Principles of Microfluidics

Viktor Shkolnikov

ISBN: 9781790217281

Contents

Preface

This book serves the goal as an introductory text to the field of microfluidics for advanced undergraduate and early graduate students, as well as a reference for practitioners in the field. The reader is expected to have prior knowledge of vector calculus, ordinary and partial differential equations, basic fluid mechanics (including some mass and heat transfer), basic electrodynamics, and basic chemistry. The book aims to present the fundamental concepts and principles needed in designing microfluidic devices as well as analyzing microfluidic phenomena.

In this text the author chose to use the editorial "we" even though there is a single author. By "we" the author means the company of the author, the reader, and the community whose work is described in this text. We hope that the use of "we" makes the text more readable.

Microfluidics, as defined by this author, encompasses flow and transport phenomena whose key feature is small length scale. We specifically do not give a specific size for what is "small" and let this size be governed by the relevant physics of the specific problem. For example, (but not always) this may mean that the length scales are such that the Reynolds number is significantly less than unity so that creeping flow model can adequately describe the flow. Or, that the device geometry is less than capillary length so that capillary forces dominate over gravitational forces. Typically, length scales in microfluidics are below 1 mm.

Applications of microfluidics abound. Microflows are found both in the natural world and industrial applications. In the natural world, we see microflows in water droplets rolling off plant leaves, in blood cells flowing in microvessels, in oil and gas percolating in pores of porous rock. In industrial applications we see microflows in microprocessor cooling (heat pipes), in printing, in chemical analysis systems, and more recently in lab-on-a-chip devices.

The text is meant to be an introductory text, and not to be a comprehensive treatment of this broad and interdisciplinary field. It is written from an industry perspective (rather than an academic one) and so attempts to emphasize immediate applicability of the concepts. However, while some concepts discussed in this text are centuries (!) old, many are still not well understood and are subjects of active research. Here we attempted to present the concepts in the simplest way possible, often presenting the simplest models that are "good enough" to describe a particular phenomena, rather than the latest and the most accurate (but much more complicated) ones. This is particularly true of subjects that are hot topics of research. This is meant to give the reader an introduction to the topic, allowing the reader to both basically understand and use the phenomena, but also to be able to read the most current literature on it as well.

Lastly, we thank Jennifer Y. Wang for assistance with cover art.

Finally, the author greatly appreciates the patience of the reader in regards to any typos, mistakes, or omissions, and encourages the reader to contact the author with any comments or suggestions.

Part 1

Hydrodynamics

Design of microfluidic devices and analysis of microfluidic phenomena requires an understanding of hydrodynamics, capillarity, electrokinetics, as well other areas pertinent to the particular device or situation. We thus begin with the hydrodynamics. In this part, we will cover the basics of motion of a single phase, incompressible fluid as it interacts with walls and particles. We will understand how the motion of the fluid can lead to useful effects such as particle sorting and separation as well as mixing, effects very useful and fundamental in many microfluidic situations.

CHAPTER 1

Creeping (low Reynolds number) flows

We begin with creeping, low Reynolds number flows by recalling the definition of the Reynolds number,

$$Re = \rho U L / \mu \tag{1.0.1}$$

where ρ is the fluid density, U is the characteristic velocity of the fluid, L is the characteristic length scale, and μ is the dynamic viscosity of the fluid. We are interested in low Reynolds numbers number flows, as microfluidic devices typically have characteristic channel dimensions of less than 1 mm and characteristic fluid velocities less 1 mm/s, typically use aqueous fluid with $\rho/\mu = 10^{-6} \, \mathrm{m^2/s}$, and so have a Reynolds numbers of less than unity. We caveat that the Reynolds numbers in microfluidic systems are often small, but not always less than unity. Microfluidic systems may also have multiple Reynolds numbers, with some Reynolds numbers quite a bit less than unity, and some, for example, order 10-100. This occurs for example, in some particle separation devices, where the Reynolds number based on particle dimension is very much less than unity, while the Reynolds number based on the channel dimensions is greater than unity. However, low Reynolds number, creeping flows occur quite frequently in microfluidic devices, and so to begin, we proceed first with creeping flows.

The basic assumption of creeping flows is that the inertia terms in the Navier-Stokes momentum equation are negligible. In such a flow pressure cannot scale with the inertial scale ρU^2 but rather must scale with a viscous scale of $\mu U/L$. This is because here, in a simple pressure driven flow, (as opposed to flow driven with other forces to be discussed later) the pressure must overcome the viscosity, and not the inertia of the flow, which is quite small relative to viscosity.

Let's begin with the continuity and the Navier-Stokes momentum equation:

$$\frac{\partial \rho}{\partial t} + \nabla \cdot (\rho \mathbf{u}) = 0 \tag{1.0.2}$$

$$\rho \left(\frac{\partial \mathbf{u}}{\partial t} + \mathbf{u} \cdot \nabla \mathbf{u} \right) = -\nabla p + \mu \nabla^2 \mathbf{u} + \frac{1}{3} \mu \nabla \left(\nabla \cdot \mathbf{u} \right) \tag{1.0.3}$$

where $\mathbf{u}$ is the velocity vector, and p is local pressure. As a reminder, the continuity equation is just a statement of conservation of mass. It states that the change in mass of fluid per time in a particular region must be balanced by the mass flow in or out of that region. (Here we neglect, for example, chemical reactions that create or destroy mass of a fluid, transforming into a different fluid, that we would have to track as well). The momentum equation, (1.0.3), is just a statement of conservation of momentum, i.e., Newton's second law. It states that the change in quantity of motion (i.e., momentum, the product of mass and its velocity) per time in a particular region must be balanced by the mass flow in or out of that region (accounted by the second term in parentheses on the left-hand side), or a force applied on this mass (accounted by the first term on the right-hand side), or diffusion of momentum (accounted by the second term on the right-hand side). (Yes, momentum, like mass, or heat can diffuse). Conservation of mass and conservation of momentum are also accompanied by conservation of energy, typically expressed as the heat equation. Conservation of mass (1.0.2) and conservation of momentum (1.0.3) are almost always mathematically coupled, and hence we must solve them together. Conservation of energy is frequently decoupled from these two (although it depends on them), and so can be frequently solved separately after we have a solution to the conservation of mass and momentum equations. Hence, we will discuss conservation of energy directly when we will need to employ it.

Firstly, we will simplify the continuity equation, noting that for majority of microfluidic flows the fluid is incompressible. The fluid is typically a liquid, or if it is a gas, its velocity is low enough such that compressibility effects are negligible. Thus, the density of the fluid is constant, and so (1.0.2) becomes

$$(1.0.4) \qquad \nabla \cdot \mathbf{u} = 0$$

$$(1.0.5) \qquad \rho \left(\frac{\partial \mathbf{u}}{\partial t} + (\mathbf{u} \cdot \nabla) \mathbf{u} \right) = -\nabla p + \mu \nabla^2 \mathbf{u}$$

Secondly, to simplify these equations accounting for the fact that Reynolds number is small, we first nondimensionalize our variables with the characteristic length, velocity, time, and pressure scales. Note, we here have used the advective time scale, U/L as the characteristic time scale. While this is a commonly useful choice, it is not the only choice. If the flow oscillates with a certain frequency, the inverse of a frequency, for example, can be used as the relevant time scale. Using these time scales, we obtain,

$$(1.0.6) \qquad \begin{aligned} x* &= x/L \\ \mathbf{u}* &= \mathbf{u}/U \\ t* &= tU/L \\ p* &= (p - p_\infty)/(\mu U/L) \end{aligned}$$

where p_∞ is free stream pressure. We designate the dimensionless variables with an asterisk. Thus we obtain the dimensionless continuity equation (1.0.4) as,

$$(1.0.7) \qquad \nabla \cdot \mathbf{u}* = 0$$

and the momentum equation (1.0.5) as:

$$(1.0.8) \qquad Re \left(\frac{\partial \mathbf{u}*}{\partial t*} + (\mathbf{u}* \cdot \nabla*) \mathbf{u}* \right) = -\nabla * p * + \nabla *^2 \mathbf{u}*$$

Often once we have transformed our variables into the dimensionless form, we will often drop the asterisks for convenience and ease of reading, and continue to work with the equations in this form, remembering which equation is in dimensionless and which is in dimensioned form. Dropping the asterisks, the dimensionless continuity and momentum equation become,

$$(1.0.9) \qquad \nabla \cdot \mathbf{u} = 0$$

$$(1.0.10) \qquad Re \left(\frac{\partial \mathbf{u}}{\partial t} + (\mathbf{u} \cdot \nabla) \mathbf{u} \right) = -\nabla p + \nabla^2 \mathbf{u}$$

Here we are concerned with flows where $Re \ll 1$. Thus the left-hand side is much smaller than the right-hand side, and so we can neglect these terms, and so obtain

$$(1.0.11) \qquad \nabla^2 \mathbf{u} - \nabla p = 0$$

Flows that satisfy this equation are termed Stokes or creeping flows. In dropping these terms we drastically change the character of the momentum equation. Firstly, the equation becomes steady. This means that any change on the boundary is immediately felt everywhere in the domain. We would like to stress this immediacy: it is as if information propagates in creeping flows at infinite speed. We know however, that information propagation speed is of course limited by the speed of light, as that is the highest possible speed for any interaction in nature. The reason why our creeping flow model predicts immediate response through the domain, i.e., infinite propagation speed, is because we have effectively set the inertia or mass of the system to zero While this is not completely true, however, on practical time scales, we expect the system to really respond instantaneously. Thus, time appears in a creeping flow solution only as a parameter that characterizes the change at the boundary. If we reverse what happens on the boundaries, the situation in the domain reverses as well. This makes it appear that the flow has some sort of memory. It doesn't. This is just the consequence of changes on the boundary immediately affecting the bulk.

Secondly, the low Reynolds number approximation and dropping of the left-hand side in (1.0.10) turned the momentum equation from being non-linear to linear, as the non-linear momentum advection term (second term on the left-hand side of (1.0.10)) is dropped. We will come back to the implications of linearity of this momentum equation shortly. Our goal was in fact to turn a non-linear equation linear, as to make it easier to solve. We didn't need to drop the non-steady term (first term on the left-hand side of (1.0.10)) to do this, and in fact if we chose a different characteristic time scale, this term may have been non-negligible. However, Reynolds number being the ratio of advection of momentum to diffusion of momentum, small Re implies that the advection of momentum is negligible. In other words, that we don't expect momentum to be advected in these flows - and that momentum primarily diffuses. Since advection and diffusion are very different transport mechanisms, with advection carrying things in the direction of the flow, and diffusion spreading things from high to low concentrations (down concentration gradient), we expect the behavior of such flows to be significantly different from advection dominated flows we experience on our macro scale.

To observe the next property of our simplified equations, let's take the curl and then the gradient of (1.0.11). We obtain that

$$(1.0.12) \qquad \begin{aligned} \nabla^2 \omega &= 0 \\ \nabla^2 p &= 0 \end{aligned}$$

where ω is the vorticity, $\omega = \nabla \times \mathbf{u}$. Notice that vorticity and pressure both satisfy Laplace's equation. Laplace's equation has an interesting property that if the domain is circular, the value of the function that satisfies Laplace's equation, in the center of the circular domain is the average of the values around the boundary. Furthermore, for a function that satisfies Laplace's equation on a general domain, the maximum (or minimum) of that function in a domain is found on the boundary of that domain. This means that the highest and lowest pressures must be on the boundaries of the domain for Stokes flows. This in turn means that Stokes flows are wall driven - whatever happens at the walls governs what happens inside the domain.

1.1. Consequences of linearity of creeping flow equations

The fact that creeping flows are governed by a linear momentum equation allows us to make remarkable predictions about these flows, just from the properties of linear PDEs, without even having to solve these PDEs.

1.1.1. Drag on a mirrored shape.
The first such remarkable property is the property of the drag of objects that are mirror images of each other with respect to the flow. Consider a wedge shaped object in Figure 1.1.1. The wedge is placed in uniform Stokes flow with free stream velocity U, either with the tip facing the flow (Figure 1.1.1a) or with the blunt rear facing the flow (Figure 1.1.1b). Is the drag on the wedge in Figure 1.1.1a greater, less than, or equal to that in Figure 1.1.1b? To obtain the drag we have to solve Stokes momentum equation, (1.0.11) and the continuity equation, (1.1.7), subject to the boundary conditions that there is no-slip on the surface of the wedge,

$$(1.1.1) \qquad \mathbf{u} = 0 \text{ on } S_{wedge}$$

and that far away from the wedge the velocity returns to its free-stream value,

$$(1.1.2) \qquad \mathbf{u} = \mathbf{U} \text{ at } \infty$$

Let

$$(1.1.3) \qquad \mathbf{u} = \mathbf{u}_a$$

and

$$(1.1.4) \qquad p = p_a$$

be the solutions for the velocity field and pressure field respectively to the situation in Figure 1.1.1a. Integrating this pressure over the surface of the wedge we obtain a net drag force,

$$(1.1.5) \qquad \mathbf{F}_D = \mathbf{F}_{D,a}$$

However, notice that the equations for the situation in Figure 1.1.1b are the same as that for Figure 1.1.1a, with the exception that (1.1.2) becomes

$$(1.1.6) \qquad \mathbf{u} = -\mathbf{U} \text{ at } \infty$$

A property of linear equations is that a linear combination of solutions to a linear equation is also a solution to the linear equation (the linear superposition principle). By reversing the sign of the boundary condition, is like multiplying everything by negative one. Hence, for the situation in Figure 1.1.1b,

$$(1.1.7) \qquad \mathbf{u} = \mathbf{u}_b = -\mathbf{u}_a$$

$$(1.1.8) \qquad p = p_b = -p_a$$

and so

$$(1.1.9) \qquad \mathbf{F}_D = \mathbf{F}_{D,b} = -\mathbf{F}_{D,a}$$

Which means that the magnitudes of the drag forces are the same for both situations! This result may be surprising because we are used to a high Reynolds number world where inertia is significant. When inertia is significant, the radial

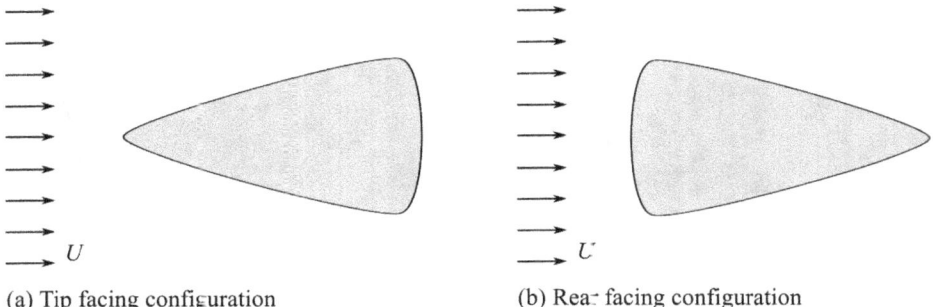

(a) Tip facing configuration   (b) Rear facing configuration

FIGURE 1.1.1. Schematic of a wedge shaped object in uniform Stokes flow. In (a) the wedge's tip is facing the flow, while in (b) the wedge's blunt rear is facing the flow. While in high Reynolds number flow the drag on the wedge in (a) would be less than the drag on the same wedge in (b), in Stokes flow the wedge experiences the same drag in both situations!

outflow from the bluff body in Figure 1.1.1b would be much harder to deflect back into the main flow direction, and so the main flow is much more disturbed than the streamlined configuration in Figure 1.1.1a. Hence the configuration in Figure 1.1.1b would have higher drag at high Reynolds numbers.

Comparing the solutions for the two situations, (1.1.3) and (1.1.7), we see that they are mirror images of one another, and so their streamline positions are identical. By just looking at the streamlines (e.g., dye streak visualization), we would not be able to tell if the flow is from left to right or right to left! This is also another important feature of Stokes flow.

1.1.2. Lift on a spinning sphere is simple shear flow. In the second example we will use the linearity of the momentum and continuity equations to obtain a lift on a spinning sphere in a simple shear flow, again without solving the equations themselves. A spinning sphere in shear flow might occur if we used, for example Quincke (electrical) rotation (to be discussed in later sections) to set a spherical bead spinning while it translates in a channel, where in its vicinity the velocity profile can be approximated as linear. A "macrofludic" (high Reynolds number) example of this is a baseball thrown with velocity U and spin Ω, in a shear flow generated by the wind (with a no-slip condition on the ground). In the macrofluidic case, Magnus lift force pushes the ball in the direction given by $\mathbf{U} \times \Omega$. We may ask, is there a similar force in creeping flow?

To answer this question, we start with the continuity and the Stokes momentum equations, (1.0.9) and (1.0.11). In this case, the no-slip boundary condition on the wall of the sphere gives,

$$(1.1.10) \qquad \mathbf{u} = \left(\frac{a}{U}\right) \Omega \times \mathbf{r} \text{ at } r = 1$$

where a is the radius of the sphere and U is the sphere's velocity. Far away from the sphere we have the shear flow, and so the velocity there is given by

$$(1.1.11) \qquad \mathbf{u} = \left(1 + \gamma \frac{a}{U} \gamma\right) \hat{\mathbf{z}} \text{ at } \infty$$

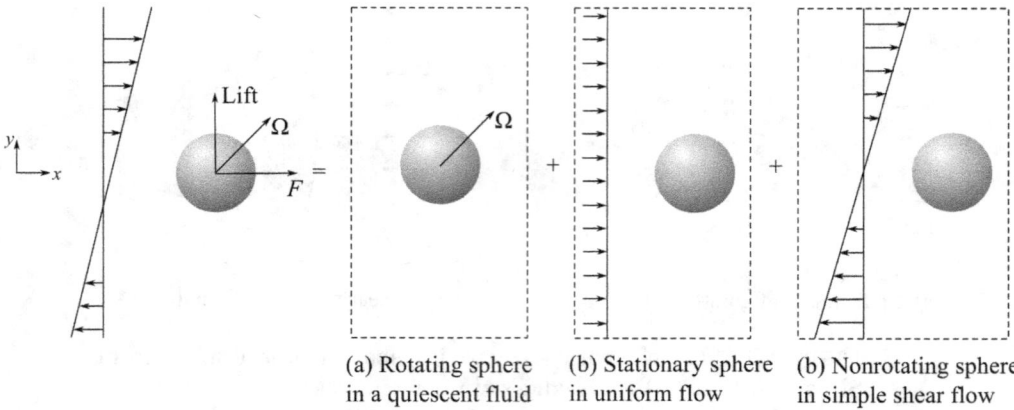

(a) Rotating sphere   (b) Stationary sphere   (b) Nonrotating sphere
in a quiescent fluid   in uniform flow   in simple shear flow

FIGURE 1.1.2. Linear decomposition of a linear problem of a spinning sphere in shear flow into three problems: (a) a sphere rotating
in a quiescent fluid, (b) a sphere held stationary in a uniform flow,
and (c) a non-rotating sphere in a simple shear flow, with zero flow
velocity coinciding with the x-axis (center of the sphere).

At the first glance the problem might look quite complicated, with the combination
of shear and spinning motion. We haven't even solved the flow profile around a
simple stationary sphere yet! However, to do this problem we take advantage of the
linearity of equations, and split the problem through the boundary conditions into
three problems (Figure 1.1.2): (1) a spinning sphere in a fluid that is stationary at
infinity:

$$(1.1.12) \qquad\qquad \mathbf{u}_1 = \left(\frac{a}{U}\right) \boldsymbol{\Omega} \times \mathbf{r} \text{ at } r = 1$$

$$(1.1.13) \qquad\qquad \mathbf{u}_2 = 0 \text{ at } \infty$$

(2) a sphere fixed in a uniform flow,

$$(1.1.14) \qquad\qquad \mathbf{u}_2 = 0 \text{ at } r = 1$$

$$(1.1.15) \qquad\qquad \mathbf{u}_2 = \hat{\mathbf{z}} \text{ at } \infty$$

(3) a sphere fixed in a simple shear flow that has zero velocity at the center of the
sphere,

$$(1.1.16) \qquad\qquad \mathbf{u}_2 = 0 \text{ at } r = 1$$

$$(1.1.17) \qquad\qquad \mathbf{u} = \left(\gamma\frac{a}{U}\gamma\right) \hat{\mathbf{z}} \text{ at } \infty$$

We see that if we add these boundary conditions up, we obtain the original boundary
conditions of (1.1.10) and (1.1.11). Now, let's examine each of these problems one
by one. In the first problem, the only direction in this problem is the axis of
rotation. There is no other direction in the plane orthogonal to the axis of rotation
that can be found in this problem. Hence no direction orthogonal to the axis of
rotation is "special" (distinguishable from another) and so with which we can take
a cross product with $\boldsymbol{\Omega}$ and expect the lift force to be in that direction. Thus we
conclude that the lift force on a spinning sphere in a stationary fluid is zero.

For the second problem, we see the same symmetry. There is no direction perpendicular to the only direction in the problem, **U** that is distinguishable from any other. Hence, again we cannot expect a force that is not in the direction of **U**, and hence we again conclude that the lift force is zero. There is indeed a force collinear with the direction of **U**, the drag force, but we shall look at this force later.

For the third problem, the only natural direction of the problem is the direction of the shear. If there was non-zero lift in the positive y-direction, then it would also be non-zero in the negative y-direction, as from the symmetry of the problem, there is no distinction between positive and negative y-direction. Once again, because we cannot find a distinguishing direction, we conclude that the lift must be zero.

Since the solutions of velocity distributions and pressure distributions for these three problems add up to the original problem, so does the force on the sphere. Since we could not find a lift force on the sphere in each of the problems, the lift force in the original problem is also zero. Notice how useful decomposition of a problem into simpler problems is. Unfortunately this is only available when the equations and boundary conditions are linear, and so is not available for the higher Reynolds number flows, where the non-linear advection term in the momentum equation prevents us from doing such decompositions.

1.1.3. Lateral migration of a sphere in creeping Poiseuille flow.
Seeing as there is no lift on a spinning sphere in shear flow, we may ask ourselves if there is a lateral force on a sphere in axisymmetric or two dimensional creeping Poiseuille flow - a very commonly occurring pressure driven flow. To understand this, consider a sphere in a center of a circular pipe in Figure 1.1.3. Let's consider if as the undisturbed flow is moving from left to right, the sphere has a force on it to move it towards the axis of the tube (Figure 1.1.3a). However, since we are considering the flow to be creeping flow, the governing equations (continuity and momentum equations) are linear and so are the boundary conditions. Thus, as before, we can change the sings of all velocities and pressure and obtain a solution with the undisturbed flow direction reversed (Figure 1.1.3b). But this would also reverse the sign of the force on our spherical particle, and so the particle now would be migrating towards the wall of the tube (Figure 1.1.3b). However, this does not make physical sense - why would a particle care if the flow was from left to right or from right to left? It's indistinguishable to the particle. If the sphere were to undergo lateral motion, it should be in the same direction in both cases. However, the linearity of the problem does not allow that. Hence the particle cannot have any lateral migration in the creeping flow limit.

However, as we shall discuss in detail in a later section, particles indeed can have lateral migration velocities, even when the particle Reynolds number (with length scale taken as a particle length scale) is small, but when for example, the channel Reynolds number (with length scale taken as a channel length scale) is large. However, as we shall see later, we need different governing equations for this to occur.

1.1.4. Resistance matrices for force and torque in creeping flow.
Lastly, we would like to use the linearity of the governing equations to predict what forces (and so motions) are allowed on bodies of various shapes (or more various symmetries) in creeping flow. This is quite powerful, firstly, because it allows us

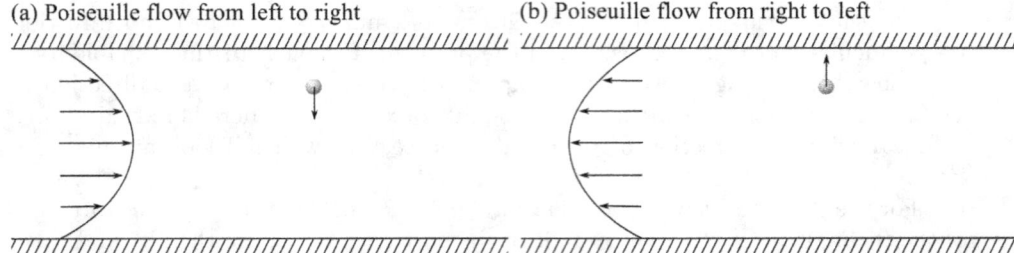

FIGURE 1.1.3. Schematic of a spherical particle in creeping Poiseuille flow as a proof that a spherical particle cannot undergo lateral migration in either axisymmetric (tube) or 2D (parallel plate) Poiseuille flow. In (a) the main flow moves from left to right, and we suppose the particle moves away from the wall. In (b) the main flow is from right to left. Since the flow is a creeping flow, and the governing equations are linear, if the direction of the main flow is reversed, the signs of all velocities are reversed, and so the velocity of the sphere is reversed, as we sketch in (b). However, since the situations in (a) and (b) are indistinguishable from the sphere's perspective, the only lateral migration velocity that can occur is a zero velocity - the sphere does not migrate laterally.

to predict behavior of objects without solving the governing equations themselves. Secondly, if we do see a behavior that is "not allowed" this hints to us that the situation is not truly governed by our creeping flow model. Hence, we either need to introduce small corrections to our model (e.g., simplifying the Navier-Stokes and keeping the creeping flow as a zeroth order equations in an asymptotic expansion, but also using higher order terms) or keep more terms in the governing equations.

To begin, let's consider a rigid particle of arbitrary shape moving with velocity $\mathbf{U}(t)$ and angular velocity $\mathbf{\Omega}(t)$ moving through an unbounded fluid, such that the Reynolds number of the particle is small, and so the creeping flow limit is justified. To calculate the force on the particle we need to solve the continuity and the Stokes momentum equations. Here we consider the problem in dimensional terms,

$$(1.1.18) \qquad\qquad \nabla \cdot \mathbf{u} = 0$$

$$(1.1.19) \qquad\qquad \mu \nabla^2 \mathbf{u} - \nabla p = 0$$

with a no-slip boundary condition on the particle surface, S,

$$(1.1.20) \qquad\qquad \mathbf{u} = \mathbf{U} + \mathbf{\Omega} \times \mathbf{x} \text{ on } S$$

and far away from the particle we have the flow return on the undisturbed flow,

$$(1.1.21) \qquad\qquad \mathbf{u} \to 0 \text{ as } |\mathbf{x}| \to \infty$$

where $\mathbf{x}$ is a position vector from the center of mass of the particle. The force on the particle is just the integral of the stress tensor $\mathbf{T}$ over the surface of the particle,

$$(1.1.22) \qquad\qquad \mathbf{F} = \int_S (\mathbf{T} \cdot \mathbf{n}) \, dS$$

and correspondingly the torque on the particle is,

$$(1.1.23) \qquad \mathbf{G} = \int\limits_{S} (\mathbf{x} \times \mathbf{T} \cdot \mathbf{n}) \, dS$$

The problem is particularly difficult because it depends on both the orientations of $\mathbf{U}$ and $\boldsymbol{\Omega}$ relative to the axis fixed on the particle, as well as the relative magnitudes of $\mathbf{U}$ and $\boldsymbol{\Omega}$. However, since the problem is linear, we split the problem into two problems: the first where $\mathbf{U}$ is arbitrary and $\boldsymbol{\Omega} = 0$ and the second problem where $\mathbf{U} = 0$, and $\boldsymbol{\Omega}$ is arbitrary. For the first problem we obtain a solution consisting of $\mathbf{u}_U, p_U, \mathbf{F}_U, \mathbf{G}_U$ and for the second problem we obtain a solution $\mathbf{u}_\Omega, p_\Omega, \mathbf{F}_\Omega, \mathbf{G}_\Omega$ and so the solution to the original problem becomes

$$(1.1.24) \qquad \begin{aligned} \mathbf{u} &= \mathbf{u}_U + \mathbf{u}_\Omega \\ p &= p_U + p_\Omega \\ \mathbf{F} &= \mathbf{F}_U + \mathbf{F}_\Omega \\ \mathbf{G} &= \mathbf{G}_U + \mathbf{G}_\Omega \end{aligned}$$

The governing equations and the boundary conditions for the first problem become,

$$(1.1.25) \qquad \nabla \cdot \mathbf{u}_U = 0$$

$$(1.1.26) \qquad \mu \nabla^2 \mathbf{u}_U - \nabla p_U = 0$$

$$(1.1.27) \qquad \mathbf{u}_U = \mathbf{U} \text{ on } S$$

$$(1.1.28) \qquad \mathbf{u}_U \to 0 \text{ as } |\mathbf{x}| \to \infty$$

We see that the problem is linear in $\mathbf{U}$. In other words, the solution $\mathbf{u}_U$ and p_U depend linearly on $\mathbf{U}$. This in turn means that $\mathbf{F}_U$ and $\mathbf{G}_U$ are also linear with $\mathbf{U}$ as their corresponding integrals are linear with $\mathbf{u}_U$ and p_U. Because $\mathbf{F}_U$ and $\mathbf{U}$ are true vectors the most general linear relationship between them is through a product with a true second order tensor (for our purposes here a 2D matrix),

$$(1.1.29) \qquad \mathbf{F}_U = \bar{\mathbf{A}} \cdot \mathbf{U}$$

where $\bar{\mathbf{A}}$ is this matrix. Notice that the ij component of $\bar{\mathbf{A}}$ is just the ith component of the force on the body for translation with unit velocity in the jth direction. Similarly, $\mathbf{G}_U$ is a pseudo-vector and so the most general relationship between it and $\mathbf{U}$ is

$$(1.1.30) \qquad \mathbf{G}_U = \bar{\mathbf{C}} \cdot \mathbf{U}$$

where $\bar{\mathbf{C}}$ is a second-order pseudo-tensor (again for our purposes here a 2D matrix). Again, the ij component of $\bar{\mathbf{C}}$ is just the ith component of the torque on the body produced by translation with unit velocity in the jth direction. (A pseudo-vector is the result of a cross product of two other vectors. It's key property that it changes sign when we invert the axis from a right- to a left-handed coordinates system, whereas a true vector is invariant to this transformation. Torque is an example of a pseudo-vector because it is a cross product of a "lever arm" and a force. Here we don't plan on doing an inversion of the handedness of the coordinate system, so we should be aware of this property of these vectors for other problems, but not worry about them). We see that if we solve the individual problems of translation with unit velocity in the three coordinate directions and calculate the components of force and torque in each of the cases, we get all nine components of $\bar{\mathbf{A}}$ and all nine components of $\bar{\mathbf{C}}$. Once we have that, we can use (1.1.29) and (1.1.30) to compute the force and torque for translation of the body in any direction we want!

Similarly, we can perform the above analysis on the second problem. We obtain

$$(1.1.31) \qquad\qquad \mathbf{F}_\Omega = \bar{\mathbf{B}} \cdot \mathbf{\Omega}$$

$$(1.1.32) \qquad\qquad \mathbf{G}_\Omega = \bar{\mathbf{D}} \cdot \mathbf{\Omega}$$

Since our original problem is just a linear combination of our two subproblems, the total force and torque on the particles can be written as

$$(1.1.33) \qquad\qquad \mathbf{F} = \mu \left(\mathbf{A} \cdot \mathbf{U} + \mathbf{B} \cdot \mathbf{\Omega} \right)$$

$$(1.1.34) \qquad\qquad \mathbf{G} = \mu \left(\mathbf{C} \cdot \mathbf{U} + \mathbf{D} \cdot \mathbf{\Omega} \right)$$

where we have factored out the viscosity from our second order tensors as the problem is also linear in viscosity. Now the second order tensors $\mathbf{A}, \mathbf{B}, \mathbf{C}, \mathbf{D}$ only depend on the geometry of the particle are independent of all other parameters. These tensors are termed resistance tensors. The most powerful property of this approach is that for a particular particle shape we can solve a total of six problems (three coordinates for the first problem, three coordinates for the second), tabulate the result, and be able to employ it in a large number of arbitrary problems. For more details on obtaining resistance tensors we refer the reader to Happel and Brenner (1983) and Leal (2007). Here describe some simple cases.

For a particle with spherical symmetry, the resistance tensors take the form of

$$(1.1.35) \qquad\qquad \mathbf{A} = a\mathbf{I}$$

$$(1.1.36) \qquad\qquad \mathbf{D} = d\mathbf{I}$$

$$(1.1.37) \qquad\qquad \mathbf{C} = \mathbf{B} = 0$$

where a and d are scalars dependent on particle geometry, and $\mathbf{I}$ is the identity matrix. For a particle that is an ellipsoid of revolution, with the coordinate system at the geometric center of the particle and the coordinate axes aligned with the principle axis of the particle the resistance tensors take the form of

$$(1.1.38) \qquad\qquad \mathbf{A} = \begin{bmatrix} a_\parallel & 0 & 0 \\ 0 & a_\perp & 0 \\ 0 & 0 & a_\perp \end{bmatrix}$$

$$(1.1.39) \qquad\qquad \mathbf{D} = \begin{bmatrix} d_\parallel & 0 & 0 \\ 0 & d_\perp & 0 \\ 0 & 0 & d_\perp \end{bmatrix}$$

$$(1.1.40) \qquad\qquad \mathbf{C} = \mathbf{B} = 0$$

For these two cases we can immediately see from the resistance tensors and (1.1.33) and (1.1.34) that a force will not cause rotational motion as $\mathbf{B} = 0$ and that translational motion will not result in a torque on the particle, as $\mathbf{C} = 0$. This is of course not generally true for other shapes. For example, a screw like structure exhibits coupling between translation and rotation: when we rotate a screw it translates in the direction of its screw axis. If we want to restrain a screw from translating, we see from (1.1.33) that we must apply a force $-\mu\mathbf{B}\cdot\boldsymbol{\Omega}$ to balance out the hydrodynamic force generated by its rotation, where

$$(1.1.41) \qquad \mathbf{B} = \begin{bmatrix} b_x & 0 & 0 \\ 0 & 0 & 0 \\ 0 & 0 & 0 \end{bmatrix}$$

with x axis coincident with the screw axis. Although we have derived our concept of resistance tensors with the boundary conditions of a particle in an unbounded fluid, we could have analogously used a boundary condition of a wall and arrived at a similar formulation. When walls are included the tensors may have factors both accounting for the presence of the wall as well as relative distance from the wall. For example, for a sphere on which a force is applied parallel to a wall, the sphere will not only translate parallel to the wall but also rotate because of the hydrodynamic interaction with this wall. This situation will have the following resistance tensors, where the coordinate axes are normal and tangential to the wall:

$$(1.1.42) \qquad \mathbf{A} = \begin{bmatrix} a_\parallel & 0 & 0 \\ 0 & a_\perp & 0 \\ 0 & 0 & a_\perp \end{bmatrix}$$

$$(1.1.43) \qquad \mathbf{D} = \begin{bmatrix} d_\parallel & 0 & 0 \\ 0 & d_\perp & 0 \\ 0 & 0 & d_\perp \end{bmatrix}$$

$$(1.1.44) \qquad \mathbf{B} = \mathbf{C}^T = \begin{bmatrix} 0 & 0 & 0 \\ 0 & 0 & b \\ 0 & b & 0 \end{bmatrix}$$

where the components of $\mathbf{A}, \mathbf{B}, \mathbf{C}, \mathbf{D}$ will dependent on the ratio of the distance between the wall and the sphere and the sphere radius.

Using resistance tensors we can obtain the angular velocity of a particle from its translation velocity, and translational velocity through its angular velocity - the translation and angular velocities are coupled through the resistance tensors and the fact that the inertia of the particle is negligible. The inertia of the particle is negligible when the density of the particle is comparable to that of the fluid, and the Reynolds number is low so that we are in the creeping flow regime.

When the inertia of the particle is negligible, by Newton's second law the particle is practically always at equilibrium (not accelerating), and so the sum of the forces and the sum of the torques on it are both zero. Say we apply an external force on a particle $\mathbf{f}$, but apply no torque on the particle. Substituting this into (1.1.33) and (1.1.34) and rearranging, we obtain

$$(1.1.45) \qquad \boldsymbol{\Omega} = -\mathbf{D}^{-1}\cdot(\mathbf{C}\cdot\mathbf{U})$$

$$(1.1.46) \qquad \mathbf{U} = \left(\mathbf{I} - \mathbf{A}^{-1}\big(\mathbf{B}\cdot(\mathbf{D}^{-1}\cdot\mathbf{C})\big)\right)^{-1}\cdot\left(\mathbf{A}^{-1}\cdot\mathbf{f}/\mu\right)$$

From (1.1.45) we see that a particle with no external torque will rotate as it translates if the coupling tensor $\mathbf{C}$ is non-zero. We can see that spherical particles and ellipsoids of revolution in an unbounded fluid (or far enough from walls) the particle will not rotate as it translates as $\mathbf{C} = 0$. The particle will translate with velocity given by (1.1.46), and interestingly the velocity is not necessarily collinear with the applied force.

1.2. Stream function for 2D and axisymmetric creeping flows

Many of the flows we will encounter can be simplified to a 2D or an axisymmetric (effectively 2D) representation. This dimensional reduction together with linearity of the continuity and momentum equations in the creeping flow regime, allows us to solve the continuity and momentum equations for a large number of problems. For this, we introduce (or remind the reader of) the concept of the stream function, a function for the streamlines of the flow. Obtaining the streamlines all by themselves is already quite powerful, as it already gives us paths of particles and reagents traveling in our flow (e.g., when their diffusion is negligible).

We begin by recalling a general representation theorem from vector calculus that states that any continuously differentiable vector field (such as velocity field), can be represented by three scales functions in the form

$$(1.2.1) \qquad \mathbf{a} = \nabla\varphi + \nabla \times (\psi\nabla\chi)$$

where φ, ψ, and, χ are the scalar functions. This representation is a decomposition of the general vector field $\mathbf{a}$ into an irrotational part, $\nabla\varphi$, and a divergence-free (solenoidal) part, $\nabla \times (\psi\nabla\chi)$. Since divergence of a curl is zero,

$$(1.2.2) \qquad \nabla \cdot (\nabla \times (\psi\nabla\chi)) = \nabla \cdot (\nabla\psi \times \nabla\chi) = 0$$

we can rewrite (1.2.1) as

$$(1.2.3) \qquad \mathbf{a} = \nabla\varphi + \nabla \times \mathbf{A}$$

where

$$(1.2.4) \qquad \nabla \cdot \mathbf{A} = 0$$

Thus, we represented the original general vector field as a sum of an irrotational part (gradient of a scalar) and a divergence free part (a divergence free vector field). We know that the velocity field is a continuously differentiable vector field. Thus, we can now write

$$(1.2.5) \qquad \mathbf{u} = \nabla\varphi + \nabla \times (\psi\nabla\chi)$$

as well as

$$(1.2.6) \qquad \mathbf{u} = \nabla\varphi + \nabla \times \mathbf{A}$$

$$(1.2.7) \qquad \nabla \cdot \mathbf{A} = 0$$

Recall that vorticity is defined as the curl of the velocity field,

$$(1.2.8) \qquad \omega = \nabla \times \mathbf{u}$$

There is a vector calculus identity for the vector Laplacian,

$$(1.2.9) \qquad \nabla^2\mathbf{A} = \nabla (\nabla \cdot \mathbf{A}) - \nabla \times (\nabla \times \mathbf{A})$$

Since $\nabla \cdot \mathbf{A} = 0$

$$(1.2.10) \qquad \nabla^2\mathbf{A} = -\nabla \times (\nabla \times \mathbf{A})$$

Since curl of a gradient is zero, i.e., $\nabla \times (\nabla\varphi) = 0$, we are justified in writing via (1.2.6),

$$(1.2.11) \qquad \nabla^2\mathbf{A} = -\nabla \times \mathbf{u} = -\nabla \times (\nabla \times \mathbf{A})$$

Or from (1.2.8) that

(1.2.12)
$$\nabla^2 \mathbf{A} = -\nabla \times \mathbf{u} = -\omega$$

Since the vector Laplacian of $\mathbf{A}$ is negative vorticity, $\mathbf{A}$ is known as vector potential for vorticity.

Now let's see where all of this gets us. Substituting our expanded form of velocity into the continuity equation for creeping flows (1.0.9), we obtain

(1.2.13)
$$\nabla^2 \varphi = 0$$

since the divergence of a curl is zero. Next we substitute our expanded form of velocity into the momentum equation for creeping flow (1.0.11), we obtain

(1.2.14)
$$\nabla^2 (\nabla \varphi) + \nabla^2 (\nabla \times \mathbf{A}) = \nabla (\nabla^2 \varphi) + \nabla (\nabla^2 \times \mathbf{A}) = \nabla p$$

and using (1.2.13) obtain

(1.2.15)
$$\nabla \times (\nabla^2 \mathbf{A}) = \nabla p$$

or

(1.2.16)
$$\nabla^4 \mathbf{A} = 0$$

We see that in general $\mathbf{A}$ must be non-zero to satisfy (1.2.15). Additionally, from (1.64) we see that $\nabla \varphi$ is non-zero only if

(1.2.17)
$$(\mathbf{u} - \nabla \times \mathbf{A}) \cdot \mathbf{n} \neq 0$$

on all boundaries. Since generally this is not the case, we can set

(1.2.18)
$$\nabla \times \mathbf{A} = \mathbf{u}$$

as we have done above to define the vector potential for vorticity. So far, we have not gotten any simplification from casting the velocity field in this expanded form. The simplification comes from the fact that it turns out that the vector potential of vorticity can be represented in terms of a single scalar function,

(1.2.19)
$$\mathbf{A} = h_3 \psi (q_1, q_2) \hat{\mathbf{i}}_3$$

where $\hat{\mathbf{i}}_3$ is the unit vector orthogonal to the plane of motion in 2D flows (typically $\hat{\mathbf{z}}$) and the azimuthal (typically $\hat{\theta}$) direction in axisymmetric flows. Here q_i are the symbols of an orthogonal curvilinear coordinate system such that

(1.2.20)
$$(q_1, q_2, q_3)$$

which for Cartesian coordinate system are

(1.2.21)
$$(x, y, z)$$

or a spherical coordinate system are

(1.2.22)
$$(r, \theta, \phi)$$

h_i are scale factors, defined such that the length of the differential line element ds is

(1.2.23)
$$(ds)^2 = \frac{1}{h_1^2}(dq_1)^2 + \frac{1}{h_2^2}(dq_2)^2 + \frac{1}{h_3^2}(dq_3)^2$$

$1/h_i$ are termed Lame coefficients. As a reminder, for a curvilinear orthogonal coordinate system (q_1, q_2, q_3) the first order differential operators are

(1.2.24)
$$\nabla \mathbf{A} = grad\mathbf{A} = \left(h_1 \frac{\partial \mathbf{A}}{\partial q_1}, h_2 \frac{\partial \mathbf{A}}{\partial q_2}, h_3 \frac{\partial \mathbf{A}}{\partial q_3} \right)$$

$$(1.2.25) \quad \nabla \cdot \mathbf{A} = div\mathbf{A} = h_1 h_2 h_3 \left(\frac{\partial}{\partial q_1} \left(\frac{A_1}{h_2 h_3} \right) + \frac{\partial}{\partial q_2} \left(\frac{A_2}{h_1 h_3} \right) + \frac{\partial}{\partial q_3} \left(\frac{A_3}{h_1 h_2} \right) \right)$$

$$(1.2.26) \qquad \nabla \times \mathbf{A} = \mathrm{rot}\, \mathbf{A} = h_1 h_2 h_3 \begin{vmatrix} \mathbf{e}_1/h_1 & \mathbf{e}_2/h_2 & \mathbf{e}_3/h_3 \\ \frac{\partial}{\partial q_1} & \frac{\partial}{\partial q_2} & \frac{\partial}{\partial q_3} \\ A_1/h_1 & A_2/h_2 & A_3/h_3 \end{vmatrix}$$

where $\mathbf{e}_1, \mathbf{e}_2, \mathbf{e}_3$ are unit vectors that are parallel to the tangents of the corresponding coordinate lines. If our coordinates (q_1, q_2, q_3) are connected with Cartesian coordinates (x, y, z) via a vector relation

$$(1.2.27) \qquad \mathbf{r} = \mathbf{r}\,(q_1, q_2, q_3)$$

where $\mathbf{r}$ is the radius vector of the point of interest, then the Lame coefficients are

$$(1.2.28) \qquad \frac{1}{h_i} = \left| \frac{\partial \mathbf{r}}{\partial q_i} \right|$$

For example, a common cylindrical coordinate system (r, φ, z) can be connected to a Cartesian system by a vector

$$(1.2.29) \qquad \mathbf{r} = \begin{pmatrix} x \\ y \\ z \end{pmatrix} = \begin{pmatrix} r\cos\varphi \\ r\sin\varphi \\ z \end{pmatrix}$$

Thus, h_3 using (1.2.28) is just

$$(1.2.30) \qquad h_3 = 1 \Big/ \sqrt{0^2 + 0^2 + 1^2} = 1$$

Similarly, for 2D flows that we will be interested in, q_3 is the z-coordinate and $h_3 = 1$. For axisymmetric flows q_3 is the azimuthal angle about the axis of symmetry (θ in cylindrical coordinates, ϕ in spherical) and h_3 is a function of q_1, q_2. For example, for spherical coordinates where (q_1, q_2, q_3) are (r, θ, ϕ) and

$$(1.2.31) \qquad \mathbf{r} = \begin{pmatrix} x \\ y \\ z \end{pmatrix} = \begin{pmatrix} r\sin\theta\cos\varphi \\ r\sin\theta\sin\varphi \\ r\cos\theta \end{pmatrix}$$

and again using (1.2.28)

$$(1.2.32) \qquad \begin{aligned} h_1 &= 1 \\ h_2 &= 1/r \\ h_3 &= 1/(r\sin\theta) \end{aligned}$$

Furthermore, in this generalized form of orthogonal curvilinear coordinate system, the continuity equation (1.0.9) is

$$(1.2.33) \qquad \left(\frac{\partial}{\partial q_1} \left(\frac{u_1}{h_2 h_3} \right) + \frac{\partial}{\partial q_2} \left(\frac{u_2}{h_1 h_3} \right) + \frac{\partial}{\partial q_3} \left(\frac{u_3}{h_1 h_2} \right) \right) = 0$$

the (non-simplified) momentum equation (1.0.10) is

(1.2.34)

$$Re \left(\begin{array}{c} \frac{\partial u_i}{\partial t} + u_1 h_1 \frac{\partial u_i}{\partial q_1} + u_2 h_2 \frac{\partial u_i}{\partial q_2} + u_3 h_3 \frac{\partial u_i}{\partial q_3} \\ + u_i h_i \left(u_1 h_1 \frac{\partial}{\partial q_1} \left(\frac{1}{h_i} \right) + u_2 h_2 \frac{\partial}{\partial q_2} \left(\frac{1}{h_i} \right) + u_3 h_3 \frac{\partial}{\partial q_3} \left(\frac{1}{h_i} \right) \right) \\ - h_i \left(u_1^2 h_1 \frac{\partial}{\partial q_i} \left(\frac{1}{h_1} \right) + u_2^2 h_2 \frac{\partial}{\partial q_i} \left(\frac{1}{h_2} \right) + u_3^2 h_3 \frac{\partial}{\partial q_i} \left(\frac{1}{h_3} \right) \right) \end{array} \right)$$
$$= h_i \frac{\partial p}{\partial q_i} + \left(\nabla^2 \mathbf{u} \right)_i$$

where $i = 1, 2, 3$ and the Laplace operator is
(1.2.35)
$$\nabla^2 = h_1 h_2 h_3 \left(\frac{\partial}{\partial q_1} \left(\frac{h_1}{h_2 h_3} \frac{\partial}{\partial q_1} \right) + \frac{\partial}{\partial q_2} \left(\frac{h_2}{h_1 h_3} \frac{\partial}{\partial q_2} \right) + \frac{\partial}{\partial q_3} \left(\frac{h_3}{h_1 h_2} \frac{\partial}{\partial q_3} \right) \right)$$

If we substitute this form of vector potential of vorticity, (1.2.19) into (1.2.18), we obtain

(1.2.36)
$$\mathbf{u} = \left(h_2 h_3 \frac{\partial \psi}{\partial q_2}, -h_1 h_3 \frac{\partial \psi}{\partial q_1} \ 0 \right)$$

and we see that the resulting velocity field, as we expect, is indeed two dimensional. Furthermore, from (1.2.36) (the gradient of the function ψ is always orthogonal to the direction of the velocity, that is

(1.2.37)
$$\frac{\mathbf{u}}{|\mathbf{u}|} \cdot \nabla \psi = 0$$

In other words, the lines of constant ψ are always tangent to the local velocity, and are thus streamlines. Because of this we call ψ the stream function. Because velocity vectors are always tangent to the streamlines, nothing flows across streamlines. This gives the stream function an interesting property: the volume flux across a curve joining any two arbitrary points is directly related to the difference in magnitude to of the stream function of the two points.

Introducing the stream function simplifies our solution of the continuity and momentum equation for creeping flows in two ways. Firstly, by substituting (1.2.36) into the continuity equation, we see that it is automatically satisfied. Secondly, we can define a derivative operator E^2 based on (1.2.16) and (1.2.19), where

(1.2.38)
$$E^2 = \frac{h_1 h_2}{h_3} \left(\frac{\partial}{\partial q_1} \left(\frac{h_1 h_3}{h_2} \frac{\partial}{\partial q_1} \right) + \frac{\partial}{\partial q_2} \left(\frac{h_2 h_3}{h_1} \frac{\partial}{\partial q_2} \right) \right)$$

This operator is analogous to, but not the same as our familiar Laplacian operator ∇^2, which is defined as

(1.2.39)
$$\nabla^2 = h_1 h_2 h_3 \left(\frac{\partial}{\partial q_1} \left(\frac{h_1}{h_2 h_3} \frac{\partial}{\partial q_1} \right) + \frac{\partial}{\partial q_2} \left(\frac{h_2}{h_1 h_3} \frac{\partial}{\partial q_2} \right) \right)$$

Substituting (1.2.19) into (1.2.16), we obtain that

(1.2.40)
$$E^4 \psi = 0$$

Thus by doing all this work to create a stream function, we converted a problem with four coupled PDEs into a problem with a single higher order PDE, and so simplified the problem. For a Cartesian coordinate and cylindrical polar coordinate systems (i.e., in our case 2D flows), $h_3 = 1$, so that for this case the operators E^2 and ∇^2 are identical. Thus for 2D problems we have

$$(1.2.41) \qquad\qquad \nabla^4 \psi = 0$$

This equation is referred to as the biharmonic equation. For completeness we may also cast the full Navier-Stokes into the form of the stream function. For this we take the curl of (1.0.10),

$$(1.2.42) \qquad Re\left(\frac{\partial \omega}{\partial t} - \nabla \times (\mathbf{u} \times \omega)\right) + \nabla \times (\nabla \times \omega) = 0$$

Recall that curl of a gradient is zero, which eliminates the pressure gradient term. Now from (1.2.19) and (1.2.12)

$$(1.2.43) \qquad\qquad \omega = \nabla \times \left(\nabla \times \left(h_3 \psi \hat{\mathbf{i}}_3\right)\right)$$

So substituting that into (1.2.42) we obtain
$$(1.2.44)$$
$$Re\left(\frac{\partial}{\partial t}\left(E^2 \psi\right) + \frac{h_1 h_2}{h_3}\left(\frac{\partial}{\partial q_1}\left(h_3^2 \frac{\partial \psi}{\partial q_2} E^2 \psi\right) - \frac{\partial}{\partial q_2}\left(h_3^2 \frac{\partial \psi}{\partial q_1} E^2 \psi\right)\right)\right) = E^4 \psi$$

and for the 2D case,

$$(1.2.45) \quad Re\left(\frac{\partial}{\partial t}\left(\nabla^2 \psi\right) + h_1 h_2\left(\frac{\partial}{\partial q_1}\left(\frac{\partial \psi}{\partial q_2} \nabla^2 \psi\right) - \frac{\partial}{\partial q_2}\left(\frac{\partial \psi}{\partial q_1} \nabla^2 \psi\right)\right)\right) = \nabla^4 \psi$$

Unsurprisingly, we see that these reduce to (1.2.40) and (1.2.41) as $Re \to 0$.

1.2.1. Solutions to the biharmonic equation: Cartesian and cylindrical polar coordinates.
Having obtained the biharmonic equation (1.2.41), now we have to solve it subject to appropriate boundary conditions. Turns out it is convenient to split the biharmonic equations into a pair of coupled PDEs

$$(1.2.46) \qquad\qquad \nabla^2 \psi = -\omega$$

$$(1.2.47) \qquad\qquad \nabla^2 \omega = 0$$

These are just the familiar Poisson's equation and Laplace's equation respectively. While the equations are coupled, the solution method is straightforward: we get to solve (1.2.47) first, and then using this solution solve (1.2.46). To solve these equations we can use standard methods such as, for example, separation of variables for bounded problems and Fourier transforms for problems with boundary conditions at infinity. We will not cover these methods here, and we refer the reader to texts such by Bland (1962).

Briefly, the separation of variables solution in Cartesian coordinates for (1.2.47) takes the form of

$$(1.2.48) \qquad\qquad \omega = X(x) Y(y)$$

substituting this back into (1.2.47), we obtain

$$(1.2.49) \qquad\qquad X''/X = -Y''/Y = \pm m^2$$

where primes designate differentiation, and m is an arbitrary complex number (the separation variable). Thus, we obtain

$$(1.2.50) \qquad\qquad X'' \pm m^2 X = 0$$

(1.2.51)
$$Y'' \mp m^2 Y = 0$$

These equations have an exponential solution, and so putting this together with (1.2.48), we obtain

(1.2.52)
$$\omega = \exp(mx) \exp(imy)$$

Now to obtain the solution to Poisson's equation (1.2.46), we set

(1.2.53)
$$\nabla^2 \psi = -\gamma_m \exp(mx) \exp(imy)$$

where γ_m is an arbitrary constant. The solution to this equation is a sum of the homogeneous part that has the form of (1.2.52) and a particular solution, and so the final solution has the form of

(1.2.54)
$$\psi_m = \alpha_m \exp(mx) \exp(imy) + \beta_m x \exp(mx) \exp(imy) + \delta_m y \exp(mx) \exp(imy)$$

and since the problem is linear, the general solution to this problem is

(1.2.55)
$$\psi = \sum_m \psi_m$$

For cylindrical polar coordinates, (1.2.46) and (1.2.47) are

(1.2.56)
$$\frac{1}{r} \frac{\partial}{\partial r} \left(r \frac{\partial \psi}{\partial r} \right) + \frac{1}{r^2} \frac{\partial^2 \psi}{\partial \theta^2} = -\omega$$

(1.2.57)
$$\frac{1}{r} \frac{\partial}{\partial r} \left(r \frac{\partial \psi}{\partial r} \right) + \frac{1}{r^2} \frac{\partial^2 \psi}{\partial \theta^2} = 0$$

and our solution methodology is similar to the one for Cartesian coordinates. The general solution for ω is

(1.2.58)
$$\omega = (a_0 + b_0 \theta)(c_0 + d_0 \ln r) + \sum_{n=1}^{\infty} (a_n \cos \lambda_n \theta + b_n \sin \lambda_n \theta) \left(c_n r^{\lambda_n} + d_n r^{-\lambda_n} \right)$$

and the general solution for ψ is

(1.2.59)
$$\psi = \left(c_0 + d_0 \ln r + \gamma_0 r^2 + \delta_0 r^2 (\ln r - 1) \right) (a_0 + b_0 \theta)$$
$$+ \left(c_1 r + d_1 r^{-1} + \gamma_1 r^3 \right) (a_1 \sin \theta + b_1 \cos \theta) + \delta_1 r (\alpha_1 \theta \sin \theta + \beta_1 \theta \cos \theta)$$
$$+ \sum_{n=2}^{\infty} \left(c_{\lambda_n} r^{\lambda_n} + c'_{\lambda_n} r^{-\lambda_n} + \gamma_{\lambda_n} r^{2+\lambda_n} + \delta_{\lambda_n} r^{2-\lambda_n} \right) (a_{\lambda_n} \sin \lambda_n \theta + b_{\lambda_n} \cos \lambda_n \theta)$$

1.2.2. Flow in a corner and Moffat vortices. We can apply the general streamline solutions that we have just obtained to investigate flow in a corner. A corner can be idealized as a wedge, or a section of a circle and thus we can use the solution for Navier-Stokes equations in polar coordinates. A set of corner flow problems were investigated by Moffatt (1964) and Taylor (1960) but here we will consider a flow in a corner that is induced by an arbitrary flow roughly orthogonal to the axis of the corner (Figure 1.2.1), as that situation frequently arises in microfluidic devices. (We will consider some of the situations in which this arises as well as experimentally obtained streamlines for the flows in these situations in Section 1.5). This problem is particularly interesting in that its solution shows that

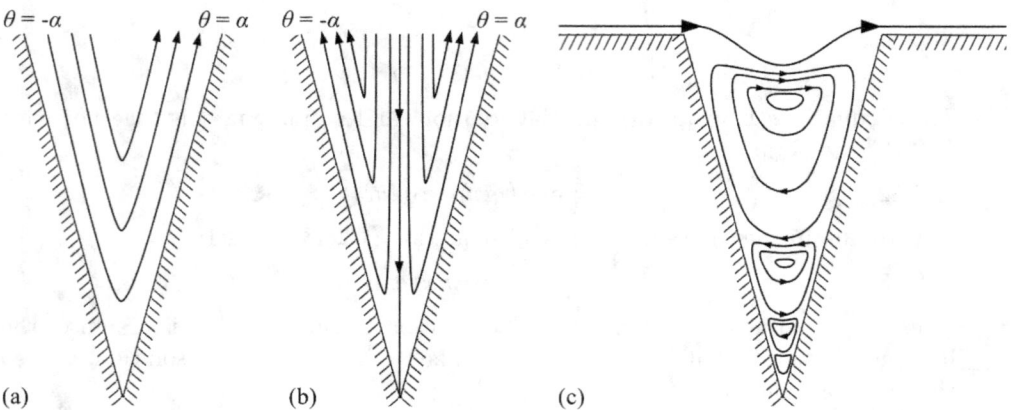

FIGURE 1.2.1. Sketch of streamlines of 2D flow in a corner, induced by arbitrary flow (e.g., stirring) far from the corner. The general direction of this flow is parallel to the top walls. (a) Sketch of antisymmetric component of the streamlines. (b) Sketch of symmetric component of the streamlines. (c) Sketch of streamlines in the corner, showing an infinite sequence of counter rotating eddies known as Moffat vortices.

there is an infinite sequence of vortices going down into the corner (Figure 1.2.1c)! (And as we later discuss, such a sequence of vortices has been observed experimentally, albeit of course only a finite number of vortices have been observed). These vortices are known as Moffat vortices.

One interesting aspect of this geometry is that it lacks a physical length scale (since we have not yet specified how deep the corner is, and it turns out that it is more interesting not to specify this, since the deeper you go into the corner, the smaller your length scale). Since we don't have a length scale, we cannot specify the Reynolds number is the usual way and so we don't know if and when the creeping flow equations apply. So we have to define Reynolds number in a new way, namely as the ratio of inertia to viscous terms in the momentum equation,

$$(1.2.60) \qquad Re = \frac{|\mathbf{u} \cdot \nabla \mathbf{u}|}{|\nu \nabla^2 \mathbf{u}|}$$

where ν is the kinematic viscosity. To obtain the magnitude of velocity, we can turn to our general solution for the stream function (1.2.59). From (1.2.59) and definition of the stream function (1.2.36), we can obtain a scaling for the magnitude of velocity as

$$(1.2.61) \qquad |\mathbf{u}| = O\left(Ar^{\xi-1}\right)$$

where ξ is the real part of λ in (1.2.59), and A is a constant of order unity with units of velocity/length$^\xi$. Substituting this scaling into (1.2.60) we obtain

$$(1.2.62) \qquad Re = \frac{O\left(A^2 r^{2\xi-3}\right)}{O\left(\nu Ar^{\xi-3}\right)} = O\left(\frac{Ar^\xi}{\nu}\right)$$

Thus for this problem the Reynolds number is based on the distance from the corner. From (1.2.62) we see that for $\xi > 0$ Reynolds number is small for small values of r. Whereas for $\xi < 0$ Reynolds number is small only if r is large. Since we typically scale the geometry such that $r < 1$, and we would like to use the creeping flow approximation, we thus only consider the case for $\xi > 0$.

The boundary conditions for this problem are that firstly there is no-slip at the walls,

$$(1.2.63) \qquad u_r = 0 \text{ at } \theta = \alpha, -\alpha$$

secondly, there is also no-penetration at the walls, hence

$$(1.2.64) \qquad u_\theta = 0 \text{ at } \theta = \alpha, -\alpha$$

thirdly, we need a boundary condition to represent the arbitrary stirring motion far from the corner. For now we hold off on defining this boundary condition. Often, to define this condition, we use the solution for the flow far from the corner and match the flow velocity field of that flow in the region closer to the corner. The velocity components are related to the stream function from definition of the stream function,

$$(1.2.65) \qquad u_r = \frac{1}{r}\frac{\partial \psi}{\partial \theta}$$

$$(1.2.66) \qquad u_r = -\frac{\partial \psi}{\partial r}$$

In general there are two types of flows that can be induced in a corner: an anti-symmetric flow (Figure 1.2.1a), and symmetric flow (Figure 1.2.1b). Since we are in a creeping flow regime and our governing equations are linear, we take the linear combination of these two flows to construct a real flow in a corner. Here we will consider the antisymmetric case, since it contributes the Moffat vortices. Thus from (1.2.59) we pick out the antisymmetric form of ψ, subject to the consideration that $\xi > 0$,

$$(1.2.67) \qquad \psi = \sum_{n=1}^{\infty} r^{\lambda_n} f_{\lambda_n}(\theta)$$

$$(1.2.68) \qquad f_{\lambda_n}(\theta) = A_n \cos \lambda_n \theta + C_n \cos(\lambda_1 - 2)\theta$$

From the boundary conditions at the walls we have

$$(1.2.69) \qquad f'(\pm\alpha) = f(\pm\alpha) = 0$$

Next, since r is small, and $\xi > 0$ and so $\text{Re}(\lambda) > 0$, as a first approximation, we only take the first (but the most dominant) term of (1.2.67),

$$(1.2.70) \qquad \psi \sim r^{\lambda_1} f_{\lambda_1}(\theta) = r^{\lambda_1}(A_1 cos\lambda_1\theta + C_1 cos(\lambda_1 - 2)\theta)$$

Substituting in the boundary conditions (1.2.69), we obtain

$$(1.2.71) \qquad A_1 \cos \lambda_1 \alpha + C_2 \cos(\lambda_1 - 2)\alpha = 0$$

$$(1.2.72) \qquad A_1 \lambda_1 \sin \lambda_1 \alpha + C_2(\lambda_1 - 2)\sin(\lambda_1 - 2)\alpha = 0$$

To solve this set of linear equations, we can use the method of determinants, where

$$(1.2.73) \qquad \det \begin{vmatrix} \cos \lambda_1 \alpha & \cos(\lambda_1 - 2)\alpha \\ \lambda_1 \sin \lambda_1 \alpha & (\lambda_1 - 2)\sin(\lambda_1 - 2)\alpha \end{vmatrix} = 0$$

Here λ_1 is known as the eigenvalue of this problem and so $f_{\lambda_1}(\theta)$ the eigenfunction. We find that

$$(1.2.74) \qquad\qquad A_1 = K \cos\left(\lambda_1 - 2\right)\alpha$$

$$(1.2.75) \qquad\qquad C_1 = -K \cos \lambda_1 \alpha$$

and so

$$(1.2.76) \qquad \psi = K r^{\lambda_1}\left(\cos\left(\lambda_1 - 2\right)\alpha \cos \lambda_1 \theta - \cos \lambda_1 \alpha \cos\left(\lambda_1 - 2\right)\theta\right)$$

The coefficient K is still undefined as we need the third boundary condition, the velocity far from the corner, to define it. From (1.2.73) we also have

$$(1.2.77) \qquad \left(\lambda_1 - 2\right)\sin\left(\lambda_1 - 2\right)\alpha \cos \lambda_1 \alpha - \lambda_1 \sin \lambda_1 \alpha \cos\left(\lambda_1 - 2\right)\alpha = 0$$

or

$$(1.2.78) \qquad \sin\left(2\alpha\left(\lambda_1 - 1\right)\right) - \left(\lambda_1 - 1\right)\sin 2\alpha = 0$$

Solving this equation numerically, we find that when $2\alpha > 146°$ this equation has a real solution for λ_1. When $0 \leqslant 2\alpha \leqslant 146°$ the solution is complex. For the complex solution, it turns out we can find that the streamline has infinitely many zeros. Zero valued streamlines ($\psi_1 = 0$) separate two counter rotating vortices. This means that this solution predicts a sequence of counter rotating vortices going down into the corner (Figure 1.2.1c). To see that the solution has these zeros, let

$$(1.2.79) \qquad\qquad \left(\lambda_n - 1\right) = p + iq$$

Then we can rewrite (1.2.78) (now for an arbitrary n) as

$$(1.2.80) \qquad\qquad \sin \xi \cosh \eta = -\frac{\sin \alpha}{\alpha}\xi$$

$$(1.2.81) \qquad\qquad \cos \xi \cosh \eta = -\frac{\sin \alpha}{\alpha}\eta$$

where $\xi = 2\alpha p$, $\eta = 2\alpha q$ and where these equations must satisfy,

$$(1.2.82) \qquad\qquad \left(2n - 1\right)\pi < \xi_n < \left(2n - 1/2\right)\pi$$

and $\sin \xi$ and $\cos \xi$ are both negative. Then

$$(1.2.83) \qquad\qquad \lambda_n = 1 + \left(2\alpha\right)^{-1}\left(\xi_n + i\eta_n\right)$$

The values of ξ_n and η_n are tabulated in Moffatt (1964). Substituting (1.2.83) into (1.2.76), we obtain

$$(1.2.84) \qquad \psi_1 = r^{(1+p)}\left(\cos\left(q \ln r\right)g_1\left(\theta\right) - \sin\left(q \ln r\right)g_2\left(\theta\right)\right)$$

where g_1 and g_2 are functions from Moffatt (1964). $\psi_1 = 0$ when

$$(1.2.85) \qquad q \ln r = \tan^{-1}\left(\frac{g_1\left(\theta\right)}{g_2\left(\theta\right)}\right) - \frac{\pi}{2} - n\pi \text{ for } n = 0, 1, 2, \ldots$$

We see that indeed there are infinitely many zeros of the stream function (implying infinitely many vortices). We also see that these zeros depend logarithmically on the distance from the tip of the corner, r. As we will discuss in a later section, such a sequence of vortices has been indeed observed experimentally by Taneda (1979), albeit of course only for a finite number of vortices.

1.2.3. Solutions to the biharmonic equation: spherical coordinates.
In this section we will discuss the last common axisymmetric geometry, the spherical axisymmetric geometry. This geometry is quite important as it gives us the solution of creeping flow around a single sphere - a first approximation model for practically any particle, from ions to biological cells (and occasionally actually spherical hard particles of reasonable dimensions). Unfortunately for spherical geometry $E^4\psi = 0$ does not simplify to $\nabla^4\psi = 0$, and so we must solve the more general equation. Here we will consider a spherical coordinate system with (r, θ, φ), where everything is uniform in the φ direction and $u_\varphi = 0$. Additionally,

$$(1.2.86) \qquad u_\theta = 0 \text{ at } \theta = 0, \pi$$

Furthermore, instead of working with θ directly, it is more convenient to introduce

$$(1.2.87) \qquad \eta = \cos\theta$$

so that the coordinate system becomes (r, η, φ), $-1 \leqslant \eta \leqslant 1$. In this coordinate system the operator E^2 becomes

$$(1.2.88) \qquad E^2 = \frac{\partial^2}{\partial r^2} + \frac{1-\eta^2}{r^2}\frac{\partial^2}{\partial\eta^2}$$

From (1.2.36) the velocity components of the flow expressed in terms of the stream function are

$$(1.2.89) \qquad u_r = \frac{1}{r^2}\frac{\partial\psi}{\partial\eta}$$

$$(1.2.90) \qquad u_\theta = -\frac{1}{\sqrt{1-\eta^2}}\frac{1}{r}\frac{\partial\psi}{\partial r}$$

Condition (1.2.86) combined with (1.2.90) requires

$$(1.2.91) \qquad \frac{\partial\psi}{\partial r} = 0 \text{ at } \eta = \pm 1$$

Typically we are interested in axisymmetric bodies with their centers at the point of $r = 0$. Typically also these bodies are impermeable. If they are, then the stream function is constant on their surface. From integrating (1.2.91) on this surface this means

$$(1.2.92) \qquad \psi = const \text{ at } \eta = \pm 1$$

It is convenient to take this constant to be zero. Since the stream function really represents a potential, we have no problem in arbitrarily setting some reference point to be zero, and the surface of our body of interest is a good point to take as a reference. Thus,

$$(1.2.93) \qquad \psi = 0 \text{ at } \eta = \pm 1$$

Now our task is to solve $E^4\psi = 0$. As before we split this fourth order equation into two second order equations,

$$(1.2.94) \qquad E^2\psi = -\omega$$

$$(1.2.95) \qquad E^2\omega = 0$$

As before, we proceed with a separation of variables solution, first for ω and then for ψ. We thus first seek a solution in the form of

$$(1.2.96) \qquad \omega = R(r)H(\eta)$$

which upon substituting into (1.2.95) via the definition (1.2.88) we obtain

$$(1.2.97) \qquad \frac{r^2}{R}\frac{d^2R}{dr^2} = -\frac{\left(1-\eta^2\right)}{H}\frac{d^2H}{d\eta^2} = n\left(n+1\right)$$

where $n\left(n+1\right)$ is our separation variable. (We could have chosen something simple as a separation variable, but from experience this form makes the algebra simpler later.) Thus, as usual, we obtain two equations

$$(1.2.98) \qquad r^2\frac{d^2R}{dr^2} - n\left(n+1\right)R = 0$$

$$(1.2.99) \qquad \left(1-\eta^2\right)\frac{d^2H}{d\eta^2} + n\left(n+1\right)H = 0$$

Equation (1.2.98) is Euler's equation, and has two independent solutions,

$$(1.2.100) \qquad \begin{aligned} R &= r^{n+1} \\ R &= r^{-n} \end{aligned}$$

To put (1.2.99) into a form of a known ODE we can let $Y = dH/d\eta$ and so rewrite (1.2.99) as

$$(1.2.101) \qquad \frac{d}{d\eta}\left(\left(1-\eta^2\right)\frac{dY}{d\eta}\right) + n\left(n+1\right)Y = 0$$

This is a Legendre's equation and its solutions are Legendre functions of the first and second kind (see Abramowitz and Stegun (1964)). With our convenient choice of separation variable $n\left(n+1\right)$, n dictates the degree of the Legendre polynomial. Legendre polynomial of the first kind is regular on our domain, while the Legendre polynomial of the second kind has a logarithmic singularity at $\eta = \pm 1$. We would like our solution to be regular in the entire domain, hence we can automatically state the coefficient for the second solution, the Legendre polynomials of the second kind is zero. Hence we focus on Legendre polynomials of the first kind, designated $P_n\left(\eta\right)$. Thus,

$$(1.2.102) \qquad Y\left(\eta\right) = P_n\left(\eta\right)$$

As an example,

$$(1.2.103) \qquad \begin{aligned} P_0 &= 1 \\ P_1 &= \eta \\ P_2 &= \left(3\eta^2 - 1\right)/2 \end{aligned}$$

Now we need to find the solution for H. To satisfy (1.2.93) we need

$$(1.2.104) \qquad H\left(\eta\right) = 0 \text{ at } \eta = \pm 1$$

Thus, integrating Y we get

$$(1.2.105) \qquad H\left(\eta\right) = \int_{-1}^{\eta} Y\left(\eta\right)d\eta$$

The condition $H\left(\eta\right) = 0$ at $\eta = -1$ is satisfied automatically. Due to the nature of Legendre polynomials, the condition $H\left(\eta\right) = 0$ at $\eta = 1$ is also satisfied for all n except $n = 0$, as Legendre polynomials are orthogonal. Turns out the integral

(1.2.105) when Legendre polynomials are substituted in evaluate to another type of named polynomials, Gegenbauer polynomials (Abramowitz and Stegun (1964)),

$$(1.2.106) \qquad Q_n(\eta) = \int_{-1}^{\eta} P_n(\eta) d\eta$$

which for example for $\eta \neq 0$ are

$$(1.2.107) \qquad \begin{aligned} Q_1(\eta) &= (\eta^2 - 1)/2 \\ Q_2(\eta) &= (\eta^3 - \eta)/2 \\ Q_3(\eta) &= (5\eta^2 - 1)(\eta^2 - 1)/8 \end{aligned}$$

Conveniently Gegenbauer polynomials are also orthogonal, with the orthogonality condition

$$(1.2.108) \qquad \int_{-1}^{1} \frac{Q_n(\eta) Q_m(\eta)}{(1 - \eta^2)} d\eta = \begin{cases} 0 \text{ for } n \neq m \\ 2/(n(n+1)(2n+1)) \text{ for } n = m \end{cases}$$

Q_n are zero at $\eta = \pm 1$ for all n except $n = 0$. Thus, $Q_n(\eta)$ for $n \geqslant 1$ are the eigenfunctions of (1.2.99) and of course the separation variable, $n(n+1) = \lambda_n$ are the eigenvalues. Thus the solution of (1.2.95) subject to the symmetry boundary condition, is

$$(1.2.109) \qquad \omega = \sum_{n=1}^{\infty} \left(\overline{A_n} r^{n+1} + \overline{C_n} r^{-n} \right) Q_n(\eta)$$

Substituting this into (1.2.94)

$$(1.2.110) \qquad E^2 \psi = -\sum_{n=1}^{\infty} \left(\overline{A_n} r^{n+1} + \overline{C_n} r^{-n} \right) Q_n(\eta)$$

Once again the solution is formed from two parts: the homogeneous part (a solution to Laplace's equation, just like the (1.2.95)),

$$(1.2.111) \qquad \psi_h = \sum_{n=1}^{\infty} \left(B_n r^{n+1} + D_n r^{-n} \right) Q_n(\eta)$$

and a particular part. From experience, we try a particular solution in the form

$$(1.2.112) \qquad \psi_p = r^\lambda Q_n(\eta)$$

and so

$$(1.2.113) \qquad E^2 \psi_p = \alpha r^{\lambda-2} Q_n(\eta)$$

where α is a constant that depends on λ and n. Comparing this with (1.2.111) we see that for one term $\lambda = n + 3$ and for another term $\lambda = 2 - n$. Thus we see that the particular solution should be

$$(1.2.114) \qquad \psi_p = \sum_{n=1}^{\infty} \left(A_n r^{n+3} + C_n r^{2-n} \right) Q_n(\eta)$$

and thus the complete solution is

$$(1.2.115) \qquad \psi = \sum_{n=1}^{\infty} \left(A_n r^{n+3} + B_n r^{n+1} + C_n r^{2-n} + D_n r^{-n} \right) Q_n(\eta)$$

where, as usual, the constants A, B, C, D are found from applying the specific boundary conditions.

1.2.4. Axial force on an arbitrary axisymmetric body. Very often we are less interested about the distribution of flow velocities around a body, and are chiefly interested in the drag experienced by the body (i.e., the force antiparallel to the body's motion). Turns out for axisymmetric bodies (Figure 1.2.2) in creeping flow we can obtain this force by just obtaining a single constant from (1.2.115), rather than all the constants, which can be quite laborious. This result applies to any creeping flow (e.g., uniform, shear, parabolic) in which the symmetry axis of the body is aligned to the general direction of the flow. To obtain this, we begin by noting that the force on an object is just

$$(1.2.116) \qquad \mathbf{F} = \mu U l_c \int_S (\mathbf{T} \cdot \mathbf{n}) dA$$

where here inside the integral the quantities are dimensionless but the force is dimensioned. Note we made the quantities inside the integral dimensionless with the viscosity μ, characteristic velocity U, and characteristic length scale l_c. To evaluate the integral we recall that the momentum equation for creeping flows, (1.0.11),

$$(1.2.117) \qquad \nabla^2 \mathbf{u} - \nabla p = 0$$

can be rewritten in terms of the stress tensor as

$$(1.2.118) \qquad \nabla \cdot \mathbf{T} = 0$$

Since

$$\mathbf{T} = p\mathbf{I} + \left(\nabla \mathbf{u} + \nabla \mathbf{u}^T\right)$$

Applying the divergence theorem, with the volume being the volume of fluid that is contained between the surface of the body of interest S and any arbitrary closed surface that encloses the body S^*,

$$(1.2.119) \qquad 0 = \int_V (\nabla \cdot \mathbf{T}) dV = \int_S (\mathbf{T} \cdot \mathbf{n}) dS - \int_{S^*} (\mathbf{T} \cdot \mathbf{n}^*) dS^*$$

where $\mathbf{n}^*$ denotes the outer normal to the surface S^*. The trick here is that we get to replace the integral over the surface of our body of interest S, which may be of complicated shape, with the integral over the surface that we get to choose S^*, which we choose to make the integral as simple as possible. A typical choice is a sphere centered with the center of mass of the body, and radius R^* that circumscribes our body of interest. With this choice of integration surface (1.2.116) becomes,

$$(1.2.120) \qquad \mathbf{F} = \mu U l_c \int_0^{2\pi} \int_0^{\pi} \mathbf{T} \cdot \mathbf{n}^* (R^*)^2 \sin\theta d\theta d\varphi$$

Since $\mathbf{n}^*$ is along the radial coordinate, we can simplify this to

$$(1.2.121) \qquad \mathbf{F} = \mu U l_c \int_0^{2\pi} \int_0^{\pi} \mathbf{T} \cdot \hat{\mathbf{r}} (R^*)^2 \sin\theta d\theta d\varphi$$

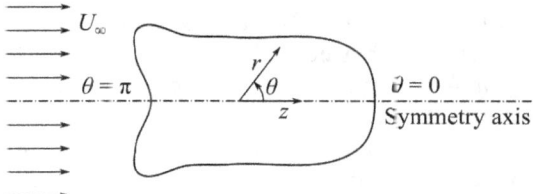

FIGURE 1.2.2. Axisymmetric body in a uniform flow with the axis of symmetry aligned with the flow.

Since we are interested in the direction along the axis of symmetry,

$$(1.2.122) \qquad F_z = \hat{\mathbf{z}} \cdot \mathbf{F} = \mu U l_c \int_0^{2\pi} \int_0^{\pi} \hat{\mathbf{z}} \cdot (\mathbf{T} \cdot \hat{\mathbf{r}}) \, (R^*)^2 \sin\theta \, d\theta \, d\varphi$$

Furthermore, since for this problem everything is uniform in the φ direction and $u_\varphi = 0$, $T_{r\varphi} = 0$. Thus,

$$(1.2.123) \qquad \mathbf{T} \cdot \hat{\mathbf{r}} = T_{rr}\hat{\mathbf{r}} + T_{\theta r}\hat{\theta}$$

Since

$$(1.2.124) \qquad \hat{\mathbf{z}} = \hat{\mathbf{r}} \cos\theta - \hat{\theta} \sin\theta$$

we obtain that

$$(1.2.125) \qquad \hat{\mathbf{z}} \cdot (\mathbf{T} \cdot \hat{\mathbf{r}}) = \left(\hat{\mathbf{r}}\cos\theta - \hat{\theta}\sin\theta\right)\left(T_{rr}\hat{\mathbf{r}} + T_{\theta r}\hat{\theta}\right) = T_{rr}\cos\theta - T_{\theta r}\sin\theta$$

since $\hat{\mathbf{r}}$ and $\hat{\theta}$ are orthogonal The relevant components of the stress tensor are,

$$(1.2.126) \qquad T_{rr} = -p + 2\frac{\partial u_r}{\partial r}$$

$$(1.2.127) \qquad T_{r\theta} = r\frac{\partial}{\partial r}\left(\frac{u_\theta}{r}\right) + \frac{1}{r}\frac{\partial u_r}{\partial \theta}$$

We can simplify (1.2.122) further noting that everything is uniform in the φ direction, and so obtain

$$(1.2.128) \qquad F_z = 2\pi \mu U l_c \int_0^{\pi} \hat{\mathbf{z}} \cdot (\mathbf{T} \cdot \hat{\mathbf{r}})\, (R^*)^2 \sin\theta \, d\theta$$

Substituting the relation for the stream function (1.2.115) into the definitions of the velocity components, (1.2.89) and (1.2.90) and then that into (1.2.126) and (1.2.127) and performing the integral (1.2.128), we obtain that

$$(1.2.129) \qquad F_z = 4\pi \mu U l_c C_1$$

This is an important result since the entire body's geometry information as it pertains to the drag force on the body is summarized in a single constant C_1 and a length scale l_c. This result allows us, for example to tabulate the drag force for axisymmetric bodies in a simple manner as we do in Table 1.

1.2.5. Flow field around an arbitrary axisymmetric body in uniform streaming flow. It turns out that not only we can obtain a simple relation for the force on an arbitrary axisymmetric body, but for an arbitrary body in a uniform streaming flow (Figure 1.2.2) we can also obtain a simple relation for the flow field contributed by that body. We begin by writing down that far from our body of interest the dimensionless velocity is uniform and directed along the axis of the body,

$$(1.2.130) \qquad \mathbf{u} = \hat{\mathbf{z}} \text{ as } r \to \infty$$

converting this to the spherical coordinates we have been using,

$$(1.2.131) \qquad \hat{\mathbf{z}} = \hat{\mathbf{r}} \cos\theta - \hat{\theta} \sin\theta = \hat{\mathbf{r}}\eta - \hat{\theta}\left(1 - \eta^2\right)^{1/2}$$

we obtain,

$$(1.2.132) \qquad u_r = \eta \text{ as } r \to \infty$$

$$(1.2.133) \qquad u_\theta = -\left(1 - \eta^2\right)^{1/2} \text{ as } r \to \infty$$

Substituting this into the definitions of the stream function, (1.2.89) and (1.2.90),

$$(1.2.134) \qquad \frac{\partial \psi}{\partial r} = r\left(1 - \eta^2\right) \text{ as } r \to \infty$$

$$(1.2.135) \qquad \frac{\partial \psi}{\partial \eta} = -\eta r^2 \text{ as } r \to \infty$$

Integrating (1.2.134) and (1.2.135) we obtain

$$(1.2.136) \qquad \psi \to \frac{r^2}{2}\left(1 - \eta^2\right) + const \text{ as } r \to \infty$$

Since the stream function is a potential function, we can arbitrarily set the arbitrary constant to zero,

$$(1.2.137) \qquad \psi \to \frac{r^2}{2}\left(1 - \eta^2\right) \text{ as } r \to \infty$$

This is the same as we have done for (1.2.93), and the condition (1.2.93) applies here as well,

$$(1.2.138) \qquad \psi = 0 \text{ at } \eta = \pm 1$$

Since our general form of the stream function (1.2.115) is expressed in terms of the Gegenbauer polynomials $Q_n(\eta)$ we would like to express this boundary condition in terms of these as well. From (1.2.107), we see that we can express (1.2.137) as

$$(1.2.139) \qquad \psi \to -r^2 Q_1(\eta) \text{ as } r \to \infty$$

Applying this boundary condition to the general form of the stream function (1.2.115) we can tell that for the boundary condition to be satisfied, $A_n = 0$ for all n; $B_1 = -1$ and $B_n = 0$ for all other n; We cannot say anything on the form of C_n and D_n at this time. Thus, this limits the form of the stream function to

$$(1.2.140) \qquad \psi = -r^2 Q_1(\eta) + \sum_{n=1}^{\infty}\left(C_n r^{2-n} + D_n r^{-n}\right) Q_n(\eta)$$

Notice, that thus far we have only applied the condition of uniform flow, and have stated nothing about the body geometry. The first term in (1.2.140) is due to the undisturbed uniform flow and the second (sum) term is due to the body disturbing the flow. We know from (1.2.129) that at least $C_1 \neq 0$. Furthermore, if the terms C_n and D_n are of equal order of magnitude, far away from the body (large r) the effect of C_n terms dominates over D_n terms. Thus the largest contribution of the body to the flow far from the body (large r) is

$$(1.2.141) \qquad \psi_d \approx C_1 r Q_1(\eta)$$

where ψ_d refers to the disturbed contribution to the flow. Integrating this according to the definitions of the stream function (1.2.89) and (1.2.90), we obtain the contributions to the velocity field from the disturbance by the body,

$$(1.2.142) \qquad u_{r,d} = -\frac{C_1}{r} P_1(\eta)$$

$$(1.2.143) \qquad u_{\theta,d} = -\frac{C_1}{r} \frac{1}{\sqrt{1-\eta^2}} Q_1(\eta)$$

We see that even without specifying the body geometry, that in creeping flow the disturbance by the body decays slowly, only as $1/r$, and so is felt quite a distance from the body. This means that even at a distance of 10 radii away from the body, the velocity is still roughly 10% below the free stream velocity magnitude. Furthermore, as we shall see later, the velocity field (1.2.142) and (1.2.143) represents the velocity field induced by a point force at the origin in creeping flow and is known as the Stokeslet velocity field. As we shall see later such velocity field is a fundamental solution, and such solutions can be additively superimposed on each other to obtain solutions for flows around complicated geometries.

1.3. Creeping flow around a single sphere

Having found relations for a stream functions for axisymmetric bodies and having discussed the particulars of flows for axisymmetric bodies, we are now ready to turn to investigating the flow around the quintessential axisymmetric body - the sphere. The motion of an isolated sphere (i.e. far from walls, or other bodies) is of fundamental importance in microfluidics, where this model of a motion of a sphere is used to model all sorts of situations, from motions of cells (typically 10s of microns in size) to motion of ions (typically, 0.1s nm)! We are often interested in a sphere moving in a quiescent fluid at constant speed at low enough Reynolds number such that the flow can be considered creeping. For now never mind what forces cause the sphere to move - this can be an electric, magnetic, or gravitational forces - these will be discussed later. To simplify the problem, and to use the result from the previous section, here we consider the problem from the point of view of the sphere: the sphere is stationary while a fluid with a stream speed U flows around it as shown in Figure 1.3.1.

The boundary conditions (in dimensionless variables) on the surface of the sphere are the no-slip condition,

$$(1.3.1) \qquad u_\theta = 0 \text{ at } r = 1$$

and the no-penetration boundary condition,

$$(1.3.2) \qquad u_r = 0 \text{ at } r = 1$$

In terms of our stream function, these boundary conditions translate to

$$(1.3.3) \qquad \frac{\partial \psi}{\partial \eta} = \frac{\partial \psi}{\partial r} = 0 \text{ at } r = 1$$

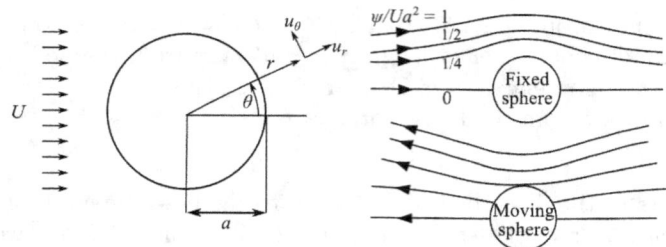

FIGURE 1.3.1. Schematic of coordinate system for calculating flow around a sphere (left). Sketch of streamlines around a fixed and moving sphere (right). The sphere in creeping flow significantly affects the flow around it. For example, for a fixed sphere even 10 radii away from the sphere, the velocities are still about 10% less than their free stream values!

As before, this implies that

(1.3.4) $\psi = const$ at $r = 1$

and as before we set this constant to zero. Therefore we have

(1.3.5) $\psi = 0$ at $r = 1$

and from (1.2.93),

(1.3.6) $\psi = 0$ at $\eta = \pm 1$

The condition (1.3.6) is a symmetry condition. Applying these conditions to (1.2.140) we obtain that $C_1 = 3/2$ and $C_n = 0$ for all other n; $D_1 = -1/2$ and $D_n = 0$ for all other n. Thus, (1.2.140) simplifies to

(1.3.7) $$\psi = -\left(r^2 - \frac{3}{2}r + \frac{1}{2r}\right) Q_1(\eta)$$

Substituting this into the definition of the stream function, and converting the velocity components to dimensional form we obtain

(1.3.8)
$$u_r = U \cos\theta \left(1 + \frac{a^3}{2r^3} - \frac{3a}{2r}\right)$$
$$u_\theta = U \sin\theta \left(-1 + \frac{a^3}{4r^3} + \frac{3a}{4r}\right)$$

Let's now study the velocity profile around the sphere. We sketch the streamlines for the sphere in the right part of Figure 1.3.1. Firstly, notice that the streamlines poses perfect fore-aft symmetry. (This is due to the symmetries of the sine squared function about the 12 o'clock - 6 o'clock axis). If you took a snapshot of a sphere and its streamlines (e.g. visualized by a dye), you could not tell if it was going forwards or backwards! This is very much unlike in higher Reynolds number flows, where a wake present at the back of the object is a dead giveaway to the direction of travel. The wake is due to the inertial forces, that are negligible here. Such similar, "wakeless" fore-aft symmetry behavior is observed for other symmetrical bodies in creeping flows.

Secondly, the effect of the sphere on the flow drops off slowly, as $1/r$ and even at a distance of 10 radii away from the sphere, the velocity is still roughly 10%

below the free stream velocity magnitude. The velocities and the streamlines are independent of the fluid viscosity (and this turns out to be true for all creeping flows), so the above holds for any fluid the particles are immersed in. This puts a constraint that spheres must be at least 10 radii from each other to be considered "isolated" and not interacting. For a quick estimate of maximum solution concentration of ideally dispersed spheres that satisfies this condition: we can assume that each sphere needs an exclusion radius 10 times its own radius, and such exclusion spheres are closed-packed, giving a concentration of 0.074% v/v. For comparison, the typical concentration of red blood cells in blood (the hematocrit) is around 40-50% for whole blood - the red blood cells are certainly interacting with each other! Lastly, the velocity is everywhere less than its free stream value, unlike in potential flows, where at the sphere shoulder the velocity is 1.5x that of the free stream.

With the velocity field known, we can use the momentum equation (1.C.11) to calculate the pressure distribution:

$$(1.3.9) \qquad p = p_\infty - \frac{3\mu a U}{2r^2} \cos\theta$$

Observe that the pressure is higher than the free stream at the front of the sphere, and is lower at the rear of the sphere. This causes pressure drag on the sphere. There is also surface shear force on the sphere. In spherical coordinates the relevant shear stress

$$(1.3.10) \qquad \tau_{r\theta} = \mu \left(\frac{1}{r} \frac{\partial u_r}{\partial \theta} + \frac{\partial u_\theta}{\partial r} - \frac{u_\theta}{r} \right) = -\frac{3}{2} \frac{\mu U \sin\theta}{r} \frac{a^3}{r^3}$$

We integrate the total pressure and shear around the surface of the sphere to calculate the total drag force:

$$(1.3.11) \qquad F = -\int_0^\pi \tau_{r\theta}|_{r=a} \sin\theta dA - \int_0^\pi p|_{r=a} \cos\theta dA$$

$$dA = 2\pi a^2 \sin\theta d\theta$$

where the sine and cosine terms in the respective integrals account for the direction of application of the respective force (pressure acts normal, whereas shear acts tangentially). After combining (1.3.11) with (1.3.10) and (1.3.9) we obtain

$$(1.3.12) \qquad F = 4\pi\mu U a + 2\pi\mu U a = 6\pi\mu U a$$

the famous Stokes' drag on a sphere formula, a result so influential, it is often referred to as "Stokes's law". Notice that 2/3 of the drag comes from shear stress - the so called viscous drag, and 1/3 comes from the pressure difference between the fore and aft surfaces of the sphere - pressure drag. Furthermore, notice that the force is proportional to velocity, not the first derivative of velocity, as it "ought" to be from Newton's second law! This is actually reminiscent of Aristotelian mechanics where the force is proportional to the velocity of the object. In creeping flows, Aristotelian mechanics, a special case of Newtonian mechanics, hold, since inertia is neglected. In fact, we derived this result using the Navier-Stokes momentum equation - Newton's second law as applied to fluids, by dropping the inertia term. This Stokes' drag result is strictly valid only for Reynolds number much less than unity, however, it agrees well with experiment up to Reynolds number of unity.

To appreciate how unimportant inertia is, let's briefly calculate deceleration time of a sphere traveling initially with velocity U to a negligible velocity, $0.01U$, by integrating Newton's second law and utilizing (1.3.12)

$$\int_0^t (F/m)dt = \int_U^{0.01U} du$$

(1.3.13)

$$\int_0^t (-6\pi\mu a/m)dt = \int_U^{0.01U} \frac{du}{u}$$

$$t = 4.61\frac{m}{6\pi\mu a} = 1.02a^2\frac{\rho_p}{\mu}$$

where m is the mass of the particle, and ρ_p is it's density. Notice that this deceleration time increases as the square of the particle size - the larger the particle, the more important the inertia, the longer the particle takes to decelerate. Note that the deceleration time is independent of the initial velocity! For a 10 μm and 1 μm radius particle with density about that of water in water, this deceleration times are 100 μs and 1 μs. To study the deceleration of the 1 μm particle, one would need a camera capable of more than 10^6 frames/s. To an ordinary observer without special tools such microscopic particles stop virtually instantaneously.

Lastly, we can use our newly derived Stokes' drag relation to obtain the settling velocity of a sphere. During settling, the drag force balances out the gravitational (buoyancy) force on the sphere. The boyancy force on the sphere is just

(1.3.14)
$$F_b = \frac{4}{3}\pi a^3 (\rho_s - \rho_f) g$$

where ρ_s is the density of the sphere. Equating this with the drag force, (1.3.12) and solving for the velocity, we obtain,

(1.3.15)
$$U_s = \frac{2}{9}\frac{a^2 (\rho_s - \rho_f) g}{\mu}$$

Note that this is the settling velocity of an isolated sphere in a dilute solution, at low Reynolds number. For a more concentrated solution where particles interact with each other, more complicated relations for settling velocity are needed.

1.3.1. Creeping flows around other common immersed bodies. Similar analysis yields analytical solutions for other simple spheroids: from slightly deformed spheres, to disks and needles. Their detailed derivations can be found in the works of Happel and Brenner (1983) as well as that of Clift et al. (1979). Here will just summarize the results in Table 1, with the appropriate geometry and orientation to the free stream described in Figure 1.3.2. We specify the orientation of these bodies to the free stream relative to the axis of rotation (either tangential or normal). However, creeping motion is linear, and so we can superimpose the normal and tangential components without accounting for any interaction - we can simply add the components. Hence we can obtain the drag on these shapes when their axis of rotation is at some arbitrary angle to the free stream. Furthermore, for objects not listed here, we can still estimate the drag thanks to result from Hill and Power (1956) - the Hill and Power theorem. This theorem states that the Stokes' drag on a body must be smaller than any circumscribed figure, and larger than any inscribed figure. This result is especially useful for estimating (bounding) drag on irregular shaped objects.

TABLE 1. Stokes' drag on spheroids

Shape	C where $F = C\mu U a$	
Sphere	6π	
Disk normal to the free stream	16	
Disk parallel to the free stream	$32/3$	
Spheroid, oblate $(E = b/a < 1)$, flow tangential to axis of revolution	$(6\pi/5)\,(4+E)$	(approx.*)
	$\dfrac{8\pi\left(1-E^2\right)}{\left[(1-2E^2)\cos^{-1}\left(E/\sqrt{1-E^2}\right)\right]+E}$	(exact)
Spheroid, oblate $(E = b/a < 1)$, flow normal to axis of revolution	$(6\pi/5)\,(3+2E)$	(approx.*)
	$\dfrac{16\pi\left(1-E^2\right)}{\left[(3-2E^2)\cos^{-1}\left(E/\sqrt{1-E^2}\right)\right]-E}$	(exact)
Spheroid, prolate $(E = b/a > 1)$, flow tangential to axis of revolution	$(6\pi/5)\,(4+E)$	(approx.*)
	$\dfrac{8\pi\left(E^2-1\right)}{\left[(2E^2-1)\ln\left(E+\sqrt{E^2-1}\right)/\sqrt{E^2-1}\right]-E}$	(exact)
Spheroid, prolate $(E = b/a > 1)$, flow normal to axis of revolution	$(6\pi/5)\,(3+2E)$	(approx.*)
	$\dfrac{16\pi\left(E^2-1\right)}{\left[(2E^2-3)\ln\left(E+\sqrt{E^2-1}\right)/\sqrt{E^2-1}\right]+E}$	(exact)
Needle $(a \gg b)$, flow tangential to axis of revolution	$\dfrac{4\pi}{\ln(2a/b)-0.5}$	
Needle $(a \gg b)$, flow normal to axis of revolution	$\dfrac{8\pi}{\ln(2a/b)-0.5}$	

* $< \pm 10\%$ error for $0 < E < 5$

1.3.2. Creeping flow around a fluid sphere. Creeping flow around an immiscible fluid sphere is of great importance in microfluidics in the area of droplet microfluidics. This problem was originally solved independently by Rybczynski (1911) and Hadamard (1911a) and the derivation for this problem can be found in Happel and Brenner (1983). Here we will list the assumptions and the final result. First, we assume that the fluid drop is spherical. Interfacial tension at the junction between the two fluids tends to maintain the spherical shape, while the shearing forces due to the motion tend to deform this shape. If the motion is slow or a particle is small (i.e. low Reynolds number) then the droplet will be sufficiently spherical. Secondly, we assume that the sphere is impenetrable and that the tangential velocity is continuous across the interface. Thirdly, we assume equilibrium theory of interfacial tension is applicable here, and so the interfacial tension manifests itself as a pressure difference across the interface. The interfacial tension does not contribute to the tangential stress across the boundary, and so tangential stress is continuous across the boundary. Using a similar method to that of a rigid sphere, we obtain the internal (subscript i) and external (subscript o) streamlines

$$(1.3.16) \qquad \begin{aligned} \psi_o &= \sin^2\theta \left(\frac{1}{2}Br + \frac{D}{r} \right), \quad r \geqslant a \\ \psi_i &= \sin^2\theta \left(\frac{1}{10}Er^4 + Gr^2 \right), \quad r < a \end{aligned}$$

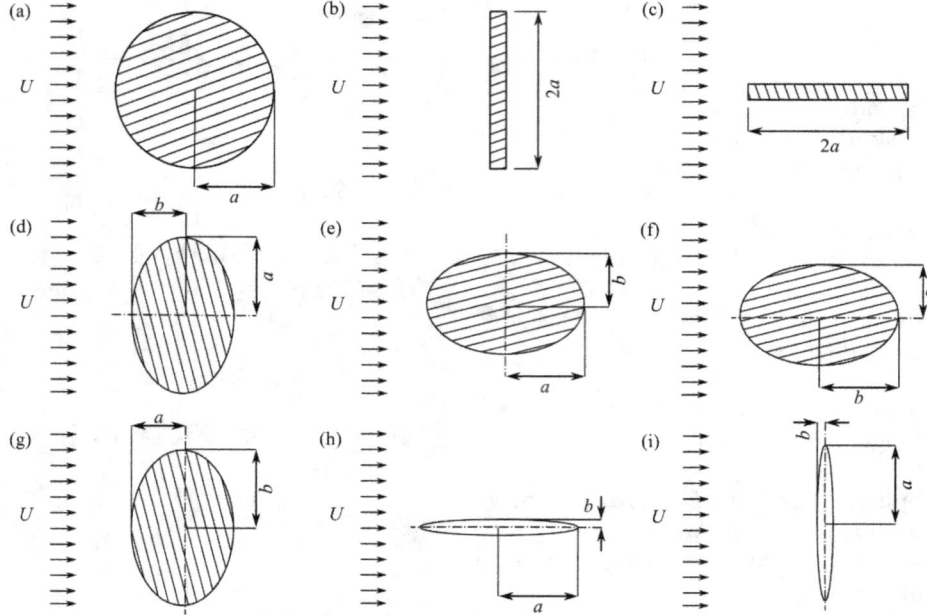

FIGURE 1.3.2. Relative dimensions and orientations of common immersed bodies: (a) sphere, (b) disk normal to the free stream, (c) disk parallel to the free stream, (d) spheroid, oblate, flow tangential to the axis of revolution, (e) spheroid, oblate, flow normal to the axis of revolution, (f) spheroid, prolate, flow tangential to the axis of revolution, (g) spheroid, prolate, flow normal to the axis of revolution, (h) needle, flow tangential to axis of revolution, (i) needle, flow normal to axis of revolution.

where

$$(1.3.17) \qquad \begin{aligned} B &= \frac{3}{2}Ua\frac{1+2\sigma/3}{1+\sigma} \\ D &= \frac{1}{4}Ua^3\frac{1}{1+\sigma} \\ E &= \frac{5}{2}\frac{U}{a^2}\frac{\sigma}{1+\sigma} \\ G &= -\frac{1}{4}U\frac{\sigma}{1+\sigma} \\ \sigma &= \mu_o/\mu_i \end{aligned}$$

A sketch of these streamlines is shown in Figure 1.3.3. We can clearly see that there is significant circulation inside the droplet. Hence the motion of a fluid droplet relative to the outside fluid can cause significant mixing of the contents of a droplet. This can either be detrimental, for example in the case where two reagents (e.g. two dyes) are deposited separately inside the droplet for creating Janus particles (e.g. a particle with a green florescent dye on one side, and a red fluorescent dye on another). This mixing can also be beneficial when we want the contents of the droplets to be well mixed - for example reagents inside the droplet undergoing a

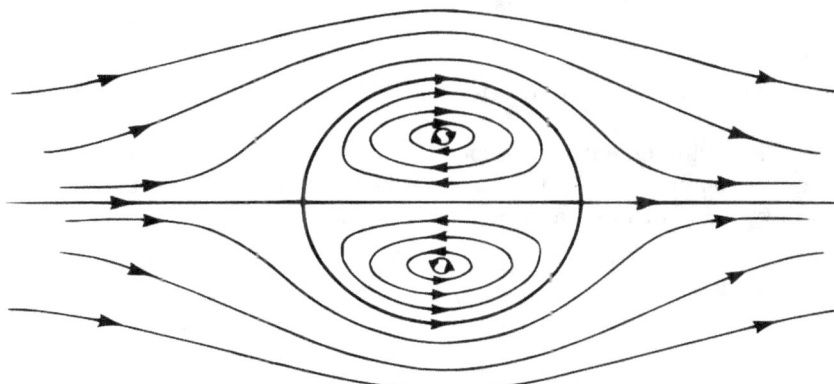

FIGURE 1.3.3. Sketch of streamlines for a fluid droplet moving in Stokes flow. We can clearly observe circulation inside the droplet.

reaction. Similarly, like in the case of a rigid sphere, we can obtain the total drag force on a fluid sphere:

$$(1.3.18) \qquad F = 6\pi\mu_o a U \frac{1 + 2\sigma/3}{1 + \sigma}$$

This force appears as a force on a rigid sphere with a correction factor for viscosity ratio of the internal and external liquids. Note that the surface tension does not contribute to the drag force. Its role is to raise internal pressure relative to the external pressure by $2\gamma/a$. For cases where $\mu_o \gg \mu_i$, e.g., an air bubble in water ($\sim 10^{-5}$ vs. $\sim 10^{-3}$ Pa s), the drag force is 2/3 of that for a rigid particle. When, $\mu_o \ll \mu_i$, the drag force tends to that of the rigid sphere, as expected.

Furthermore, we can find the velocity components of the flow inside and outside the drop by substituting the streamlines from (1.3.16) into the definition of the stream function (here in dimensional terms)

$$(1.3.19) \qquad \begin{aligned} u_r &= \frac{1}{r^2 \sin\theta} \frac{\partial \psi}{\partial \theta} \\ u_\theta &= -\frac{1}{r \sin\theta} \frac{\partial \psi}{\partial r} \end{aligned}$$

We obtain

$$(1.3.20) \qquad u_{r,o} = 2\left(\frac{D}{r} + \frac{Br}{2}\right)\frac{\cos\theta}{r^2}$$

$$(1.3.21) \qquad u_{r,i} = 2\left(\frac{G}{r^2} + \frac{Er^4}{10}\right)\frac{\cos\theta}{r^2}$$

$$(1.3.22) \qquad u_{\theta,o} = -\left(\frac{B}{2} - \frac{D}{r^2}\right)\frac{\sin\theta}{r}$$

$$(1.3.23) \qquad u_{\theta,i} = -\left(-\frac{2G}{r^3} + \frac{2Er^3}{5}\right)\frac{\sin\theta}{r}$$

The corresponding streamlines of this flow are shown in Figure 1.3.3.

1.3.3. Sphere in an axisymmetric extensional flow. After uniform flow, the next interesting flow is the axisymmetric extensional flow (Figure 1.3.4). Such flow is an approximation for the flow at a crossing of two channels, where for example the fluid from the vertical channel flows into the horizontal channel. This flow provides a simple and reproducible way of subjecting objects to a controlled shear at the center of crossed channels. This purely straining flow field can be written in Cartesian coordinates as

$$(1.3.24) \qquad \mathbf{u}_\infty = \mathbf{E} \cdot \mathbf{x} = -E\left(x\hat{\mathbf{x}} + y\hat{\mathbf{y}} - 2z\hat{\mathbf{z}}\right)$$

where

$$(1.3.25) \qquad \mathbf{E} = E \begin{bmatrix} 1 & 0 & 0 \\ 0 & 1 & 0 \\ 0 & 0 & -2 \end{bmatrix}$$

where E is the extensional flow parameter. When $E > 0$ the flow is outward away from the sphere parallel to the axis of symmetry and towards the sphere on the plane orthogonal to the axis of symmetry (Figure 1.3.4). This flow is called uniaxial extensional flow. When $E < 0$ the flow direction is reversed relative to the uniaxial extension flow, and this flow is called biaxial extensional flow. In both cases the velocity field is axisymmetric, and so we will use the stream function from (1.2.115). To do this we will just need to convert the undisturbed velocity (1.3.24) into spherical coordinate system. For this the standard relations are

$$(1.3.26) \qquad \begin{aligned} \hat{\mathbf{x}} &= \hat{\mathbf{r}}\sin\theta\cos\varphi + \hat{\theta}\cos\theta\cos\varphi - \hat{\varphi}\sin\varphi \\ \hat{\mathbf{y}} &= \hat{\mathbf{r}}\sin\theta\sin\varphi + \hat{\theta}\cos\theta\sin\varphi - \hat{\varphi}\cos\varphi \\ \hat{\mathbf{z}} &= \hat{\mathbf{r}}\cos\theta - \hat{\theta}\sin\theta \end{aligned}$$

and

$$(1.3.27) \qquad \begin{aligned} x &= r\sin\theta\cos\varphi \\ y &= r\sin\theta\sin\varphi \\ z &= r\cos\theta \end{aligned}$$

Using these relations we find that the dimensioned form of the undisturbed stream velocity is

$$(1.3.28) \qquad \mathbf{u}_\infty = -Er\left(1 - 3\cos^2\theta\right)\hat{\mathbf{r}} - Er\left(3\cos\theta\sin\theta\right)\hat{\theta}$$

We see that relevant velocity scale for this problem is

$$(1.3.29) \qquad u_c = Ea$$

Thus the dimensionless undisturbed stream velocity is

$$(1.3.30) \qquad \mathbf{u}_\infty = -r\left(1 - 3\cos^2\theta\right)\hat{\mathbf{r}} - r\left(3\cos\theta\sin\theta\right)\hat{\theta}$$

where $r = 1$ at the sphere's surface. Far away from the sphere (i.e. at infinity) the velocity must approach this velocity. Substituting this result into the definition of the stream function (1.2.89) and (1.2.90), we obtain

$$(1.3.31) \qquad \psi \to r^3\eta\left(1 - \eta^2\right) = -2r^3 Q_2\left(\eta\right) \text{ as } r \to \infty$$

Applying this to the general form of the stream function (1.2.115), we obtain that $A_n = 0$ for all n; $B_1 = 0$, $B_2 = -2$, and $B_n = 0$ for all other n. Thus (1.2.115)

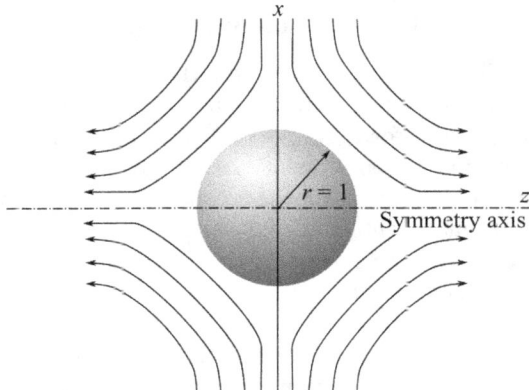

FIGURE 1.3.4. Sketch of streamlines around a sphere for an axisymmetric uniaxial extensional flow. If the flow direction is reversed, this flow is termed biaxial extensional flow.

simplifies to,

$$(1.3.32) \qquad \psi = -2r^3 Q_2(\eta) + \sum_{n=1}^{\infty} \left(C_n r^{2-n} + D_n r^{-n} \right)$$

Now we apply the standard boundary conditions at the surface of the sphere, the no-slip,

$$(1.3.33) \qquad u_\theta = 0 \text{ at } r = 1$$

and the no-penetration boundary condition,

$$(1.3.34) \qquad u_r = 0 \text{ at } r = 1$$

which as before translate to

$$(1.3.35) \qquad \frac{\partial \psi}{\partial r} = \psi = 0 \text{ at } r = 1$$

Applying this to (1.3.32) we obtain $C_2 = 5$ and $C_n = 0$ for all other n; $D_2 = -3$ and $D_n = 0$ for all other n. Thus we obtain the stream function for a sphere in axisymmetric extensional flow to be

$$\psi = 2 \left(-r^3 + \frac{5}{2} - \frac{3}{2} \frac{1}{r^2} \right) Q_2(\eta)$$

Interestingly we see that $C_1 = 0$. This means, from (1.2.129) that there is no axial force (and in fact any hydrodynamic force) on the sphere in axisymmetric extensional flow. This implies that if a sphere is carefully placed in such a flow will not be transported anywhere by the flow. In practice however, if another force (e.g., imperfections in the flow even Brownian motion) knocks the sphere out of the origin position, the force on the sphere will not necessarily be zero and the sphere will be transported by the flow.

1.4. Modern methods for solving creeping flow equations

So far we have solved the creeping flow equations by constructing a stream function and solving the resulting equation via separation of variables or via transform methods such as Fourier transform. While these traditional methods are quite powerful and many useful, fundamental solutions are obtained with them, their main weakness is that they require us to choose a coordinate system that coincides with the geometry. If we want to obtain a solution for flow around a sphere, we need to choose a spherical coordinate system, if we would like to find the flow field in a curved pipe, we need to choose a toroidal coordinate system (as we shall see later). However, for complex geometries, and especially 3D geometries, no convenient coordinate systems exist and so we need different methods. The methods that we describe here can be described as "pseudo-analytical" in a sense that for realistic problems we cannot obtain a full analytical solution and obtain part of the solution analytically and then evaluate the rest numerically. A scientist or engineer using results from such solutions often has to re-implement the numerical portion of the calculation to obtain results for their particular problem. Thus in this text we present these modern methods for solving creeping flow equations maybe not in sufficient detail for the engineer to obtain novel solutions themselves, but to be able to be familiar enough to follow the method and re-implement pertinent portions.

1.4.1. Mathematical Preliminaries.
In order to develop the modern methods of solving creeping flow equations, we need to review the Fourier transforms, the Dirac delta distribution, and harmonic functions. We begin by reviewing the basics of Fourier transforms. Recall that the Fourier transform of a function $f(x)$ is defined as

$$(1.4.1) \qquad \tilde{f}(k) = \int_{-\infty}^{\infty} f(x) \exp(ikx)\, dx$$

where the transform variable k is real. It can be shown that $\tilde{f}(k)$ will exist for every real value of k if

$$(1.4.2) \qquad \int_{-\infty}^{\infty} |f(x)|^2 dx < \infty$$

If f is at least piecewise smooth, then we can recover f from $\tilde{f}$ via

$$(1.4.3) \qquad f(x) = \lim_{\lambda \to \infty} \frac{1}{2\pi} \int_{-\lambda}^{\lambda} \tilde{f}(k) \exp(-ikx)\, dk$$

Here we will be focusing on 3D problems, and so in 3D we have a position vector

$$(1.4.4) \qquad \mathbf{r} = x_1 \hat{\mathbf{e}}_1 + x_2 \hat{\mathbf{e}}_2 + x_3 \hat{\mathbf{e}}_3$$

and so we have $f(\mathbf{r})$ for which we have a corresponding Fourier transform $\tilde{f}(\mathbf{k})$ where

$$(1.4.5) \qquad \mathbf{k} = k_1 \hat{\mathbf{e}}_1 + k_2 \hat{\mathbf{e}}_2 + k_3 \hat{\mathbf{e}}_3$$

$$(1.4.6) \qquad \tilde{f}(\mathbf{k}) = \int_{-\infty}^{\infty} f(\mathbf{r}) \exp(i\mathbf{k} \cdot \mathbf{r}) \, d\mathbf{r}$$

$$(1.4.7) \qquad f(\mathbf{r}) = \frac{1}{(2\pi)^3} \int_{-\infty}^{\infty} \tilde{f}(\mathbf{k}) \exp(-i\mathbf{k} \cdot \mathbf{r}) \, d\mathbf{k}$$

where the integrals are triple integrals.

The Dirac delta distribution (or a symbolic function) represents a point force and we will later use it to construct solutions consisting of a distribution of point forces equivalent to the effect of presence of a body. Dirac delta distribution has the following basic properties:

$$(1.4.8) \qquad \begin{aligned} \delta(x) &= 0 \text{ at } x \neq 0 \\ \delta(x) &= \infty \text{ at } x = 0 \end{aligned}$$

$$(1.4.9) \qquad \int_{-\infty}^{\infty} \delta(x) \, dx = 1$$

Dirac delta distribution exhibits a shifting (or selecting) property

$$(1.4.10) \qquad \int_{-\infty}^{\infty} f(x) \delta(x-a) \, dx = f(a)$$

Using (1.4.10) we can obtain the scaling property,

$$(1.4.11) \qquad \delta(cx) = \delta(x)/|c|$$

Where c is a constant not equal to zero. The integral of the delta function is the Heavyside step function

$$(1.4.12) \qquad H(x) = \int_{-\infty}^{\infty} \delta(x) \, dx = \begin{cases} 0 \text{ for } x < 0 \\ 1 \text{ for } x \geqslant 0 \end{cases}$$

Heavyside step function is a piecewise function. The derivative of the Heavyside function is the Dirac delta function,

$$(1.4.13) \qquad \frac{dH(x)}{dx} = \delta(x)$$

The Dirac delta function has other properties, such as

$$(1.4.14) \qquad \int_{-\infty}^{\infty} \delta(x) \exp(ikx) \, dx = \exp(ik0) = 1$$

Combining this with (1.4.3)

$$(1.4.15) \qquad \delta(x) = \frac{1}{2\pi} \int_{-\infty}^{\infty} \exp(-ikx) \, dk$$

we see that the inverse Fourier transform of unity is the Dirac delta function. Finally the derivative of the Dirac delta function has the following properties:

$$(1.4.16) \qquad \frac{d}{dx}\delta(-x) = \frac{d}{dx}\delta(x)$$

$$(1.4.17) \qquad \frac{d}{dx}\delta(-x) = -\frac{d}{dx}\delta(x)$$

$$(1.4.18) \qquad x\frac{d}{dx}\delta(x) = -\delta(x)$$

In three dimensions the Dirac delta function is defined as

$$(1.4.19) \qquad \begin{aligned} \delta(x_1, x_2, x_3) &= 0 \text{ at } x_1^2 + x_2^2 + x_3^2 \neq 0 \\ \delta(x_1, x_2, x_3) &= \infty \text{ at } x_1^2 + x_2^2 + x_3^2 = 0 \end{aligned}$$

and

$$(1.4.20) \qquad \int_{-\infty}^{\infty} \delta(x_1, x_2, x_3) dx_1 dx_2 dx_3 = 1$$

Additionally,

$$(1.4.21) \qquad \delta(x_1, x_2, x_3) = \delta(x_1)\,\delta(x_2)\,\delta(x_3)$$

Once again the inverse Fourier transform of unity in three dimensions is the Dirac delta function,

$$(1.4.22) \qquad \delta(\mathbf{r}) = \delta(x_1, x_2, x_3) = \frac{1}{(2\pi)^3} \int_{-\infty}^{\infty} \exp(-i\mathbf{k}\cdot\mathbf{r})\, d\mathbf{k}$$

Now let's turn to reviewing harmonic function solutions to the Laplace's equation. We need these because we can decompose the creeping flow equations,

$$(1.4.23) \qquad \nabla \cdot \mathbf{u} = 0$$

$$(1.4.24) \qquad \mu\nabla^2\mathbf{u} - \nabla p = 0$$

into Laplace's equation. Taking the divergence of the momentum equation,

$$(1.4.25) \qquad \nabla \cdot (\mu\nabla^2\mathbf{u} - \nabla p) = 0$$

we see that

$$(1.4.26) \qquad \nabla^2 p = 0$$

As from the continuity equation (1.4.23) we see that the velocity is divergence free (solenoidal). It can be also shown that the velocity can be split into a part due to pressure and a part that is a harmonic function,

$$(1.4.27) \qquad \mathbf{u} = \frac{\mathbf{r}}{2\mu}p + \mathbf{u}_H$$

where

$$(1.4.28) \qquad \nabla^2\mathbf{u}_H = 0$$

which we can check by substituting (1.4.27) into the continuity and momentum equations (1.4.23) and (1.4.24). We also see that by substituting (1.4.27) into the continuity equation, we obtain

$$(1.4.29) \qquad \nabla \cdot \mathbf{u}_H = -\frac{1}{2\mu} \left(3p + \mathbf{r} \cdot \nabla p \right)$$

From (1.4.27) together with (1.4.26) and (1.4.28) we can see that we can obtain the solution to the creeping flow equations by solving two Laplace's equations. This is quite a powerful observation because we can now construct a solution from solutions of Laplace's equations – harmonic functions. A harmonic function is defined as a function that has continuous second partial derivatives and solves Laplace's equation. Because of the linear nature of the Laplace's equation harmonic functions have a property that if φ and ψ are harmonic functions, then $\varphi + c\psi$ is also a harmonic function.

There are three main classes of harmonic functions: (a) homogeneous polynomials in two variables; (b) exponentially growing or decaying oscillations; and (c) functions with radial symmetry. From the first class, firstly, all constants (degree 0) are harmonic; secondly, all linear polynomials (degree 1) $ax + by$ are harmonic; thirdly, quadratic (degree 2) polynomials $x^2 - y^2$ and xy and their linear combinations are harmonic; fourthly, (degree n) real and imaginary parts of the complex polynomial $(x + iy)^n$ are harmonic. For the second class, functions such as $\exp(kx)\sin(ky)$ are harmonic. For the third class, functions such as $1/r$, $\ln r$ where $r = |\mathbf{r}|$ are harmonic. We can group harmonic functions into two categories: growing harmonics, whose magnitude increases with increasing r and decaying harmonics, whose magnitude decreases with increasing r. We will be most interested here in decaying harmonics and the most convenient form on those is $1/r$ and its higher order derivatives. For example, the first decaying harmonic past $1/r$ is

$$(1.4.30) \qquad \nabla \left(\frac{1}{r} \right) = \frac{x_i}{r^3} = \frac{\mathbf{r}}{r^3}$$

and the second is

$$(1.4.31) \qquad \nabla \left(\frac{\mathbf{r}}{r^3} \right) = \frac{x_i x_j}{r^5} - \frac{\delta_{ij}}{3r^3}$$

The third harmonic is

$$(1.4.32) \qquad \frac{x_i x_j x_k}{r^7} - \frac{x_i \delta_{jk} + x_j \delta_{ki} + x_k \delta_{ij}}{5r^5}$$

The general formula for this series of decaying harmonics is

$$(1.4.33) \qquad \varphi_{-(n+1)} = \frac{(-1)^n}{1 \cdot 3 \cdot 5 \cdot \ldots \cdot (2n-1)} \frac{\partial^{n-1}}{\partial x_i \partial x_j \partial x_k \ldots} \left(\frac{1}{r} \right)$$

for $n = 0, 1, 2, \ldots$. From this formula we can also obtain a formula for growing harmonics,

$$(1.4.34) \qquad \psi_n = r^{2n+1} \varphi_{-(n+1)}$$

Note that for the third class of harmonic functions, we have written them so that they take the general position vector $\mathbf{r}$ and its magnitude $r = |\mathbf{r}|$ and so they can be represented in any coordinate system.

1.4.2. Green's function for creeping flow: the Stokeslet. We are now ready to derive the Green's function for the creeping flow equations. As a reminder, Green's function is a solution to a differential equation where the inhomogeneous part is an impulse (i.e. the Dirac delta distribution) for a particular set of initial and boundary conditions. For linear problems, when we need to obtain the solution for an arbitrary inhomogeneous part, $f(x)$, we simplify convolve $f(x)$ with the Green's function. Hence having a Green's function to an equation is almost like having a general solution to it.

To begin, let's consider the continuity and momentum equations of creeping flow but with a condition that a point force $\mathbf{F_c}$ is applied at the origin, $\mathbf{r} = 0$

$$(1.4.35) \qquad \nabla \cdot \mathbf{u} = 0$$

$$(1.4.36) \qquad \nabla \cdot \mathbf{T} = \mu \nabla^2 \mathbf{u} - \nabla p = -\mathbf{F_c} \delta(\mathbf{r})$$

From our discussions of the delta function we see that (1.4.36) can be interpreted as two equations

$$(1.4.37) \qquad \nabla \cdot \mathbf{T} = 0 \text{ for } \mathbf{r} \neq 0$$

and for any volume V that encloses the point $\mathbf{r} = 0$,

$$(1.4.38) \qquad \iiint\limits_V \nabla \cdot \mathbf{T} d\tau = -\mathbf{F_c}$$

To solve this version of the creeping flow equations, we employ the Fourier transform. Taking the Fourier transform of (1.4.35) and (1.4.36)

$$(1.4.39) \qquad \mathbf{k} \cdot \tilde{\mathbf{u}} = 0$$

$$(1.4.40) \qquad -\mu k^2 \tilde{\mathbf{u}} + i \mathbf{k} \tilde{p} = -\mathbf{F_c}$$

where $k = |\mathbf{k}|$. Taking the dot product of (1.4.40) with $\mathbf{k}$ and employing (1.4.39), we obtain

$$(1.4.41) \qquad \tilde{p} = \frac{i \mathbf{k} \cdot \mathbf{F_c}}{k^2}$$

Substituting this result into (1.4.40)

$$(1.4.42) \qquad \mu k^2 \tilde{\mathbf{u}} + \mathbf{k} \left(\frac{\mathbf{k} \cdot \mathbf{F_c}}{k^2} \right) = \mathbf{F_c}$$

or

$$(1.4.43) \qquad \tilde{\mathbf{u}} = \frac{1}{\mu k^2} \left(\mathbf{F_c} - \mathbf{k} \left(\frac{\mathbf{k} \cdot \mathbf{F_c}}{k^2} \right) \right)$$

Thus we have obtained the solutions for the pressure (1.4.41) and the velocity field (1.4.43) in Fourier space. Now we need to transform these back into our physical space. Taking the Fourier transform of (1.4.41) and (1.4.43)

$$(1.4.44) \qquad p(\mathbf{r}) = \frac{1}{(2\pi)^3} i \int\limits_{-\infty}^{\infty} \frac{i \mathbf{k} \cdot \mathbf{F_c}}{k^2} \exp(-i\mathbf{k} \cdot \mathbf{r}) d\mathbf{k}$$

$$(1.4.45) \qquad \mathbf{u}(\mathbf{r}) = \frac{1}{(2\pi)^3} \frac{1}{\mu} \int\limits_{-\infty}^{\infty} \frac{1}{k^2} \left(\mathbf{F_c} - \mathbf{k} \left(\frac{\mathbf{k} \cdot \mathbf{F_c}}{k^2} \right) \right) \exp(-i\mathbf{k} \cdot \mathbf{r}) d\mathbf{k}$$

where the integrals are triple integrals. Evaluating these integrals is a bit tricky, so we need to develop a few tricks to proceed with this integration. Firstly, recall from the previous section that $1/r$ is a harmonic function, and so

$$(1.4.46) \qquad \nabla^2 (1/r) = -4\pi\delta(\mathbf{r})$$

is true. From (1.4.22)

$$(1.4.47) \qquad \nabla^2\left(\frac{1}{4\pi r}\right) = -\delta(\mathbf{r}) = -\frac{1}{(2\pi)^3}\int\limits_{-\infty}^{\infty}\exp(-i\mathbf{k}\cdot\mathbf{r})\,d\mathbf{k}$$

from which by successive integration, we obtain that

$$(1.4.48) \qquad \nabla\left(\frac{1}{4\pi r}\right) = -\frac{1}{(2\pi)^3}i\int\limits_{-\infty}^{\infty}\frac{\mathbf{k}}{k^2}\exp(-i\mathbf{k}\cdot\mathbf{r})\,d\mathbf{k}$$

$$(1.4.49) \qquad \frac{1}{4\pi r} = \frac{1}{(2\pi)^3}\int\limits_{-\infty}^{\infty}\frac{1}{k^2}\exp(-i\mathbf{k}\cdot\mathbf{r})\,d\mathbf{k}$$

Taking the dot product of (1.4.48) with $\mathbf{F_c}$ we obtain

$$(1.4.50) \qquad \mathbf{F_c}\cdot\nabla\left(\frac{1}{4\pi r}\right) = -\frac{1}{(2\pi)^3}i\int\limits_{-\infty}^{\infty}\frac{(\mathbf{F_c}\cdot\mathbf{k})}{k^2}\exp(-i\mathbf{k}\cdot\mathbf{r})\,d\mathbf{k}$$

comparing this with (1.4.44) we see that we have a solution for the pressure distribution,

$$(1.4.51) \qquad p(\mathbf{r}) = \mathbf{F_c}\cdot\nabla\left(\frac{1}{4\pi r}\right) = \frac{\mathbf{F_c}\cdot\mathbf{r}}{4\pi r^3}$$

Now we find the solution to the velocity distribution. Firstly we multiply (1.4.49) by $\mathbf{F_c}$ and obtain

$$(1.4.52) \qquad \frac{\mathbf{F_c}}{4\pi r} = \frac{1}{(2\pi)^3}\int\limits_{-\infty}^{\infty}\frac{\mathbf{F_c}}{k^2}\exp(-i\mathbf{k}\cdot\mathbf{r})\,d\mathbf{k}$$

to obtain the first part of the integral in (1.4.45). Now it can be shown that r is a fundamental solution to the biharmonic equation and so that

$$(1.4.53) \qquad \nabla^4\left(\frac{r}{8\pi}\right) = -\delta(\mathbf{r})$$

From (1.4.22)

$$(1.4.54) \qquad \nabla^4\left(\frac{r}{8\pi}\right) = -\delta(\mathbf{r}) = -\frac{1}{(2\pi)^3}\int\limits_{-\infty}^{\infty}\exp(-i\mathbf{k}\cdot\mathbf{r})\,d\mathbf{k}$$

via successive integration we can obtain

$$(1.4.55) \qquad \frac{r}{8\pi} = -\frac{1}{(2\pi)^3}\int\limits_{-\infty}^{\infty}\frac{1}{k^4}\exp(-i\mathbf{k}\cdot\mathbf{r})\,d\mathbf{k}$$

(1.4.56)
$$\nabla\nabla\left(\frac{r}{8\pi}\right) = \frac{1}{(2\pi)^3} \int\limits_{-\infty}^{\infty} \frac{\mathbf{kk}}{k^4} \exp\left(-i\mathbf{k}\cdot\mathbf{r}\right) d\mathbf{k}$$

Comparing (1.4.45) to (1.4.52) and (1.4.56), we obtain the velocity distribution as

(1.4.57)
$$\mathbf{u}\left(\mathbf{r}\right) = \frac{\mathbf{I}\cdot\mathbf{F_c}}{4\pi\mu r} - \frac{\mathbf{F_c}}{\mu}\cdot\nabla\nabla\left(\frac{r}{8\pi}\right)$$

where $\mathbf{I}$ is the unit tensor (identity matrix). Since $\nabla\left(r\right) = \mathbf{r}/r$ and $\nabla\left(\mathbf{r}\right) = \mathbf{I}$

(1.4.58)
$$\nabla\nabla\left(\frac{r}{8\pi}\right) = -\frac{1}{8\pi}\frac{\mathbf{rr}}{r^3} + \frac{1}{8\pi}\frac{\mathbf{I}}{r}$$

and so

(1.4.59)
$$\mathbf{u}\left(\mathbf{r}\right) = \frac{1}{8\pi\mu}\left(\frac{\mathbf{I}}{r} + \frac{\mathbf{rr}}{r^3}\right)\cdot\mathbf{F_c}$$

The tensor

(1.4.60)
$$\mathbf{B} = \left(\frac{\mathbf{I}}{r} + \frac{\mathbf{rr}}{r^3}\right)$$

or

(1.4.61)
$$B_{ij} = \left(\frac{\delta_{ij}}{r} + \frac{x_i x_j}{r^3}\right) \text{ for } i, j = 1, 2, 3$$

is called the Oseen-Burgers tensor. The velocity field in (1.4.59) is termed the Stokeslet of strength

(1.4.62)
$$\alpha = \frac{\mathbf{F_c}}{8\pi\mu}$$

The tensor

(1.4.63)
$$\mathbf{G} = \frac{\mathbf{B}}{8\pi\mu}$$

is known as the free space Green's function for creeping flow for infinite unbounded flow. We can rewrite the pressure and velocity distributions as

(1.4.64)
$$p\left(\mathbf{r}\right) = \frac{\mathbf{P}\left(\mathbf{r}\right)}{8\pi\mu}\cdot\mathbf{F_c}$$

(1.4.65)
$$\mathbf{u}\left(\mathbf{r}\right) = \frac{\mathbf{B}}{8\pi\mu}\cdot\mathbf{F_c}$$

where

(1.4.66)
$$P_j = 2\mu\frac{x_j}{r^3} + P_{\infty,j}$$

where $P_{\infty,j}$ is the pressure field far from the origin. If the point force is applied at point $\mathbf{r} = \rho$ the Oseen-Burgers takes the form of

(1.4.67)
$$\mathbf{B}\left(\mathbf{r} - \rho\right) = \left(\frac{\mathbf{I}}{|\mathbf{r} - \rho|} + \frac{\left(\mathbf{r} - \rho\right)\left(\mathbf{r} - \rho\right)}{|\mathbf{r} - \rho|^3}\right)$$

and the Stokeslet becomes

(1.4.68)
$$\mathbf{u}\left(\mathbf{r}\right) = \frac{1}{8\pi\mu}\left(\frac{\mathbf{I}}{|\mathbf{r} - \rho|} + \frac{\left(\mathbf{r} - \rho\right)\left(\mathbf{r} - \rho\right)}{|\mathbf{r} - \rho|^3}\right)\cdot\mathbf{F_c}$$

Green's functions can be categorized into three categories: (a) the free space Green's function for infinite unbounded flow (which we just derived); (b) the Green's function for infinite or semi-infinite flow that is bounded by a rigid wall or fluid/fluid interface; and (c) Green's function for internal flow that is completely bounded by rigid walls.

1.4.3. Lorentz reciprocal theorem. To use the result we have obtained in the previous section, i.e., calculate the velocity field around a body by calculating the velocity field from a distribution of point forces, we need to translate the geometry of the body into a distribution of point forces. The basic machinery for this is the Lorentz reciprocal theorem (see Lorentz (1907), and Hasimoto (1996)). To obtain this theorem we first need to derive a general integral theorem for creeping flow equations reminiscent of Green's theorem in vector calculus. Let's consider a space outside of a closed surface ∂D where ∂D may represent the surface of a body. In this space let $\mathbf{u}_1$ and $\mathbf{u}_2$ be any smooth vector fields (in our case they will be velocity fields) that satisfy

$$(1.4.69) \qquad \nabla \cdot \mathbf{u}_1 = 0$$

$$(1.4.70) \qquad \nabla \cdot \mathbf{u}_2 = 0$$

In other words, the vector fields are divergence free (solenoidal). Let's also define corresponding tensors

$$(1.4.71) \qquad \mathbf{T}_1 = -p_1\mathbf{I} + \mu\left(\nabla\mathbf{u}_1 + \nabla\mathbf{u}_1^T\right)$$

$$(1.4.72) \qquad \mathbf{T}_2 = -p_2\mathbf{I} + \mu\left(\nabla\mathbf{u}_2 + \nabla\mathbf{u}_2^T\right)$$

where p_1 and p_2 are any smooth scalar fields (in our case they will be pressure fields). Finally, the divergence of the tensors is also zero, (which for us represents the creeping flow momentum equation),

$$(1.4.73) \qquad \nabla \cdot \mathbf{T}_1 = 0$$

$$(1.4.74) \qquad \nabla \cdot \mathbf{T}_2 = 0$$

Here we will prove the Lorentz reciprocal theorem using Cartesian tensor methods, but it can be equivalently proven using general tensor methods for any coordinate system. In terms of Cartesian components the continuity equation (1.4.69) and (1.4.70) is

$$(1.4.75) \qquad \frac{\partial u_{1,i}}{\partial x_i} = 0$$

$$(1.4.76) \qquad \frac{\partial u_{2,i}}{\partial x_i} = 0$$

where we are using the Einstein index notation, where repeat over the index indicates summation. The momentum equation (1.4.73) and (1.4.74) in this notation becomes

$$(1.4.77) \qquad \frac{\partial T_{1,ij}}{\partial x_j} = 0$$

$$(1.4.78) \qquad \frac{\partial T_{2,ij}}{\partial x_j} = 0$$

The definitions of the stress tensor in this notation become

(1.4.79) $$T_{1,ij} = -p_1\delta_{ij} + 2\mu D_{1,ij}$$

(1.4.80) $$T_{2,ij} = -p_2\delta_{ij} + 2\mu D_{2,ij}$$

Where δ_{ij} is the Kronecker delta and

(1.4.81) $$D_{1,ij} = \frac{1}{2}\left(\frac{\partial u_{1,i}}{\partial x_j} + \frac{\partial u_{1,j}}{\partial x_i}\right)$$

(1.4.82) $$D_{2,ij} = \frac{1}{2}\left(\frac{\partial u_{2,i}}{\partial x_j} + \frac{\partial u_{2,j}}{\partial x_i}\right)$$

Now we multiply $T_{1,ij}$ by $D_{2,ij}$ and obtain

(1.4.83) $$T_{1,ij}D_{2,ij} = -p_1\delta_{ij}D_{2,ij} + 2\mu D_{1,ij}D_{2,ij}$$

Now

(1.4.84) $$\delta_{ij}D_{2,ij} = D_{2,ii} = \frac{\partial u_{2,i}}{\partial x_i} = 0$$

from the continuity equation (1.4.76), and so

(1.4.85) $$T_{1,ij}D_{2,ij} = 2\mu D_{1,ij}D_{2,ij}$$

Similarly, we multiply $T_{2,ij}$ by $D_{1,ij}$ and simplify, and obtain

(1.4.86) $$T_{2,ij}D_{1,ij} = 2\mu D_{1,ij}D_{2,ij}$$

Comparing (1.4.85) and (1.4.86) we obtain

(1.4.87) $$T_{1,ij}D_{2,ij} = T_{1,ij}D_{2,ij}$$

Now we expand both sides, keeping in mind that the stress tensor is symmetric, $T_{ij} = T_{ji}$ and using the product rule,

(1.4.88)
$$T_{1,ij}D_{2,ij} = \frac{1}{2}T_{1,ij}\frac{\partial u_{2,i}}{\partial x_j} + \frac{1}{2}T_{1,ij}\frac{\partial u_{2,j}}{\partial x_i} = T_{1,ij}\frac{\partial u_{2,i}}{\partial x_j} = \frac{\partial}{\partial x_j}(u_{2,i}T_{1,ij}) - u_{2,i}\frac{\partial T_{1,ij}}{\partial x_j}$$

(1.4.89)
$$T_{2,ij}D_{1,ij} = \frac{1}{2}T_{2,ij}\frac{\partial u_{1,i}}{\partial x_j} + \frac{1}{2}T_{2,ij}\frac{\partial u_{1,j}}{\partial x_i} = T_{2,ij}\frac{\partial u_{1,i}}{\partial x_j} = \frac{\partial}{\partial x_j}(u_{1,i}T_{2,ij}) - u_{1,i}\frac{\partial T_{2,ij}}{\partial x_j}$$

From (1.4.77) and (1.4.78) we have that $\partial T_{ij}/\partial x_j = 0$ and so

(1.4.90) $$T_{1,ij}D_{2,ij} = \frac{\partial}{\partial x_j}(u_{2,i}T_{1,ij})$$

(1.4.91) $$T_{2,ij}D_{1,ij} = \frac{\partial}{\partial x_j}(u_{1,i}T_{2,ij})$$

and from (1.4.87) we have

(1.4.92) $$\frac{\partial}{\partial x_j}(u_{2,i}T_{1,ij}) = \frac{\partial}{\partial x_j}(u_{1,i}T_{2,ij})$$

Converting this back to general tensor form,

(1.4.93) $$\nabla \cdot (\mathbf{T}_1 \cdot \mathbf{u}_2) = \nabla \cdot (\mathbf{T}_2 \cdot \mathbf{u}_1)$$

Integrating this over a volume V

$$(1.4.94) \qquad \int_V \nabla \cdot (\mathbf{T}_1 \cdot \mathbf{u}_2)\, dV = \int_V \nabla \cdot (\mathbf{T}_2 \cdot \mathbf{u}_1)\, dV$$

or

$$(1.4.95) \qquad \int_V \nabla \cdot (\mathbf{T}_1 \cdot \mathbf{u}_2) - \nabla \cdot (\mathbf{T}_2 \cdot \mathbf{u}_1)\, dV = 0$$

Using the divergence theorem we convert the volume integral into the surface integral. Because we have an inner surface ∂D and the outer surface S (enveloping the entire volume),

$$(1.4.96) \qquad \begin{aligned} &\int_V \nabla \cdot (\mathbf{T}_1 \cdot \mathbf{u}_2) - \nabla \cdot (\mathbf{T}_2 \cdot \mathbf{u}_1)\, dV = \\ &\int_S \mathbf{n} \cdot (\mathbf{T}_1 \cdot \mathbf{u}_2) - \mathbf{n} \cdot (\mathbf{T}_2 \cdot \mathbf{u}_1)\, dS - \int_{\partial D} \mathbf{n} \cdot (\mathbf{T}_1 \cdot \mathbf{u}_2) - \mathbf{n} \cdot (\mathbf{T}_2 \cdot \mathbf{u}_1)\, dS = 0 \end{aligned}$$

However, if $\mathbf{u}_1$ and $\mathbf{u}_2$ decay as $1/r$ or faster as $r \to \infty$ and if p_1 and p_2 decay as $1/r^2$ or faster as $r \to \infty$, then the integral over the outer surface S is zero, and so
(1.4.97)

$$\int_V \nabla \cdot (\mathbf{T}_1 \cdot \mathbf{u}_2) - \nabla \cdot (\mathbf{T}_2 \cdot \mathbf{u}_1)\, dV = -\int_{\partial D} \mathbf{n} \cdot (\mathbf{T}_1 \cdot \mathbf{u}_2) - \mathbf{n} \cdot (\mathbf{T}_2 \cdot \mathbf{u}_1)\, dS = 0$$

The right-hand side can be rewritten as

$$(1.4.98) \qquad \int_{\partial D} \mathbf{F}_1 \cdot \mathbf{u}_2\, dS = \int_{\partial D} \mathbf{F}_2 \cdot \mathbf{u}_1\, dS$$

The result (1.4.97) is often referred to as the general integral relation for creeping flow, and the result (1.4.98) is known as Lorentz reciprocal theorem. These are very powerful results that will enable us to obtain solutions of creeping flow equations via distributions of singularities either on the surface of the object around which the flow interests us, or in the center of this object.

1.4.4. Ladyzhenskaya representation and boundary integral method of solution.
Firstly, the Lorentz reciprocal theorem leads us to a general integral representation of the creeping flow equations (originally due to Ladyzhenskaya). To obtain this, we firstly let $\mathbf{u}_1$ and $\mathbf{T}_1$ represent an arbitrary solution of creeping flow equation (with the exception that $\mathbf{u}_1$ decays as $1/r$ or faster and p_1 decays as $1/r^2$ or faster as $r \to \infty$). Secondly, we let $\mathbf{u}_2$ and $\mathbf{T}_2$ represent the Stokeslet solution we obtained in the previous section ((1.4.51) and (1.4.59)), just with a unity point force. For the Stokeslet the velocity field is

$$(1.4.99) \qquad \mathbf{u}_2 = \frac{1}{8\pi} \left(\frac{(\mathbf{r} - \rho)(\mathbf{r} - \rho)}{R^3} + \frac{\mathbf{I}}{R} \right) \cdot \mathbf{f}$$

where $R = |\mathbf{r} - \rho|$ and here $\mathbf{f}$ is a point force of unit magnitude. The corresponding stress tensor is

$$(1.4.100) \qquad \mathbf{T}_2 = -\frac{3}{4\pi} \frac{(\mathbf{r} - \rho)(\mathbf{r} - \rho)(\mathbf{r} - \rho)}{R^5} \cdot \mathbf{f}$$

Now we can substitute (1.4.99) and (1.4.100) into (1.4.97) together with the fact that

$$(1.4.101) \qquad \nabla \cdot \mathbf{T}_1 = 0$$

$$(1.4.102) \qquad \nabla \cdot \mathbf{T}_2 = \delta \left(\mathbf{r} - \rho \right) \mathbf{f}$$

and obtain

$$(1.4.103) \qquad \int_V \left(\delta \left(\mathbf{r} - \rho \right) \mathbf{u}_1 \cdot \mathbf{f} \right) dV_\rho = - \int_{\partial D} \mathbf{n} \cdot \left(\left(\mathbf{U}_2 \cdot \mathbf{f} \right) \cdot \mathbf{T}_1 - \mathbf{u}_1 \cdot \left(\mathbf{\Lambda} \cdot \mathbf{f} \right) \right) dS_\rho$$

where the subscript ρ indicates that the variable of integration is ρ and the fixed point is $\mathbf{r}$ and where we have defined

$$(1.4.104) \qquad \mathbf{u}_2 = \mathbf{U}_2 \cdot \mathbf{f}$$

$$(1.4.105) \qquad \mathbf{T}_2 = \mathbf{\Lambda} \cdot \mathbf{f}$$

We have made these definitions so that we can factor out the point force $\mathbf{f}$. Note that the tensor component of $\mathbf{U}_2$, $U_{2,ik}$ is the i component of the velocity field generated by the point force of unit magnitude in the k direction. Likewise, the tensor component of $\mathbf{\Lambda}$, Λ_{ijk} is the ij component of the stress tensor produced by the point force of unit magnitude in the k direction. Factoring out the point force $\mathbf{f}$,

$$(1.4.106) \qquad \int_V \left(\delta \left(\mathbf{r} - \rho \right) \mathbf{u}_1 \right) dV_\rho = - \int_{\partial D} \mathbf{n} \cdot \left(\mathbf{U}_2 \cdot \mathbf{T}_1 - \mathbf{u}_1 \cdot \mathbf{\Lambda} \right) dS_\rho$$

and evaluating the left-hand side integral via the property of the Dirac delta function,

$$(1.4.107) \qquad \mathbf{u}_1 = - \int_{\partial D} \mathbf{n} \cdot \left(\mathbf{U}_2 \cdot \mathbf{T}_1 - \mathbf{u}_1 \cdot \mathbf{\Lambda} \right) dS_\rho$$

and substituting in (1.4.99) and (1.4.100) via (1.4.104) and (1.4.105), we obtain

$$
\begin{aligned}
\mathbf{u}_1 = & -\frac{3}{4\pi} \int_{\partial D} \left(\frac{\left(\mathbf{r} - \rho \right)\left(\mathbf{r} - \rho \right)\left(\mathbf{r} - \rho \right)}{R^5} \cdot \mathbf{u}_1 \left(\rho \right) \right) \cdot \mathbf{n} dS_\rho \\
& + \frac{1}{8\pi} \int_{\partial D} \left(\frac{\left(\mathbf{r} - \rho \right)\left(\mathbf{r} - \rho \right)}{R^3} + \frac{\mathbf{I}}{R} \right) \cdot \mathbf{T}_1 \left(\rho \right) \cdot \mathbf{n} dS_\rho
\end{aligned}
$$

(1.4.108)

and correspondingly the pressure,
(1.4.109)

$$
p_1 = \frac{1}{2\pi} \int_{\partial D} \left(\frac{\mathbf{I}}{R} - \frac{3 \left(\mathbf{r} - \rho \right)\left(\mathbf{r} - \rho \right)}{R^3} \cdot \mathbf{u}_1 \left(\rho \right) \right) \cdot \mathbf{n} dS_\rho + \frac{1}{4\pi} \int_{\partial D} \frac{\left(\mathbf{r} - \rho \right)}{R^3} \cdot \mathbf{T}_1 \left(\rho \right) \cdot \mathbf{n} dS_\rho
$$

This is the integral representation of the creeping flow equations due to Ladyzhenskaya. Because our derivation required that the velocity field and pressure field vanished at infinity, $\mathbf{u}_1$ should be used as a disturbance velocity (i.e., the actual velocity field minus the velocity at infinity). In fact we can rewrite (1.4.108) as

(1.4.110)
$$\mathbf{u}\left(\mathbf{r}\right) = \mathbf{u}_\infty\left(\mathbf{r}\right) - \frac{3}{4\pi} \int\limits_{\partial D} \left(\frac{\left(\mathbf{r} - \rho\right)\left(\mathbf{r} - \rho\right)\left(\mathbf{r} - \rho\right)}{R^5} \cdot \mathbf{u}\left(\rho\right) \right) \cdot \mathbf{n}dS_\rho$$
$$+ \frac{1}{8\pi} \int\limits_{\partial D} \left(\frac{\left(\mathbf{r} - \rho\right)\left(\mathbf{r} - \rho\right)}{R^3} + \frac{\mathbf{I}}{R} \right) \cdot \mathbf{T}\left(\rho\right) \cdot \mathbf{n}dS_\rho$$

since

(1.4.111)
$$\int\limits_{\partial D} \mathbf{n} \cdot \left(\mathbf{u}_2 \cdot \mathbf{T}_\infty - \mathbf{u}_\infty \cdot \mathbf{\Lambda}\right) dS_\rho = 0$$

The first integral on the right-hand side of (1.4.108) is often referred to as the double layer potential (not to be confused with electrical double layer, which we will discuss much later). The double layer potential has a density function that is the (disturbance) velocity $\mathbf{u}_1$ on the boundary ∂D of the domain, which typically corresponds to the surface of the body. The second integral on the right-hand side, is referred to as the single layer potential and its density function is the surface stress tensor $\mathbf{T}_1 \cdot \mathbf{n}$ on the boundary ∂D of the domain, which again typically corresponds to the surface of the body of interest. What we have done in deriving (1.4.108) and (1.4.109) is obtained an integral formulation of the velocity equivalent to the differential form of the creeping flow equations. This allows us to solve the creeping flow equations via the boundary integral method. This method is especially useful for bodies with complicated geometries ∂D. It is especially useful for geometries where there is no coordinate system available to map to the bodies geometry to, as we have with a sphere and the spherical coordinate system. To obtain a solution via this method, we determine the density functions so that the resulting (disturbance) velocity $\mathbf{u}_1$ satisfies the boundary conditions on ∂D. Typically this is done numerically. This is the first of what we call "modern" solution methods to creeping flow equations.

1.4.5. Solution by distribution of internal singularities. The second "modern" solution of creeping flow equation that we consider is the solution by distribution of singularities. This approach tackles an important class of creeping flow problems – the problems of motion past a stationary solid body. This method has been especially effective in finding solutions near a sphere or spheroidal shapes. For the stationary solid body problems the no-slip and no-penetration boundary condition dictates that

(1.4.112)
$$\mathbf{u} = 0 \text{ for all } \mathbf{x} \text{ on } \partial D$$

Applying these boundary conditions to (1.4.110) we obtain

(1.4.113)
$$\mathbf{u}_\infty\left(\mathbf{r}\right) = -\frac{1}{8\pi} \int\limits_{\partial D} \left(\frac{\left(\mathbf{r} - \rho\right)\left(\mathbf{r} - \rho\right)}{R^3} + \frac{\mathbf{I}}{R} \right) \cdot \mathbf{T}\left(\rho\right) \cdot \mathbf{n}dS_\rho$$

and so the general velocity distribution for this class of problems is

(1.4.114)
$$\mathbf{u}\left(\mathbf{r}\right) = \mathbf{u}_\infty\left(\mathbf{r}\right) + \frac{1}{8\pi} \int\limits_{\partial D} \left(\frac{\left(\mathbf{r} - \rho\right)\left(\mathbf{r} - \rho\right)}{R^3} + \frac{\mathbf{I}}{R} \right) \cdot \mathbf{T}\left(\rho\right) \cdot \mathbf{n}dS_\rho$$

The meaning of (1.4.114) is that we can represent an arbitrary solution around a rigid body by a distribution of Stokeslets at some positions ρ on the boundary of the body ∂D. Effectively the distribution of Stokeslets replaces the no-slip and no-penetration condition. We obtain these by solving (1.4.113) for $\mathbf{T}(\rho) \cdot \mathbf{n}$ from a known (prescribed) $\mathbf{u}_\infty$. (Typically, this is done numerically). While we have obtained (1.4.114) for a rigid body, we can obtain similar relations even when the body is not rigid and its boundaries are not stationary. Thus, we can obtain a solution to creeping flow equations around these bodies via a distribution (even a moving distribution) of Stokeslets.

Here we will show that for the flow past a rigid body, as we can find a solution for creeping flow equations where the no-slip and no-penetration condition is replaced by an internal distribution of singularities – a point force (the Stokeslet), a "point torque" (a rotlet), a "point strain" (the stresslet), and even higher order singularities. We will derive the point torque and point strain constructs shortly. We may ask if this approach – placing singularities of various orders inside the body to represent the boundary conditions, is better than the approach of placing the singularity of only a single order (the Stokeslet) in a distribution at the boundary of the object. The short answer is that for certain problems one is more advantageous than others, and this is in fact a subject of current research. We can imagine that for a bumpy object with spherical symmetry (and so a complicated surface) may be better solved via the method of internal singularities as placing singularities on the complicated surface would be quite complicated.

To develop the method of internal singularities, we begin with the Stokeslet solutions ((1.4.51) and (1.4.59)), which we write as

$$(1.4.115) \qquad \mathbf{u}_S(\mathbf{r};\alpha) = \frac{\alpha}{r} + \frac{(\alpha \cdot \mathbf{r})\,\mathbf{r}}{r^3}$$

$$(1.4.116) \qquad p_S(\mathbf{r};\alpha) = 2\frac{\alpha \cdot \mathbf{r}}{r^3}$$

where we mean that the velocity and pressure is evaluated at a point $\mathbf{r}$ and the variables after the semicolon are auxiliary variables. Here, for example $\alpha = \mathbf{F_c}/8\pi$. The subscript S stands for Stokeslet. Recall that the derivatives of the creeping flow equations are also solutions of the creeping flow equations (Section 1.4.1). Thus the derivatives of $\mathbf{u}_S$ and p_S represent a solution of creeping flow equations, and specifically one corresponding to a point singularity that is a derivative of the same order of the point force $\mathbf{F_c}$. Furthermore, we can express the Stokeslet solution for a point force at a position ρ as a Taylor series expansion (multipole expansion) about $\mathbf{r}$ as

$$(1.4.117) \qquad \mathbf{u}_S(\mathbf{r} - \rho) = \mathbf{u}_S(\mathbf{r}) - (\rho \cdot \nabla)\,\mathbf{u}_S(\mathbf{r}) + \frac{1}{2}(\rho \cdot \nabla)^2 \mathbf{u}_S(\mathbf{r}) + \ldots$$

$$(1.4.118) \qquad p_S(\mathbf{r} - \rho) = p_S(\mathbf{r}) - (\rho \cdot \nabla)\,p_S(\mathbf{r}) + \frac{1}{2}(\rho \cdot \nabla)^2 p_S(\mathbf{r}) + \ldots$$

The term

$$(1.4.119) \qquad (\rho \cdot \nabla)\,\mathbf{u}_S(\mathbf{r})$$

represents the contribution to the velocity field generated by a force dipole $(\rho \cdot \nabla) \mathbf{F_c}$ (two point forces infinitesimally close together pointing in opposite directions). Similarly the term $(\rho \cdot \nabla)^2 \mathbf{u}_S (\mathbf{r})$ represents the contribution to the velocity field generated by a force quadrupole, $(\rho \cdot \nabla)^2 \mathbf{F_c}$. Since we can express the Stokeslet solution for a point force at a position ρ as a multipole expansions of various singularities, we should be able to substitute (1.4.117) into a general solution in terms of distribution of Stokeslets such as (1.4.110). Thus we should be able to represent a solution to the creeping flow equations in terms of singularities of various order near the origin (inside the body) rather than the distribution of Stokeslets at the boundary of the body ∂D. Substituting (1.4.115) into the dipole term $(\rho \cdot \nabla) \mathbf{u}_S (\mathbf{r})$, we obtain

$$(1.4.120) \quad \begin{aligned} \mathbf{u}_{SD} (\mathbf{r}; \alpha, \beta) &= (\beta \cdot \nabla) \mathbf{u}_S (\mathbf{r}; \alpha) \\ &= \frac{(\beta \times \alpha) \times \mathbf{r}}{r^3} - \left(\frac{(\alpha \cdot \beta) \mathbf{r}}{r^3} - 3 \frac{(\alpha \cdot \mathbf{r})(\beta \cdot \mathbf{r}) \mathbf{r}}{r^5} \right) \end{aligned}$$

$$(1.4.121) \quad \begin{aligned} p_{SD} (\mathbf{r}; \alpha, \beta) &= (\beta \cdot \nabla) p_S (\mathbf{r}; \alpha) \\ &= \frac{(\beta \times \alpha) \times \mathbf{r}}{r^3} - \left(\frac{(\alpha \cdot \beta)}{r^3} - 3 \frac{(\alpha \cdot \mathbf{r})(\beta \cdot \mathbf{r})}{r^5} \right) \end{aligned}$$

where once again the parameter α is the strength of the point force, and β can be thought of as the position of the point force locations. The subscript SD refers to stokes dipole. $\mathbf{u}_{SD}$ formally corresponds to a velocity field produced by a pair of forces, at $\mathbf{r} = \pm \beta/2$, with a strength of α and $-\alpha$ respectively, in the limit of $|\beta| \to \infty$. We can similarly arrive at a quadruple,

$$(1.4.122) \quad \mathbf{u}_{SQ} (\mathbf{r}; \alpha, \beta, \delta) = (\delta \cdot \nabla)(\beta \cdot \nabla) \mathbf{u}_S (\mathbf{r}; \alpha)$$

$$(1.4.123) \quad p_{SQ} (\mathbf{r}; \alpha, \beta, \delta) = (\delta \cdot \nabla)(\beta \cdot \nabla) p_S (\mathbf{r}; \alpha)$$

It is useful to split the dipole into a sum of two parts, the rotlet and the stresslet, as each have a physical significance. We define the rotlet as the antisymmetric part of the dipole, and so its velocity and pressure is

$$(1.4.124) \quad \begin{aligned} \mathbf{u}_R (\mathbf{r}; \alpha, \beta) &= \frac{(\beta \times \alpha) \times \mathbf{r}}{r^3} \\ p_R (\mathbf{r}; \alpha, \beta) &= 0 \end{aligned}$$

The rotlet represents the contribution of a singular, point torque at the origin. Similarly, we define the stresslet as the symmetric part of the dipole, and so its velocity and pressure is

$$(1.4.125) \quad \mathbf{u}_{SD} (\mathbf{r}; \alpha, \beta) = \left(-\frac{(\alpha \cdot \beta) \mathbf{r}}{r^3} + 3 \frac{(\alpha \cdot \mathbf{r})(\beta \cdot \mathbf{r}) \mathbf{r}}{r^5} \right)$$

$$(1.4.126) \quad p_{SS} (\mathbf{r}; \alpha, \beta) = \left(-\frac{(\alpha \cdot \beta)}{r^3} + 3 \frac{(\alpha \cdot \mathbf{r})(\beta \cdot \mathbf{r})}{r^5} \right)$$

The stresslet represents the contribution of a straining motion of a fluid that is symmetric about the α, β plane with the principal axes of stain being in the direction of $\alpha/|\alpha| + \beta/|\beta|$, $\alpha/|\alpha| - \beta/|\beta|$, and $\alpha \times \beta$. Interestingly the stresslet exerts zero net force or torque on the fluid.

The quadrupole solutions are more complicated, and only one of them, the potential dipole is commonly used. The potential dipole is defined as

$$(1.4.127) \qquad \mathbf{u}_D\left(\mathbf{r};\delta\right) = -\frac{1}{2}\nabla^2\mathbf{u}_S\left(\mathbf{r};\delta\right)$$

and its velocity and pressure is

$$(1.4.128) \qquad \mathbf{u}_D\left(\mathbf{r};\delta\right) = \left(-\frac{\delta}{r^3} + \frac{3\left(\delta\cdot\mathbf{r}\right)\mathbf{r}}{r^5}\right)$$

$$(1.4.129) \qquad p_D\left(\mathbf{r};\delta\right) = 0$$

The potential dipole represents the flow generated by a mass source at $\mathbf{r} = \delta/2$ and a mass sink of the same magnitude at $\mathbf{r} = -\delta/2$, in the limit of $|\delta| \to \infty$. The potential dipole exerts zero net force fluid.

1.4.6. Demonstration of the method of internal singularities: sphere moving in quiescent fluid. To get a feel for how we construct solutions via the method of internal singularities, let's construct our familiar solution for a sphere moving through a quiescent fluid. For this problem we nondimensionalize the velocities with the sphere velocity U and the length scales with the sphere radius a. Thus the boundary condition on the surface of the sphere

$$(1.4.130) \qquad \mathbf{u} = \mathbf{i}_u \text{ at } r = 1$$

where here we use the symbol $\mathbf{i}$ to indicate that the vector has been nondimensionalized and has a magnitude of order unity. With this nondimensionalization the Stokeslet velocity is

$$(1.4.131) \qquad \mathbf{u}_S = \alpha + \left(\alpha\cdot\mathbf{i}_r\right)\mathbf{i}_r$$

We see that the Stokeslet (alone) cannot satisfy the boundary condition, as there is nothing to balance the $\left(\alpha\cdot\mathbf{i}_r\right)\mathbf{i}_r$ term. Looking at our (small) catalog of singularities, we can guess that maybe a potential dipole, which with this nondimensionalization has a velocity of

$$(1.4.132) \qquad \mathbf{u}_D = -\delta + 3\left(\delta\cdot\mathbf{i}_r\right)\mathbf{i}_r$$

together with the Stokeslet might balance the $\left(\alpha\cdot\mathbf{i}_r\right)\mathbf{i}_r$ term. Letting

$$(1.4.133) \qquad \mathbf{u} = \mathbf{u}_S + \mathbf{u}_D = \alpha - \delta + \left(\alpha\cdot\mathbf{i}_r\right)\mathbf{i}_r + 3\left(\delta\cdot\mathbf{i}_r\right)\mathbf{i}_r$$

we see that we have a way to satisfy the boundary condition (1.4.130), namely through

$$(1.4.134) \qquad \alpha - \delta = \mathbf{i}_u$$

$$(1.4.135) \qquad \alpha + 3\delta = 0$$

Thus, $\alpha = 3\mathbf{i}_u/4$, $\delta = -\mathbf{i}_u/4$. Since the potential dipole applies no force on the fluid, the only force here is from the Stokeslet. From Section 1.4.2 the force is just $\mathbf{F}^* = 6\pi\mathbf{i}_u$ or in dimensional terms, $\mathbf{F} = 6\pi\mu Ua\mathbf{i}_u$. Hence we rederived Stokes's law.

1.4.7. Faxen's law: force on a body in unbounded fluid in any creeping flow. In prior sections we derived the force on a sphere in uniform creeping flow – the so called "Stokes's law" or "Stokes' drag law". Stokes' drag is used almost universally as the first approximation to a drag on a spherical particle in practically any creeping flow. (As a mater of fact, Stokes' drag is used as a first approximation for the drag on non-completely spherical, non-rigid particles, in low (but not very low, or effectively zero) Reynolds number as well and with some success.) As we will see shortly, the use of Stokes' drag as the drag on a sphere for any creeping flow is just an approximation. Here we will derive the exact result for drag on a sphere in any creeping flow – Faxen's law.

To begin, let's consider a particle that translates with a velocity $-\mathbf{U}$ in a quiescent fluid. (Notice that if we have the flow field for this problem we can also simply calculate the flow field for a stationary particle in uniform flow with velocity $\mathbf{U}$). We denote the flow field around this translating particle as $\mathbf{u}$. We place the particle near the origin of the coordinate system so that the flow field $\mathbf{u}$ vanishes as $r \to \infty$. We denote the corresponding surface force vector on the surface of the particle ∂D as $\mathbf{f}$. Applying the Lorentz reciprocal theorem (1.4.98),

$$(1.4.136) \qquad \mathbf{U} \cdot \int_{\partial D} \mathbf{f}_2 dS = \mathbf{U} \cdot \mathbf{F}_2 = \int_{\partial D} \mathbf{u}_2 \cdot \mathbf{f} dS$$

We can let the second velocity field $\mathbf{u}_2$ be the undisturbed flow around the fixed body $\mathbf{u}_\infty$, so that

$$(1.4.137) \qquad \mathbf{U} \cdot \mathbf{F}_\infty = \int_{\partial D} \mathbf{u}_\infty \cdot \mathbf{f} dS$$

Now we can obtain the force on the particle in an undisturbed flow $\mathbf{F}_\infty$, by integrating the right-hand side of (1.4.137) as we know $\mathbf{f}$ from the solution of the simpler uniform flow problem. To obtain all three components of $\mathbf{F}_\infty$ we need solutions for uniform flow around a body, directed at the body in three directions (e.g., x, y, and z directions). Thus from initially solving three simpler problems of a particle in uniform flow, we can obtain, via the Lorentz reciprocal theorem, the force on a particle in any arbitrary flow (given that both the uniform flow and the arbitrary flow satisfy the creeping flow equations).

Let's now apply this to find the force (drag) on a sphere. From Section 1.3 the force density on a sphere in uniform flow is,

$$(1.4.138) \qquad \mathbf{f} = \frac{3}{2} \frac{\mu \mathbf{U}}{a}$$

Since the sphere is spherically symmetric, this force density of course does not depend on the direction of the flow. Thus,

$$(1.4.139) \qquad \mathbf{U} \cdot \mathbf{F}_\infty = \int_{\partial D} \mathbf{u}_\infty \cdot \frac{3}{2} \frac{\mu \mathbf{U}}{a} dS$$

Because $\mathbf{U}$ is arbitrary, we can factor it, and multiply by the inverse on both sides, obtaining

$$(1.4.140) \qquad \mathbf{F}_\infty = \frac{3}{2} \frac{\mu}{a} \int_{\partial D} \mathbf{u}_\infty dS$$

Now we need and expression for $\mathbf{u}_\infty$ to calculate the force, but we want to avoid calculating $\mathbf{u}_\infty$. Thus we can expand $\mathbf{u}_\infty$ about the point of the center of the sphere (the origin) as a multipole expansion,

$$(1.4.141) \qquad \mathbf{u}_\infty\left(\mathbf{x}\right) = \mathbf{u}_\infty\left(0\right) + \mathbf{x} \cdot \left(\nabla \mathbf{u}_\infty\right)_0 + \frac{\mathbf{xx}}{2!} : \left(\nabla\left(\nabla \mathbf{u}_\infty\right)\right)_0 + \ldots$$

where : is the double dot product. We substitute this into (1.4.140) and obtain

$$(1.4.142) \qquad \mathbf{F}_\infty = \frac{3}{2}\frac{\mu}{a} \int\limits_{\partial D} \left(\mathbf{u}_\infty\left(0\right) + \mathbf{x} \cdot \left(\nabla \mathbf{u}_\infty\right)_0 + \frac{\mathbf{xx}}{2!} : \left(\nabla\left(\nabla \mathbf{u}_\infty\right)\right)_0 + \ldots\right) dS$$

Noting that because of spherical symmetry the odd derivatives vanish at the surface of the sphere ∂D, so that

$$(1.4.143) \qquad \mathbf{F}_\infty = 6\pi\mu a \left(\mathbf{u}_\infty\left(0\right) + \frac{a^2}{6}\left(\nabla^2 \mathbf{u}_\infty\right)_0 + k_4\left(\nabla^4 \mathbf{u}_\infty\right)_0 + \ldots\right)$$

where k_4 is a constant. However, since $\mathbf{u}_\infty$ satisfies the creeping flow equations, $\nabla^4 \mathbf{u}_\infty = 0$. In fact it can be shown if a velocity distribution satisfies the creeping flow equations, $\nabla^{2n}\mathbf{u}_\infty = 0$ for $n \geqslant 2$. Thus all the higher order terms are exactly zero in (1.4.143)! So instead of getting an approximate result for the force we get an exact result,

$$(1.4.144) \qquad \mathbf{F}_\infty = 6\pi\mu a \left(\mathbf{u}_\infty\left(0\right) + \frac{a^2}{6}\left(\nabla^2 \mathbf{u}_\infty\right)_0\right)$$

This result is called Faxen's law for a sphere. This is a powerful result that allows us to calculate the drag force on a sphere by just specifying the undisturbed velocity field (the velocity field in the absence of the sphere) and its Laplacian at the origin (where the sphere will be located). However, the only requirement is that the undisturbed flow field be unbounded. Since an unbounded flow field is a flow field that is that without walls or wall interactions, the only unbounded flow field that is easy to produce in practice is the uniform flow field. We can immediately see that for a uniform flow field, Faxen's law collapses to Stokes's law,

$$(1.4.145) \qquad \mathbf{F}_\infty = 6\pi\mu a \mathbf{u}_\infty\left(0\right)$$

since the second derivative of a constant is zero. The next realistic, but slightly more difficult "unbounded" flow fields that we can produce in practice is a linear flow field, and a parabolic flow field. In practice, we can produce a linear flow field in Couette flow (flow between parallel plates with one plate moving relative to another). We can produce a parabolic flow field via Poiseulle flow. While these fields are not truly unbounded, when the particle is small relative to the channel dimensions, (but not so small that the field local to the particle appears uniform) we can model these fields as unbounded, especially for a particle far from the walls. Other unbounded flow fields are possible to generate in practice but are significantly more difficult to produce. One example of producing these is via a Newtonian ferrofluid driven by a shaped magnet field, with a particle that does not respond to a magnetic field. Another example is the flow field in a closed tube driven by non-uniform electroosmotic flow (e.g., due to non-uniformly coated, e.g. patterned walls).

Let's consider an arbitrary linear flow,

$$(1.4.146) \qquad \mathbf{u}_\infty = \mathbf{U}_\infty + \mathbf{\Gamma}_\infty \cdot \mathbf{x}$$

where $\mathbf{\Gamma}_\infty$ is a velocity gradient tensor. If our spherical particle is moving with velocity $\mathbf{U}_p$ then the force on the particle depends on the relative velocity between the body and the undisturbed fluid, and so

(1.4.147)
$$\mathbf{F}_\infty = 6\pi\mu a \left(\mathbf{U}_\infty + \mathbf{\Gamma}_\infty \cdot \mathbf{x}_p - \mathbf{U}_p \right)$$

where $\mathbf{x}_p$ is the location of the particle center. Notice that the Laplacian term does not come into play, since the second derivative of a linear function is zero. For a neutrally buoyant particle (and a particle not experiencing electrical, magnetic or other forces), the force on it is zero, and so

(1.4.148)
$$\mathbf{U}_p = \mathbf{U}_\infty + \mathbf{\Gamma}_\infty \cdot \mathbf{x}_p$$

We see that in a linear flow there is no difference between Stokes' law and Faxen's law. Now let's consider a particle in a parabolic flow, for example a 2D Poiseuille flow between parallel plates separated by a distance d,

(1.4.149)
$$\mathbf{u}_\infty = \frac{G}{2\mu} y \left(d - y \right) \hat{\mathbf{x}}$$

and where G is the negative of the pressure gradient applied to generate the flow. Substituting this into Faxen's laws

$$\mathbf{F}_\infty = 6\tau\mu a \left(\frac{G}{2\mu} y \left(d - y \right) \hat{\mathbf{x}} - \mathbf{U}_p - \frac{Ga^2}{6\mu} \hat{\mathbf{x}} \right)$$

where the last term on the right-hand side comes from the second derivative of the flow profile (the Laplacian). Once again if the particle experiences no forces on it, then

(1.4.150)
$$\mathbf{U}_p = \frac{G}{2\mu} \left(\frac{y}{d} - \left(\frac{y}{d} \right)^2 - \frac{1}{3} \left(\frac{a^2}{d^2} \right) \right) \hat{\mathbf{x}}$$

We see something very interesting – the particle's velocity is slower than the flow velocity by $a^2/3d^2$! The particle is not "faithful" – it does not actually follow the flow. In practice however, our derivation is applicable only for unbounded flows, so where the size of the particle is small compared to the size of the channel - $a/d \ll 1$. Hence, for most practical situations the correction from the $a^2/3d^2$ is small.

1.5. Separated flows around immersed bodies

So far we have considered creeping flows where the fluid remains attached the body interest. In this section we will consider flows where the fluid detaches from the body that it is flowing around, and causes vortices to form. These vortices can provide additional local mixing and sometimes can entrain particles - sometimes desired and sometimes undesired effects. These vortices are often misinterpreted by those unfamiliar with fluid phenomena as "turbulence". We would like to emphasize that these are not an example of turbulence, because, for one, inertia of the flow is negligible. In this section we will consider a number of flows past objects commonly encountered in microfluidic devices and discuss the associated streamlines of the flow based on experimental visualization of these streamlines by Taneda (1979) and numerical predictions of these streamlines by Shen and Floryan (1985). In these flows Reynolds number was order 10^{-2}.

Let's begin with the flow around two equal spheres placed streamwise in the flow (Figure 1.5.1). When the spheres are spaced apart more than one sphere diameter no separation occurs. Flow appears to begin to separate when the ratio of separation to sphere diameter $\varepsilon/d = 1$. When $\varepsilon/d = 0.7$ vortex rings were clearly observed in the gap between the two spheres. When the spheres are brought together and are touching the fluid in the two gaps rotates in an infinite set of ring vortices. Theoretical predictions of this flow are given in Davis et al. (1976). These predictions agree well with behavior observed by Taneda at $Re = 10^{-2}$ (based on sphere diameter). These streamlines are also similar to the streamlines between to bubbles in a bubble train (which also have spherical end caps) as we will see later.

A similar flow to the flow around equal spheres is the flow around equal cylinders. Such flow is commonly encountered in microfluidic devices, as equal cylinders often form pillar arrays, whose behavior might be loosely inferred from the behavior of a pair of cylinders. Taneda observed at $Re = 10^{-2}$ that when $\varepsilon/d = 1.5$ no separation occurred. Clear vortex pairs occurred at $\varepsilon/d = 1$ and these vortex pairs merge into one vortex pair at $\varepsilon/d = 0.5$. When the cylinders are set to touch, once again an infinite set of ring vortices occurs. This series of vortices of decreasing size and strength are known as a type of Moffatt vortices (Moffatt (1964)). Moffatt vortices are found in viscous flows over cavities where the cavity (corner) angle is less than the critical angle and consist of a sequence of vortices of decreasing size and rapidly decreasing intensity. Because the intensity of the vortices rapidly decreases as we get further into the corner, the fluid is virtually stagnant in the corners.

When the cylinders are placed side by side instead of behind each other in a uniform flow, no separation of flow is observed (Figure 1.5.3).

For the flow around a cylinder placed near a plane (Figure 1.5.4) no flow separation is observed for the gap to diameter ratio, $\varepsilon/d = 0.6$. For smaller gap ratios such as $\varepsilon/d = 0.25, 0.2,$ and 0 flow separation is clearly observed. Davis and O'Neill (1977) have theoretically studied the structure of Stokes flow past a cylinder near a plane. They have found that separation should occur starting from $\varepsilon/d = 0.34$. At $\varepsilon/d = 0.34$ the flow separates from the plane, and at $\varepsilon/d = 0.015$ the flow separates from the cylinder. Furthermore, they predict that there is always a small flux of fluid between the plane and the cylinder, even when the gap is small.

Another fundamental structure found in microfluidic devices is a "fence" - a thin rib protruding from a plane into the flow (Figure 1.5.5). For a symmetric 90° fence the flow separates and forms two symmetrical vortices. The distance between the center of the vortex and the corner point was observed to be 0.54 times the height of the fence. For the 105° fence the two vortices formed are not symmetric. However, the streamline pattern remains the same even if the direction of flow across the fence is reversed! (This is a characteristic property of Stokes flows).

Other very common features found in microfluidic devices are a square bank and (especially) a step (Figure 1.5.6). Flow separates both in front and behind a square bank, and a symmetric pair of vortices develop. The distance between the vortex center and the corner point was observed to be 0.46 the bank height. For a step, the flow structure is very similar to that downstream of a square bank - a single vortex develops there. The distance between the vortex center and the corner point was observed to be 0.47 the bank height. Notice that the flows around a 90° (virtually very thin square bank), a square bank, and a step (virtually a very long square bank) are quite similar. Even the center of the vortex found downstream of each structure is roughly a distance of 0.5 times the structure's height away from the corner point. The surfaces of the corners in which these vortices are found don't interact with the rest of the flow, as it is very difficult for particles to move across

(a) $\varepsilon/d = 1$

(b) $\varepsilon/d = 0.7$

(c) $\varepsilon/d = 0$

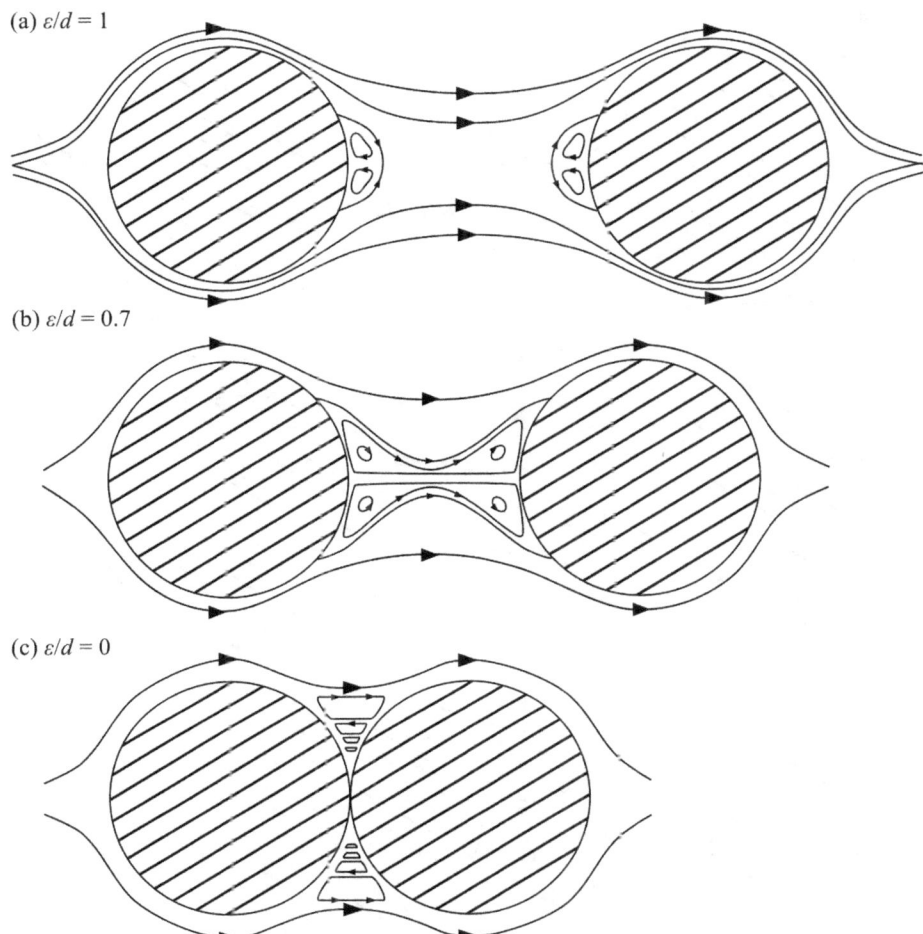

FIGURE 1.5.1. Schematic of flow between two spheres diameter d separated with spacing ε. (a) For $\varepsilon/d = 1$ flow separation was not observed experimentally but small amount of flow separation was predicted theoretically. (b) At $\varepsilon/d = 0.7$ flow was observed to separate from the spheres and form vortex rings in the gap as shown. (c) When the spheres are touching an infinite sequence of vortices occurs.

streamlines in Stokes flow (Bretherton (1962)). (In fact, motion of objects across streamlines in Stokes flow occurs either by diffusion or if external forces are applied to the objects). Thus, an object traveling on a streamline that is near a vortex is unlikely to land in a vortex, and an object trapped in a vortex is unlikely to come out. Hence, such corners can be often treated as virtual "dead zones", as surface sensors (e.g., antibodies, electrochemical sensors) will have poor interaction with the rest of the flow.

An inverse of a bank is a cavity (Figure 1.5.7) and this shape is also very commonly encountered in microfluidic flows. A cavity with an aspect ratio of $w/h = 1$ is observed to house a single large vortex. Shen and Floryan (1985) also predict

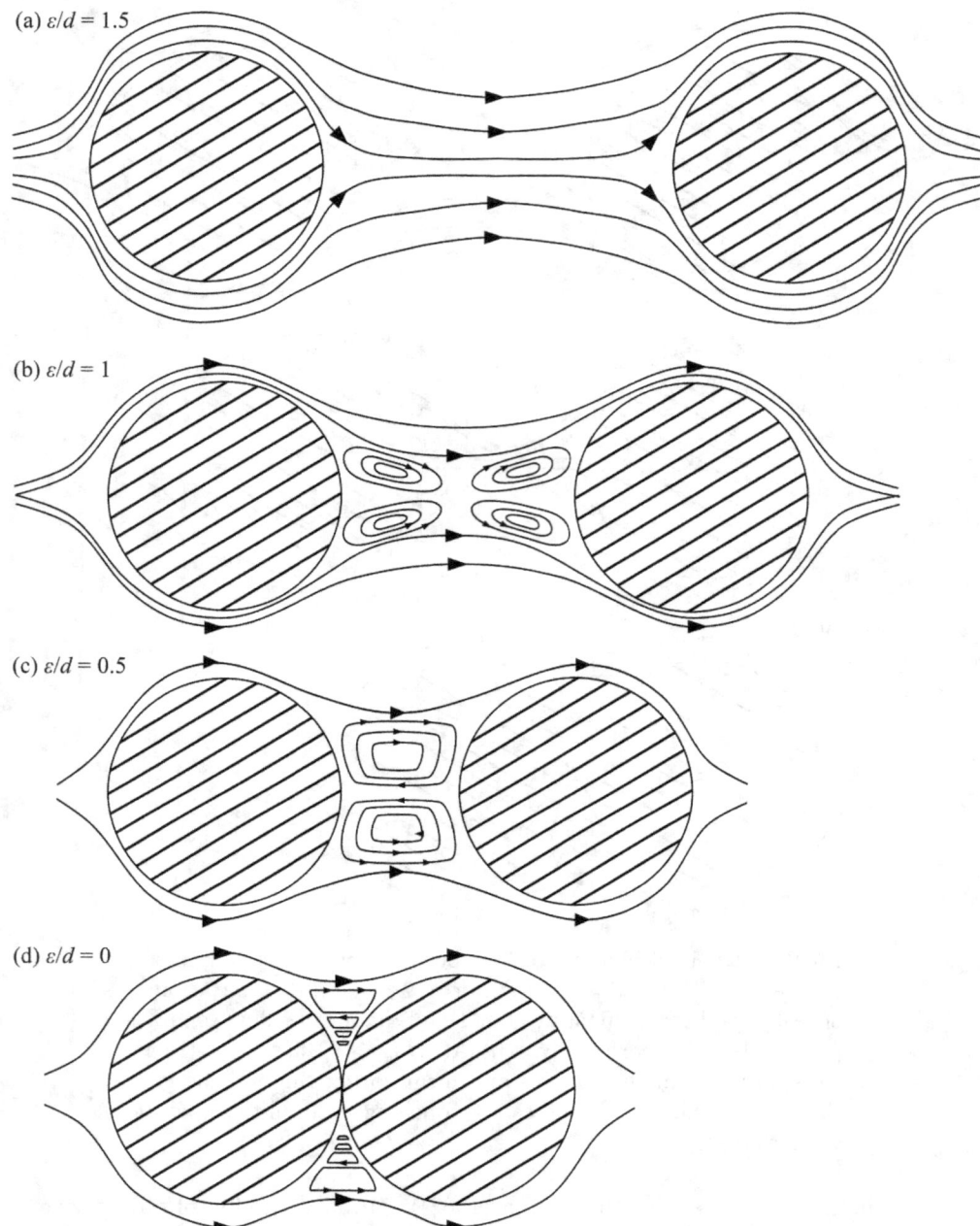

(a) $\varepsilon/d = 1.5$

(b) $\varepsilon/d = 1$

(c) $\varepsilon/d = 0.5$

(d) $\varepsilon/d = 0$

FIGURE 1.5.2. Schematic of two dimensional flow between circular cylinders. (a) For $\varepsilon/d = 1.5$ no flow separation occurs. (b) At $\varepsilon/d = 1$ flow separation is observed and two vortex pairs are observed in the gap between cylinders. (c) As the gap is narrowed to $\varepsilon/d = 0.5$ these vortex pairs collapse into two vortex pairs. (d) As the gap between the cylinders is closed, $\varepsilon/d = 0$, an infinite sequence of vortices occurs in the two gaps.

$\varepsilon/d = 0.2$

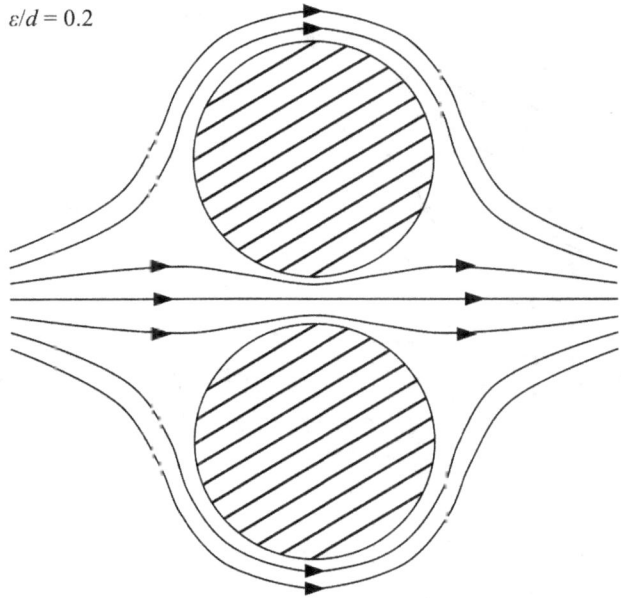

FIGURE 1.5.3. Sketch of a streamline pattern around two equal circular cylinders, placed side by side relative to the flow, with gap to diameter ratio, $\varepsilon/d = 0.2$, and Reynolds number order 10^{-2}. No flow separation was observed in this configuration.

that the very corners of the cavity also house a cascade of weak Moffatt type vortices. A cavity with an aspect ratio of $w/h = 0.5$ (a deep cavity), is observed to house a pair of large vortices (with the bottom one being smaller), as well as weak Moffatt type vortices in the corners. A cavity with an aspect ratio of $w/h = 2$ (a wide cavity) is observed to house a large flat vortex (as well as weak Moffatt vortices in the corners). As the aspect ratio is increased to $w/h = 3$ the large flat vortex breaks up into two smaller triangular vortices. Another type of cavity is a wedge (Figure 1.5.8). The wedge is clearly observed to house a cascade of Moffatt type vortices (as predicted by Moffatt (1964)). As predicted, the vortices appear self similar, and are of decreasing strength the deeper you go into the wedge.

$\varepsilon/d = 0.6$ $\varepsilon/d = 0.25$

$\varepsilon/d = 0.1$ $\varepsilon/d = 0$

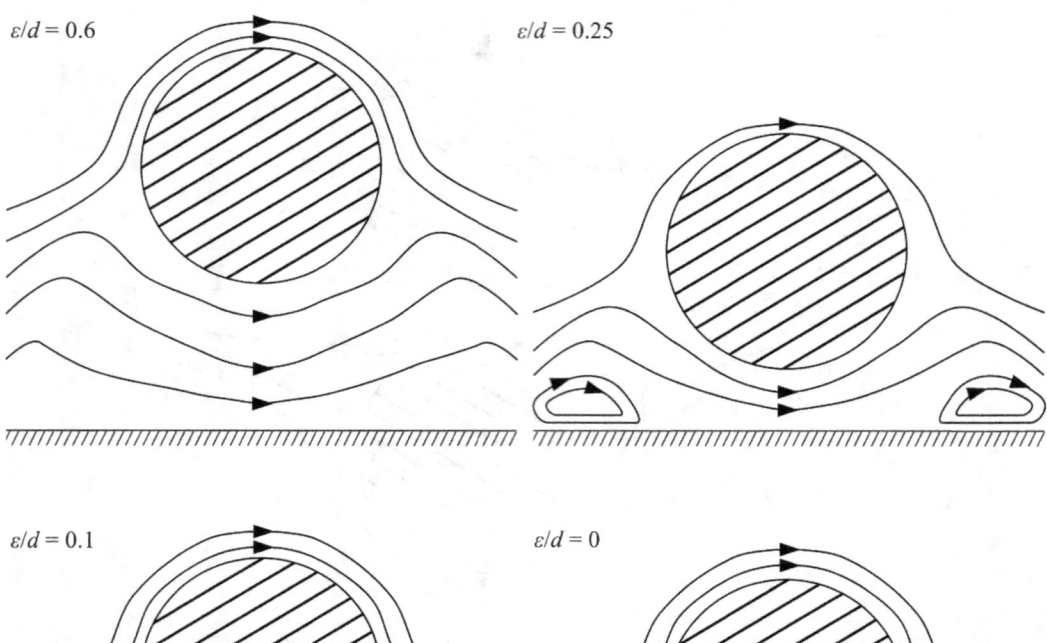

FIGURE 1.5.4. Sketch of streamline patterns around a circular cylinder placed near a flat plate. Reynolds number was order 10^{-2}. When gap to diameter ratio, $\varepsilon/d = 0.6$ no flow separation was observed. Flow separation was observed at $\varepsilon/d = 0.25$, 0.2, and 0.

(a) 90° fence   (b) 105° fence

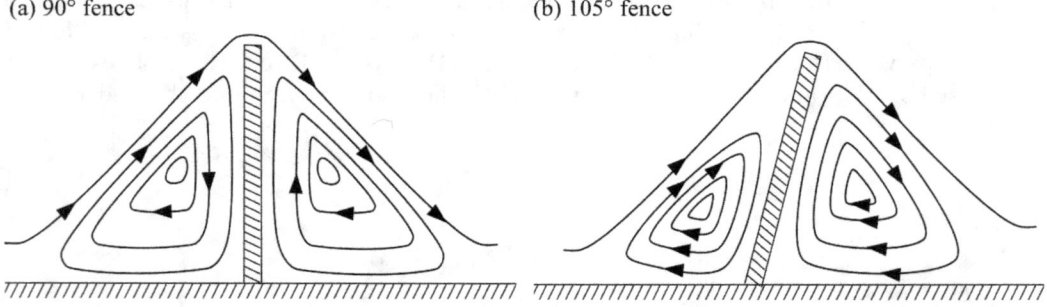

FIGURE 1.5.5. Sketch of a streamline pattern past a 90° and 105° fence, Reynolds number order 10^{-2}.

(a) Square bank   (b) Step

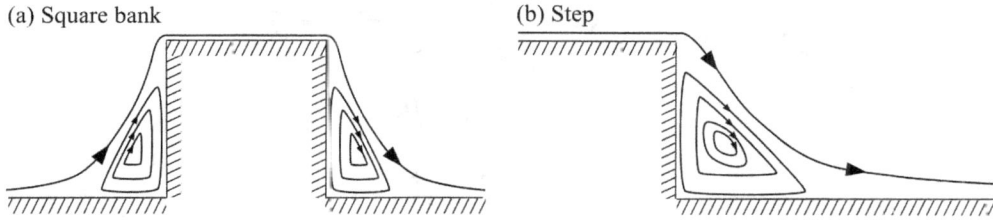

FIGURE 1.5.6. Sketch of streamlines around (a) a square bank, and (b) a 90° step. (a) In the flow around the square bank, the distance between the vortex center and the corner point was 0.46 the bank height. (b) Similarly, in the flow around the step, the distance between the vortex center and the corner point was 0.47 the step height. Furthermore, the flow pattern around a step was the same when the direction of flow was reversed (as expected for a low Reynolds number flow).

(a) $w/h = 1$ (b) $w/h = 0.5$

(c) $w/h = 2$

(d) $w/h = 3$

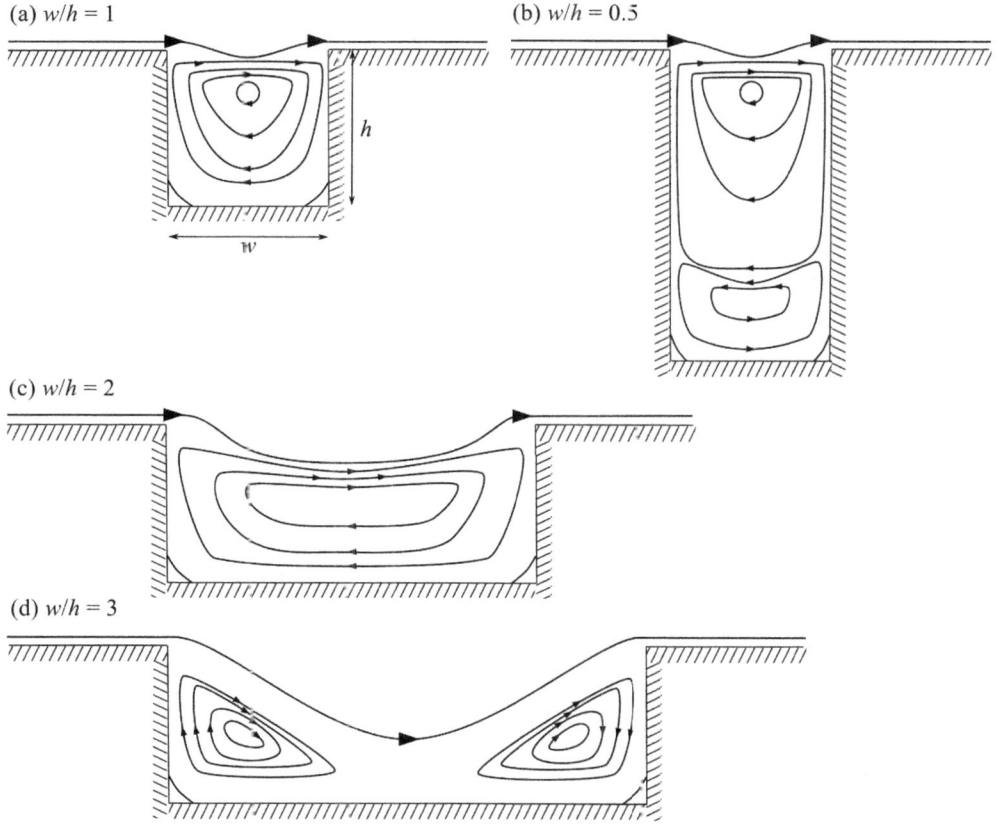

FIGURE 1.5.7. Sketch of streamlines of a flow past a cavity. Single or multiple vortices are observed. The corners often posses Moffat type vorticies.

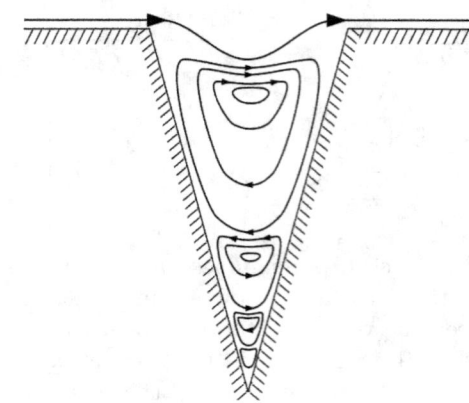

FIGURE 1.5.8. Sketch of a streamline pattern in a wedge shaped region (wedge angle 28.5°). A cascade of Moffatt type vortices is observed.

CHAPTER 2

Thin geometry flows: Hele-Shaw and Lubrication flows

Very high aspect ratio geometries, such as a flat thin chamber, a thin chamber with a taper, or a thin film of liquid bounded between a solid and a gas occurs quite frequently in microfludics. We here will examine two models for such cases: Hele-Shaw model and Lubrication theory model. These models drastically simplify the Navier-Stokes equations and often permit obtaining exact analytical solutions for the flow. Furthermore, they accurately model realistic microfluidic devices and situations.

2.1. Hele-Shaw flows

Hele-Shaw flow refers to flow around objects sandwiched between flat plates, where the gap between the plates is much smaller than the characteristic length scale of the flow (typically length scale of the object). The Reynolds number (based on the gap) of such flows is also small. The flow is named after Henry Hele-Shaw (1898) who studied this flow. This flow is particularly interesting because of a peculiar property that the streamlines in this flow reproduce exactly the streamlines of a potential flow (i.e , infinite Reynolds number flow). The flow is practically two dimensional! Since a Hele-Shaw chambers (or cells) are fairly easy to fabricate - e.g., sandwich a rubber gasket cutout between transparent plastic plates, these were used for obtaining potential flow streamlines around unusual geometries. However, now it is easy to obtain potential flow solutions for many geometries numerically, and so now we can use these solutions to predict the flow in the Hele-Shaw resembling microfluidic chambers.

Here to understand the simplifications of Navier-Stokes equations for Hele-Shaw flows we consider the steady flow around a cylinder between flat parallel plates (Figure 2.1.1). The plates are located at $x = \pm b$ and we define our Reynolds number as

$$(2.1.1) \qquad\qquad Re = U_0 b/\nu$$

Between the plates and at the center of the coordinate system is a cylinder of radius $2a$ and height $2b$. Far upstream of the cylinder the flow has a uniform velocity U_0. Since this is a Hele-Shaw cell, the gap must be small compared to the characteristic length scale of the cylinder. Thus,

$$(2.1.2) \qquad\qquad b/a = \varepsilon \ll 1$$

In fact, we define a Hele-Shaw limit as where

$$(2.1.3) \qquad\qquad Re \ll \varepsilon^2 \ll 1$$

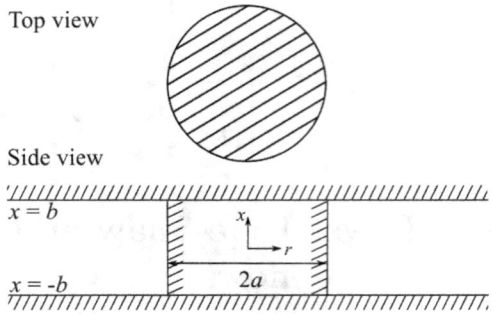

FIGURE 2.1.1. Schematic of a cylinder sandwiched between flat parallel plates, with the gap between the plates exaggerated for clarity. In reality, the diameter of the cylinder is much larger than the gap between the plates, $b/a \ll 1$. The configuration of a flow around an object sandwich between parallel plates, when the objects characteristic length scale is much larger than the gap is often referred to Hele-Shaw flow and the apparatus referred to as a Hele-Shaw cell.

Since we are working with a flow around a cylinder, we will use a cylindrical coordinate system. To help us simplify the Navier-Stokes equations we introduce the following dimensionless variables. We scale the vertical coordinate by the characteristic dimension of the gap:

(2.1.4) $$x^* = x/b$$

We scale the radial coordinate by the characteristic dimension of the object of interest,

(2.1.5) $$r^* = r/a$$

We scale the velocity vector by the upstream velocity U_0,

(2.1.6) $$\mathbf{v}^* = \mathbf{v}/U_0$$

Lastly, we scale the pressure by the viscous pressure scale $\mu U_0/b$, since we expect the Reynolds number to be small,

(2.1.7) $$p^* = \frac{p - p_\infty}{\mu U_0/b}$$

where p_∞ is a reference pressure, for example atmospheric pressure. Dropping the asterisks for convenience of readability, the continuity equation (taking into account the flow is steady and incompressible) is

(2.1.8) $$\frac{\partial u_x}{\partial x} + \varepsilon \left(\frac{1}{r} \frac{\partial}{\partial r}(r u_r) + \frac{1}{r} \frac{\partial u_\theta}{\partial \theta} \right) = 0$$

and the momentum equations are:

(2.1.9)
$$Re \left(u_x \frac{\partial u_r}{\partial x} + \varepsilon \left(u_r \frac{\partial u_r}{\partial r} + \frac{u_\theta}{r} \frac{\partial u_r}{\partial \theta} - \frac{u_\theta^2}{r} \right) \right)$$
$$= -\frac{\partial p}{\partial r} + \frac{\partial^2 u_r}{\partial x^2} + \varepsilon^2 \left(\frac{\partial^2 u_r}{\partial r^2} + \frac{1}{r} \frac{\partial u_r}{\partial r} + \frac{1}{r^2} \frac{\partial^2 u_r}{\partial \theta^2} - \frac{u_r}{r^2} - \frac{2}{r} \frac{\partial u_\theta}{\partial \theta} \right)$$

(2.1.10)
$$Re\left(u_x\frac{\partial u_\theta}{\partial x} + \varepsilon\left(u_r\frac{\partial u_\theta}{\partial r} + \frac{u_\theta}{r}\frac{\partial u_\theta}{\partial\theta} - \frac{u_r u_\theta}{r}\right)\right)$$
$$= -\frac{1}{r}\frac{\partial p}{\partial\theta} + \frac{\partial^2 u_\theta}{\partial x^2} + \varepsilon^2\left(\frac{\partial^2 u_\theta}{\partial r^2} + \frac{1}{r}\frac{\partial u_\theta}{\partial r} + \frac{1}{r^2}\frac{\partial^2 u_\theta}{\partial\theta^2} - \frac{u_\theta}{r^2} - \frac{2}{r^2}\frac{\partial u_r}{\partial\theta}\right)$$

(2.1.11)
$$Re\left(u_x\frac{\partial u_x}{\partial x} + \varepsilon\left(u_r\frac{\partial u_x}{\partial r} + \frac{u_\theta}{r}\frac{\partial u_x}{\partial\theta}\right)\right)$$
$$= -\frac{\partial p}{\partial x} + \varepsilon^2\left(\frac{\partial^2 u_x}{\partial r^2} + \frac{1}{r}\frac{\partial u_x}{\partial r} + \frac{1}{r^2}\frac{\partial^2 u_x}{\partial\theta^2}\right)$$

Because the condition for Hele-Shaw flow is that $Re \ll \varepsilon^2 \ll 1$, we take the limit $Re \to 0$ first and drop the convective acceleration terms that multiply Re. We then take the limit $\varepsilon \to 0$ and drop the corresponding terms. The continuity equation gives us

(2.1.12)
$$\frac{\partial u_x}{\partial x} = O(\varepsilon) \to 0$$

Since the no-penetration boundary condition specifies that at each of the parallel plates $u_x = 0$, combining this with (2.1.12), we obtain that $u_x = 0$ everywhere in our domain. This is not surprising, as we expected the flow to be practically two dimensional. Next the momentum equations give us

(2.1.13)
$$\frac{\partial^2 u_r}{\partial x^2} = \frac{\partial p}{\partial r} + O(\varepsilon^2)$$

(2.1.14)
$$\frac{\partial^2 u_\theta}{\partial x^2} = \frac{1}{r}\frac{\partial p}{\partial\theta} + O(\varepsilon^2)$$

(2.1.15)
$$0 = \frac{\partial p}{\partial x} + O(\varepsilon^2)$$

Equation (2.1.15) tells us that $p = p(r,\theta)$ and is independent of x. This is again not surprising as we expected the flow to be practically two dimensional. Integrating (2.1.13) and (2.1.14) with respect to x, and applying the no-slip boundary condition that at $x = \pm 1$, $u_r = u_\theta = 0$ we obtain,

(2.1.16)
$$u_r = -\frac{\partial}{\partial r}\left(\frac{1}{2}p\left(1 - x^2\right)\right)$$

(2.1.17)
$$u_\theta = -\frac{1}{r}\frac{\partial}{\partial\theta}\left(\frac{1}{2}p\left(1 - x^2\right)\right)$$

We can now define a two dimensional potential function φ, such that the two dimensional velocity $\mathbf{u} = \nabla\varphi$, or

(2.1.18)
$$u_r = \frac{\partial\varphi}{\partial r}$$

(2.1.19)
$$u_\theta = \frac{1}{r}\frac{\partial\varphi}{\partial\theta}$$

and so we see that for our case,

(2.1.20)
$$\varphi = -\frac{1}{2}p\left(1 - x^2\right)$$

To solve this, we take a look at some of the terms we have dropped in the continuity equation. When simplifying the continuity equation we stated that the term

(2.1.21)
$$\varepsilon \left(\frac{1}{r} \frac{\partial}{\partial r} (r u_r) + \frac{1}{r} \frac{\partial u_\theta}{\partial \theta} \right)$$

is negligible because ε is very small, and so because of our scaling,

(2.1.22)
$$\frac{1}{r} \frac{\partial}{\partial r} (r u_r) + \frac{1}{r} \frac{\partial u_\theta}{\partial \theta} = O(1)$$

However, if $u_x = 0$, then in fact (2.1.8) tells us that

(2.1.23)
$$\frac{1}{r} \frac{\partial}{\partial r} (r u_r) + \frac{1}{r} \frac{\partial u_\theta}{\partial \theta} = 0$$

Substituting in the velocities in terms of the potential function (2.1.18) and (2.1.19) we obtain

(2.1.24)
$$\nabla^2 \varphi = 0$$

where ∇^2 is the two dimensional Laplacian in r, θ. Translating the boundary conditions into the potential function, (with the help of (2.1.18) and (2.1.19)) of we obtain that at the cylinder wall, $r = 1$, we have a no-penetration boundary condition and so

(2.1.25)
$$\frac{\partial \varphi}{\partial r} = 0 \text{ at } r = 1$$

and we have uniform flow at infinity,

$$\varphi \to \frac{1}{2} \left(1 - x^2 \right) r \cos \theta \text{ at } r \to \infty$$

Solving Laplace's equation (2.1.24) subject to these boundary conditions, we obtain

(2.1.26)
$$\varphi = r \cos \theta \left(1 + \frac{1}{r^2} \right) \frac{\left(1 - x^2 \right)}{2}$$

Substituting this result into (2.1.18) and (2.1.19), we obtain,

(2.1.27)
$$u_r = \cos \theta \left(1 - \frac{1}{r^2} \right) \frac{\left(1 - x^2 \right)}{2}$$

(2.1.28)
$$u_\theta = -\sin \theta \left(1 + \frac{1}{r^2} \right) \frac{\left(1 - x^2 \right)}{2}$$

To find the pressure, we substitute (2.1.27) and (2.1.28) into (2.1.18) and (2.1.19), and integrate to solve the ODE, to obtain,

(2.1.29)
$$p = -R \cos \theta \left(1 + \frac{1}{r^2} \right)$$

To check our obtained velocity profile, let's see what occurs at the cylinder surface, $r = 1$. We see that

(2.1.30)
$$u_r (r = 1) = 0$$

so, as expected the no-penetration boundary condition is satisfied. However,

(2.1.31)
$$u_\theta (r = 1) = -2 \sin \theta \frac{\left(1 - x^2 \right)}{2}$$

and so the no-slip condition is not satisfied - there is slip at the surface of the cylinder. We must then conclude that what we just obtained is the outer solution for the flow (the flow outside the boundary layer) and that there is a boundary later near the cylinder (just like with high Reynolds number flows). In this boundary layer $u_\theta = 0$ when $r = 1$. Since this boundary layer must be thin, we can assume that throughout this boundary layer, $u_r \approx 0$ and $\partial p/\partial r \approx 0$. Using our relation for pressure (2.1.29) for $r \approx 1$, we obtain

$$(2.1.32) \qquad \frac{1}{r}\frac{\partial p}{\partial \theta} \approx 2\sin\theta$$

Notice that in this boundary layer u_θ has to go from zero at the cylinder wall to a value given by (2.1.28) in a very small distance (thickness of the boundary layer). Hence for the θ momentum equation (2.1.10) the derivatives must be very large. There the dominant terms are

$$(2.1.33) \qquad \frac{\partial^2 u_\theta}{\partial x^2} + \varepsilon^2 \frac{\partial^2 u_\theta}{\partial r^2} = \frac{1}{r}\frac{\partial p}{\partial \theta}$$

Substituting in (2.1.32), we obtain

$$(2.1.34) \qquad \frac{\partial^2 u_\theta}{\partial x^2} + \varepsilon^2 \frac{\partial^2 u_\theta}{\partial r^2} = 2\sin\theta$$

Because the boundary layer is very thin in the r direction, and this scale is very much off compared to for example, the x direction. Therefore, let's stretch this direction so that the scales match better. To do this we define

$$(2.1.35) \qquad \hat{r} = (r - 1)/\varepsilon$$

Applying this stretched coordinate system to (2.1.34),

$$(2.1.36) \qquad \frac{\partial^2 u_\theta}{\partial x^2} + \frac{\partial^2 u_\theta}{\partial \hat{r}^2} = 2\sin\theta$$

In this coordinate system the cylinder surface is at $\hat{r} = 0$, and we must apply the no-slip condition, $u_\theta = 0$ there. In this coordinate system the boundary layer ends at $\hat{r} \to \infty$, where it must match the outer flow. Here u_θ must be given by (2.1.31). Equation (2.1.36) is just Poisson's equation, and we solve it using standard methods subject to these boundary conditions. We obtain,

$$(2.1.37) \qquad u_\theta(\hat{r}, \theta, x) = -(1 - x^2)\sin\theta + \sum_{n=0}^{\infty} A_n \cos(k_n x)\exp(-k_n\hat{r})\sin\theta$$

$$(2.1.38) \qquad A_n = \int_{-1}^{1} (1 - x^2)\cos(k_n x)dx$$

$$(2.1.39) \qquad k_n = \left(n + \frac{1}{2}\right)\pi$$

We thus obtained the fluid velocity in the boundary layer. We see that even for such a simple flow as flow around a cylinder we have a solution for an outer flow and a solution for a boundary layer. Very often we are not interested in the flow very near the object itself (in the boundary layer), unless we care for example about the mass transport to a surface sensor on the object's surface. Thus, we may only solve just for the outer flow, simply assuming that the flow in the direction of the gap,

u_x is simply equal to zero and so assume that the flow is two dimensional. This allows us to immediately construct a potential function such as (2.1.18), (2.1.19), and (2.1.24) and obtain a potential flow solution.

Further, notice that for the outer flow, the both u_r and u_θ decay as $1/r^2$. This means that even a few radii distance from the object the disturbance in the flow field due to the object vanishes, as if the object is not there. For example, a distance of 3 radii away from the cylinder, u_θ is roughly 10% below that infinitely far from the cylinder. Contrast this with a sphere in stokes flow (1.3.8). Even at a distance of 10 radii away from the sphere, the velocity is still roughly 10% below the free stream velocity magnitude! Thus, we may observe that for large flat objects, for which Hele-Shaw flow applies, the objects don't influence the flow very far from themselves. Small objects, for which Stokes flow applies, on the other hand, have a much greater reach and influence on the flow, compared to their size.

2.2. Lubrication theory flows

Another important flow model that is frequently encountered as a model for microfluidic flows is the steady incompressible flow in a thin, slowly tapering gap - lubrication flows. Lubrication (i.e. friction reduction between two solids) is often accomplished by having a viscous fluid move through a narrow but variable thickness gap. Typically one of the bodies is moving. The theory for this flow was originally developed by Reynolds (1886). Such flows are found ubiquitously, from journal bearings, to disk drives, to flow of liquid around bubbles.

To understand general lubrication theory flow and understand the simplifying assumptions, let's analyze a typical lubrication problem: low Reynolds number Couette flow in a varying gap (Figure 2.2.1). The top wall is stationary and the bottom wall moves at a velocity U relative to the top wall, creating Couette flow in the gap. The key to lubrication flows is that the gap is very narrow, $h(x) \ll L$. In this simple problem (and many that this problem models), the gap decreases linearly from h_0 to h_L over the distance L. The second key assumption to lubrication theory is that the flow is a Stokes flow, i.e., it has negligible inertia, or viscous terms dominate the inertial terms,

$$(2.2.1) \qquad \rho u_x \frac{\partial u_x}{\partial x} \ll \mu \left(\frac{\partial^2 u_x}{\partial y^2} \right)$$

or using the characteristic length scales,

$$(2.2.2) \qquad \rho U \frac{U}{L} \ll \frac{\mu U}{h^2}$$

which rewritten is,

$$\frac{\rho U L}{\mu} \frac{h^2}{L^2} \ll 1$$

From this we see that the Reynolds number based on the gap must be small, but the Reynolds number based on the length of the liquid film can be quite large. Because the Reynolds number based on the gap is small it is acceptable to neglect the inertia of the fluid.

The third usual simplifying assumption is that the velocity in the direction of the gap, u_y is small compared to u_x. Most of the flow is along the film thickness, not perpendicular to it. The forth often (but not always) used simplifying assumption of lubrication flows is that the flow is two dimensional, i.e., that there is no variation in

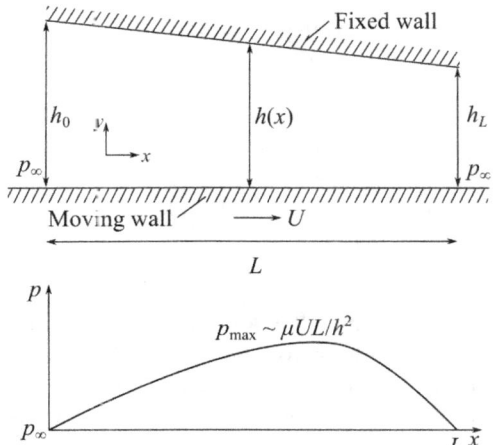

FIGURE 2.2.1. Schematic of a gap (top) in which a low Reynolds number Couette-Poiseuille type flow occurs. The gap is greatly exaggerated for clarity, while in reality $h/L \ll 1$. We sketch the resulting representative pressure distribution in the bottom part of the figure.

the z-direction (here into the page), $\partial/\partial z = 0$ and $u_z = 0$. With these assumptions we simplify the continuity equation to

$$(2.2.3) \qquad \frac{\partial u_x}{\partial x} + \frac{\partial u_y}{\partial y} = 0$$

and the momentum equations simplify to

$$(2.2.4) \qquad 0 = -\frac{\partial p}{\partial x} + \mu \frac{\partial^2 u_x}{\partial y^2}$$

$$(2.2.5) \qquad 0 = -\frac{\partial p}{\partial y}$$

$$(2.2.6) \qquad 0 = -\frac{\partial p}{\partial z}$$

Integrating the momentum equation (2.2.4) with no-slip and moving boundary conditions, we obtain the Couette-Poiseuille velocity profile,

$$(2.2.7) \qquad u_x = \frac{1}{2\mu}\frac{\partial p}{\partial x} y\,(y - h) + U\left(1 - \frac{y}{h}\right)$$

Integrating the continuity equation (2.2.3), we obtain

$$(2.2.8) \qquad \int_0^h \frac{\partial u_x}{\partial x}\,dy = -\int_0^h \frac{\partial u_y}{\partial y}\,dy = -\left(u_y\,(h) - u_y\,(0)\right)$$

Since we have no-penetration condition on the walls, both $u_y(h)$ and $u_y(0)$ are zero. Therefore,

$$(2.2.9) \qquad \int_0^h \frac{\partial u_x}{\partial x} dy = 0$$

Substituting in (2.2.7) and integrating we obtain

$$(2.2.10) \qquad \frac{\partial}{\partial x}\left(h^3 \frac{\partial p}{\partial x}\right) = 6\mu U \frac{\partial h}{\partial x}$$

a second order ODE for pressure. To solve it we need the relation $h(x)$, which should be known from the geometry of the device. In principle, we can solve this equation (at least numerically) for $h(x)$ having any form, and then use that pressure distribution to find the u_x distribution via (2.2.7). Here we give the result for the simplest $h(x)$,

$$(2.2.11) \qquad h(x) = h_0 + (h_L - h_0)\frac{x}{L}$$

For this distribution,

$$(2.2.12) \qquad \frac{p - p_\infty}{\mu U L / h_0^2} = \frac{6(x/L)(1 - x/L)(1 - h_L/h_0)}{(1 + h_L/h_0)(1 - (1 - h_L/h_0)(x/L))^2}$$

and

$$(2.2.13) \qquad \frac{\partial p}{\partial x} = \frac{\mu U L}{h_0^2}\frac{6 L h_0^2 (h_0 - h_L)((L - x)h_0 - x h_L)}{(h_0 + h_L)((L - x)h_0 + x h_L)^3}$$

where p_∞ is the reference (atmospheric) pressure at the entrance and exit of the lubrication film. We see that the maximum pressure in the gap is order $\mu U L / h_0^2$. For a water having a viscosity of $10^{-3} Pa\,s$, the wall moving at a brisk but reasonable 10 m/s (think journal bearing), 5 cm length, and 100 µm nominal gap, we obtain a pressure of 50 kPa or about 0.5 atm. Replacing that water with a common motor oil, SAE 0W-30, having a viscosity of $250 \times 10^{-3} Pa\,s$, we obtain a pressure of 123 atm! We can see now how journal bearings can easily support very large loads.

In this example we have analyzed a flow of liquid into a converging gap. However, since Stokes flows are linear, they are reversible. If we reverse the flow, i.e. set, $U < 0$, the pressure (correctly) predicted by (2.2.12) will be negative. The liquid will attempt to reach these pressures, but before it can attain such large negative pressures, it will outgas (dissolved gasses in the liquid will no longer be dissolved and will form gas bubbles) and will boil (again producing gas bubbles). In other words, the liquid will cavitate. The flow into an expanding gap thus does not provide good load bearing or lubrication. However, in a circular geometry, e.g., a journal bearing, we have both an expanding and a contracting gap. Thus this effect is unavoidable in journal bearings.

CHAPTER 3

Particle filtration in low Reynolds number flows

An important unit operation in many industries, from ore extraction to lab-on-a-chip, is particle separation. These particles can be spherical and irregular, rigid, and viscoelastic. In this section we will concentrate on how some of these can be separated at low Reynolds numbers, mostly based on their size. In the later sections we will also consider particle separations based on lift (at higher Reynolds numbers), density, and later on their electric properties.

3.1. Traditional filtration

Here we will briefly take a look at the mechanisms of particle separation via traditional filtration, both from the point of view of implementing this in microfluidic devices (especially lab-on-a-chip devices), as well as competitor technology to non-filtration particle separators. Traditional filtration can be separated into two types: cake filtration, and depth filtration (Figure 3.1.1). In cake filtration, the particles of interest are never meant to go into the filter media, and are meant to be kept above the filter media. Here the surface of the filter performs the separation (Figure 3.1.1a). Cake filtration can be operated in either "dead end" mode, where the flow direction is normal to the plane of the filter (as shown in Figure 3.1.1a) or in "crossflow" mode where the direction of the flow is parallel to the plane of the filter. In both modes particles accumulate above the filter and form their own porous structure – the so called "cake" after which the mode is named. Right below the cake is the working layer of the membrane, typically quite thin. This working layer is supported by a mechanical supporting layer, which provides mechanical support to the separation layer, is much thickener, and has bigger pores (for resistance reduction). In depth filtration on the other hand, it is the bulk (volume) of the filter performs the actual separation (Figure 3.1.1b).

Cake and depth filtration use different mechanisms to perform particle separation (Figure 3.1.2). Cake filtration relies on either complete blocking filtration (Figure 3.1.2a) or bridging filtration (Figure 3.1.2b). In complete blocking filtration the pore size is smaller than the size of the particle of interest, and particles are captured by sieving or screening process. We typically use low to medium particle concentrations with this filtration process. In bridging filtration the pore size is actually larger than the particle size. Particles are retained at the entrance of the filter when they being to clump together and form stable, but permeable bridges. For the filter to operate in this mechanism, the particles need to have propensity to clump together as to initiate the initial clumping and so the initial cake formation at the entrance of the filter, that actually performs the sieving and sorting. Since high concentration of particles increases the chances that they clump together, we see this mechanism when feed particle concentration is high. Complete blocking

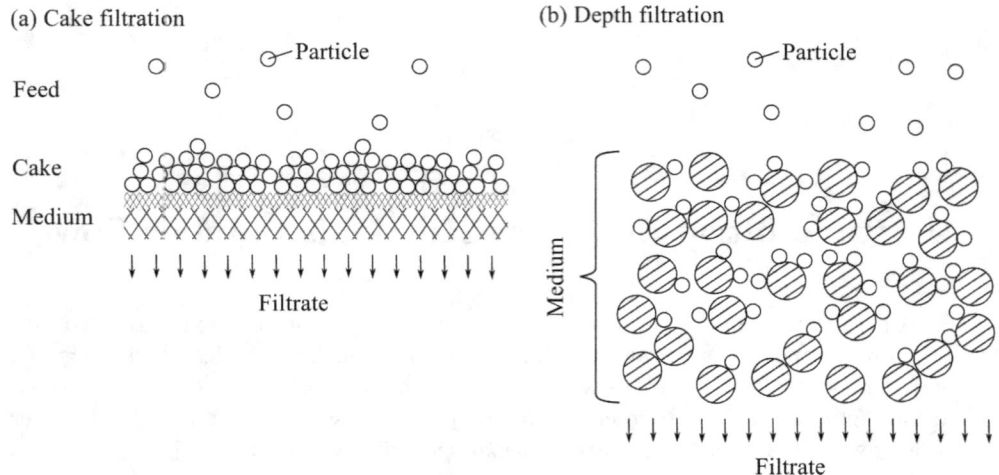

FIGURE 3.1.1. Schematic of filtration process in dead end cake filtration (a) and depth filtration (b). Notice that in cake filtration, after significant cake has built up on the filtration medium, it is both the cake and the filter media that sets the size cut off for the particles to be filtered. For depth filtration we draw the medium to be disordered as is typical of natural filters (e.g., packed beads, fiber meshes). However, for certain microfluidic devices the media can be ordered. For example, it can be a lithographically defined ordered post array.

filtration is the desirable mechanism when particle sizing is desired. Bridging filtration is undesired for particle sizing, since it is difficult to control the particle sieving in the developing cake. Bridging filtration can be desired when we need coarsely remove solids from a liquid, especially when we don't have access to smaller pore filtration media (e.g., limitation of lithography for a lab-on-a-chip device). However, complete blocking filtration allows us only to filter dilute solutions, and accurate particle sizing occurs only when no significant cake has developed (so that the sieving is done by the pores, not the cake). Thus, complete blocking filtration is typically a lower throughput filtration than other filtration modes.

Depth filtration typically uses standard blocking filtration mode (Figure 3.1.2c), where the pore size is larger than the size of the particle of interest. These particles get captured in the filter via a combination of hydrodynamic and physicochemical (e.g., affinity) based mechanisms. The hydrodynamic mechanisms bring the particle near the wall of the filter media, so that the physicochemical mechanisms can actually retain the particle on the wall. Hydrodynamic mechanisms include straining (Figure 3.1.3a), inertial impaction (Figure 3.1.3b), diffusion (Figure 3.1.3c), and interception (Figure 3.1.3d), as well as several others such as sedimentation, eddy interaction, and electrostatic interactions.

Straining capture occurs when a particle encounters a pore or constriction in the bulk of the filter that is smaller than its size. While it may seem that this is mechanism should belong to complete blocking filtration of the cake filtration, note that real filter media has a distribution of pore sizes, and so the particle may pass

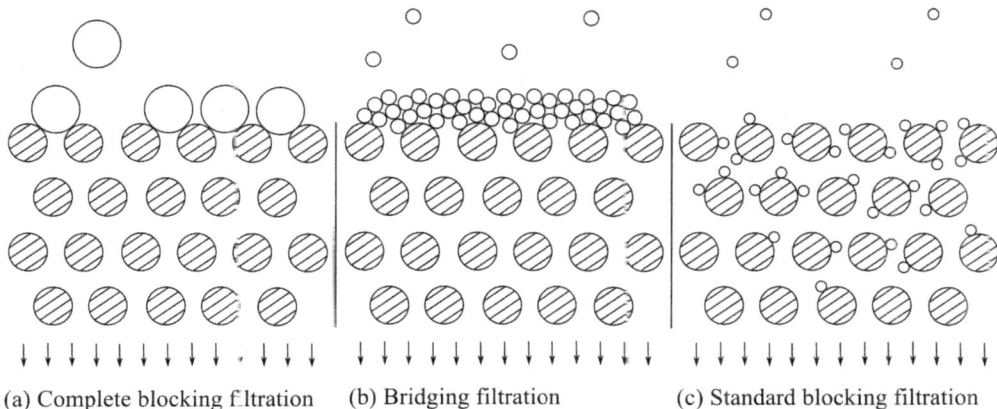

(a) Complete blocking filtration   (b) Bridging filtration   (c) Standard blocking filtration

FIGURE 3.1.2. Schematic of the three most common filtration mechanisms: (a) complete blocking and (b) bridging filtration that occurs in cake filtration, and (c) standard blocking filtration that occurs in depth filtration. In complete blocking filtration the particle size is larger than the media pore size and typically particle concentration in the feed is low. In bridging filtration particle size is smaller than the media pore size, but because the particles form bridges they are captured at the surface of the media. Typically particle concentration in the feed is high, to increase the likelihood of bridge formation. In standard blocking filtration, particle size is smaller than the media pore size, but the particle concentration in the feed is low, so that bridges are unlikely to form, and particle capture occurs predominantly in the filter media.

through larger pores at the entrance of the filter but be captured by a small pore in the bulk of the filter. Hence straining capture occurs even in depth filtration.

Inertial impaction occurs when a particle has a density significantly different from that of the fluid and so is unable to follow the streamlines of the fluid as the fluid curves around the walls of the filter media (the collectors). The ability of a particle to follow streamlines of the fluid, especially around an obstacle, is characterized by Stokes number,

$$(3.1.1) \qquad St = \frac{U}{l_o} \frac{\rho_p l_p^2}{18\mu}$$

Here U is the characteristic velocity of the flow, typically the flow velocity away from the obstacle; l_o is the characteristic length scale of the obstacle (e.g., collector diameter); l_p is the characteristic length scale of the particle (e.g., particle diameter); ρ_p is the particle density; and μ is the viscosity of the fluid. For $St \ll 1$, particles closely follow the fluid streamlines and so do not impact on the collectors. For $St \gg 1$ the particles detach from the flow, especially if the flow changes direction rapidly. Hence, inertial impaction occurs for particles with $St \gg 1$. In principle, inertial impactor mode of filtration can be used to sort particles not only by size but by density too. However, for common particles in aqueous solutions, Stokes numbers are typically 10^{-9} to 10^{-3} and so inertial impaction is not useful for particle separation from aqueous solutions. For example, for a 10 µm particle

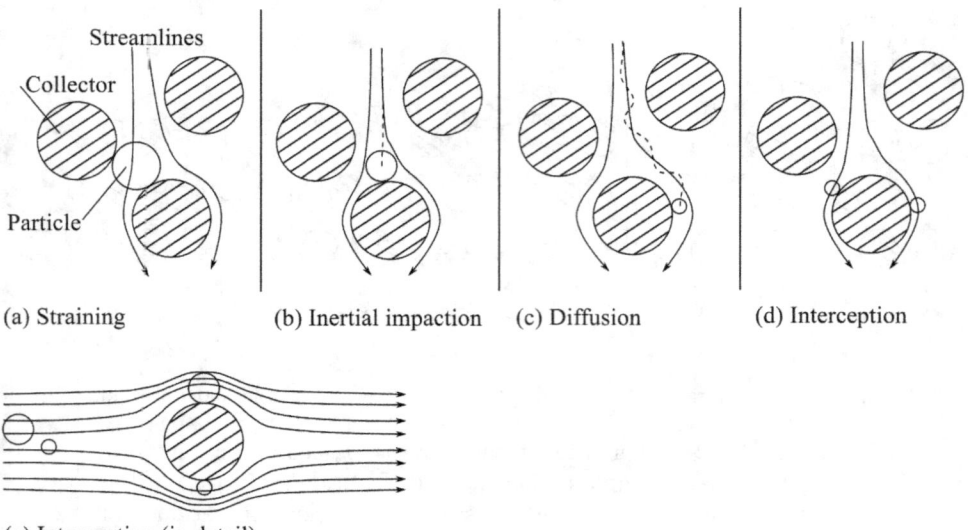

(a) Straining   (b) Inertial impaction   (c) Diffusion   (d) Interception

(e) Interception (in detail)

FIGURE 3.1.3. Sketch of common filtration mechanisms in depth filtration: (a) straining, (b), inertial impaction (c) diffusional caputure, and (d) interception. Solid lines with arrows are the fluid streamlines around the solid walls of the filter media (the collectors). Dashed lines in (b) and (c) show the particle path. In (e) we show streamlines around a collector and the effect of particle size on capture efficiency in interception filtration mechanism.

with a density of 2 g/ml, in a filter with 20 μm pore size and 1 mm/s fluid velocity, Stokes number is 6×10^{-4}. Inertial impaction is much more useful for sorting particles in a gas, where Stokes numbers are larger.

It may seem that for a low Stokes number situation a particle that is on a streamline far away from a collector will never touch the collector. In practice this is not true. The particle may diffuse across streamlines, and which is the mechanism of diffusional capture. For this to occur, particle diffusion must be high enough relative to its advection through the filter. Peclet number represents the ratio of importance of advection to diffusion,

$$(3.1.2) \qquad Pe = \frac{l_p U}{D}$$

where l_p is the characteristic particle length scale (particle diameter), and also the characteristic length scale over which the particle gets transported, U is the characteristic particle velocity, and D is the particle's diffusion coefficient. Thus, diffusional capture mechanism occurs when $Pe \ll 1$. For a spherical particle, we can estimate the particle's diffusion coefficient via the Stokes-Einstein relation (we will discuss it in detail later),

$$(3.1.3) \qquad D = \frac{k_B T}{3\pi \mu l_p}$$

Substituting this into (3.1.2) gives us a Peclet number for a spherical particle,

$$(3.1.4) \qquad Pe = 3\pi \frac{\mu l_p^2 U}{k_B T}$$

From (3.1.4) we can tell the particle size and velocity necessary for the diffusional capture mechanism to be significant. Peclet number in (3.1.2) is one way to define the Peclet number. Another way is to define it, is to compare the time the particle has to advect through the entire filter to the time scale it needs to diffuse over from a streamline where it will not meet a collector to one where it will. Assuming the length scale to diffuse over from a streamline where the particle never meets a collector to one where it will is the same order as the particle length scale, and the length scale to advect through the entire thickness of the filter is d, we can define another Peclet number as

$$(3.1.5) \qquad Pe = \frac{l_p^2/D}{d/U} = \frac{l_p}{d} \frac{l_p U}{D}$$

Since typically $l_p \ll d$, this Peclet number is a lot smaller than the Peclet number in (3.1.2). While the Peclet number in (3.1.2) represents the chance that diffusional capture mechanism occurs locally, at the scale of the collector, the Peclet number in (3.1.5) represents the chance of diffusional capture occurring anywhere in the filter. Using Stokes-Einstein relation (3.1.3), we can rewrite (3.1.5) for a spherical particle,

$$(3.1.6) \qquad Pe = \frac{\mu l_p^3 U}{k_B T d}$$

The interception capture mechanism (Figure 3.1.3d and e) represents the fact that since particles have finite size, there are always streamlines that will carry a particle into a collector. As we can see from Figure 3.1.3e, the larger the particle the more streamlines are in its cross-section area, and so there are more streamlines that will carry it right into the collector. The consequence of this is that larger particles, are more likely to be captured via this mechanism, and so get captured closer to the entrance of the filter, while the smaller particles would be found more downstream.

In real depth filtration media several mechanisms typically work together to achieve filtration. For example, diffusional capture may work with interception capture for the capture to eventually occur. Furthermore, there are other intricacies to depth filtration that we have not touched upon, such as reduction in permeability of the filter as the particles fill the filter, and increase in interstitial flow velocities due to a reduction in gap (and so higher shear rates, potentially leading to particle detachment). For more information on traditional filtration and design of filtration systems, we refer the reader to Tien (2012), Tarleton and Wakeman (2005), and Tien and Ramarao (2011).

3.2. Wall streamline displacement filtration

Having discussed mechanisms of traditional filtration which typically use irregular structures often with wide distribution of parameters, we now turn to "structured filtration" methods that specifically rely on highly ordered structure to function. The first class of structured filtration methods that we will discuss is the wall streamline displacement filtration methods. In what we call "wall streamline displacement" a particle comes in contact with a wall and the wall pushes the particle from its streamline to another streamline, thus eventually aligning the particles according to size. Wall streamline displacement filtration methods include (a) pinched flow fractionation and (b) deterministic lateral displacement filtration.

Pinched flow fractionators are very simple devices consisting of inlets for feed flow and sheath flow, a constriction where the feed flow meets the sheath flow, an expansion section, and a number of outlets (Figure 3.2.1). In pinched flow fractionation we introduce a stream containing particles of various sizes (feed flow) through one of the channels into a constriction and at the same time introduce a stream of solution without particles (sheath flow) from another channel into the constriction. In the feed flow the particles are distributed randomly across the streamlines, and follow the streamline running through their center. (In typical operation of pinched flow fractionators the Stokes number is low). In the constriction, the distance between adjacent streamlines narrows. The sheath flow, by adding additional volume of fluid through the constriction, adds to the narrowing of the distance between adjacent streamlines. In fact, by controlling the flow rate of the sheath flow relative to the feed flow, we can control the change in distance between the streamlines as they enter into the constriction (Figure 3.2.1b and c). As the distance between streamlines narrows, some of the larger particles may encounter that the streamline that they were on places them right in contact with the wall. Thus, the wall pushes these particles from their native streamline closer to the wall to a new streamline that is closer to the center of the channel. The larger the particle, the further the wall pushes them towards the constriction center. Hence for good particle separation, we like the streamlines of the feed flow to be compressed into a width on the order of the radius of the smallest particle we are interested in separating. This way all of the particles to be separated contact the wall and are pushed to new locations based on their size, with the smallest particles occupying streamlines nearest to the wall and the largest particles occupying streamlines towards the center. After the constriction, is the expansion section. Here the streamlines spread apart, and so do the particles of different sizes. If we direct the streamlines into different outlets, we will also direct particles into different outlets, and thus achieve particle separation. Practically, for this separation to occur as we describe, the concentration of particles in the feed must be low enough so that they do not interact and most importantly do not stick together. Secondly, the flow velocity and channel Reynolds number must be low enough so that the wall lift and shear lift on the particles is negligible, so that these forces do move the particles across streamlines. (We will discuss wall lift and shear lift in a separate section). Thirdly, however, the flow velocity must be high enough so that diffusion across streamlines relative to the advection through the constriction is negligible. In other words, Peclet number such as that in (3.1.5), where now d is the length of the constriction, should be high. From the Peclet number for a spherical particle, (3.1.6), we can see that due to the l_p^3 scaling that the last two conditions put a severe restriction on the size of the particle that can be separated in this separator. For more details on design of pinched flow fractionation devices see Yamada et al. (2004).

Deterministic lateral displacement (DLD) separation operates similarly to pinched flow fractionation in that again a wall pushes particles off their native streamlines. In DLD there are many constriction walls in parallel as well as in series, pushing the particles across the streamlines (Figure 3.2.2). In DLD if a center of the particle is less than its radius away from a post when the streamlines of the flow come

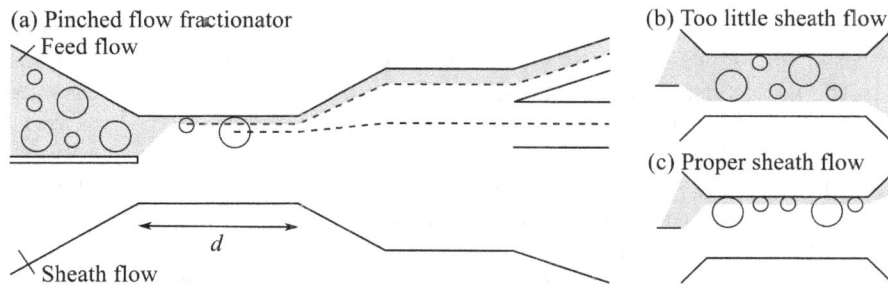

(a) Pinched flow fractionator
Feed flow

d

Sheath flow

(b) Too little sheath flow

(c) Proper sheath flow

FIGURE 3.2.1. Schematic of a pinched flow fractionator consisting of a feed flow inlet, a sheath flow inlet, a pinched (constricted) section, an expanding section, and a series of outlets. Dashed lines show the streamlines of the small and large particles to be separated. (b) shows a case where there is not enough sheath flow, and so pinching is ineffective. When pinching is ineffective the particles are not pushed against the upper wall and hence and their location on the streamlines past the constriction is poorly defined. (c) shows properly tuned sheath flow, where the feed flow streamlines are compacted into a space on the order of the smallest particle's radius. This ensures that all particles are pushed against the upper wall, which defines their location on the streamlines past the constriction.

together in a constriction between the posts, the particle is pushed of its streamline towards the center streamline. The posts in a DLD array are arranged such that the particle gets pushed towards the center streamline in this way if its diameter is greater than a critical diameter, D_c. Since the array of posts is arranged at an angle, all the larger particles will be deviated in the direction of the array (Figure 3.2.2) while the particles with diameters smaller than the critical diameter will continue to migrate on their original streamlines (but zigzag). The particles with diameters smaller than D_c are termed to travel is "zigzag mode", while the particles with diameter larger than D_c are termed to travel in "bump mode."

In our discussion of DLD separation, we will analyze DLD in a rhombic array of circular posts (Figure 3 2.2a) and follow the analysis of Inglis et al. (2006). We note that many different array shapes and post shapes have been used for DLD separations, but circular posts are the most common shaped used (McGrath et al. (2014)) and similar analysis can be applied to those devices. We begin by considering the flow though the gap between the posts (Figure 3.2.2a). The fluid flux through each gap can be divided into n flow streams (stream tubes), each carrying equal flow rate. From the geometry of the post array we can define a row shift fraction, ε as the ratio of the horizontal distance that each subsequent row is shifted, $\varepsilon\lambda$, to the array period λ. For convenience, we set the row shift fraction to be $\varepsilon = 1/n$. In this flow the stream tubes are separated by stall lines, each begging and ending on a post. The geometry, specifically ε, determines the number of stall lines. The stream tubes shift their positions cyclically relative to the left post in each constrictions, so that after n rows the position of the stream tube relative to the left post in its constriction returns to its original relationship. We can see in Figure 3.2.2a where we chose $n = 3$, that the stream tube position in the gap repeats each third row. To see this, let's observe the stream tube starting in the upper left constriction in Figure 3.2.2a. It starts being closest to the left post, i.e., position (1). In the next constriction, in the second row (only partially shown), it is closest to the right post, i.e. position (3). In the third row, (also only partially shown) it is in the center between the posts, i.e., position (2). In the fourth row it returns to being in position (1). Now consider a particle placed into this stream tube. If its radius is larger than the width of this stream tube, the particle will bump against the wall of the post and be forced to move to the next (higher numbered) stream tube (Figure 3.2.2b). This repeats from row to row, with the particle shifting by $\varepsilon\lambda$ each row. This travel is termed "bumped mode". If the particle's radius is less than the width of the stream tube, it will continue in its stream tube, zigzaging as it travels. This travel is termed "zigzag mode". Thus, the critical particle radius which separates the two modes is the width of the stream tube, β in Figure 3.2.2. Thus, $D_c = 2\beta$. Since we divided the flow in the constriction between the posts into $n = 1/\varepsilon$ equal parts, then we can write that,

$$(3.2.1) \qquad \int_0^\beta u(x)\, dx = \int_0^g u(x)\, dx$$

where $u(x)$ is the velocity profile in the gap. As a first approximation, we assume a 2D parabolic profile in the gap,

$$(3.2.2) \qquad u(x) = \frac{g^2}{4} - \left(x - \frac{g}{2}\right)^2$$

Substituting this into (3.2.1) and integrating we obtain,

$$(3.2.3) \qquad (\beta/g)^3 - (3/2)(\beta/g)^2 + \varepsilon/2 = 0$$

which has a solution,

$$(3.2.4) \qquad \beta/g = (1 + 2w + 1/(2w))/2$$

where

$$(3.2.5) \qquad w^3 = \frac{1}{8} - \frac{\varepsilon}{4} \pm \sqrt{\frac{\varepsilon(\varepsilon - 1)}{16}}$$

and with the physical root of w

$$(3.2.6) \qquad w = \left(\frac{1}{8} - \frac{\varepsilon}{4} \pm \sqrt{\frac{\varepsilon(\varepsilon - 1)}{16}}\right)^{\frac{1}{3}} \left(-\frac{1}{2} - i\frac{\sqrt{2}}{3}\right)$$

   Which gives us a relation for the critical particle diameter that the particular DLD array can separate. Notice that in this analysis we have ignored diffusion or lift forces, as well as physicochemical interactions that the particles may have with the posts. Davis (2008) empirically found a relation for the critical diameter to be

$$(3.2.7) \qquad D_c = 1.4g\varepsilon^{0.48}$$

DLD devices have been used to separate beads from order 500 nm up to order 500 μm; white blood cells and other blood components from red blood cells; suspensions of cultured cells; and even droplets (McGrath et al. (2014)). To achieve fractionation of multiple particle sizes, several DLD arrays with different gap and row shift fraction are put in series. Since the array angle θ is typically small, 1-6°, DLD arrays are typically much longer in the flow direction than in width.

(a) Section of a DLD array

(b) Particles navigating post

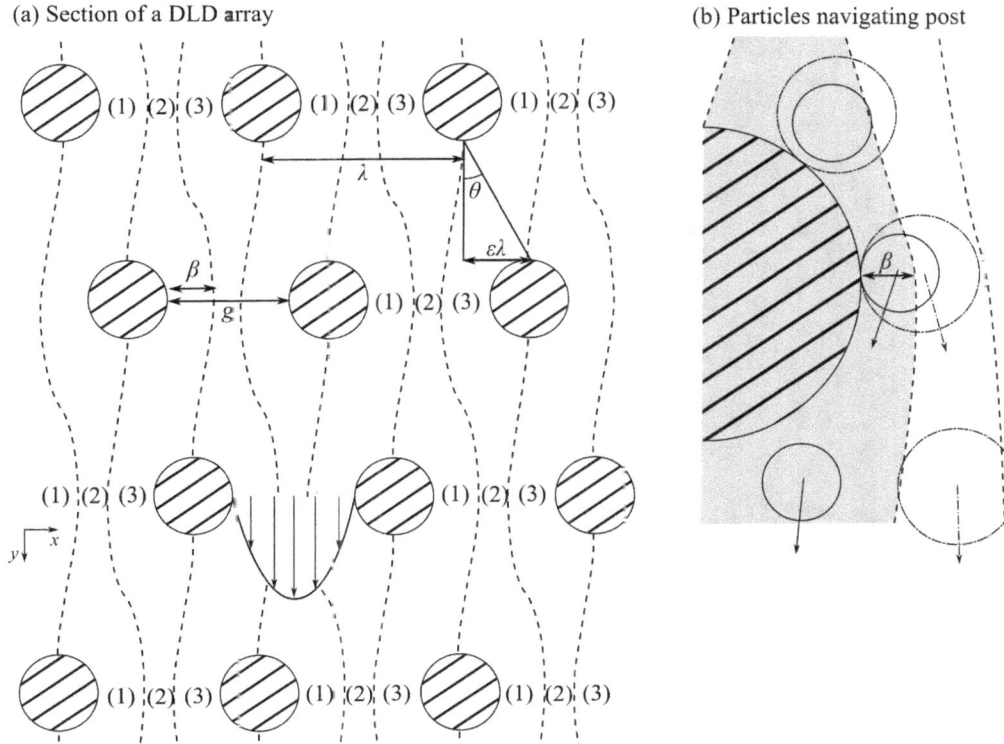

FIGURE 3.2.2. (a) Schematic of flow in a rhombic array of circular posts forming part of a deterministic lateral displacement (DLD) device. We assume that there is parabolic flow profile between each of the posts. Dashed lines here represent the stall lines, with each stall line beginning and ending on a post. The stall lines divide the flow into regions of equal flow rate (stream tubes). We divide the flow between the posts into streams of equal flow rate, in this case into three streams, which we label (1), (2), (3). Stream (1) is always closest to the left post, while stream (3) is closest to the right post in the constriction between the posts. λ is the distance between the post centers, while g is the spacing between the posts. The array angle θ is defined as shown. This angle is typically small, 1-6°. ε is the so called row shift fraction, the ratio of the horizontal distance that each subsequent row is shifted, $\varepsilon\lambda$, to the array period λ. (b) Close up of two particles, one larger and one smaller than a critical diameter navigating a post, with the first stream tube (1) highlighted in gray. The larger particle's radius is just larger than the width of the stream tube and so it is pushed into the neighboring, higher numbered stream tube (as shown by arrows). Meanwhile, the smaller particle's radius is smaller than the width of the stream tube, and so it continues in its stream tube.

CHAPTER 4

Higher Reynolds number flows

Up till now, for the most part, we concerned ourselves with creeping flow, where the most fundamental assumption was $Re \to 0$. However, in practical microfluidic devices, such as for example, field flow fractionation devices, this is not always the case. In fact there are many flows of interest where the Reynolds number is indeed small for one length scale, but is not small for a second length scale. It would seem that inertia is thus not completely negligible for this class of problems. (While the Reynolds number is not small, here we only consider cases where even our "higher Reynolds numbers" are still small enough so that the flow is not turbulent). As we will see shortly, inertia indeed plays a role and produces interesting phenomena that cannot be predicted based on creeping flow assumptions alone. In addition to inertia, curvature of channels introduces additional flow and forces on particles. Since realistic channels are curved, it is important to understand these effects. Thus we will extend our standard creeping flow analysis (and approaches) that we have developed in the past few sections, first to describe effects of inertia for flow in straight channels, and then to describe flow in curved channels.

4.1. Flow in straight channels

In this section we will first consider the two "paradoxes" that arise by neglecting the inertia terms (the non-linear terms) in the momentum equation - the Whitehead paradox and the Stokes paradox. We will see that by "partially" including a little bit of inertia into the momentum equation we can overcome the problems caused by these paradoxes and obtain more accurate solutions. Lastly, we will consider a problem where a "little bit of momentum" gives rise to an entire new phenomena - lift on particles in parabolic flow, the Segre-Silberberg effect. Recall that particles in creeping flow will not experience a lift force. However, it turns out that even when the particle length scale Reynolds number is low but the channel length scale Reynolds number is high enough, a particle experiences a non-negligible lift both away from the channel center line (shear lift), and away from the wall (wall lift), and positions itself at a particular location from the centerline of the channel.

4.1.1. Whitehead paradox and Stokes paradox. The Whitehead paradox (after Alfred North Whitehead, Whitehead (1889)) arises if we consider that the creeping flow equations are the first order approximation in a regular asymptotic series for the solution to the momentum equation. We will investigate this for the problem of flow around a sphere. To see this we begin by considering the full momentum equation expressed in terms of the stream function, where as usual the velocity is nondimensionalized with the free stream velocity, and length is nondimensionalized with the radius of our sphere. We had already expressed the full momentum equation as a stream function in (2.2.44), and here we rewrite this expression in axisymmetric (and so 2D) spherical coordinates (with as before, $\eta = \cos\theta$), since we are dealing with a sphere.

$$(4.1.1) \qquad E^4\psi = Re\left(\frac{1}{r^2}\left(\frac{\partial\psi}{\partial r}\frac{\partial}{\partial\eta}\left(E^2\psi\right)\right) + \frac{2}{r^2}E^2\psi\left(\frac{\partial\psi}{\partial r}\frac{\eta}{1-\eta^2} + \frac{1}{r}\frac{\partial\psi}{\partial\eta}\right)\right)$$

Once again we have the same boundary conditions as before (Section 1.3): the no-penetration and no-slip conditions, and the far-field condition (flow far from the sphere):

$$(4.1.2) \qquad \frac{\partial\psi}{\partial r} = \psi = 0 \text{ at } r = 1$$

$$(4.1.3) \qquad \psi = 0 \text{ at } \eta = \pm 1$$

$$(4.1.4) \qquad \psi \to -r^2 Q_1(\eta) \text{ as } r \to \infty$$

Naively, we seek a regular perturbation expansion to the stream function, with the Reynolds number as the smallness parameter,

$$(4.1.5) \qquad \psi(r,\eta;Re) = \psi_0(r,\eta) + \psi_1(r,\eta)Re + \psi_2(r,\eta)Re^2 + \dots$$

This was roughly the approach of Whitehead, however when he was working on this problem around 1889 the modern methods of asymptotic expansions (including methods of multiple scales, various regularization methods, and methods of matched asymptotic expansions) were not yet developed. Whitehead thus used a related but simpler method of successive approximation, which is equivalent to our naive approach of regular perturbation expansion. Substituting (4.1.5) into (4.1.1) and into the boundary conditions (4.1.2) - (4.1.4) and keeping the zeroth and first order terms, we obtain from the main equation

$(4.1.6)$
$$E^4\psi_0 + ReE^4\psi_1 + O\left(Re^2\right) =$$
$$Re\left(\frac{1}{r^2}\left(\frac{\partial\psi_0}{\partial r}\frac{\partial}{\partial\eta}\left(E^2\psi_0\right) - \frac{\partial\psi_0}{\partial\eta}\frac{\partial}{\partial r}\left(E^2\psi_0\right)\right) + \frac{2E^2\psi_0}{r^2}\left(\frac{\partial\psi_0}{\partial r}\frac{\eta}{1-\eta^2} + \frac{1}{r}\frac{\partial\psi_0}{\partial\eta}\right)\right)$$
$$+O\left(Re^2\right)$$

and from the boundary conditions

$$(4.1.7) \qquad \frac{\partial\psi_0}{\partial r} + Re\frac{\partial\psi_1}{\partial r} + O\left(Re^2\right) = \psi_0 + Re\psi_1 + O\left(Re^2\right) = 0 \text{ at } r = 1$$

$$(4.1.8) \qquad \psi_0 + Re\psi_1 + O\left(Re^2\right) = 0 \text{ at } \eta = \pm 1$$

$$(4.1.9) \qquad \psi_0 + Re\psi_1 + O\left(Re^2\right) \to -r^2 Q_1(\eta) \text{ as } r \to \infty$$

Collecting the terms by each order, we obtain that the zeroth order terms (i.e., where the smallness parameter appears to the zeroth power, Re^0) are,

$$(4.1.10) \qquad E^4\psi_0 = 0$$

$$(4.1.11) \qquad \frac{\partial\psi_0}{\partial r} = \psi_0 = 0 \text{ at } r = 1$$

$$(4.1.12) \qquad \psi_0 = 0 \text{ at } \eta = \pm 1$$

$$(4.1.13) \qquad \psi_0 \to -r^2 Q_1(\eta) \text{ as } r \to \infty$$

We see that we have our familiar equations for the creeping flow around a sphere from Section 1.3. Things appear so far so good. Now let's obtain the first order terms (i.e., where the smallness parameter appears to the first power, Re^1),

(4.1.14)
$$E^4\psi_1 = \frac{1}{r^2}\left(\frac{\partial\psi_0}{\partial r}\frac{\partial}{\partial\eta}\left(E^2\psi_0\right) - \frac{\partial\psi_0}{\partial\eta}\frac{\partial}{\partial r}\left(E^2\psi_0\right)\right) + \frac{2E^2\psi_0}{r^2}\left(\frac{\partial\psi_0}{\partial r}\frac{\eta}{1-\eta^2} + \frac{1}{r}\frac{\partial\psi_0}{\partial\eta}\right)$$

(4.1.15)
$$\frac{\partial\psi_1}{\partial r} = \psi_1 = 0 \text{ at } r = 1$$

(4.1.16)
$$\psi_1 = 0 \text{ at } \eta = \pm 1$$

(4.1.17)
$$\psi_1 \to 0 \text{ as } r \to \infty$$
$$\frac{1}{r}\frac{\partial\psi_1}{\partial r} \to 0 \text{ as } r \to \infty$$
$$\frac{1}{r^2}\frac{\partial\psi_1}{\partial\eta} \to 0 \text{ as } r \to \infty$$

We converted the (4.1.17) into the form of the velocity components (from the definition of this stream function) as when $r \to \infty$, $\mathbf{u} = 1$ so $\mathbf{u}_0 = 1$ and $\mathbf{u}_1 = 0$. These equations are more complicated, and (as expected) depend on the zeroth order solution. Conveniently, we already have the zeroth order solution from Section 1.3. Substituting that solution, (1.3.7), into (4.1.14) we obtain

(4.1.18)
$$E^4\psi_1 = \frac{9}{2}\left(\frac{2}{r^2} - \frac{3}{r^3} + \frac{1}{r^5}\right)G_2(\eta)$$

The general solution to this problem, the homogeneous (1.2.115) plus the particular solution, is

(4.1.19)
$$\psi_1 = \sum_{n=1}^{\infty}\left(A_n r^{n+3} + B_n r^{n+1} + C_n r^{2-n} + D_n r^{-n}\right)Q_n(\eta)$$
$$+ \frac{3}{16}\left(2r^2 - 3r - \frac{1}{r}\right)Q_2(\eta)$$

where we have yet to apply the boundary conditions (4.1.15) - (4.1.17) to obtain the constants A_n, B_n, C_n, and D_n. However, if we try to apply the boundary conditions, specifically (4.1.17), we see that no choice of these constants will satisfy this condition! This implies that there is no solution for ψ_1. Since higher order terms will depend on the first order term, the situation is quite bad - regular perturbation expansion for this problem fails. This is the crux of the Whitehead paradox. However, careful examination of the situation hints why the regular perturbation expansion (4.1.5) fails. Note, for (4.1.5) to work, we need that each higher term in the expansion be smaller than the previous one - we use this fact to justify truncating off the higher order terms. If the Reynolds number is really small, each higher order term should indeed be smaller than the previous one, if each of the ψ_i are order unity or decreasing with increasing i. However, because we have an infinite domain (because of a boundary condition at infinity) we really have two Reynolds numbers: one based on particle radius, and one based on domain dimension - a paradoxical one, which approaches infinity. This suggests that we should break our domain really into two domains: one domain on the order of particle size, and one domain on the order of the original domain size, but rescaled. We call the first "small" domain the inner region (because its closer to the particle), and the second domain the outer region (because its further from the particle). However, before we proceed with this two domain approach, let's also take a look at the original Navier-Stokes momentum equation to see if it can give us more clues, as did Carl Wilhelm Oseen (1927) when he explained and resolved this paradox. In deriving the creeping flow solution we assumed that Reynolds number is small over the entire domain, even as $r \to \infty$. However, Oseen noted that from the creeping flow solution, the inertia term

$$(4.1.20) \qquad\qquad Re\left(u_r \frac{\partial u_r}{\partial r}\right) \sim O\left(\frac{Re}{r^2}\right)$$

while a viscous term

$$(4.1.21) \qquad\qquad \frac{\partial^2 u_r}{\partial r^2} \sim O\left(\frac{Re}{r^3}\right)$$

Comparing the two we see that the inertia term is smaller than the viscous term only when

$$(4.1.22) \qquad\qquad r < O\left(\frac{1}{Re}\right)$$

where the Reynolds number is based on particle radius. Hence we are not justified in dropping the inertia terms far away from the particle. Thus it is not surprising that there was no solution to the first order term in the regular perturbation expansion. What is actually surprising, and very fortunate that we were able to find a zeroth order solution.

In our next problem, 2D creeping flow around a cylinder, the Stokes paradox, we are not even able to find the zeroth order solution! Consider 2D flow around a circular cylinder with the axis of the cylinder perpendicular to the direction of the flow. The flow far from the cylinder is uniform. Using the stream function form of the 2D Navier-Stokes from (1.2.45), we can similarly substitute in (4.1.5) and collect terms according to their order, like we did above. For the first order we obtain

$$(4.1.23) \qquad\qquad \nabla^4 \psi_0 = 0$$

We can express the uniform flow far away from the cylinder boundary condition as

$$(4.1.24) \qquad\qquad \psi_0 \to r \sin\theta \text{ as } r \to \theta$$

The no-slip, no-penetration, and symmetry conditions as before produce

$$(4.1.25) \qquad\qquad \frac{\partial \psi_0}{\partial r} = \psi_0 = 0 \text{ at } r = 1$$

$$(4.1.26) \qquad\qquad \psi_0 = 0 \text{ at } \eta = \pm 1 \text{ or } \theta = 0, \pi$$

Solving (4.1.23) and applying the boundary conditions (4.1.25) and (4.1.26), we obtain

$$(4.1.27) \qquad\qquad \psi_0 = A\left(\frac{1}{r} - r + r \ln r\right) \sin\theta$$

where A is a coefficient to be determined from (4.1.24) but as we see there is no choice of A that will satisfy this boundary condition. Hence, we cannot even find the zeroth order solution - the creeping flow solution for seemingly benign (and certainly commonly encountered) problem of flow around a cylinder.

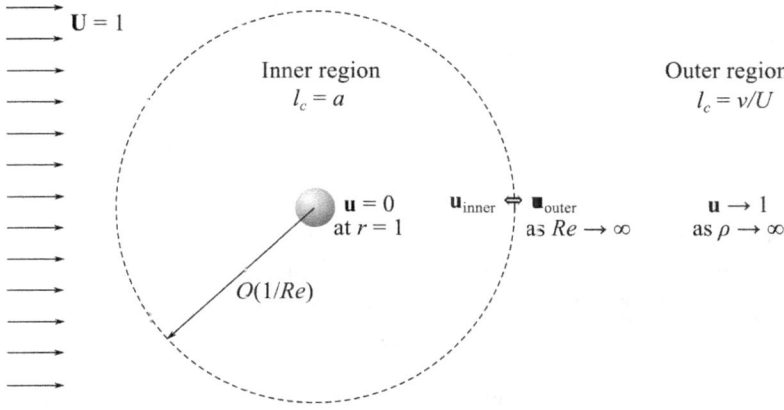

FIGURE 4.1.1. Schematic of the two domains used to solve the flow around a sphere at finite Reynolds numbers. The inner region has a characteristic length scale of the radius of the sphere, while the characteristic length scale in the outer region is ν/U. In the inner region inertia is negligible, but in the outer region the inertia is non-negligible. We find the solutions to the respective equations in each region, and then match the solutions at the boundary.

Seeing that these problems really have two length scales - one small one near the body and one very large one at infinity, we will attempt this problem via the method of matched asymptotic expansions. The method involves splitting the domain into two regions - the inner region near the body, and the outer region, far away from the body (Figure 4.1.1). We will rescale the length in the outer region, so that it is more convenient to work with. For the inner region we have a boundary conditions at the body and at the outside border of the inner region. For the outer region we have a boundary conditions at the outer border of the outer region - the boundary condition corresponding to "infinity", and also at the inner border of the outer region. Since outer border of the inner region borders the inner border of the outer region the solutions must agree with each other there - they must match. Hence the name for this method - matched asymptotic expansions.

Let's begin with the solution for the inner region. We still would like to use the method of asymptotic expansions, but instead of assuming a form of regular expansion, like (4.1.5) where we have a power series in Reynolds number, let's not assume the form of the gauge functions $f_n(Re)$ and take the stream function to be

$$(4.1.28) \qquad \psi = \sum_{n=0}^{N} f_n(Re)\, \psi_n(r, \eta)$$

However, since we would like to truncate the series at a finite (and small) number of terms, we require that $f_n(Re)$ decreases as n increases or

$$(4.1.29) \qquad \lim_{Re \to 0} \frac{f_{n+1}(Re)}{f_n(Re)} \to 0$$

Finally, for the inner region we have the usual boundary conditions at the body: the no-slip, no-penetration, and symmetry:

$$(4.1.30) \qquad \frac{\partial \psi}{\partial r} = \psi = 0 \text{ at } r = 1$$

$$(4.1.31) \qquad \psi = 0 \text{ at } \eta = \pm 1$$

Having setup the inner region, let's now set up the outer region. Firstly we need to rescale the length scale in this region. A simple choice for this is Reynolds number to an unknown power, so that our new radial coordinate ρ is scaled to be

$$(4.1.32) \qquad \rho = rRe^m$$

(Note that we are not using density in this problem, hence we are conveniently reusing the variable ρ). Thus the new characteristic length scale of the problem is

$$(4.1.33) \qquad l_c = aRe^{-m}$$

Now we must rescale the stream function (4.1.1) and the velocities from (4.1.2) and (4.1.3)

$$(4.1.34) \qquad u_r = \frac{1}{r^2} \frac{\partial \psi}{\partial \eta}$$

$$(4.1.35) \qquad u_\theta = -\frac{1}{\sqrt{1-\eta^2}} \frac{1}{r} \frac{\partial \psi}{\partial r}$$

We would like the rescaled velocities to be still order unity when nondimensionalized with respect to the free stream velocity. Substituting in (4.1.32) into (4.1.34) and (4.1.35)

$$(4.1.36) \qquad u_r = \frac{Re^{2m}}{\rho^2} \frac{\partial \psi}{\partial \eta}$$

$$(4.1.37) \qquad u_\theta = -\frac{1}{\sqrt{1-\eta^2}} \frac{Re^{2m}}{\rho} \frac{\partial \psi}{\partial \rho}$$

Since we already have that u_r and u_θ be order unity (as they were already scaled by the free stream velocity), we can leave them as is, and scale the stream function in the outer region as

$$(4.1.38) \qquad \Psi = Re^{2m}\psi$$

Now we can substitute this into (4.1.1) along with (4.1.32), and obtain

$$(4.1.39) \quad Re^{2m} E_\rho^4 \Psi = \frac{Re^{m+1}}{\rho^2} \left(\frac{\partial \Psi}{\partial \rho} \frac{\partial}{\partial \eta} \left(E_\rho^2 \Psi \right) - \frac{\partial \Psi}{\partial \eta} \frac{\partial}{\partial \rho} \left(E_\rho^2 \Psi \right) \right) +$$
$$\frac{2Re^{m+1}}{\rho^2} E_\rho^2 \Psi \left(\frac{\partial \Psi}{\partial \rho} \frac{\eta}{1-\eta^2} + \frac{1}{r} \frac{\partial \Psi}{\partial \eta} \right)$$

where we now have a new operator based on ρ

$$(4.1.40) \qquad E_\rho^2 = \frac{\partial}{\partial \rho^2} + \frac{1-\eta^2}{\rho^2} \frac{\partial}{\partial \eta^2}$$

Since, as indicated by Oseen, the viscous and inertial terms are important in the outer region we must keep both the left- and the right-hand side. Hence to make this work

(4.1.41) $$2m = m + 1 \text{ or } m = 1$$

And so our rescaling becomes

(4.1.42) $$\rho = rRe$$

(4.1.43) $$\Psi = Re^2 \psi$$

Now, just like in the inner region we would like an asymptotic expansion for the stream function, and once again we do not assume the form of the gauge functions and so have

(4.1.44) $$\Psi = \sum_{n=0}^{N} F_n(Re) \Psi_n(\rho, \eta)$$

we just require that $F_n(Re)$ decreases with increasing order, or

(4.1.45) $$\lim_{Re \to 0} \frac{F_{n+1}(Re)}{F_n(Re)} \to 0$$

Now we need the boundary conditions for this domain. We still expect that the solution is symmetric and so

(4.1.46) $$\Psi = 0 \text{ at } \eta = \pm 1$$

We also have a condition at the outer border of the outer region, which again happens to be at infinity,

(4.1.47) $$\Psi \to -\rho^2 Q_1(\eta) \text{ as } \rho \to \infty$$

Finally, we also have the matching condition between the regions - the stream function on the outer border of the inner region must match the stream function on the inner border of the outer region. The matching condition can be written as

(4.1.48) $$Re^2 \psi \big|_{r \gg 1} \Leftrightarrow \Psi \big|_{\rho \ll 1} \text{ as } Re \to \infty$$

where $\Leftrightarrow$ is the matching symbol, and works similarly to equality, asking us to make both sides equal to each distinct order (e.g., make equal at order $O(1)$, then at $O(Re)$, then at $O(Re^2)$ and so on).

Now let's try to obtain the solution for the outer region. As a first try we can try the boundary condition $\Psi_0 = -\rho^2 Q_1(\eta)$ itself. The intuition to try this stems from the physical observation of the problem - the streamlines far enough from the sphere are just barely perturbed by the presence of the sphere and practically correspond to an undisturbed uniform flow, so why not try that as a first approximation? Substituting

(4.1.49)
$$\Psi_0 = -\rho^2 Q_1(\eta)$$
$$F_0(Re) = 1$$

into (4.1.39) we see that this works as a solution and of course it satisfies the boundary conditions. Now let's try to obtain the solution for the inner region. Recall that in the inner region the inertia terms are negligible and the viscous terms dominate. Hence the governing equation for the stream function is

(4.1.50) $$E^4 \psi = 0$$

Substituting (4.1.28) into (4.1.50) we see that for any choice of gauge function $f_0\left(Re\right)$ the zeroth order solution is

(4.1.51) $$E^4\psi_0 = 0$$

which as we saw from Section 1.3 has a solution that already satisfies the symmetry boundary condition,

(4.1.52) $$\psi_0 = \sum_{n=1}^{\infty}\left(A_n r^{n+3} + B_n r^{n+1} + C_n r^{2-n} + D_n r^{-n}\right) Q_n\left(\eta\right)$$

Let's first apply the matching condition, (4.1.48). To do this we convert this stream function into the variables of the outer region via (4.1.42), and so obtain

(4.1.53) $\quad -\rho^2 Q_1\left(\eta\right)$

$$\Leftrightarrow f_0\left(Re\right)\sum_{n=1}^{\infty}\left(A_n\frac{\rho^{n+3}}{Re^{n+1}} + B_n\frac{\rho^{n+1}}{Re^{n-1}} + C_n\rho^{2-n}Re^n + D_n\rho^{-n}Re^{n+2}\right) Q_n\left(\eta\right)$$

We see that the only term that exhibits ρ^2 dependence is the B_n term. Namely, the right-hand side should have a

(4.1.54) $$f_0\left(Re\right)B_1\rho^2 Q_1\left(\eta\right)$$

where

(4.1.55) $$f_0\left(Re\right) = 1$$

(4.1.56) $$B_1 = -1$$

We see that the terms with coefficients of A_n are all of $O\left(Re^{-2}\right)$ or larger. Since the biggest term in the outer region is $O\left(1\right)$ we have nothing to match these terms with, and so we set $A_n = 0$ for all n. Similarly, we see that the terms with coefficients of B_n for $n \geqslant 2$ are $O\left(Re^{-1}\right)$ or larger. Since, again, we have nothing to match these terms with, and so we set $B_n = 0$ for $n \geqslant 2$. We also see that the terms with coefficients of C_n are $O\left(Re\right)$ or smaller. Similarly the terms with coefficients of D_n are $O\left(Re^3\right)$ or smaller. Hence these cannot be expected to match at this level of approximation. We find these coefficients from the boundary conditions (4.1.30) and (4.1.31). Applying (4.1.30) and (4.1.31) to (4.1.52), we obtain that $C_1 = 3/2$ and $C_n = 0$ for all other n; $D_1 = -1/2$ and $D_n = 0$ for all other n. Thus, we finally obtain

(4.1.57) $$\psi_0 = -\left(r^2 - \frac{3}{2}r + \frac{1}{2r}\right) Q_1\left(\eta\right)$$

This is just the familiar creeping flow solution (1.3.7). Notice that this solution is just the zeroth order term of a larger expansion and only for the inner region - we still have yet to obtain the solution for the outer region. Previously we naively expected that this is the solution valid everywhere, but as we see now this is not necessarily so. Notice also that the terms we just obtained C_1 and D_1, cause a small mismatch with the outer region. To see this we convert (4.1.57) into the outer variables,

(4.1.58) $$Re^2\psi_0 = -\left(\rho^2 - \frac{3}{2}\rho Re + \frac{1}{2\rho}Re^2\right) Q_1\left(\eta\right) \Leftrightarrow -\rho^2 Q_1\left(\eta\right)$$

This hints that the Re and Re^2 must come from a higher order terms in the outer region expansion. We do not provide the details of seeking these terms here as the derivation is laborious, and refer the reader to Proudman and Pearson (1957). Proudman and Pearson found that the expansion for the outer region stream function to be

$$(4.1.59) \quad \Psi = -\rho^2 Q_1(\eta) - Re\left(\frac{3}{2}(1+\eta)(1-\exp(-\rho(1-\eta)/2))\right) + O(F_2(Re))$$

and the corresponding inner region solution to be
(4.1.60)
$$\psi = -\left(r^2 - \frac{3}{2}r + \frac{1}{2r}\right)Q_1(\eta)\left(1 + \frac{3}{8}Re\right) + \frac{3Re}{16}Q_2\left(2r^2 - 3r + 1 - \frac{1}{r} + \frac{1}{r^2}\right)$$

Substituting this result into the definitions of stream function (4.1.34) and (4.1.35) and then into (1.2.122) we obtain

$$(4.1.61) \qquad F_{drag} = 6\pi a\mu U\left(1 + \frac{3}{8}Re + O(Re)\right)$$

The extra term that we obtained does indeed improve the accuracy of the drag force relation extending its use even past Reynolds number of unity. Proudman and Pearson (1957) and later Chester et al. (1969) found higher order terms in the drag equation. Unfortunately these terms did not add much accuracy to the drag force. In fact the efforts of using creeping flow to expand into higher Reynolds number flows has not been fruitful. While we rarely encounter very high Reynolds numbers in microfluidic applications, since a sphere is such an important geometry, we give an empirical fit for drag on a sphere valid up to Reynolds $O(10^5)$,

$$(4.1.62) \qquad F_{drag} = \frac{\rho_f U^2 \pi a}{2}\left(\frac{12}{Re} + \frac{6}{1 + \sqrt{2Re}} + 0.4\right)$$

4.1.2. Lift on particles at finite Reynolds numbers. In previous section we observed that the drag on a particle at finite (non-zero, but less than unity) Reynolds number differs from the drag predicted by the Stokes' law. The smaller the Reynolds number, the smaller this difference. Furthermore, we have seen that in creeping flow (zero Reynolds number) a particle does not experience a lift (a force orthogonal to the main direction of motion, see Section 1.1.3). Similarly, at finite Reynolds numbers this is not exactly the case. We will see in this section that particles in bounded (channel flow) do indeed experience lift. For this to be observable, the Reynolds number based on the channel dimension (e.g., width) has to be typically greater than unity, while the Reynolds number based on particle size can be less than unity. While this lift is generally small, it leads to interesting, and surprising results, some of which we can harness for, for example, particle separation. One of such results is the so called "tubular pinch effect" (Figure 4.1.2), observed by Segre and Silberberg (1962). In the tubular pinch effect, if you load a circular pipe uniformly with neutrally buoyant, rigid spherical particles in which there is a pressure driven (Poiseuille) flow (Figure 4.1.2, t_0) after some time downstream they will segregate themselves into a ring in a pipe (Figure 4.1.2, t_2). For a continuous supply of particles into the pipe, sometime downstream there will be a tube of particles, as if the particles were pinched into a tube shape (narrow annulus). For a wide range of channel Reynolds numbers investigated (2-700), Segre and Silberberg found that the particles concentrated themselves at a radius of 0.6

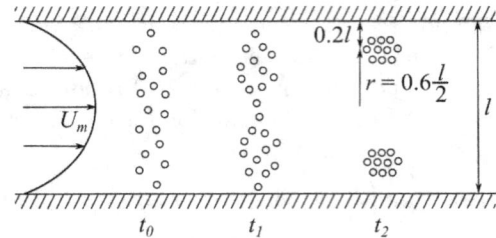

FIGURE 4.1.2. Schematic of the Segre-Silberberg tubular pinch effect. Particles initially injected uniformly across the cross-section of the pipe with Poiseuille flow (t_0) will focus themselves into a tight annulus at a radius of 0.6 the radius of the pipe (t_2). This focusing is due two oppositely directed forces: a wall interaction lift and a shear gradient lift.

the radius of the pipe. The particles experience two lift forces – the first lift force from the interaction with the wall where the wall "pushes" the particles towards the center of the pipe, and a second lift force from the gradient in the shear due to a parabolic (undisturbed) velocity profile in the pipe.

Non-naturally buoyant spherical particles, also exhibit lift, but their behavior depends on the direction of gravity relative to the axis of the pipe. When the gravity vector is orthogonal to the axis of the pipe, even creeping flow equations predict that the particles will undergo settling. Hogg (1994) showed that the inertial (finite Reynolds number) correction to the settling velocity predicted by Stokes' law is small, and of order $aBRe_c$ where a is the particle radius, Re_c is the Reynolds number based on the channel width as the length scale and B is the buoyancy parameter,

$$(4.1.63) \qquad\qquad B = \frac{2}{9}\frac{a^2\Delta\rho g}{\mu U_m}$$

where U_m is the peak undisturbed velocity (center velocity in our parabolic flow). When the gravity vector is aligned with the axis of the pipe, the contribution of finite inertia is less obvious and more significant. Jeffrey and Pearson (1965) observed that for a particle that travels faster than the undisturbed velocity (leads the flow due to the buoyancy effect) migrates towards the walls. Meanwhile a particle that lags the flow travels towards the centerline. This effect is due to the fact that when the particle travels parallel to the walls, it pushes the fluid around it laterally. Even though the particle is typically small, and so the amount of fluid it pushes is small, the effect can be felt quite far away. Even creeping flow equations predict that particle motion can be felt order 10 particle radii away (Section 1.2.5). Due to inertia, which is significant over large distances, this displacement becomes irreversible. When this displaced fluid encounters the walls, it pushes on the walls, and the walls push back on the fluid. The force is transmitted and results in the force on the particle, in the direction towards the center line. This is the wall interaction lift. However, the displaced fluid also interacts with the velocity gradient, and the resulting shear. When the particle lags the fluid, the difference between the velocity of the displaced fluid and background flow is greater in the direction of increasing velocity. This leads to a lateral pressure gradient that moves the particle in the

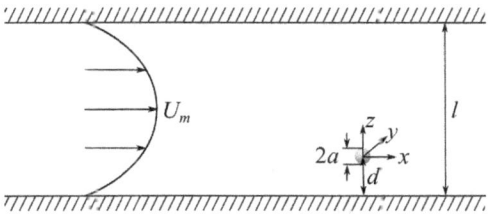

FIGURE 4.1.3. Schematic of a spherical particle, radius a, in Poiseuille flow between parallel plates separated by a distance l. The center particle is located a distance d from the nearest wall.

direction of increasing velocity (centerline). Conversely, when the particle leads the fluid, the difference between the velocity of the displaced fluid and background flow is greater in the direction of decreasing velocity. This sets up a pressure gradient so that the particle migrates in the direction of decreasing velocity (towards the walls). This force competes with the wall interaction lift. Hogg (1994) found that wall interaction lift dominates when $Re_c B \gg 1$ and the lateral migration towards the wall is order $(a/l) B^2 Re_c$.

Segre and Silberberg investigated a dilute suspension of spheres (volume fraction, $\phi < 1\%$). More recently Han et al. (1999) investigated the effect of volume fraction on the tubular pinch effect for neutrally buoyant spheres. They found again that when $\phi \leqslant 0.1$, but Re_p Reynolds number based on the particle radius as the length scale is significant, $Re_p \geqslant 0.2$, the particles segregated themselves to a radius 0.5-0.6 of the pipe radius. For this particle flow, the velocity profile was still parabolic. At the other limit when $\phi = 0.4$, the spheres always migrated towards the center of the pipe. For this particle flow, the velocity profile was blunted (as is typical of high volume fraction suspension flows). (Compare this to volume fraction of red blood cells, the hematocrit, in healthy humans, which is 0.4-0.5). At the intermediate ranges of volume fraction, sphere behavior depended on Re_p. For low Re_p particles migrated to the pipe axis. For $Re_p \geqslant 0.2$ the particles equilibrated into two bands: one at the center of the pipe, and one between the center and the pipe wall. The latter behavior was also observed by Martel and Toner (2013).

4.1.3. Quantitative analysis of lift at finite Reynolds numbers.

Unfortunately, there are not many quantitative analytical models of lift on particles at finite Reynolds numbers. Most analyses attempt to analyze this lift via asymptotic expansions of creeping flow equations, often with $Re_p (a/l)$ as the smallness parameter, where l is the width of the channel. These analyses typically only consider a single particle and ignore particle-particle interactions. Here we will discuss the approach and the results of the currently more advanced of these analytical approaches, that of Asmolov (1999). Asmolov's analysis considers both neutrally buoyant and non-neutrally buoyant particles, and is valid up to a channel Reynolds number of order 2000 (up to transition to turbulence). Asmolov's analysis builds upon the analysis of Schonberg and Hinch (1989) and Hogg (1994). This analysis considers Poiseuille flow between two flat plates separated by distance l in which a spherical particle radius a travels. For this analysis we pin the origin of the coordinate system at the center of the sphere, moving with the sphere (Figure 4.1.3).

For this analysis we define the channel Reynolds number as

$$(4.1.64) \qquad Re_c = \frac{U_m l}{\nu}$$

while we define a modified particle Reynolds number, following Asmolov, as

$$(4.1.65) \qquad Re_{p,A} = \frac{U_m a}{\nu}\frac{a}{l} = Re_p \frac{a}{l}$$

To solve the problem, we follow the example of Oseen's correction to Stokes' law (Section 4.1.1) and split the domain of the problem into two sub domains. The first domain is in the local vicinity of the particle (the inner solution), and has the characteristic length scale of the radius of the particle. The second domain is the region far from the particle, and has the characteristic length scale

$$(4.1.66) \qquad l_p = a Re_{p,A}^{-1/2} = \sqrt{\nu l/U_m} = l Re_c^{-1/2}$$

As we have done in deriving Oseen's correction to Stokes' law, we then match the two domains. For the governing equations in both domains we begin with the assumptions that the flow is steady and incompressible. The undisturbed flow is just a Poiseuille flow between flat plates,

$$(4.1.67) \qquad \bar{V}(d) = 4 U_m \frac{d\,(l-d)}{l^2}$$

where here the overbar will imply undisturbed flow variables. For generality, the particle moves at a velocity different from the flow velocity, due to the buoyancy force. (We are considering the neutrally buoyant and the non-neutrally buoyant particles together). Thus the difference between the undisturbed velocity and the particle's velocity, the slip velocity is,

$$(4.1.68) \qquad V = 4 U_m \frac{d\,(l-d)}{l^2} - U_p$$

In the frame of reference of the particle, the undisturbed flow is then

$$(4.1.69) \qquad \bar{\mathbf{v}} = \left(V - U_m \left(\gamma \frac{z}{l} - 4\frac{z^2}{l} \right) \right) \hat{\mathbf{x}}$$

where γ is the dimensionless shear rate at the location of the center of the particle,

$$(4.1.70) \qquad \gamma = 4 - 8d/l$$

Subtracting the undisturbed flow (4.1.69) from the steady momentum equation (which includes inertial terms), we obtain the disturbance flow,

$$(4.1.71) \qquad \mathbf{u} \cdot \nabla \mathbf{u} + \bar{\mathbf{v}} \cdot \nabla \mathbf{u} + \mathbf{u} \cdot \nabla \bar{\mathbf{v}} = \frac{-\nabla p}{\rho} + \nu \nabla^2 \mathbf{u}$$

The continuity equation for the disturbance flow is still,

$$(4.1.72) \qquad \nabla \cdot \mathbf{u} = 0$$

Our goal will be to find the solution to the disturbance flow, and from that to find the lift force on the particle. On the particle surface we have the no-slip and no-penetration boundary conditions. In the frame of reference traveling with the particle these conditions are

$$(4.1.73) \qquad \mathbf{u} = \mathbf{\Omega}_p \times \mathbf{r} - \bar{\mathbf{v}} \text{ at } r = a$$

where $\mathbf{\Omega}_p$ is the sphere's rotational velocity, and $\mathbf{\Omega}_p \times \mathbf{r}$ represents the rotation of the sphere. In the frame of reference traveling with the sphere, the no-slip and no-penetration condition on the particle walls are represented by

(4.1.74) $\mathbf{u} = 0$ at $z = -d$ and $z = l - d$

In the axial direction of the channel, the disturbance velocity decays to zero, and this is represented by

$$\mathbf{u} \to 0 \text{ as } |x| \to \infty$$

We nondimensionalize the variables in the following way,

(4.1.75) $\mathbf{r}^* = \mathbf{r}/a$

(4.1.76) $\mathbf{u}^* = \mathbf{u}/U_m$

(4.1.77) $\mathbf{\Omega}_p^* = \mathbf{\Omega}_p l / U_m$

(4.1.78) $\mathbf{F}^* = \dfrac{\mathbf{F}}{\mu a U_m}$

where $\mathbf{F}$ is the force of the sphere. The dimensionless undisturbed velocity becomes

(4.1.79) $\bar{\mathbf{v}} = \left(V - \varepsilon z \dfrac{\gamma}{\sqrt{Re_c}} - 4\varepsilon^2 z^2 \dfrac{1}{Re_c} \right) \hat{\mathbf{x}}$

where we have dropped the asterisks designating dimensionless variables for ease of readability. Here we also introduced ε, the smallness parameter that we will use in our asymptotic expansions,

(4.1.80) $\varepsilon = \sqrt{Re_{p,A}} = \sqrt{\dfrac{U_m a}{\nu} \dfrac{a}{l}}$

The dimensionless momentum and continuity equations become,

(4.1.81) $\varepsilon \sqrt{Re_c} \left(\mathbf{u} \cdot \nabla \mathbf{u} + \bar{\mathbf{v}} \cdot \nabla \mathbf{u} + \mathbf{u} \cdot \nabla \bar{\mathbf{v}} \right) = -\nabla p + \nabla^2 \mathbf{u}$

$$\nabla \cdot \mathbf{u} = 0$$

Notice that the zeroth order equations are just the creeping flow equations. The inertia term is introduced only in the first order. The respective boundary conditions become,

(4.1.82) $\mathbf{u} = -V\hat{\mathbf{x}} + \dfrac{\varepsilon}{\sqrt{Re_c}} \left(\mathbf{\Omega}_p \times \mathbf{r} - \gamma z\hat{\mathbf{x}} \right) + 4\varepsilon^2 z^2 \dfrac{1}{Re_c}\hat{\mathbf{x}}$ at $r = 1$

(4.1.83) $\mathbf{u} = 0$ at $z = -\dfrac{\sqrt{Re_c}}{\varepsilon} \dfrac{d}{l1.4}$ and $z = \dfrac{\sqrt{Re_c}}{\varepsilon} \dfrac{l - d}{l}$

(4.1.84) $\mathbf{u} \to 0$ as $|x| \to \infty$

To solve these equations we begin with the inner solution, and seek a regular asymptotic expansion of the form,

(4.1.85) $\mathbf{u} = \mathbf{u}_0 + \varepsilon \mathbf{u}_1 + O(\varepsilon)$

(4.1.86) $p = p_0 + \varepsilon p_1 + O(\varepsilon)$

(4.1.87) $\mathbf{F} = \mathbf{F}_0 + \varepsilon \mathbf{F}_1 + O(\varepsilon)$

Substituting these into (4.1.81) through (4.1.84), and collecting the like powers of the smallness parameter ε, we obtain for zeroth order the creeping flow equations,

$$(4.1.88) \qquad \nabla^2 \mathbf{u}_0 - \nabla p_0 = 0$$

$$(4.1.89) \qquad \nabla \cdot \mathbf{u}_0 = 0$$

with the boundary conditions,

$$(4.1.90) \qquad \mathbf{u}_0 = -V\hat{\mathbf{x}} \text{ at } r = 1$$

$$\mathbf{u}_0 \to 0 \text{ as } |x| \to \infty$$

The solution to the creeping flow equations with this boundary condition, is

$$(4.1.91) \qquad \mathbf{u}_0 = -V\left(\frac{1}{4}\left(\frac{3}{r} + \frac{1}{r^3} \right)\hat{\mathbf{x}} + \frac{3}{4}\left(\frac{1}{r} + \frac{1}{r^3} \right)\frac{x\mathbf{r}}{r^2} \right)$$

with the force on the sphere being that given by Stokes' law,

$$(4.1.92) \qquad \mathbf{F}_0 = 6\pi V\hat{\mathbf{x}}$$

Notice, that, as we of course expected, the creeping flow equations do not yield a lift force.

The first order equations are

$$(4.1.93) \qquad \nabla^2 \mathbf{u}_1 - \nabla p_1 = \sqrt{Re_c}\,(\mathbf{u}_0 + V\hat{\mathbf{x}}) \cdot \nabla \mathbf{u}_0$$

$$(4.1.94) \qquad \nabla \cdot \mathbf{u}_1 = 0$$

with the boundary condition,

$$(4.1.95) \qquad \mathbf{u}_1 = \frac{1}{\sqrt{Re_c}}\,(\mathbf{\Omega}_p \times \mathbf{r} - \gamma z\hat{\mathbf{x}}) \text{ at } r = 1$$

The disturbance velocity of the first order, $\mathbf{u}_1$ does not decay to zero at infinity, and so we must replace the boundary condition at infinity with the matching condition to the outer flow. Asmolov (1999) gives the solution to this set of equations and boundary condition as

$$(4.1.96) \qquad \mathbf{u}_1 = \mathbf{u}_{0,nb} + \mathbf{u}_{1,pp}\sqrt{Re_c} + \mathbf{V}_1$$

where

$$(4.1.97)$$
$$\mathbf{u}_{0,nb} = \frac{1}{\sqrt{Re_c}}\left(\left(\mathbf{\Omega}_p + \frac{\gamma}{2}\hat{\mathbf{y}} \right) \times \frac{\mathbf{r}}{r^3} + \frac{5}{2}\gamma r x z \left(\frac{1}{r^7} - \frac{1}{r^5} \right) - \frac{\gamma}{2r^5}\,(z\hat{\mathbf{x}} + x\hat{\mathbf{z}}) \right)$$

$$(4.1.98) \quad \mathbf{u}_{1,pp} = \frac{3}{32}V^2\left(2 - \frac{3}{r} + \frac{1}{r^2} - \frac{1}{r^3} + \frac{1}{r^4} \right)\left(1 - \frac{3x^2}{r^2} \right)\frac{\mathbf{r}}{r} +$$
$$\frac{3}{32}V^2\left(4 - \frac{3}{r} + \frac{1}{r^3} - \frac{2}{r^4} \right)\left(\frac{x^2\mathbf{r}}{r^3} - \frac{x\hat{\mathbf{x}}}{r} \right)$$

The term $\mathbf{V}_1$ is found from the matching condition. Asmolov solves the first order equations by noticing that they are linear, and splitting the problem into three problems (see Section 1.1.2 for a similar approach). $\mathbf{u}_{0,nb}$ is the solution to

$$(4.1.99) \qquad \nabla^2 \mathbf{u}_1 - \nabla p_1 = 0$$

$$(4.1.100) \qquad \mathbf{u}_1 = \frac{1}{\sqrt{Re_c}}\,(\mathbf{\Omega}_p \times \mathbf{r} - \gamma z\hat{\mathbf{x}}) \text{ at } r = 1$$

$\mathbf{u}_{1,pp}\sqrt{Re_c}$ is the solution to

(4.1.101) $$\nabla^2\mathbf{u}_1 - \nabla p_1 = \sqrt{Re_c}\,(\mathbf{u}_0 + V\hat{\mathbf{x}})\cdot\nabla\mathbf{u}_0$$

(4.1.102) $$\mathbf{u}_1 = 0 \text{ at } r = 1$$

and $\mathbf{V}_1$ is the solution to

(4.1.103) $$\nabla^2\mathbf{u}_1 - \nabla p_1 = 0$$

(4.1.104) $$\mathbf{V}_1 = 0 \text{ at } r = 1$$

as well as a matching condition at infinity with the outer solution.

For the outer solution we scale the coordinates with our smallness parameter ε, such that

(4.1.105) $$\mathbf{R} = (X, Y, Z) = \varepsilon\mathbf{r} = \varepsilon\,(x, y, z)$$

The zeroth order solution (4.1.91) written in outer variables is

$$\mathbf{u}_0 = \varepsilon\mathbf{U}_s + O\left(\varepsilon^3\right)$$

where

$$\mathbf{U}_s = -\frac{3}{4}V\left(\frac{1}{R}\hat{\mathbf{x}} + \frac{X}{R^3}\mathbf{R}\right)$$

is the Stokeslet velocity field. The outer solution equations are obtained via a regular asymptotic expansion similar to the inner solution, which due to the scaling (4.1.105) become,

(4.1.106) $$\mathbf{u} = \varepsilon\mathbf{U} + O\left(\varepsilon\right)$$

(4.1.107) $$p = \varepsilon^2 P + O\left(\varepsilon^2\right)$$

The first order outer equations can be written as

(4.1.108) $$\nabla^2\mathbf{U} - \nabla P - \bar{V}_x\frac{\partial\mathbf{U}}{\partial\mathbf{X}} - \frac{\partial\bar{V}_x}{\partial\mathbf{X}}U_z\hat{\mathbf{x}} = 6\pi V\hat{\mathbf{x}}\delta\left(\mathbf{R}\right)$$

where

(4.1.109) $$\bar{V}_x = v + \gamma Z - \frac{4}{\sqrt{Re_c}}Z^2$$

(4.1.110) $$v = V\sqrt{Re_c}$$

(4.1.111) $$\nabla\cdot\mathbf{U} = 0$$

along with the boundary conditions,

(4.1.112) $$\mathbf{U} = 0 \text{ at } Z = -\sqrt{Re_c}\frac{d}{l} \text{ and } Z = \sqrt{Re_c}\frac{l-d}{l}$$

(4.1.113) $$\mathbf{U} = 0 \text{ at } X \to \infty$$

The solution to (4.1.108) to (4.1.113) are quite involved and are detailed in Asmolov (1999). It consists of taking a two dimensional Fourier transform of (4.1.108) to (4.1.113), which results in a forth order ordinary differential equation; then solving this equation and finally numerically taking the inverse Fourier transform, via an orthonormalization method to obtain the solution. Solving this equation and matching with inner solution allows us to find the first order lateral migration velocity, V_{1z} and from it the first order lift force F_{1z}. We plot this lift force as a function of the difference between the particle velocity and the undisturbed flow, V and channel Reynolds number Re_c in Figure 4.1.4 and Figure 4.1.5. We observe that for the majority of the channel the lift force is positive, which means that if the particle leads the undisturbed flow (i.e., $V < 0$) the particle is pushed towards the walls and conversely if the particle lags the undisturbed flow (i.e., $V > 0$) the particle is pushed towards the center. This qualitatively agrees well with the observations of Jeffrey and Pearson (1965). The maximum value of the lift force increase rapidly between Re_c 0.1 and 100 and increases more slowly for Re_c greater than 100. Particles lagging the flow (Figure 4.1.4b and Figure 4.1.5b) only experience a force directed towards the center of the channel and so reach an equilibrium position there - particles become focused into a tight stream near the channel center. Particles leading the flow (Figure 4.1.4a and Figure 4.1.5a) experience two forces: near the channel center a shear gradient force directed towards the wall, and secondly, near the wall the particle experiences a wall interaction force directed towards the center of the channel. The wall lift force becomes significant when the particle approaches the wall at a distance of the outer region length scale, $l_p = l/\sqrt{Re_c}$. At a point closer to the channel wall these two forces equally oppose each other and so the particle finds an equilibrium position. The more the particle leads the flow, the closer this position is towards the center of the channel.

Now let's consider the neutrally buoyant particle case. For neutrally buoyant particle, which we define as one with $|V| \ll \varepsilon^2/Re_c$, the flow disturbance is caused by the shear on the sphere. We expand the disturbance velocity and pressure on the sphere as

(4.1.114)
$$\mathbf{u} = \varepsilon\mathbf{u}_{0,nb} + O(\varepsilon)$$

(4.1.115)
$$p = \varepsilon p_{0,nb} + O(\varepsilon)$$

Substituting this into the governing equations (4.1.81) through (4.1.84), we obtain

(4.1.116)
$$\nabla^2\mathbf{u}_{0,nb} - \nabla p_{0,nb} = 0$$

(4.1.117)
$$\nabla \cdot \mathbf{u}_{0,nb} = 0$$

with the boundary conditions on the particle

(4.1.118)
$$\mathbf{u}_{0,nb} = \frac{1}{\sqrt{Re_c}}\left(\mathbf{\Omega}_p \times \mathbf{r} - \gamma z\hat{\mathbf{x}}\right) \text{ at } r = 1$$

(4.1.119)
$$\gamma = 4 - 8d/l$$

(4.1.120)
$$\mathbf{u}_{0,nb} \to 0 \text{ at } r \to \infty$$

Asmolov found that the solution to this is
(4.1.121)
$$\mathbf{u}_{0,nb} = \frac{1}{\sqrt{Re_c}}\left(\left(\mathbf{\Omega}_p + \frac{\gamma}{2}\hat{\mathbf{y}}\right) \times \frac{\mathbf{r}}{r^3} + \frac{5}{2}\gamma rxz\left(\frac{1}{r^7} - \frac{1}{r^5}\right) - \frac{\gamma}{2r^5}\left(z\hat{\mathbf{x}} + x\hat{\mathbf{z}}\right)\right)$$

This velocity field produces no force on the sphere. It does produce a couple on the sphere of

(4.1.122)
$$\frac{8\pi}{\sqrt{Re_c}}\left(\mathbf{\Omega}_p \times \frac{\gamma}{2}\hat{\mathbf{y}}\right)$$

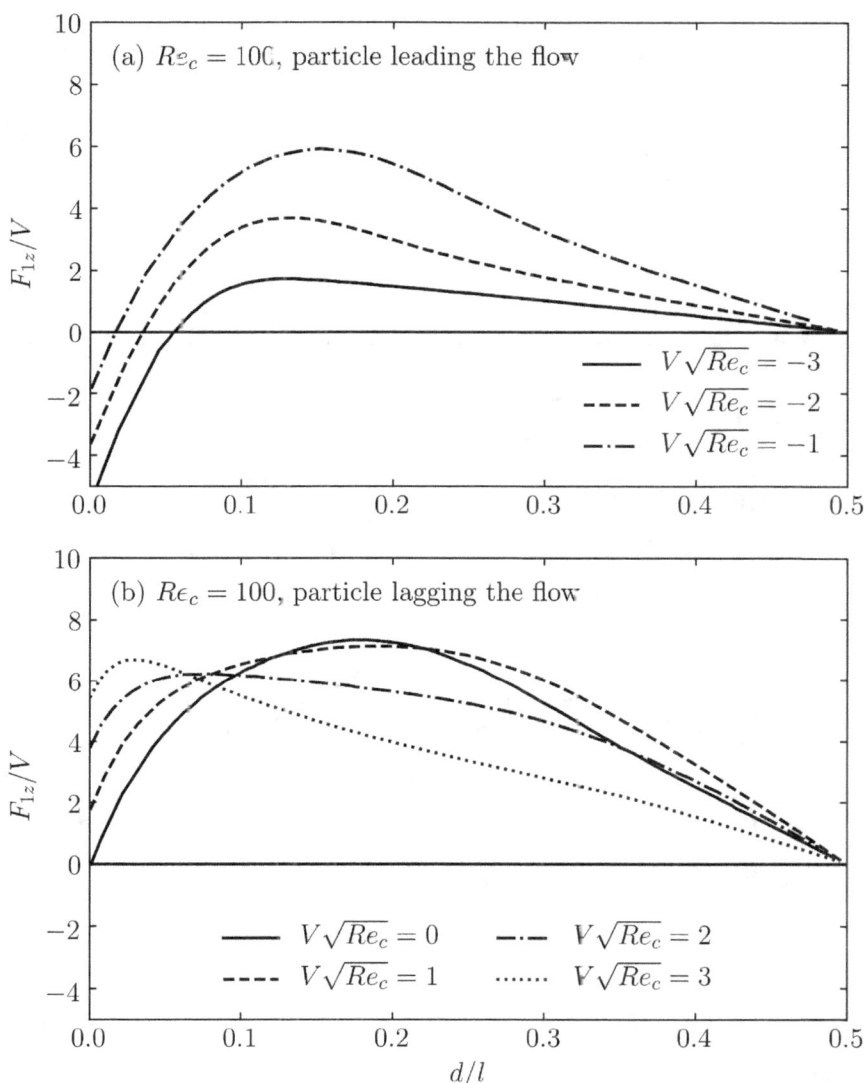

FIGURE 4.1.4. Scaled lift force on a non-neutrally buoyant spherical particle in Poiseuille flow between parallel plates with channel Reynolds numbers of 100 as calculated by Asmolov (1999). When the particle is leading the flow ($V < 0$) the particle experiences two lift forces: firstly, in the majority of the channel it experiences a shear gradient force directed towards the wall, and secondly, near the wall the particle experiences a wall interaction force directed towards the center of the channel. At some point, these two forces equally oppose each other and a particle attains an equilibrium position. When the particle lags the flow ($V > 0$) the lift force is only directed towards the center of the channel, and so the particle attains an equilibrium position at the center of the channel.

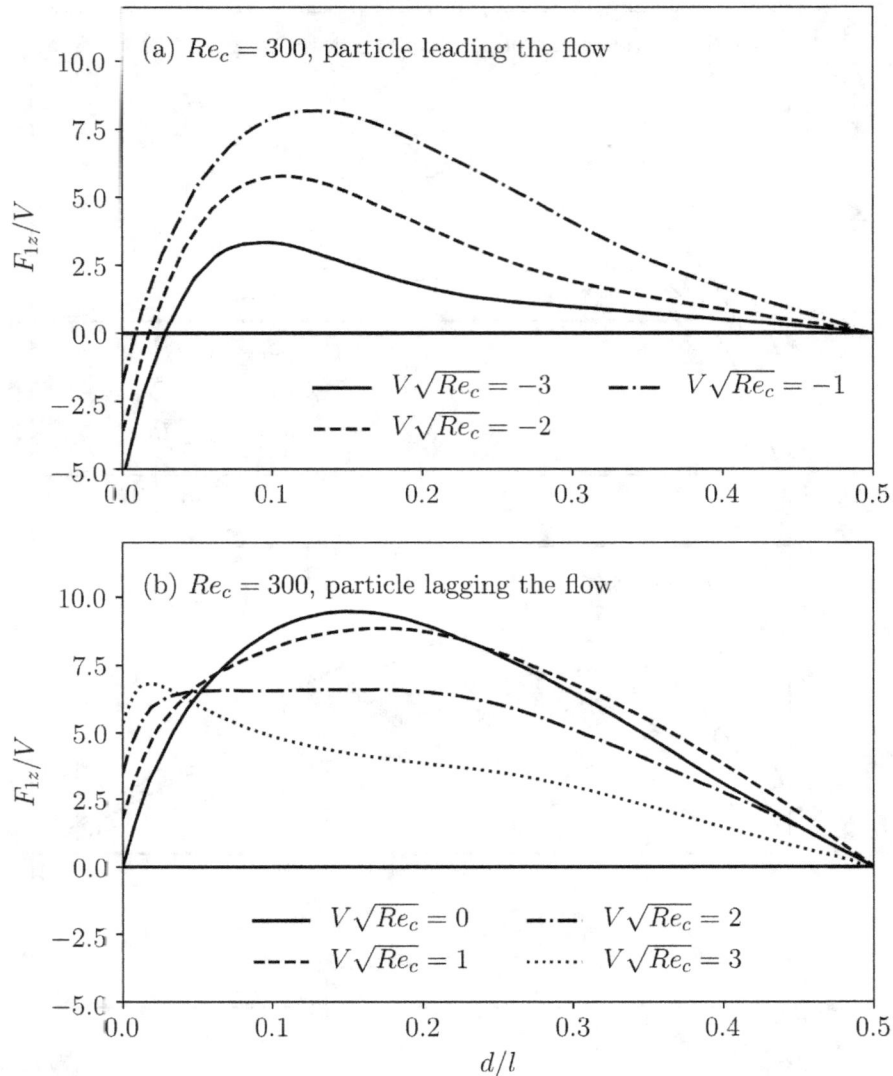

FIGURE 4.1.5. Scaled lift force on a non-neutrally buoyant spherical particle in Poiseuille flow between parallel plates with channel Reynolds numbers of 300 as calculated by Asmolov (1999). Just like in Figure 4.1.4, when the particle is leading the flow ($V < 0$) the particle experiences two lift forces: firstly, in the majority of the channel it experiences a shear gradient force directed towards the wall, and secondly, near the wall the particle experiences a wall interaction force directed towards the center of the channel. At some point, these two forces equally oppose each other and a particle attains an equilibrium position. When the particle lags the flow ($V > 0$) the lift force is only directed towards the center of the channel, and so the particle attains an equilibrium position at the center of the channel.

Since there are no other torques on the sphere other this couple from the fluid, the particles angular velocity is

$$(4.1.123) \qquad \mathbf{\Omega}_p = -\frac{\gamma}{2}\hat{\mathbf{y}}$$

and so,

$$(4.1.124) \qquad \mathbf{u}_{0,nb} = \frac{1}{\sqrt{Re_c}}\left(\frac{5}{2}\gamma \mathbf{r}xz\left(\frac{1}{r^7}-\frac{1}{r^5}\right) - \frac{\gamma}{2r^5}\left(z\hat{\mathbf{x}}+x\hat{\mathbf{z}}\right)\right)$$

Since we will need to match this solution with the outer region, we also transform this solution into the outer variables, and so

$$(4.1.125) \qquad \mathbf{u}_{0,nt} = -\varepsilon^2\frac{5}{2}\frac{1}{\sqrt{Re_c}}\gamma\frac{\mathbf{R}XZ}{R^5} + O\left(\varepsilon^4\right)$$

The first term in this expression is the strainlet velocity field, corresponding to flow created by a symmetric force dipole (see Section 1.4.5).

We can expand the outer region flow disturbance velocity and pressure as

$$(4.1.126) \qquad \mathbf{u} = \varepsilon^3\mathbf{U} + O\left(\varepsilon^3\right)$$
$$p = \varepsilon^3 P + O\left(\varepsilon^3\right)$$

The momentum equation for the outer region becomes,
(4.1.127)
$$\nabla^2\mathbf{U} - \nabla P - \bar{V}_x\frac{\partial\mathbf{U}}{\partial\mathbf{X}} - \frac{\partial\bar{V}_x}{\partial\mathbf{X}}U_z\hat{\mathbf{x}} = \frac{10}{3}\pi\gamma\frac{1}{\sqrt{Re_c}}\left(\left(\frac{\partial}{\partial X}\left(\delta\left(\mathbf{R}\right)\right)\hat{\mathbf{x}} + \frac{\partial}{\partial Z}\left(\delta\left(\mathbf{R}\right)\right)\hat{\mathbf{z}}\right)\right)$$

where we encapsulated the momentum equation into this equation by introducing a singularity corresponding to a symmetric force dipole (right-hand side, see Asmolov for details). Here

$$(4.1.128) \qquad \bar{V}_x = \gamma Z - \frac{4}{\sqrt{Re_c}}Z^2$$

since for a neutrally buoyant particle, V and so v are zero. The continuity equation is still

$$(4.1.129) \qquad \nabla\cdot\mathbf{U} = 0$$

and the boundary conditions are still,

$$(4.1.130) \qquad \mathbf{U} = 0 \text{ at } Z = -\sqrt{Re_c}\frac{d}{l} \text{ and } Z = \sqrt{Re_c}\frac{l-d}{l}$$

$$(4.1.131) \qquad \mathbf{U} = 0 \text{ at } X \to \infty$$

As before, Asmolov solves this equation by taking a two dimensional Fourier transform of (4.1.127) through (4.1.131), which results in a forth order ordinary differential equation; then solving this equation and finally numerically taking the inverse Fourier transform to obtain the solution. We plot the resulting solution in Figure 4.1.6. For $Re_c \leqslant 15$ the curves nearly collapse onto the $Re_c = 15$ curve. We see that the force is negative near the center of the channel, indicating that the force is directed towards the wall. This is the shear gradient force. Near the wall the force is positive and so is directed towards the center. This is the wall interaction force. When these two forces act on the particle equally, the particle achieves an equilibrium position. For $Re_c \leqslant 15$ this occurs around $d/l = 0.2$ which agrees with original observations of Segre and Silberberg (1962). At higher Reynolds numbers this position shifts towards the walls with increasing Reynolds number.

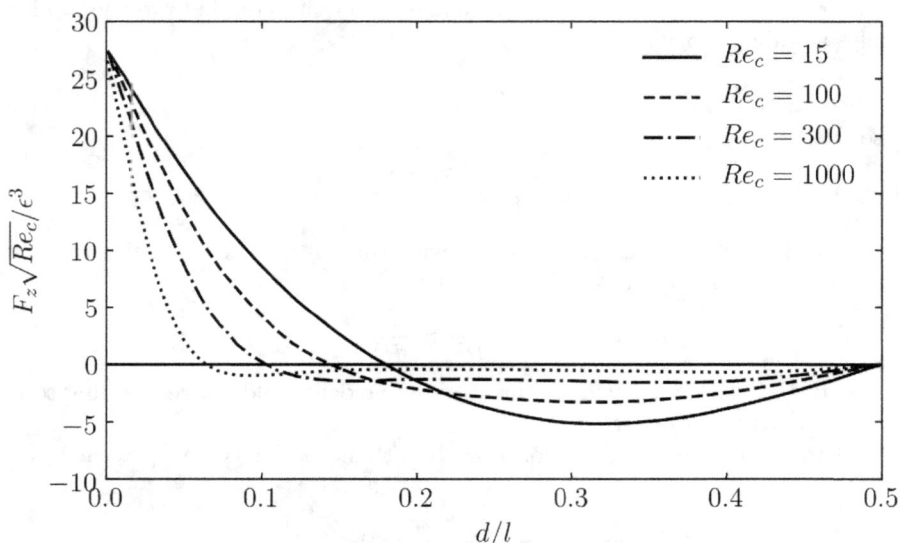

FIGURE 4.1.6. Scaled lift force on a neutrally buoyant spherical particle traveling in Poiseuille flow between flat plates. Positive values indicate a force directed towards the center of the channel, while negative values indicate a force directed towards the walls. The particle experiences two forces: firstly, a shear gradient force directed towards the wall, and secondly, near the wall the particle experiences a wall interaction force directed towards the center of the channel. At some point, these two forces equally oppose each other and a particle attains an equilibrium position. This equilibrium position shifts toward the wall as the channel Reynolds number increases. For $Re_c < 15$ the equilibrium position is roughly around $d/l = 0.2$. For $Re_c \leqslant 15$ the scaled force can be well approximated by the $Re_c = 15$ curve.

While an analytic expression for the force on the particle is not available, we fit the numerical results for $Re_c = 15$ to a polynomial and obtain (to less than 5% error) the dimensionless lift force as

(4.1.132) $$F_z = \frac{\varepsilon^3}{\sqrt{Re_c}} \left(27.79 - 241.6 \left(\frac{d}{l}\right) + 543.3 \left(\frac{d}{l}\right)^2 - 342.3 \left(\frac{d}{l}\right)^3 \right)$$

or in dimensional terms,
(4.1.133)

$$F_z = \mu a U_m \left(\frac{a}{l}\right)^2 \left(\frac{U_m a}{\nu}\right) \left(27.79 - 241.6 \left(\frac{d}{l}\right) + 543.3 \left(\frac{d}{l}\right)^2 - 342.3 \left(\frac{d}{l}\right)^3 \right)$$

With the dimensional form we can clearly see the scaling. The lift force scales as U_m/l to the second power, and the particle radius to the fourth power. This latter dependence may be used to separate particles based on their size.

To estimate the time necessary to reach the equilibrium position we equate this lift force with Stokes' drag,

$$(4.1.134) \qquad 6\pi\mu a l \frac{d}{dt}\left(\frac{d}{l}\right) = \mu a U_m \left(\frac{a}{l}\right)^2 \left(\frac{U_m a}{\nu}\right) f\left(\frac{d}{l}\right)$$

where

$$(4.1.135) \qquad f\left(\frac{d}{l}\right) = \left(27.79 - 241.6\left(\frac{d}{l}\right) + 543.3\left(\frac{d}{l}\right)^2 - 342.3\left(\frac{d}{l}\right)^3\right)$$

Simplifying and integrating to solve this differential equation,

$$(4.1.136) \qquad \int_{0.5}^{0.2} \frac{d\,(d/l)}{f\,(d/l)} = \int_0^{t_{eq}} \frac{1}{6\pi} \frac{a^3}{l^3} \frac{U_m^2}{\nu}\,dt$$

where, for this estimate, we are having the particle migrate from the channel centerline ($d/l = 0.5$) to an equilibrium position, $d/l = 0.2$. Evaluating the integrals we obtain,

$$(4.1.137) \qquad t_{eq} = 0.5388 \frac{l^3}{a^3} \frac{\nu}{U_m^2}$$

We see that while the equilibrium position of the particle does not depend on particle size, the time scale to reach equilibrium strongly depends on particle size. We see that larger particles reach the equilibrium positions much faster. From this we can conclude that passing different sized particles through a channel with Poiseuille flow and then skimming a particular layer can be used to enrich (or deplete) the suspension with a particular particle size. We also observe that the velocity of the flow matters - for slow flows (low U_m) the equilibrium time is long, and so we may never observe the tubular pinch effect.

4.2. Flow in curved channels - Dean flow

So far when discussing flow in a channel, we have discussed flow and flow profiles in straight channels. However, many real microfluidic devices have channels that are curved, sometimes as a necessity to fit a long channel into a compact space, and sometimes on purpose - to effect centrifugal separations in a channel. Having a curve in a channel imposes a centrifugal force on the flow and introduces vortices in the cross-section plane of the channel. This type of flow is often referred to as Dean flow, after W.R. Dean who analyzed it around 1927. Here we will mainly discuss flow in a slightly curved tube with a circular cross-section, following the analysis of Dean (1927) and Leal (2007). A general review of flow in a curved tubes is given by Berger et al. (1983).

To study this flow we will use a slightly unusual coordinate system, the toroidal coordinate system (Figure 4.2.1). Here the tube radius is a, while the coil radius is R. Since we are considering slightly curved tubes, $\varepsilon = a/R \ll 1$. Here ε is the smallness parameter which we will employ in our asymptotic expansion solution for the velocity profile. Note that as the curved tube approaches a straight tube, $R \to \infty$ and $\varepsilon \to 0$. The orthogonal coordinates of this coordinate system are (r, ψ, θ) as shown in Figure 4.2.1 and we will designate the corresponding velocity components as (u, v, w). Note that a distance along the tube, corresponding to the axial coordinate z in a straight tube is measured by $R\theta$. To study this flow,

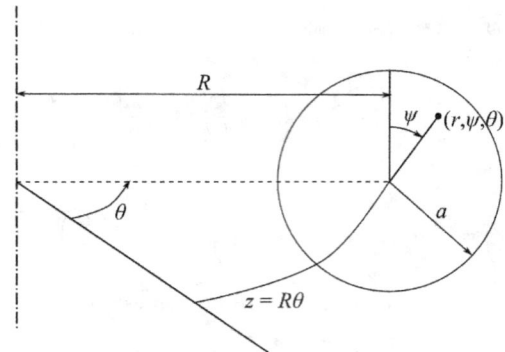

FIGURE 4.2.1. Sketch of a toroidal coordinate system we use for deriving Dean flow profile. The tube radius a is exaggerated for clarity relative to the curve radius R. In the situation we consider $a/R \ll 1$.

we firstly assume that the flow is steady and therefore ignore any entrance or exit effects. Thus the steady continuity equation in these coordinates is

$$(4.2.1) \qquad \frac{\partial u}{\partial r} + \frac{u}{r} + \frac{u \sin \psi}{R + r \sin \psi} + \frac{1}{r} \frac{\partial v}{\partial \psi} + \frac{v \cos \psi}{R + r \sin \psi} = 0$$

and the momentum equations in these coordinates are:

$$(4.2.2) \qquad \begin{aligned} &\rho \left(u \frac{\partial u}{\partial r} + \frac{v}{r} \frac{\partial u}{\partial \psi} - \frac{v^2}{r} - \frac{w^2 \sin \psi}{R + r \sin \psi} \right) \\ &= -\frac{\partial p}{\partial r} + \mu \left(\left(\frac{1}{r} \frac{\partial}{\partial \psi} + \frac{\cos \psi}{R + r \sin \psi} \right) \left(\frac{\partial v}{\partial r} + \frac{v}{r} - \frac{1}{r} \frac{\partial u}{\partial \psi} \right) \right) \end{aligned}$$

$$(4.2.3) \qquad \begin{aligned} &\rho \left(u \frac{\partial v}{\partial r} + \frac{v}{r} \frac{\partial v}{\partial \psi} + \frac{uv}{r} - \frac{w^2 \cos \psi}{R + r \sin \psi} \right) \\ &= -\frac{1}{r} \frac{\partial p}{\partial \psi} + \mu \left(\left(\frac{\partial}{\partial r} + \frac{\sin \psi}{R + r \sin \psi} \right) \left(\frac{\partial v}{\partial r} + \frac{v}{r} - \frac{1}{r} \frac{\partial u}{\partial \psi} \right) \right) \end{aligned}$$

$$(4.2.4) \qquad \begin{aligned} &\rho \left(u \frac{\partial w}{\partial r} + \frac{v}{r} \frac{\partial w}{\partial \psi} + \frac{uw \sin \psi}{R + r \sin \psi} - \frac{vw \cos \psi}{R + r \sin \psi} \right) \\ &= -\frac{1}{R + r \sin \psi} \frac{\partial p}{\partial \theta} \\ &+ \mu \left(\left(\frac{\partial}{\partial r} + \frac{1}{r} \right) \left(\frac{\partial w}{\partial r} + \frac{w \sin \psi}{R + r \sin \psi} \right) + \frac{1}{r} \frac{\partial}{\partial \psi} \left(\frac{1}{r} \frac{\partial w}{\partial \psi} + \frac{w \cos \psi}{R + r \sin \psi} \right) \right) \end{aligned}$$

We now nondimensionalize these equations. We nondimensionalize all the length scales with the tube radius a. We nondimensionalize all the velocities with a characteristic velocity

$$(4.2.5) \qquad u_c = \frac{Ga^2}{4\mu}$$

where G is the axial pressure gradient driving the flow, $G = \partial p/\partial z$. We nondimensionalize the pressure with a viscous pressure scale

$$(4.2.6) \qquad p_c = \mu u_c/a$$

Applying the nondimensionalization on (4.2.1) to (4.2.4), and rewriting them in $a/R = \varepsilon$, dropping higher order terms, and dropping the asterisks on the dimensionless equations for convenience, we obtain

$$(4.2.7) \qquad \frac{\partial u}{\partial r} + \frac{u}{r} + \frac{1}{r}\frac{\partial v}{\partial \psi} + \varepsilon \left(u \sin \psi + v \cos \psi \right) \left(1 - \varepsilon r \sin \psi + O\left(\varepsilon^2\right) \right) = 0$$

$$
(4.2.8) \qquad
\begin{aligned}
&Re\left(u\frac{\partial u}{\partial r} + \frac{v}{r}\frac{\partial u}{\partial \psi} - \frac{v^2}{r} - \varepsilon w^2 \sin \psi \left(1 - \varepsilon r \sin \psi + O\left(\varepsilon^2\right)\right) \right) \\
&= -\frac{\partial p}{\partial r} + \frac{1}{r}\frac{\partial}{\partial \psi}\left(\frac{\partial v}{\partial r} + \frac{v}{r} - \frac{1}{r}\frac{\partial u}{\partial \psi} \right) \\
&\quad + \varepsilon \cos \psi \left(1 - \varepsilon r \sin \psi + O\left(\varepsilon^2\right)\right)\left(\frac{\partial v}{\partial r} + \frac{v}{r} - \frac{1}{r}\frac{\partial u}{\partial \psi} \right)
\end{aligned}
$$

$$
(4.2.9) \qquad
\begin{aligned}
&Re\left(u\frac{\partial v}{\partial r} - \frac{v}{r}\frac{\partial v}{\partial \psi} + \frac{uv}{r} - \varepsilon w^2 \sin \psi \left(1 - \varepsilon r \sin \psi + O\left(\varepsilon^2\right)\right) \right) \\
&= -\frac{1}{r}\frac{\partial p}{\partial \psi} - \left(\frac{\partial}{\partial r} + \varepsilon \sin \psi \left(1 - \varepsilon r \sin \psi + O\left(\varepsilon^2\right)\right) \right)\left(\frac{\partial v}{\partial r} + \frac{v}{r} - \frac{1}{r}\frac{\partial u}{\partial \psi} \right)
\end{aligned}
$$

$$
(4.2.10) \qquad
\begin{aligned}
&Re\left(u\frac{\partial u}{\partial r} + \frac{v}{r}\frac{\partial u}{\partial \psi} + \varepsilon \left(uw \sin \psi + wv \cos \psi \right)\left(1 - \varepsilon r \sin \psi + O\left(\varepsilon^2\right)\right) \right) \\
&= -\left(1 - \varepsilon r \sin \psi - O\left(\varepsilon^2\right)\right)\frac{1}{r}\frac{\partial p}{\partial \theta} \\
&\quad + \left(\frac{\partial}{\partial r} + \frac{1}{r} \right)\left(\frac{\partial w}{\partial r} + \varepsilon w \sin \psi \left(1 - \varepsilon r \sin \psi + O\left(\varepsilon^2\right)\right) \right) \\
&\quad + \frac{1}{r}\frac{\partial}{\partial \psi}\left(\frac{1}{r}\frac{\partial w}{\partial \psi} + \varepsilon w \cos \psi \left(1 - \varepsilon r \sin \psi + O\left(\varepsilon^2\right)\right) \right)
\end{aligned}
$$

Our task now is to solve these four equations subject firstly to the no-slip boundary condition:

$$(4.2.11) \qquad v = 0 \text{ at } r = 1$$

$$(4.2.12) \qquad w = 0 \text{ at } r = 1$$

and the no-penetration condition,

$$(4.2.13) \qquad u = 0 \text{ at } r = 1$$

We observe that for the PDEs that we have to solve for non-zero values of ε are highly non-linear and thus are difficult to solve. Thus we employ the fact that $\varepsilon = a/R \ll 1$, and seek the simplest asymptotic expansion method, a regular asymptotic expansion in terms of ε. That is, we seek a solution in the form of

$$(4.2.14) \qquad u = u_0\varepsilon^0 + u_1\varepsilon^1 + u_2\varepsilon^2 + \dots$$

$$(4.2.15) \qquad v = v_0\varepsilon^0 + v_1\varepsilon^1 + v_2\varepsilon^2 + \dots$$

$$(4.2.16) \qquad w = w_0\varepsilon^0 + w_1\varepsilon^1 + w_2\varepsilon^2 + \dots$$

(4.2.17) $$p = p_0 \varepsilon^0 + p_1 \varepsilon^1 + p_2 \varepsilon^2 + \ldots$$

Note that the coefficients of u_i, v_i, w_i, p_i can be functions of (r, ψ) but must be independent of ε. To obtain the governing equations for these coefficients we substitute (4.2.14) through (4.2.17) into equations (4.2.7) through (4.2.10) and collect all the terms together at each order of ε. A result is a set of four equations with boundary conditions (the continuity equation and three momentum equations), each taking a form of

(4.2.18) $$f_0(r, \psi) \varepsilon^0 + f_1(r, \psi) \varepsilon^1 + f_2(r, \psi) \varepsilon^2 + \ldots = 0$$

Since ε is arbitrary, each of the coefficients $f_i(r, \psi)$ must independently equal to zero. Setting these coefficients to zero forms the equations for u_i, v_i, w_i, p_i. Performing this procedure, we obtain the zeroth order equations,

(4.2.19) $$\frac{\partial u_0}{\partial r} + \frac{u_0}{r} + \frac{1}{r} \frac{\partial v_0}{\partial \psi} = 0$$

(4.2.20) $$Re \left(u_0 \frac{\partial u_0}{\partial r} + \frac{v_0}{r} \frac{\partial u_0}{\partial \psi} - \frac{v_0^2}{r} \right) = -\frac{\partial p_0}{\partial r} + \frac{1}{r} \frac{\partial}{\partial \psi} \left(\frac{\partial v_0}{\partial r} + \frac{v_0}{r} - \frac{1}{r} \frac{\partial u_0}{\partial \psi} \right)$$

(4.2.21) $$Re \left(u_0 \frac{\partial v_0}{\partial r} + \frac{v_0}{r} \frac{\partial v_0}{\partial \psi} + \frac{u_0 v_0}{r} \right) = -\frac{1}{r} \frac{\partial p_0}{\partial \psi} + \frac{\partial}{\partial r} \left(\frac{\partial v_0}{\partial r} + \frac{v_0}{r} - \frac{1}{r} \frac{\partial u_0}{\partial \psi} \right)$$

(4.2.22) $$Re \left(u_0 \frac{\partial w_0}{\partial r} + \frac{v_0}{r} \frac{\partial w_0}{\partial \psi} \right) = -\frac{\partial p_0}{\partial z} + \frac{\partial^2 w_0}{\partial r^2} + \frac{1}{r} \frac{\partial w_0}{\partial r} + \frac{1}{r} \frac{\partial^2 w_0}{\partial \psi^2}$$

where we have substituted in

(4.2.23) $$\frac{\partial}{\partial z} \approx \frac{1}{R} \frac{\partial}{\partial \theta}$$

into the last equation, as to specify the pressure gradient in terms of the axial pressure gradient. We substitute in (4.2.14) through (4.2.17) into the boundary conditions (4.2.11) through (4.2.13) as well, (we treat boundary conditions just like any other equation) and obtain that

(4.2.24) $$u_0 = v_0 = w_0 = 0 \text{ at } r = 1$$

Let's now attempt to solve (4.2.19) through (4.2.22) subject to the boundary condition (4.2.24). From the definition of the characteristic velocity scale, (4.2.5), we have

(4.2.25) $$-\frac{\partial p_0}{\partial z} = 4$$

From this and (4.2.22) we see that $w_0 \neq 0$. On the other hand, the continuity equation as well as the other two momentum equations are homogeneous and independent of w_0. Since $u_0 = v_0 = 0$ on the boundary, $u_0 = v_0 = 0$ is also a solution in the domain. Furthermore, we assume that w_0 is axisymmetric. Thus we simplify equations (4.2.19) through (4.2.21) to

(4.2.26) $$u_0 = v_0 = 0$$

and (4.2.22) to

(4.2.27) $$\frac{1}{r} \frac{d}{dr} \left(r \frac{dw_0}{dr} \right) = -4$$

We can simply integrate this twice subject to (4.2.24) and the fact that at the center of the channel, $r = 0$, the solution must be regular (i.e., not blow up to infinity). We obtain

$$(4.2.28) \qquad\qquad w_0 = 1 - r^2$$

This is just the familiar Poiseuille flow solution subject to our particular nondimensionalization. We see that the effect of curvature is not captured in the zeroth order solution, and so it must appear in the higher order solutions. Therefore, now let's move onto the first order equations,

$$(4.2.29) \qquad\qquad \frac{\partial u_1}{\partial r} + \frac{u_1}{r} + \frac{1}{r}\frac{\partial v_1}{\partial \psi} = 0$$

$$(4.2.30) \qquad -Re\left(w_0^2 \sin\psi\right) = -\frac{\partial p_1}{\partial r} + \frac{1}{r}\frac{\partial}{\partial \psi}\left(\frac{\partial v_1}{\partial r} + \frac{v_1}{r} - \frac{1}{r}\frac{\partial u_1}{\partial \psi}\right)$$

$$(4.2.31) \qquad -Re\left(w_0^2 \cos\psi\right) = -\frac{1}{r}\frac{\partial p_1}{\partial \psi} + \frac{\partial}{\partial r}\left(\frac{\partial v_1}{\partial r} + \frac{v_1}{r} - \frac{1}{r}\frac{\partial u_1}{\partial \psi}\right)$$

$$(4.2.32) \qquad
\begin{aligned}
Re\left(u_1\frac{\partial w_0}{\partial r}\right) &= \frac{\partial p_0}{\partial z}r\sin\psi - \frac{\partial p_1}{\partial z} + \left(\frac{\partial^2 v_1}{\partial r^2} + \frac{1}{r}\frac{\partial w_1}{\partial r} + \frac{1}{r^2}\frac{\partial^2 w_1}{\partial \psi^2}\right) \\
&\quad + \left(\frac{\partial}{\partial r} - \frac{1}{r}\right)w_0 \sin\psi + \frac{1}{r}\frac{\partial}{\partial \psi}\left(w_0 \cos\psi\right)
\end{aligned}$$

And the boundary conditions are again,

$$(4.2.33) \qquad\qquad u_1 = v_1 = w_1 = 0 \text{ at } r = 1$$

While the equations now may look more intimidating, many of the terms are known because they rely on the solution of the previous order. In fact, while the original set of PDEs ((4.2.7) through (4.2.10)) was non-linear, this set of PDEs is linear. Additionally, notice that the momentum equations are now non-homogeneous and therefore all three velocity components are non-zero. Thus, the effect of tube curvature is already felt in the first order solution. To solve this set of equations, let's begin with pressure. From experience, the most general possible form for

$$(4.2.34) \qquad\qquad p_1 = H\left(r, \psi\right)z + B\left(r, \psi\right)$$

However, from (4.2.30) and (4.2.31) we see that $\partial p_1/\partial r$ and $\partial p_1/\partial \psi$ are at most functions of (r, ψ) and are not functions of z. This implies that $H\left(r, \psi\right) = 0$, and so

$$(4.2.35) \qquad\qquad p_1 = B\left(r, \psi\right)z$$

which also implies that $\partial p_1/\partial z = 0$. To solve (4.2.30) and (4.2.31), we introduce a stream function g_1 defined as

$$(4.2.36) \qquad
\begin{aligned}
ru_1 &= \frac{\partial g_1}{\partial \psi} \\
v_1 &= \frac{\partial g_1}{\partial r}
\end{aligned}$$

Our choice of this definition is inspired by the continuity equation, (4.2.29) - we want a function that satisfies the continuity equation automatically. We can see that with this definition, (4.2.29) is indeed satisfied. Substituting these as well as the solution (4.2.28) into (4.2.30) and (4.2.31), we obtain

$$(4.2.37) \qquad -Re(1-r^2)^2 \sin\psi = -\frac{\partial p_1}{\partial r} - \frac{1}{r}\frac{\partial}{\partial\psi}\left(\nabla_1^2 g_1\right)$$

$$(4.2.38) \qquad -Re(1-r^2)^2 \cos\psi = -\frac{1}{r}\frac{\partial p_1}{\partial\psi} + \frac{\partial}{\partial r}\left(\nabla_1^2 g_1\right)$$

where

$$(4.2.39) \qquad \nabla_1^2 = \frac{\partial^2}{\partial r^2} + \frac{1}{r}\frac{\partial}{\partial r} + \frac{1}{r^2}\frac{\partial^2}{\partial\psi^2}$$

We can eliminate p_1 by differentiating (4.2.37) with respect to ψ, differentiating (4.2.38) with respect to r and combining the two equations,

$$(4.2.40) \qquad r\nabla_1^2\left(\nabla_1^2 g_1\right) = 4Re\left(1-r^2\right)r^2 \cos\psi$$

Next, we simplify (4.2.32) by substituting in (4.2.28), and (4.2.25)

$$(4.2.41) \qquad \nabla_1^2 w_1 = 6r \sin\psi - (2ru_1)\,Re$$

Translating the boundary conditions (4.2.33) into our stream function, we obtain

$$(4.2.42) \qquad \frac{\partial g_1}{\partial r} = \frac{\partial g_1}{\partial\psi} = 0$$

Subject to these boundary conditions the inhomogeneous biharmonic equation (4.2.40) has a solution

$$(4.2.43) \qquad g_1 = \frac{Re}{288}r\left(1-r^2\right)^2\left(4-r^2\right)\cos\psi$$

Applying (4.2.36) to this solution, we obtain

$$(4.2.44) \qquad u_1 = \frac{Re}{288}\left(1-r^2\right)^2\left(4-r^2\right)\sin\psi$$

$$(4.2.45) \qquad v_1 = -\frac{Re}{288}\left(7r^4 - 23r^2 + 4\right)\left(1-r^2\right)\cos\psi$$

Now that we know u_1, we can solve (4.2.41) and obtain

$$(4.2.46) \quad w_1 = -\frac{3}{4}\sin\psi r\left(1-r^2\right) + Re^2\frac{\sin\psi}{11520}r\left(1-r^2\right)\left(19 - 21r^2 + 9r^4 - r^6\right)$$

And so we have obtained a solution for the velocity profile in a curved tube accurate to the first order,

$$(4.2.47) \qquad u = u_0 + u_1\varepsilon^1 + O\left(\varepsilon^2\right)$$

$$(4.2.48) \qquad v = v_0 + v_1\varepsilon^1 + O\left(\varepsilon^2\right)$$

$$(4.2.49) \qquad w = w_0 + w_1\varepsilon^1 + O\left(\varepsilon^2\right)$$

We plot the streamlines of the secondary flow (contours of constant g_1, (4.2.43)) in Figure 4.2.2. We plot the contours of axial velocity as predicted by (4.2.49) in Figure 4.2.3. We plot these contours as a function of Dean number, a dimensionless

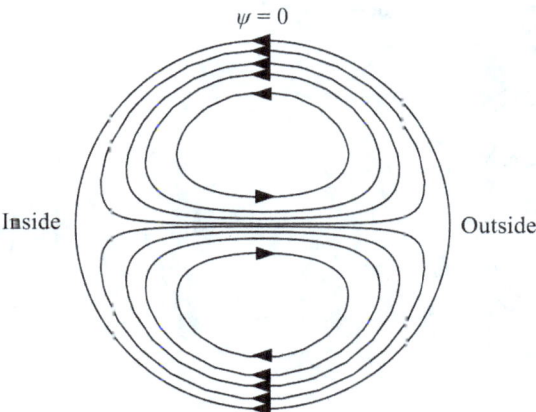

FIGURE 4.2.2. Contour plot of streamlines for the secondary flow (the function g_1) for flow in curved tube with a small tube radius compared with the curve radius. "Inside" refers to the inside of the curve. Notice that in the center the fluid moves from the inside part of the tube to outside, while near the wall the flow is from the outside, towards the inside.

number describing the ratio of inertial and centripetal forces to viscous forces. For this problem we define the Dean number as

$$(4.2.50) \qquad De = \frac{\rho u_c a}{\mu} \sqrt{\frac{a}{R}} = Re \sqrt{\frac{a}{R}} = Re \sqrt{\varepsilon}$$

In Figure 4.2.2, we see that the presence of even small amount of curvature of a pipe is already enough to produce a secondary flow, even at the leading order correction, $O(\varepsilon)$. We see that unlike in a straight pipe, the motion is no longer just in the axial direction, but there is motion in the cross-section plane as well. However, the motion is in cross-section plane is fairly small compared to the motion in the axial direction, since it is given by the $O(\varepsilon)$ term. (In a well behaved asymptotic expansion the higher the order of the term, the less it should contribute to the overall solution - this is how we justify dropping higher order terms). We see that the motion at the center of the tube is from the inside of the curve to the outside. This flow is driven by the centrifugal forces exerted on the fluid as the fluid is forced through the bend. The return flows are next to the walls. Superimposing this motion in the cross-section plane with the axial travel, we would observe that a particle in the fluid moves in a helical motion in the tube.

In Figure 4.2.3, we see that at high enough Dean numbers, the secondary flow also modifies the contours of the axial flow. The secondary flow is asymmetric with slightly higher flow in the region closer to the outside of the curve, and with slightly lower flow in the region closer to the inside of the curve. Note that our $O(\varepsilon)$ does not capture the change in volumetric flow rate through the tube, a small change that is observed experimentally. To predict this we need to calculate the next term of the expansion, which we do not do here.

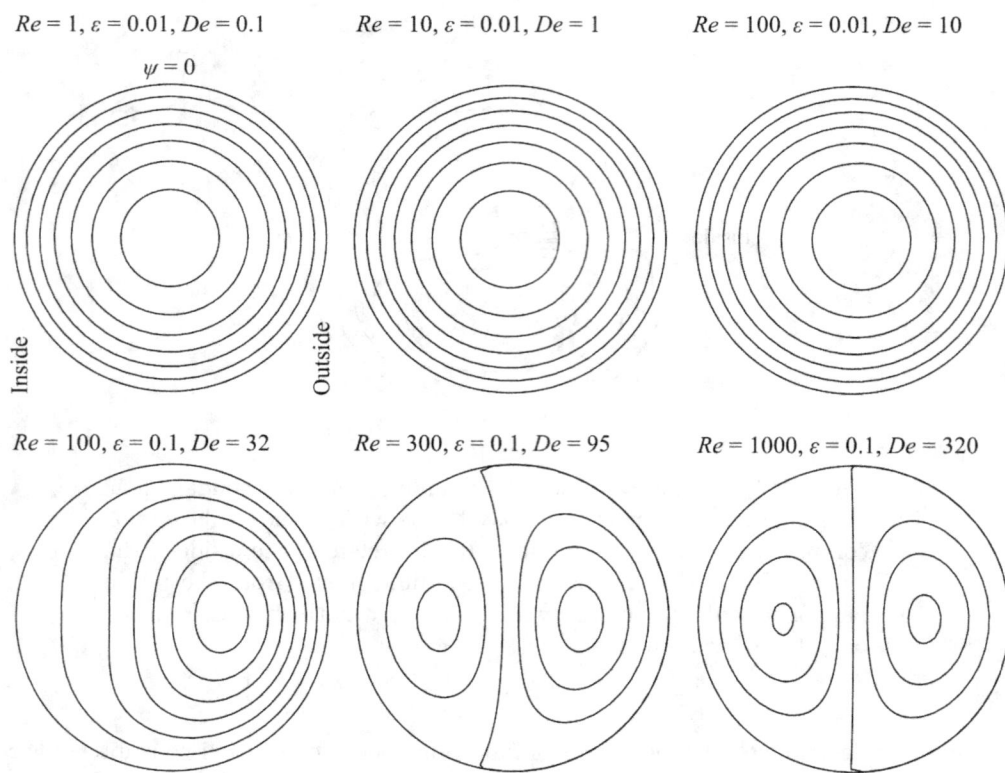

FIGURE 4.2.3. Plots of contours of axial velocity as a function of Dean number, for ψ from 0 to 2π , r from 0 to 1, as predicted by (4.2.49). At Dean numbers below unity contours of axial velocity remain symmetric about the center of the pipe, just like in a straight pipe (the contribution from secondary flow is negligible). As Dean numbers increase, the point of maximum velocity first shifts towards the outside of the pipe. At even higher velocities, a minimum develops on the inside side of the pipe.

4.2.1. Particles in Dean flow. In the past section we saw that secondary flow (Dean flow) occurs at moderate channel Reynolds numbers and at Dean numbers greater than unity. From Section 4.1.3, we saw that particles at such channel Reynolds numbers exhibit lift and migration lateral to the main flow direction. Thus, we expect particles placed in Dean flow to both experience the influence of the secondary flow as well as the influence of the lift forces. Indeed, the combination of secondary Dean flow with inertial lift, can for example cause particle focusing and can even be used for particle separation by size (e.g., Martel and Toner (2013)). Transporting particles through curved channels is a fairly common, both in lab-on-chip devices, as well as tubing in all sort of particle handling devices, and so we explore this combination to understand both the intended and unintended phenomena that can result from a simple transport of particles through a curved tube. To estimate the combined effect of Dean flow with inertial lift, we here will pursue a scaling approach. Firstly, we assume that the presence of Dean flow does not influence the scaling of the lift force, which from Section 4.1.3, (4.1.133) scales as

$$(4.2.51) \qquad F_L \sim \rho a^4 \left(\frac{U_m}{l} \right)^2 f \left(\frac{d}{l} \right)$$

where, as a reminder, ρ is the density of the fluid, a is the radius of the particle, U_m is the maximum of the Poiseuille velocity in the channel, l is the diameter or width of the channel, and d is the location of particle in the channel. This scaling is valid for $Re_c \leqslant 15$. For higher Reynolds numbers,

$$(4.2.52) \qquad F_L \sim \rho a^4 \left(\frac{U_m}{l} \right)^2 f \left(\frac{d}{l}, Re_c \right)$$

Effectively, we assume that no new lift force is generated due to the presence of the secondary flow in the channel, as the magnitude of the secondary flow is very much smaller than that of the primary flow. Secondly, the secondary flow applies a force on the particle equal to the Stokes' drag. This again can be justified by the fact that the magnitude of the secondary flow is small, and so Reynold's number based on it is small, and so this part of the flow can be approximated by creeping flow. Thirdly, we assume that the situation is linear, and that the lift force can be superimposed by the force generated by the secondary Dean flow. To estimate the force from the secondary Dean flow, we observe from equations (4.2.44), (4.2.45) and (4.2.47), (4.2.48) that the magnitude of the secondary flow scales as

$$(4.2.53) \qquad U_D \sim \varepsilon Re_c U_m = \frac{l}{R} \frac{l U_m}{\nu} U_m = De^2 \frac{\nu}{l}$$

where here we define $Re_c = lU_m/\nu$, the Dean number as $De = Re_c\sqrt{l/R}$ and $\varepsilon = l/R$. Here R is the curve radius, and ν is the liquid kinematic viscosity. Substituting this into Stokes' drag relation, $F_{SD} = 6\pi\mu aU$, we obtain the scaling for Dean drag

$$(4.2.54) \qquad F_D \sim 6\pi\rho a \frac{l^2 U_m^2}{R}$$

The balance between the inertial lift forces and Dean drag determine the equilibrium position (or the approach to equilibrium position) of the particles. While in many simple flows the equilibrium position can be found by just balancing the lift and drag forces, in our system the Dean drag is not necessarily in the same direction as the lift forces. In fact in many positions in the channel, the two are orthogonal. Thus, to understand the relative effects of these forces, we compare their relative magnitudes,

$$(4.2.55) \qquad \frac{F_L}{F_D} \sim \frac{1}{6\pi} \left(\frac{a}{l} \right)^3 \frac{R}{l} f \left(\frac{d}{l} \right) = \frac{1}{6\pi} \frac{(a/l)^3}{\varepsilon} f \left(\frac{d}{l} \right)$$

If $F_L \ll F_D$ then the Dean vortices mix the particles and no focusing occurs. $F_L \gg F_D$ then the focusing will come from inertial lift, and Dean's flow contribution to focusing will be negligible. Thus, we are interested in a regime where the two are of the same order. Specifically, our interest lies where the lift force interacts with the Dean drag at the channel midline between the two counter rotating vortices. Here the Dean drag and the lift force are collinear and their opposing each other should allow differential focusing of particles based on their size. In other locations, where the Dean drag and the lift force are orthogonal, the Dean drag causes the particle focusing in these locations to be unstable and so it eliminates potential focusing locations. From (4.2.55) we see that both Dean flow and inertial lift are important when the ratio of particle size to channel size balances out the ratio of channel radius of curvature to channel size.

Martel and Toner (2013) investigated particle focusing with combined Dean flow and inertial lift in a rectangular channel, for particle size to channel size ratio λ of 0.066 to 0.255, channel size to radius of curvature ratio ε of 0 to 0.0166, for Re_c from 9 to 425. For their rectangular channel, they take the characteristic channel size l as the hydraulic diameter of the channel, and define $\varepsilon = l/2R$. They found that for the lowest λ, $\lambda = 0.066$, they see poor particle focusing for $\varepsilon < 0.0034$, with particles being found in multiple streams at all Reynolds numbers investigated. For $\varepsilon \geqslant 0.0034$ particles focused into a single stream. The location of that stream starts closer to the inner wall of the curved channel, and moves towards the outer wall as Re_c increases. For $\lambda \geqslant 0.149$ the focusing behavior is different. Even for $\varepsilon \geqslant 0$ the particles focus practically into a single stream. For low Reynolds numbers this stream starts near the middle of the channel and moves towards the inner wall with increasing Reynolds number. We see that the focusing behavior resulting from the coupling of Dean flow and inertial lift is complex and multiple focusing modes exist. However, it shows promise for particle separation.

4.3. Lift velocity of a deformable drop in creeping Poiseuille flow

So far we have considered the lift on rigid particles at finite Reynolds numbers and found that the presence of inertia causes the particles to experience a force orthogonal to their main direction of motion (lift). Here we will see that deformability of a particle can also generate a lift force on the particle, even when the overall flow is still a creeping flow (and so the inertia is negligible). In our discussion we will follow the analysis of Chan and Leal (1979). Chan and Leal examined the lift on a deformable particle modeled as a second order fluid traveling in another second order fluid (as a generalization beyond the standard Newtonian fluid). (A second order fluid is a fluid where the stress tensor is a sum of all tensors that can be formed from the velocity field up to two derivatives. A Newtonian fluid is a first order fluid, so its stress tensor is a sum of all tensors that can be formed from the velocity field up to a single derivative). Here we will not discuss Chan and Leal's analysis in detail and just briefly summarize their approach, and discuss their results. Chan and Leal considered a neutrally buoyant drop in a quadratic shearing two dimensional flow (modeling Poiseuille flow). They assumed that the flow is dominated by viscosity and pressure and that inertia both of the fluid and the drop is negligible. Hence they used the creeping flow equations for fluid motion outside and inside the drop. They solved the equations of motion via a double asymptotic expansion where one of the smallness parameters was the deviation from "Newtonianness" of the fluid (the effective ratio of the intrinsic relaxation time scale to the convective time scale) and the second smallness parameter was the capillary number (see Section 7.8). We note that in most problems of interest these are indeed small.

In Chan and Leal's model we consider a particle radius a traveling in a Poiseuille flow either between flat plates or in a circular pipe. The spacing between plates and the diameter of the pipe is d. For a given position in this channel, we can introduce a dimensionless distance from a wall s, which is just the distance from a

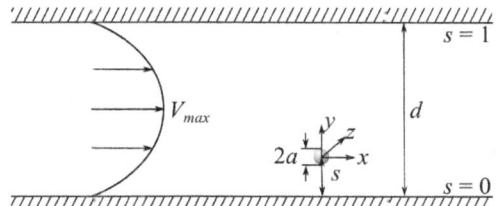

FIGURE 4.3.1. Schematic of a particle radius a in Poiseuille flow either between flat plates or in a circular pipe. The spacing between plates and the diameter of the pipe is d. The vertical location of the particle is defined by dimensionless distance from the wall s, which is the dimensioned distance from the wall scaled by d.

wall normalized by d, such that $0 \leqslant s \leqslant 1$ (Figure 4.3.1). We also introduce the ratio of drop size to channel size, $\varepsilon = a/d$. This ratio is assumed to be small and is indeed small for most problems of interest. For Poiseuille flow the undisturbed flow relative to a coordinate system fixed at the centroid of the drop can be written as

$$(4.3.1) \qquad \mathbf{V} = \left(\alpha + \beta y + \gamma y^2\right) \hat{\mathbf{x}} - \mathbf{U}_p$$

where $\mathbf{U}_p$ is the particle velocity and where for Poiseuille flow,

$$(4.3.2) \qquad \begin{aligned} \alpha &= 4V_{\max} s\,(1 - s) \\ \beta &= 4V_{\max}\,(1 - 2s)\,\varepsilon \\ \gamma &= -4V_{\max} s^2 \varepsilon^2 \end{aligned}$$

where $V_{\max}$ is the centerline velocity in Poiseuille flow.

Chan and Leal found that for Poiseuille flow between flat plates, where the flow can be treated as unbounded ($\varepsilon \to 0$), the droplet lift velocity to be
$$(4.3.3)$$

$$U_{p,y} = -\frac{\beta \gamma Ca}{(1+\kappa)^2 (2+3\kappa)} \left(\left(\frac{16 + 19\kappa}{42\,(2 + 3\kappa)\,(4 + \kappa)}\left(13 - 36\kappa - 73\kappa^2 - 24\kappa^3\right)\right) + \frac{10 + 11\kappa}{105}\left(8 - \kappa + 3\kappa^2\right) + O\left(\varepsilon\right) \right)$$

where

$$(4.3.4) \qquad \kappa = \mu_i / \mu_o$$

is the ratio of the Newtonian viscosity inside the drop to that outside the drop and the capillary number

$$(4.3.5) \qquad Ca = \frac{a \mu_o G}{\sigma}$$

where σ is the surface tension, and G is the mean shear rate of the bulk flow.

$$(4.3.6) \qquad G = \beta + \gamma$$

We plot $U_{p,y}$ in Figure 4.3.2. We see that this model predicts a lift velocity only if the velocity profile has curvature, i.e., $\gamma \neq 0$. For Couette flows, where $\gamma = 0$, the model predicts that there is no lift at this order (however, Chan and Leal's model predicts lift velocity in Couette flows at higher orders). Secondly, the migration velocity scales with the capillary number. The smaller the surface tension pulling the droplet into a sphere, the more deformable the droplet. Thus, the larger the

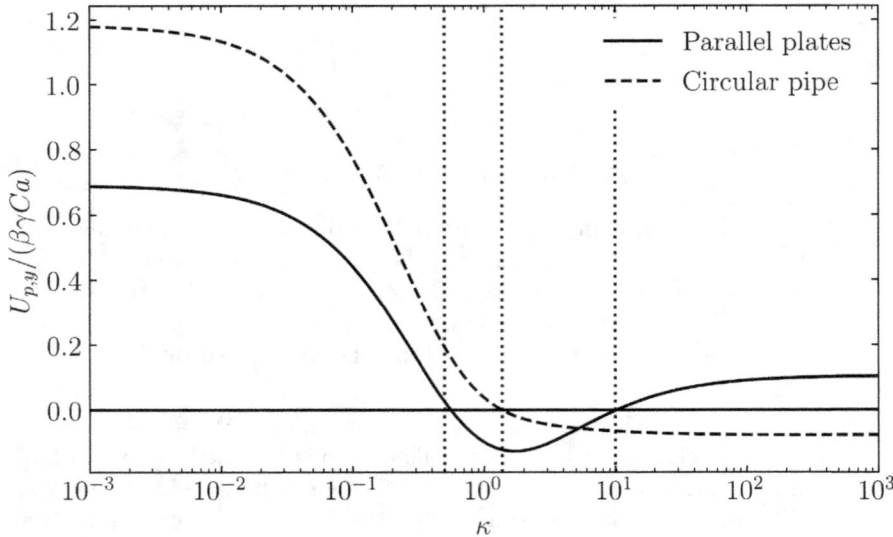

FIGURE 4.3.2. Lift velocity scaled by the capillary number and parameters β and γ as a function of the viscosity ratio κ, for both Poiseuille flow between parallel plates and in a circular pipe.

capillary number the more deformable the droplet. Thus, the more deformable the droplets the higher their lift velocity. Thirdly, and most interestingly, the direction of migration velocity depends on the viscosity ratio κ. For $\kappa < 0.5$ or $\kappa > 10$, $U_{p,y}$ is positive and so the droplets migrate towards the centerline. Meanwhile for $0.5 < \kappa < 10$ $U_{p,y}$ is negative and so the droplets migrate towards the walls of the channel. While only the migration towards the centerline indeed has been observed, the investigation of droplet motion in $0.5 < \kappa < 10$ has not been reported. This prediction is especially interesting in that it predicts that simple migration in a tube may be used to sort particles based on their viscoelastic properties, and the sorting may be tuned by selecting a suspending fluid with appropriate viscosity. This migration velocity is due to an asymmetrical deformation of the drop, since a symmetrical drop will not migrate in creeping flow (see Section 1.1). Chan and Leal found that for Poiseuille flow in a circular pipe, where the flow can be treated as unbounded ($\varepsilon \to 0$), the droplet lift velocity behaves similar to that of Poiseuille flow between flat plates, and is

$$(4.3.7) \qquad U_{p,y} = -\frac{\beta\gamma Ca}{(1+\kappa)^2 (2+3\kappa)} \left(\begin{array}{c} \left(\dfrac{3}{14} \dfrac{16+19\kappa}{(2+3\kappa)} \left(1 - \kappa - 2\kappa^2 \right) \right) \\ + \dfrac{10 + 11\kappa^2}{140} \left(8 - \kappa + 3\kappa^2 \right) + O\left(\varepsilon \right) \end{array} \right)$$

We plot $U_{p,y}$ in Figure 4.3.2. Droplet in a circular pipe has a slightly different lift velocity behavior. For $\kappa < 1.4$, $U_{p,y}$ is positive and so the droplets migrate towards the centerline, while for $\kappa > 1.4$ $U_{p,y}$ is negative and so the droplets migrate towards the walls of the channel.

CHAPTER 5

Slurry flow in microchannels

So far we have considered the flow of single, isolated particles and saw that at high enough Reynolds numbers they experience lift and may segregate themselves by size. However, we also frequently encounter situations where the volume fractions are high enough so that the particles in the flow cannot be treated as single and isolated. As observed by Leighton and Acrivos (1987) particles in this situation also experience a force orthogonal to their main motion (an effective lift force) and thus migrate towards regions of low shear rate (channel center). The same effect can be observed not only for rigid spherical particles but also for flexible discoid particles – red blood cells. Red blood cells (erythrocytes) for example migrate towards the center of the tube and away from tube walls. Firstly, this causes the Fahraeus effect (Fahraeus (1928)), where the volume fraction of red blood cells (hematocrit) is lower in a small capillary (<300 µm) than the hematocrit in a larger channel that feeds it! (Note that the effect is not purely mechanical sieving, as there are still cells in the small capillary). Secondly, this causes the Fahraeus-Lindqvist effect (Fahraeus and Lindqvist (1931)) where blood viscosity decreasing rapidly with tube diameter for capillaries less than 300 µm in diameter. Here we will first consider the segregation of particles in the high volume fraction (slurry) flow of neutrally buoyant rigid spherical particles, then consider sedimenting rigid particles, and lastly consider deformable particles such as red blood cells.

5.1. Rheology of slurries of neutrally buoyant spheres

In general, the viscosity of a slurry μ is dependent on the set of the following parameters: particle radius, a, particle density ρ_p, number density of particles n, suspending medium viscosity μ_0, suspending medium density ρ_0, thermal energy $k_B T$, the shear rate $\dot{\gamma}$, and time. All of these quantities can be expressed in units of mass, length, and time. Using nondimensionalization (Buckingham π theorem) we can reduce these nine variables to a total of six, going from

$$(5.1.1) \qquad \mu = f\left(a, \rho_p, n, \mu_0, \rho_0, kT, \dot{\gamma}, t\right)$$

to

$$(5.1.2) \qquad \mu_r = f\left(\phi, \rho_r, Pe_{\dot{\gamma}}, Re_{\dot{\gamma}}, t_r\right)$$

where

$$(5.1.3) \qquad \mu_r = \mu/\mu_0$$

is the reduced viscosity;

$$(5.1.4) \qquad \phi = n\frac{4}{3}\pi a^3$$

113

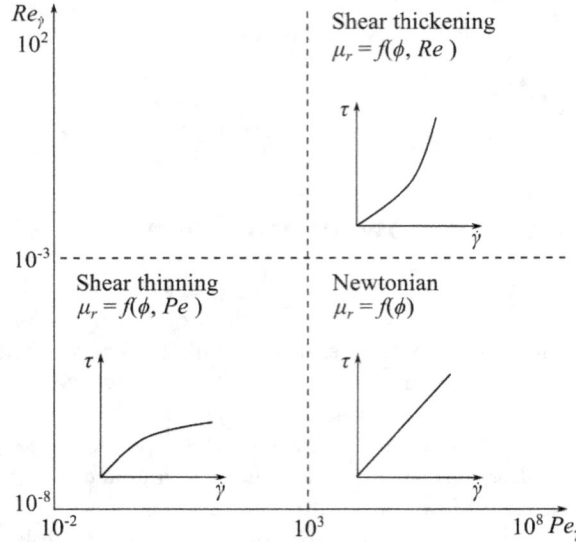

FIGURE 5.1.1. Rheological regimes of dense particle suspensions as a function of Reynolds and Peclet numbers based on flow shear rate.

is the particle volume fraction;

$$(5.1.5) \qquad \rho_r = \rho_p/\rho_0$$

is the relative density;

$$(5.1.6) \qquad t_r = \frac{tkT}{\mu_0 a^3}$$

is the dimensionless time;

$$(5.1.7) \qquad Re_{\dot\gamma} = \frac{\rho_0 a^2 \dot\gamma}{\mu_0}$$

is the shear rate based Reynolds number, and μ_0

$$(5.1.8) \qquad Pe_{\dot\gamma} = \frac{6\pi\mu_0 a^3 \dot\gamma}{kT}$$

is the shear rate based Peclet number, (and where we have used the Einstein-Smoluchowski and Einstein-Stokes relation to relate particle Brownian diffusivity to solution temperature and particle dimension; see Section 16.7.6). Here the Peclet number represents the ratio of importance of advection to Brownian diffusion of the beads (with high Peclet number indicating that advection dominates over Brownian diffusion). For a steady state situation of naturally buoyant system ρ_r and t_r become not important and so

$$(5.1.9) \qquad \mu_r = f\left(\phi, Pe_{\dot\gamma}, Re_{\dot\gamma}\right)$$

In general rheological behavior of particle suspensions depends strongly on the Reynolds and Peclet numbers. For example, when Reynolds number and Peclet number are small, the suspension exhibits shear thinning behavior. On the other hand, when Reynolds numbers are significant but Peclet numbers are large, the suspension exhibits shear thickening behavior. Newtonian behavior can be observed when Reynolds number is small and Peclet number is large (see Figure 5.1.1). We are most interested in this regime. In this regime, the viscosity is only a function of particle volume fraction. The simplest relation for viscosity as a function of volume fraction in this regime is due to Einstein (1906),

$$(5.1.10) \qquad \mu_r = 1 + \frac{5}{2}\phi + O\left(\phi^2\right)$$

This relation is only valid for dilute suspensions $\phi < 0.01$ and for rigid spheres. Batchelor extended this relation to

$$(5.1.11) \qquad \mu_r = 1 + \frac{5}{2}\phi + 6.2\phi^2 + O\left(\phi^3\right)$$

which again is only valid for dilute suspensions. The two most commonly used practical relations for viscosity are the Eilers equation (Ferrini et al. (1979)),

$$(5.1.12) \qquad \mu_r = \left(1 + \frac{5\phi/4}{1 - \phi/\phi_m}\right)$$

and the Krieger-Dougherty equation (Krieger and Dougherty (1959)),

$$(5.1.13) \qquad \mu_r = (1 - \phi/\phi_m)^{-5\phi_m/2}$$

where ϕ_m is the maximum attainable volume fraction. Typically, this is considered to be the volume fraction at which the spheres pack. Spheres can pack in a fairly low density packing, the simple cubic arrangement, $\phi = 0.524$ to a very high density packing, the hexagonal closed packed, $\phi = 0.740$. It reasonable that a well-mixed suspension does not self-assemble into any of these theoretical packing forms, but instead forms a random closed packed arrangement, for which $\phi = 0.63$ (McGeary (1961)). Typically we choose ϕ_m to be around this value. Both the Eilers equation and the Kreieger-Dougherty equation agree very well with experiments for suspensions of spherical particles for over three orders of magnitude in reduced viscosity, and for both low particle volume fractions and volume fractions approaching ϕ_m (Chang and Powell (1994)). Here we will use a simpler form of Krieger-Dougherty equation (Krieger (1972)),

$$(5.1.14) \qquad \mu_r = (1 - \phi/\phi_m)^{-1.82}$$

with $\phi_m = 0.68$, which was obtained by fitting data for suspensions with volume fractions in the range $0.01 < \phi < 0.5$.

5.2. Slurries of neutrally buoyant hard spheres

Around 1980 Gadala-Maria and Acrivos (1980) observed for a high concentration of spherical particles ($\phi > 0.4$) the viscosity of the suspensions was not constant, but instead decreased from an initial value to an equilibrium value when sheared for a prolonged time in a standard Couette viscometer. More careful observations later by Leighton and Acrivos (1987) revealed that particles slowly escaped from the high shear regions of the Couette viscometer into the viscometer's reservoir (low shear region), leaving a significantly lower volume fraction suspension in the gap. Leighton and Acrivos modeled this migration from high shear to low shear regions as shear induced diffusion. Here discuss this phenomena following the analysis of Phillips et al. (1992) who attribute this behavior to a combined effect of spatially varying interaction frequency of the particles and the effect of spatially varying viscosity.

We are interested in the migration of spheres in a suspension of hard spheres with radius a in a Newtonian liquid with a viscosity μ_0. Furthermore, we are interested in situations where the shear based Peclet number, (5.1.8), is high (i.e., Brownian diffusion of particles is negligible compared to their migration via other means). Secondly we are interested in the situation where the Reynolds number is low, so that creeping flow equations apply. For this regime we use the Kreieger's relation, (5.1.14), for suspension viscosity as a function of particle volume fraction. We also assume the that suspension stress tensor relates linearly to shear rate as for a Newtonian fluid,

$$(5.2.1) \qquad\qquad \tau = -\mu\left(\phi\right)\dot{\gamma}$$

where

$$(5.2.2) \qquad\qquad \dot{\gamma} = \nabla\mathbf{u} + \nabla\mathbf{u}^T$$

is the rate of strain tensor. For a given distribution of particle volume fraction, we find the velocity and pressure fields via the creeping flow equations ((1.0.9) and (1.0.11)) which we rewrite here as

$$(5.2.3) \qquad\qquad \nabla\cdot\mathbf{u} = 0$$

$$(5.2.4) \qquad\qquad \nabla\cdot\tau + \nabla p = 0$$

These equations combined with the diffusion equation for the particles allow us to obtain the full distribution of particles, velocity, and pressure in the system. Note that we model particle migration as a diffusion rather than accounting for individual forces from particle to particle because in a concentrated suspension tracking multiparticle interactions becomes computationally impractical as the number of interactions is high. Interestingly, while the trajectories of two isolated particles in creeping flow are time symmetric and reversible (see Section 1.1) due to the linearity of creeping flow equations, three or more particles can have irreversible and asymmetric net displacements (Drazer et al. (2002)). In fact Drazer et al. showed that trajectories of three or more particles are very sensitive to perturbations in initial conditions and exhibit chaotic behavior. This would indicate that direct tracking of particles in a suspension is not productive. Thus, we describe the migration of particles in a suspension via an averaging approach, the diffusion based approach. This particle diffusion is due to a linearly combined effect of spatially varying interaction frequency of the particles and the effect of spatially varying viscosity.

Let's first consider the diffusive flux due to spatially interaction frequency. Imagine that a shear flow (e.g., Couette flow or Poiseuille flow) as a stack of shearing planes sliding past each other, like pages in a book. In such a flow, a collision between particles occurs when two particles embedded in adjacent shearing planes slide past each other. Such collisions can cause a particle (or both particles) to be irreversibly moved from their original streamlines. Notice that a particle experiencing higher frequency of collisions from one direction than the opposite direction will generally be "bumped" but the net effect of many collisions is in the direction of lower collision frequency. (In fact, we may observe that this description of motion is reminiscent of statistical mechanics description of thermophoresis (Soret effect)). We assume that the movement of particles in a concentrated suspension is an affine transformation. In other words, the movement is such that the points (locations of particles), lines (imaginary lines between particle centers), and planes are preserved. While this motion does not preserve the distance between the points (particles) or the distance between the lines, it does preserve the ratios of distances between points of a straight line. Thus, the number of collisions experienced by a test particle will scale as proportional to the shear rate $\dot{\gamma}$, and to the number of particles in the vicinity, and so scale as $\dot{\gamma}\phi$. The higher the shear rate, the higher the movement of adjacent shearing surfaces, and so the larger the number of collisions. Notice that we need to have a significant number of particles per unit volume (high ϕ) for this phenomena to occur. If the suspension is so dilute that particle collisions are rare, the change in collision frequency over a distance of order of particle radius, a is $a\nabla(\dot{\gamma}\phi)$. We assume that the particle migration velocity due to these collisions is proportional to this variation in collision frequency, and that each collision gives a displacement a distance of order particle radius a. This gives us an estimate for the flux, which we call the collision flux,

$$(5.2.5) \qquad \mathbf{N}_c = -K_c a^2 \phi \left(\nabla \left(\dot{\gamma}\phi \right) \right) = -K_c a^2 \left(\phi^2 \nabla \dot{\gamma} + \phi \dot{\gamma} \nabla \phi \right)$$

where K_c is a "fudge factor" order unity accounting for our assumptions above (e.g., the displacement not being exactly the particle radius). This factor is found experimentally. Notice that the flux can be split into two parts: a flux due to gradient in shear rate, and a flux due to a gradient in particle volume fraction. The gradient in shear rate causes particle migration even when the particle volume fraction is uniform. Initially when we inject particles into flow, particles are often well mixed and their distribution is uniform. It this shear rate gradient driven flux that begins their segregation from high shear areas to low shear areas. The gradient in particle volume fraction typically opposes this motion. This flux is more reminiscent of traditional diffusion where species migrate from high concentration to low concentration via a random walk process.

A gradient in volume fraction also produces a gradient in suspension viscosity, since suspension viscosity depends on the local volume fraction. This gradient in viscosity in turn causes an uneven resistance to motion experienced by the two spheres undergoing a collision. The sphere that feels higher viscosity feels more resistance to motion and moves less, while the sphere that feels a slightly lower viscosity feels less resistance and so moves more in a collision. Thus the particles tend to move in the direction of lower viscosity. We assume that the particle velocity due to this phenomena is proportional to the variation in viscosity over the distance of order particle radius, relative to the overall suspension velocity and is so given by $(a/\mu)\nabla\mu$. We again assume each collision causes a displacement of order particle radius. Since the interaction frequency scales as $\dot{\gamma}\phi$, the migration velocity due to this phenomena scales as $\dot{\gamma}\phi(a^2/\mu)\nabla\mu$. To obtain the flux due to this phenomena in terms of particle volume fraction, we multiply this velocity by particle volume fraction ϕ, and use Krieger's relation (5.1.14) for viscosity as a function of volume fraction, and obtain,

$$(5.2.6) \qquad \mathbf{N}_\mu = -K_\mu \dot{\gamma}\phi^2 a^2 \frac{1}{\mu}\frac{d\mu}{d\phi}\nabla\phi$$

where K_μ is again a "fudge factor" order unity accounting for our assumptions above, and is found experimentally. $d\mu/d\phi$ is taken from Krieger's relation (5.1.14). Notice that the viscosity sensitivity to volume fraction $d\mu/d\phi$ is normalized by the magnitude of the viscosity μ. Thus, the viscosity gradient induced migration flux is actually independent of the magnitude of the viscosity. This agrees with experimental observation of Abbott et al. (1991).

Putting the these two fluxes together, we can write a conservation of equation for particles in the frame of reference moving with the fluid (the Lagrangian frame),

$$(5.2.7) \qquad \frac{\partial \phi}{\partial t} = -\nabla \cdot (\mathbf{N}_c + \mathbf{N}_\mu)$$

Converting from a Lagrangian reference frame to a Eulerian reference frame,

$$(5.2.8) \qquad \frac{\partial \phi}{\partial t} + \mathbf{u} \cdot \nabla \phi = -\nabla \cdot (\mathbf{N}_c + \mathbf{N}_\mu)$$

and substituting in (5.2.5) and (5.2.6) we obtain

$$(5.2.9) \qquad \frac{\partial \phi}{\partial t} + \mathbf{u} \cdot \nabla \phi = a^2 K_c \nabla \cdot \left(\phi^2 \nabla \dot{\gamma} + \phi \dot{\gamma} \nabla \phi\right) + a^2 K_\mu \nabla \cdot \left(\dot{\gamma} \phi^2 \frac{1}{\mu} \frac{d\mu}{d\phi} \nabla \phi\right)$$

Note that in deriving (5.2.7) and so (5.2.9) we assumed that migration takes place normal to the shearing surfaces, and so it is only valid for situations where we are interested in predicting migration in unidirectional shear flows, such as Couette flow or Poiseuille flow. Secondly, in deriving (5.2.7) we have assumed that the overall flow takes place on a length scale much larger than the particle length scale (particle radius) so that the large number of movements of order a can take place. Thus, we require that the diameter of the channel or gap between plates to be much larger than the particle radius. Lastly, we account for the fact that particles cannot overlap by prescribing a maximum volume fraction and via Krieger's relation (5.1.14), prescribing that as $\phi \to \phi_m$, $\mu_r \to \infty$.

Here we are specifically interested in Couette flow and Poiseuille flow. For these flows, the gradient in volumetric density is orthogonal to the main flow direction. (In Poiseuille flow in a pipe the main flow is in the axial direction of the pipe, while the gradient in particle volumetric density is in the radial direction). Thus, for these flows we can immediately simplify (5.2.9) to

$$(5.2.10) \qquad \frac{\partial \phi}{\partial t} = a^2 K_c \nabla \cdot \left(\phi^2 \nabla \dot{\gamma} + \phi \dot{\gamma} \nabla \phi\right) + a^2 K_\mu \nabla \cdot \left(\dot{\gamma} \phi^2 \frac{1}{\mu} \frac{d\mu}{d\phi} \nabla \phi\right)$$

Furthermore, the Couette flow we are most interested in is the Couette viscometer (flow in a thin gap between two rotating cylinders) and so even for this problem $\phi = \phi(r, t)$. Similarly, for the Poiseuille flow in a circular pipe that we are interested in $\phi = \phi(r, t)$. Thus we can simplify (5.2.10) to

$$(5.2.11) \qquad \frac{\partial \phi}{\partial t} = \frac{a^2}{r} \frac{\partial}{\partial r} \left(r \left(K_c \left(\phi^2 \frac{\partial \dot{\gamma}}{\partial r} + \phi \dot{\gamma} \frac{\partial \phi}{\partial r} \right) + K_\mu \dot{\gamma} \phi^2 \frac{1}{\mu} \frac{\partial \mu}{\partial \phi} \frac{\partial \phi}{\partial r} \right) \right)$$

where here $\dot{\gamma} = \dot{\gamma}(r)$ is the local shear rate, which we obtain from (5.2.4) and (5.2.1). For the Couette viscometer, $\dot{\gamma} = |\dot{\gamma}_{r\theta}|$ and for Poiseuille pipe flow, $\dot{\gamma} = |\dot{\gamma}_{rz}|$. In both cases the boundary condition on the advection diffusion equation is the no-penetration condition, i.e., no flux at the solid boundaries:

$$(5.2.12) \qquad \mathbf{n} \cdot \left(K_c \phi \nabla (\dot{\gamma} \phi) + K_\mu \dot{\gamma} \phi \frac{1}{\mu} \frac{\partial \mu}{\partial \phi} \nabla \phi \right) = 0$$

where here the unit normal to the boundary $\mathbf{n}$ is purely radial. For these one dimensional shear flows, we solve the continuity and momentum equations to obtain an expression for $\dot{\gamma}$ in terms of $\mu(\phi)$. We then solve (5.2.11) to obtain $\phi = \phi(r, t)$. The time dependent equations can be solved numerically (see Phillips et al. (1992)

for details). Steady state concentration profiles can be obtained analytically. Integrating (5.2.11) with $\partial\phi/\partial t = 0$, we obtain

$$(5.2.13) \qquad K_c\left(\phi^2\frac{\partial\dot{\gamma}}{\partial r} + \phi\dot{\gamma}\frac{\partial\phi}{\partial r}\right) + K_\mu\dot{\gamma}\phi^2\frac{1}{\mu}\frac{\partial\mu}{\partial\phi}\frac{\partial\phi}{\partial r} = 0$$

This equation implies that the total flux everywhere is zero in this suspension, which is not surprising since we are at steady state. Rearranging (5.2.13) we obtain

$$(5.2.14) \qquad \frac{1}{\dot{\gamma}}\frac{d\dot{\gamma}}{dr} + \frac{1}{\phi}\frac{d\phi}{dr} + \frac{K_\mu}{K_c}\frac{1}{\mu}\frac{d\mu}{dr} = 0$$

Integrating this with respect to r we obtain

$$(5.2.15) \qquad \frac{\dot{\gamma}\phi}{\dot{\gamma}_w\phi_w} = \left(\frac{\mu_w}{\mu}\right)^{K_\mu/K_c}$$

where the subscript w designates the value at the wall: the pipe wall for Poiseuille pipe flow, and the inner cylinder wall for the Couette viscometer. While experimental observations of ϕ_w, for example of Hampton et al. (1997), show that ϕ_w can approach zero, we see that this model cannot exactly reproduce this. Thus, it is best to use ϕ_w as a fitting parameter to ensure that cross-sectional average volume fraction agrees with that experimentally observed. Then, as shown by Hampton et al. (1997) this model predicts both the volume fraction for most of the channel correctly, except at the walls. The model also correctly predicts the experimentally observed blunted velocity profile in slurry flow. The discrepancy at the wall may be due to the fact that since volume fraction the wall is quite low, the mechanism of particle transport there is different from the collisional particle transport this model is based on. For example, transport in this region may be more akin to single particle dynamics discussed in Section 4.1. K_μ/K_c is another fitting parameter to be experimentally determined. Phillips et al. (1992) found $K_\mu/K_c = 0.66$ for both Couette and Poiseuille flow. Individual values of K_μ and K_c can be obtained from analysis of the flow transient (numerically calculated) and we refer the reader to Phillips et al. (1992) for details.

To find the explicit relations for volume fraction profiles we need to obtain the relevant shear rates by solving the momentum equations. We begin with the Couette viscometer flow (flow in the gap between concentric rotating cylinders). The simplified momentum equation, accounting for the fact that the flow is steady and one dimensional (in the ϑ direction), becomes

$$(5.2.16) \qquad \frac{1}{r^2}\frac{\partial}{\partial r}\left(r^2\tau_{r\theta}\right) = 0$$

Substituting in (5.2.1) and integrating we obtain

$$(5.2.17) \qquad \dot{\gamma}_{r\theta} = \frac{C_1}{r^2\mu}$$

where C_1 is a constant of integration. Since

$$(5.2.18) \qquad \dot{\gamma}_{r\theta} = r\frac{\partial}{\partial r}\left(\frac{u_\theta}{r}\right)$$

we can set

$$(5.2.19) \qquad r\frac{\partial}{\partial r}\left(\frac{u_\theta}{r}\right) = \frac{C_1}{r^2\mu}$$

Integrating this equation and imposing the boundary condition that the inner cylinder with radius κR spins with angular velocity Ω, while there is no-slip at the outer cylinder radius R, we obtain

$$(5.2.20) \qquad \frac{C}{\Omega} = -\left(\int_{\kappa R}^{R} \frac{1}{r^3 \mu} dr \right)^{-1}$$

Substituting this back into (5.2.17) we obtain

$$(5.2.21) \qquad \frac{\dot{\gamma}_{r\theta}}{\Omega} = -\left(\left(\frac{r}{R}\right)^2 \mu_r \int_{\kappa}^{1} \frac{1}{r^3 \mu_r} dr \right)^{-1}$$

Now we are ready to obtain the steady state distribution for the volume fraction. We substitute (5.2.17) and (5.1.14) into (5.2.15) and obtain

$$(5.2.22) \qquad \frac{\phi}{\phi_w} = \left(\frac{r}{R}\right)^2 \frac{1}{\kappa^2} \left(\frac{1 - \phi_w/\phi_m}{1 - \phi/\phi_m} \right)^{1.82(1 - K_\mu/K_c)}$$

Phillips et al. (1992) found $K_\mu/K_c = 0.66$ for Couette flow, and so we can approximate the exponent $1.82\left(1 - K_\mu/K_c\right)$ as -1, and so obtain an explicit expression for the volume fraction,

$$(5.2.23) \qquad \phi = \frac{\phi_m (r/R)^2}{(r/R)^2 + \alpha \kappa^2}$$

where

$$(5.2.24) \qquad \alpha = \frac{\phi_m - \phi_w}{\phi_w}$$

Integrating (5.2.23) we find the cross-sectional area averaged volume fraction,

$$(5.2.25) \qquad \bar{\phi} = \frac{2\phi_m}{1 - \kappa^2} \left(\frac{1 - \kappa^2}{2} - \frac{\alpha \kappa^2}{2} \ln\left(\frac{1 + \alpha \kappa^2}{\kappa^2 (1 + \alpha)} \right) \right)$$

Typically we use this equation to find α and so ϕ_w from some experimental $\bar{\phi}$ to match the experimental data to the predicted profile.

Now let's obtain the volume fraction profile for the Poiseuille pipe flow. The momentum equation, due to the fact that the flow is steady and one dimensional (in the z direction), becomes

$$(5.2.26) \qquad \frac{1}{r} \frac{d}{dr} (r\tau_{rz}) = -\frac{dp}{dz}$$

Integrating this equation via (5.2.1) we obtain

$$(5.2.27) \qquad \dot{\gamma}_{rz} \left(\frac{2\mu_r}{R(dp/dz)} \right) = \frac{r}{R} \frac{1}{\mu_r}$$

Once again substituting (5.2.27) and (5.1.14) into (5.2.15), we obtain

$$(5.2.28) \qquad \frac{\phi}{\phi_w} = \left(\frac{r}{R}\right)^{-1} \left(\frac{1 - \phi_w/\phi_m}{1 - \phi/\phi_m} \right)^{1.82(1 - K_\mu/K_c)}$$

Once again Phillips et al. (1992) found $K_\mu/K_c = 0.66$ for Poiseuille flow as well. and so we can approximate the exponent $1.82\,(1 - K_\mu/K_c)$ as -1, and so obtain an explicit expression for the volume fraction,

$$(5.2.29) \qquad \phi = \frac{\phi_m}{1 + \alpha\,(r/R)}$$

where

$$(5.2.30) \qquad \alpha = \frac{\phi_m - \phi_w}{\phi_w}$$

Once again integrating (5.2.29) find the cross-sectional area averaged volume fraction,

$$(5.2.31) \qquad \bar{\phi} = \frac{2\phi_m}{\alpha}\left(1 - \frac{1}{\alpha}\ln\left(1 + \alpha\right)\right)$$

We plot the volume fraction profile for several initial volume fractions for Poiseuille flow along with experimental data from Hampton et al. (1997) in Figure 5.2.1.

Notice that at the center of the channel the volume fraction practically equals the maximum volume fraction ϕ_m and a peak of volume fraction surrounds the center of the channel, in both theoretical predictions and Hampton et al. (1997) results. The overall width of this peak expands as the initial volume fraction increases. This peaking phenomena can be explained by the fact the variation in shear rate in Poiseuille pipe flow is such that it always forces particles towards the center of the pipe. Away from the center, this flux is mostly canceled by an opposing flux driven by a gradient in volume fraction, proportional to $\dot{\gamma}_{rz}\,(d\phi/dr)$ as per (5.2.5). However, at the center of the pipe $\dot{\gamma}_{rz}$ is equal to zero, so this opposing flux goes to zero. This results in particle flux towards the center of the pipe and accumulation of particles there. The increased viscosity at the center of the pipe leads to blunting of the velocity profile. For a model and experimental observations of volume fraction profiles in a rectangular channel see Shauly et al. (1997).

5.3. Slurries of non-neutrally buoyant hard spheres

Flowing slurries of non-neutrally buoyant spheres experience a gravity driven flux in addition to the flux driven by particle interaction frequency gradient and the viscosity gradient that we have discussed in the previous section. In this section, we discuss a model for the flow of non-neutrally buoyant hard sphere slurry in Poiseuille flow in a circular pipe due to Zhang and Acrivos (1994). Specifically we discuss Zhang and Acrivos' extension of the Phillips et al. (1992), model that we have used in the previous section.

The specific problem that we are interested in is the equilibrium (steady state) distribution of particles flowing in a pipe. The particles enter the pipe well mixed and so their distribution is uniform across the cross-section of the pipe. As the particles move downstream, gravity, acting perpendicular to the axis of the pipe causes the particles to settle (or float up, depending on their density). As they settle, their concentration at the bottom of the pipe increases, which firstly, creates a particle gradient that creates particle diffusion towards the center of the pipe; secondly, this increases the number of particles in the high shear region, which again creates a particle gradient towards the center (low shear region) of the pipe; thirdly, this increases the viscosity at the bottom of the pipe, and so again generates a viscosity gradient, which again moves particles toward the center of the pipe. It

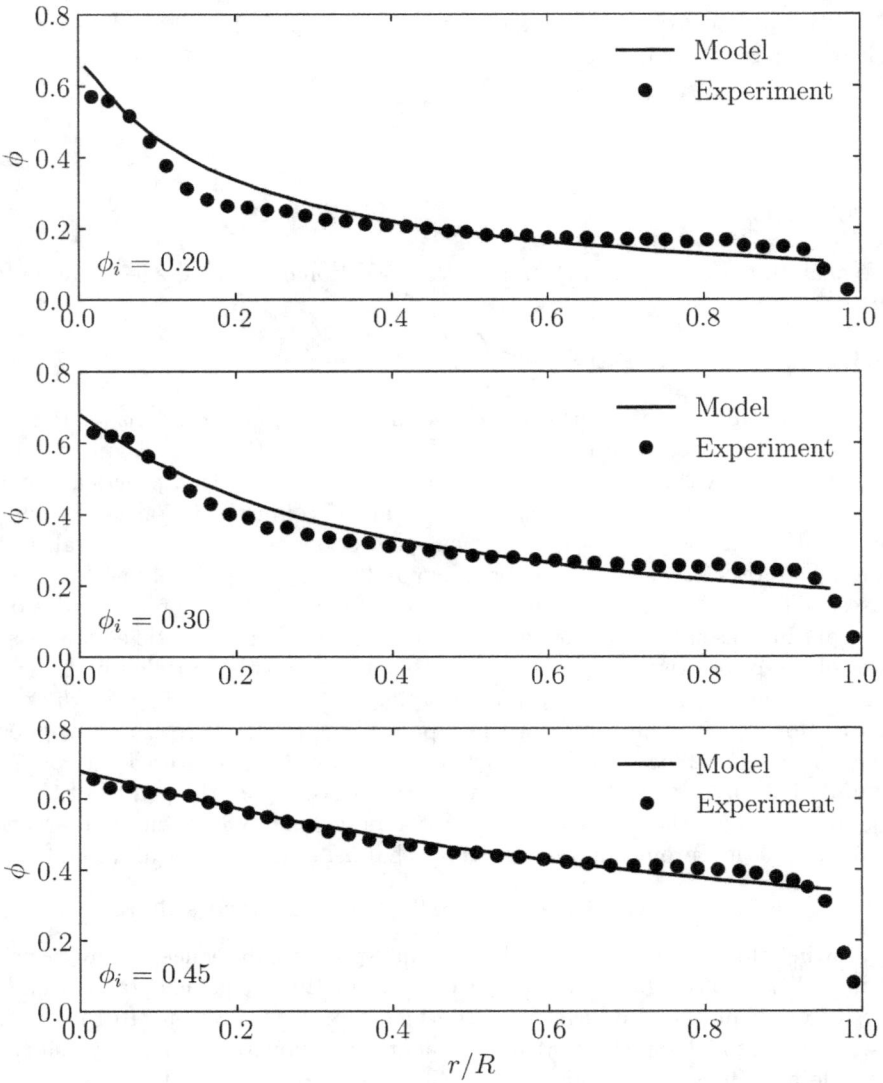

FIGURE 5.2.1. Equilibrium (fully developed) volume fraction distributions predicted by Phillips et al. (1992) model (solid line) compared with data from Hampton et al. (1997), for initial volume fractions of 0.2, 0.3, and 0.45. The particle radius to pipe radius in this flow was 0.0256. There is good agreement between model and experiment except near the pipe wall, where volume fraction is especially low.

appears that the gravity driven flux of particles is opposed by all the three fluxes we considered in the previous section. Thus we expect a steady state volume fraction profile to develop in a pipe. Indeed such a profile is observed experimentally: at the bottom of the pipe there is a layer of high concentration of particles; above it is a second, more diffuse layer of particles; and above that is a third layer that is clear of particles.

To understand how quickly the steady state is reached, we apply the scaling approach proposed by Nott and Brady (1994). The length scale for a particle to travel via a diffusive process scales as

$$(5.3.1) \qquad\qquad y \sim \sqrt{Dt}$$

where D is the diffusivity of the particle. We can claim that steady state is reached once the particles had a chance to diffuse a distance that scales with the radius of the pipe, R. Hence,

$$(5.3.2) \qquad\qquad t_{ss} \sim R^2/D$$

From the previous section (5.2.10), we saw that the particle diffusivity scales as

$$(5.3.3) \qquad\qquad D \sim \dot{\gamma}a^2$$

The shear rate in turn scales as

$$(5.3.4) \qquad\qquad \dot{\gamma} \sim U/R$$

where U is the characteristic velocity of the flow. Substituting (5.3.4) and (5.3.3) into (5.3.2), we obtain,

$$(5.3.5) \qquad\qquad t_{ss} \sim \frac{R^3}{a^3}\frac{a}{U}$$

Thus the time scale to reach steady state scales as the advective time scale, a/U times the cube of the ratio of particle size to channel size. Since the particle has to be smaller than the channel (typically much smaller) the time scale to reach steady state is much, much larger than the advective time scale. To look at it from another point of view, we can rewrite (5.3.5) as the length the particle has to travel to reach steady state, the equilibration length (or "entrance" length),

$$(5.3.6) \qquad\qquad \frac{L_{ss}}{R} \sim \frac{R^2}{a^2}$$

Thus we see that equilibration can take many tube radii. For example, for 10 µm particles traveling in a 500 µm radii tube, would only reach steady state after 1.25 m!

The combined particle collision gradient and viscosity gradient driven fluxes, (5.2.5) and (5.2.6) can be rewritten as

$$(5.3.7) \qquad\qquad \mathbf{N}_d = -D_c\nabla\phi - D_s\nabla\dot{\gamma}$$

where D_c is the volume fraction gradient diffusion coefficient,

$$(5.3.8) \qquad\qquad D_c = a^2\dot{\gamma}\left(0.43\phi + 0.65\phi^2\frac{1}{\mu}\frac{d\mu}{d\phi}\right)$$

and D_s is the shear gradient diffusion coefficient,

$$(5.3.9) \qquad\qquad D_s = 0.43a^2\phi^2$$

Here we take these diffusion coefficients as scalars, for simplicity. However, in reality these should be second order tensors, as it is has been experimentally shown by Leighton and Acrivos (1987) that even in unidirectional flows the diffusion coefficient is different if the diffusion is orthogonal to the plane of shear or parallel to it. As before, the shear rate $\dot{\gamma}$ is defined as the square root of the second invariant of the rate of deformation tensor $\mathbf{d}$, such that

$$(5.3.10) \qquad \dot{\gamma} = \sqrt{2\mathbf{d} : \mathbf{d}}$$

$$(5.3.11) \qquad \mathbf{d} = \left(\nabla\mathbf{u} + \nabla\mathbf{u}^T\right)/2$$

where, as a reminder, : is the double dot product (see, for example, Appendix A).

The gravity driven flux is given by the product of the particle volume and the Stokes settling velocity of an isolated sphere and "particle hindrance" function, which accounts for the reduction of settling velocity due to particle interactions. As a reminder, Stokes settling velocity is obtained from setting equal the force of gravity on the particle to Stokes' drag on the particle (see Section 1.3). Note that the hindrance function is not supposed to account for the interactions described by the particle collision gradient and viscosity gradient driven fluxes. Thus, the gravity driven flux can be written as

$$(5.3.12) \qquad \mathbf{N}_g = \frac{2}{9}\frac{a^2\left(\rho_p - \rho_l\right)}{\mu_0}\phi\left[\frac{1-\phi}{\mu}\right]\mathbf{g}$$

where ρ_p is the density of the particle, ρ_l and μ_0 is the density and viscosity of the liquid without particles, $\mathbf{g}$ is the gravity vector, and the term in the square brackets is the particle hindrance function.

Here we are interested in Poiseuille flows where the particle Reynolds number is relatively small and where the suspension can be modeled as a continuum Newtonian fluid, but with particle concentration dependent properties. Thus we model the flow with incompressible, creeping flow equations. Furthermore, we take advantage of the fact that Poiseuille flow is predominantly two dimensional and so consider the flow velocity and particle concentration variations in two dimensions. We denote the two dimensions in the cross-section of the pipe as x_1 and x_2, where x_2 is the coordinate aligned with gravity (vertical direction). x_3 is the direction aligned with the axis of the pipe.

In this flow, we have L as the characteristic length scale, and U as the characteristic velocity scale. Thus, we can write the continuity equation as

$$(5.3.13) \qquad \frac{\partial u_j}{\partial x_j} = 0$$

where we are employing the Einstein summation notation. We can write the momentum equation for the first direction x_1 as

$$(5.3.14) \qquad Re\left(1 + \varepsilon\phi\right)\left(\frac{\partial u_1}{\partial t} + u_j\frac{\partial u_1}{\partial x_j}\right) = \frac{\partial}{\partial x_j}\left(p\delta_{1j} + 2\mu d_{1j}\right)$$

and the momentum in the vertical direction (alighted with gravity) as

$$(5.3.15) \qquad Re\left(1 + \varepsilon\phi\right)\left(\frac{\partial u_2}{\partial t} + u_j\frac{\partial u_2}{\partial x_j}\right) = \frac{\partial}{\partial x_j}\left(p\delta_{2j} + 2\mu d_{2j}\right) - \frac{\phi}{Fr}$$

Lastly, the momentum equation into the axial direction is

$$(5.3.16) \qquad Re\left(1 + \varepsilon\phi\right)\left(\frac{\partial u_3}{\partial t} + u_j \frac{\partial u_3}{\partial x_j}\right) = K + \frac{\partial}{\partial x_j}\left(2\mu d_{3j}\right)$$

Here δ_{ij} is the Kronecker delta and K is the dimensionless pressure drop per unit length of the pipe. The advection diffusion equation is then

$$(5.3.17) \qquad \frac{\partial \phi}{\partial t} + u_j \frac{\partial \phi}{\partial x_j} = \lambda \frac{\partial}{\partial x_j}\left(\frac{D_c}{a^2}\frac{\partial \phi}{\partial x_j} + \frac{D_s}{a^2}\frac{\partial \dot{\gamma}}{\partial x_j} + \frac{2}{9}\frac{1}{Fr}\phi\left[\frac{1-\phi}{\mu}\right]\delta_{j2}\right)$$

This system of equations is governed by four dimensionless parameters: the pipe scale Reynolds number,

$$(5.3.18) \qquad Re = \frac{\rho_l U L}{\mu_0}$$

the modified Frounde number,

$$(5.3.19) \qquad Fr = \frac{\mu_0 U}{\left(\rho_p - \rho_l\right) g L^2}$$

the relative density ratio of the particles to that of the pure liquid,

$$(5.3.20) \qquad \varepsilon = \frac{\rho_p - \rho_l}{\rho_l}$$

and the square of the ratio of particle size to the characteristic length scale,

$$(5.3.21) \qquad \lambda = \left(a/L\right)^2$$

Lastly, we impose the no-slip and no-penetration condition at the pipe walls, and no particle flux on wall,

$$(5.3.22) \qquad \left(\mathbf{N}_g + \mathbf{N}_d\right)_j \mathbf{n}_j = 0$$

Zhang and Acrivos solved these equations numerically using the finite element method, for $\varepsilon = 0.2$ and λ. In Figure 5.3.1 we plot the normalized axial velocity ($u_3/u_{3,\mathrm{max}}$) as well as the particle volume fraction ϕ for cross-sectional averaged (i.e. initial) volume fraction $\bar{\phi}$ of 0.1, 0.2, 0.3 and channel Reynolds number of 5, 10, 20 (Frounde number 0.25, 0.5, 1). We see that in most cases a region of clear fluid (devoid of particles) appears at the top of the pipe when the flow rate (i.e. Reynolds and Frounde numbers) are low. When flow rate increases the region that is devoid of particles shrinks Increasing the initial volume fraction of particles also decreases the region devoid of particles. We see in Figure 5.3.1 also that the velocity profiles are blunted in the center. The blunting is especially obvious the higher the initial particle concentration. The higher the initial particle volume fraction, the higher the peak of axial velocity shifts above the center line. This can be explained by the fact that the higher initial volume fraction, the higher the concentration of particles at the bottom of the pipe, and so there is higher resistance towards the bottom of the pipe, and so the majority of the flow shifts upward. We can also observe that when $\bar{\phi}$ is high particles form a concentrated core centered slightly above the pipe centerline and roughly coincident with the point of highest velocity. This effect is enhanced at high flow rates. This effect can be explained by the fact that at higher particle volume fractions and higher flow rates (and so higher shear rates) the combined particle collision gradient and viscosity gradient driven fluxes dominate the gravity driven flux and so work to concentrate the particles at the point of highest velocity. When the mean particle volume fraction is too low to

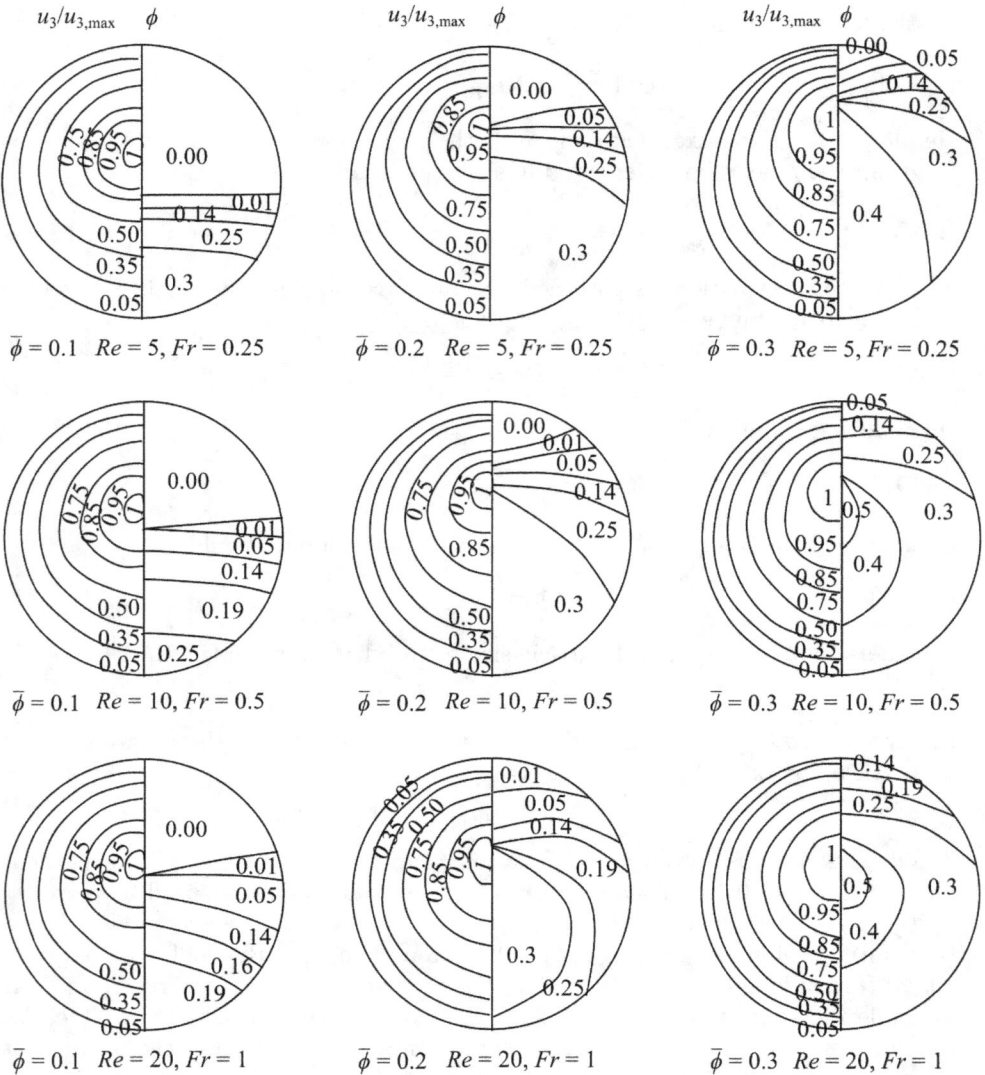

FIGURE 5.3.1. Contours of axial velocity magnitude scaled by the maximum velocity (left part of the circle) and contours of particle volume fraction (right part of the circle) as a function of Reynolds and Frounde numbers and initial (mean) volume fraction. As the flow rate increases (increasing Reynolds and Frounde numbers proportionally) the portion of the pipe devoid of particles decreases. Notice from the contours of the velocity that the velocity profile is blunted in the center.

support adequate particle collision gradient and viscosity gradient driven fluxes the equilibrium particle concentration decreases monotonically from the bottom to the top of the pipe. The results of Zhang and Acrivos's model agree well qualitatively with experimental results, such as those of Altobelli et al. (1991).

5.4. Multicomponent slurries with neutrally buoyant particles

So far, we have discussed particle migration in a slurry with one type of particle. However, practical slurries are often multicomponent – they have particles of different sizes. Here we discuss the model of Shauly et al. (1998) that extends Phillips et al. (1992) model to a slurry with multiple particle sizes. In additional to the particle collision gradient and viscosity gradient driven fluxes Shauly et al. incorporated a streamline curvature induced flux proposed by Krishnan et al. (1996). While streamlines do not have curvature in Poiseuille flow in straight pipes (the primary flow of interest to us), they do curve in curved pipes as well as other geometries, hence we discuss it here for generality.

We can understand how the streamline curvature induced flux arises by carefully examining the collision that gives rise to the other two fluxes. Under purely creeping flow conditions, when two perfectly smooth particles go for a collision, they never actually touch each other. As they approach each other, a film of liquid must be squeezed out form the space between them. From lubrication theory (Section 2.2) we know that the thinner the gap, the larger the pressure in the gap. As the gap approaches zero (particle contact) the pressure in the gap approaches infinity. Hence this model predicts that particles will never contact, and rather push each other apart as they approach. However, in fact this pushing apart is all we need for a "collision". Real particles have roughness. Experimental observations of Arp and Mason (1977) and Leighton and Rampall (1993) suggest that presence of roughness also tends to keep particles apart when two of them approach in a collision. In curved geometry, we may visualize this repulsion force acting on a line between the centers of the two particles. This force can be broken into two components: one acting along the direction of the local streamlines (tangent to the streamline) and one perpendicular to this direction (i.e. radially outward from the streamline) (see Figure 5.4.1). Once the collision is complete, the particle float away from each other on their new streamlines. However, the component of the interaction force that was radially outward from the streamline causes the radially outward migration in a curved streamline shear flow. This component of the force is responsible for the streamline curvature induced flux. Just like the collision gradient induced flux, this flux scales as the collision frequency $\dot{\gamma}\phi$, as the greater the number of collisions, the greater the number of times the particles feel this radial force. Secondly, just like for the collision gradient induced flux, we expect the distance the particle moves is order particle radius a. Thirdly, we also expect the distance the particle moves scale as the ratio of streamline curvature to particle curvature, a/r_s where r_s is the local radius of curvature of the streamlines. Finally this flux should be proportional to the particle volume fraction. Putting these scaling arguments together, we obtain that the streamline curvature induced flux is

$$(5.4.1) \qquad \mathbf{N}_r = K_r \dot{\gamma} a \phi^2 \frac{a}{r_s} \mathbf{n}_s$$

where K_r is another "fudge" factor to be experimentally determined, and $\mathbf{n}_s$ is a unit vector pointing in the radially outward direction relative to the streamlines. We can denote the ratio of streamline curvature to particle curvature a/r_s as R_s and write (5.4.1) in gradient form as

$$(5.4.2) \qquad \mathbf{N}_r = -K_r \dot{\gamma} a \phi^2 \nabla \left(\ln R_s \right)$$

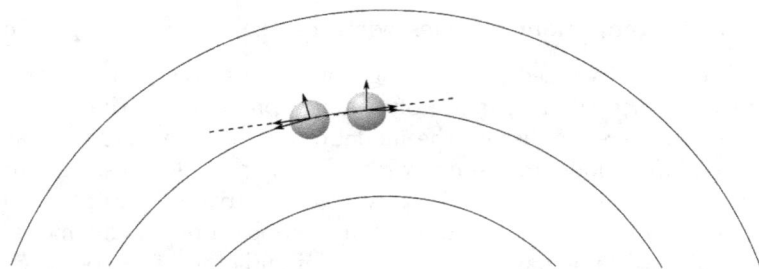

FIGURE 5.4.1. Schematic of particles colliding while traveling on curving streamlines. Because of the curvature of the streamlines, particles experience an additional force in the direction radially outward from the streamline. The dashed line is a line connecting the particle centers. Arrow sizes are exaggerated for clarity.

Combining the particle collision gradient driven flux, the viscosity gradient driven flux, and the streamline curvature induced flux together we obtain the total flux

$$(5.4.3) \qquad \mathbf{N} = -K_c a^2 \phi \nabla \left(\dot{\gamma} \phi \right) - K_\mu \dot{\gamma} \phi^2 a^2 \frac{1}{\mu} \frac{d\mu}{d\phi} \nabla \phi - K_r \dot{\gamma} a \phi^2 \nabla \left(\ln R_s \right)$$

Krishnan et al. (1996) found that $K_r \approx K_c$, and we can use this fact to combine the terms in (5.4.3) as

$$(5.4.4) \qquad \mathbf{N} = -K a^2 \dot{\gamma} \phi^2 \nabla \left(\ln \left(\phi \dot{\gamma} \mu^\lambda R_s \right) \right) = -D \nabla P$$

$$(5.4.5) \qquad\qquad D = K a^2 \dot{\gamma} \phi^2$$

$$(5.4.6) \qquad\qquad P = \ln \left(\phi \dot{\gamma} \mu^\lambda R_s \right)$$

where $K = K_r \approx K_c$, $\lambda = K_\mu / K$ and P is the so called "particle migration potential". After re-examining Phillips et al. data Sahuly et al. found that $\lambda \approx 2$.

Let's now derive the three flux relations for when particles are of different sizes, beginning with the collision gradient driven flux. The frequency of interaction of particles of species i is now $\dot{\gamma} \phi_i$. Once again we take the interaction frequency over distance $O\left(a_i \right)$ to scale as $a_i \nabla \left(\dot{\gamma} \phi_i \right)$. We take the flux resulting from the interaction of particle of species i with particles of any species to be $a_i a_j \phi_j \nabla \left(\dot{\gamma} \phi_i \right)$. Since particles of species i interact with all particles in the flow, we must sum this flux over all species j

$$(5.4.7) \qquad\qquad \mathbf{N}_{c,i} = -K_c a_i \sum_j a_j \phi_j \nabla \left(\dot{\gamma} \phi_i \right)$$

For the viscosity gradient driven flux, the relevant length scale associated with variation in viscosity and particle displacement is the weighted average particle length scale, given by

$$(5.4.8) \qquad\qquad \bar{a} = \frac{\displaystyle\sum_j a_j \phi_j}{\phi_\Sigma}$$

$$(5.4.9) \qquad\qquad \phi_\Sigma = \sum_j \phi_j$$

Following the arguments that we used to derive (5.2.6), we can similarly obtain that the viscosity gradient driven flux for a particular particle species i is

$$(5.4.10) \qquad \mathbf{N}_{\mu,i} = -K_\mu \bar{a}^2 \dot{\gamma} \phi_\Sigma \phi_i \frac{1}{\mu} \nabla \mu$$

Similarly, following the arguments that we used to derive (5.4.2), we can derive the expression for streamline curvature induced flux for a particular particle species i to be

$$(5.4.11) \qquad \mathbf{N}_{r,i} = -K_r \dot{\gamma} \bar{a} a_i \phi_\Sigma \phi_i \nabla \left(\ln R_s\right) g\left(\frac{a_i}{\bar{a}}\right)$$

where $g(x)$ is an experimentally determined function of the form $g(x) = x^q$ for $q > 0$ as suggested by Shauly et al. Furthermore, Shauly et al. suggests that the coefficients K_c, K_μ, K_r are independent of the particle properties and are in fact as those in a slurry with a single particle type. Once again we can combine these three fluxes into a single equation for the total flux of a particular particle species i

$$(5.4.12) \quad \mathbf{N}_i = -K\dot{\gamma} \bar{a} a_i \phi_\Sigma \phi_i \left(\nabla \left(\ln \left(\dot{\gamma}\phi_i\right)\right) + \frac{\bar{a}}{a_i} \nabla \left(\ln \left(\mu^\lambda\right)\right) + \left(\frac{a_i}{\bar{a}}\right)^q \nabla \left(\ln R_s\right) \right)$$

We can rewrite this in terms of a diffusion coefficient and a migration potential as

$$(5.4.13) \qquad \mathbf{N}_i = D_i \left(\nabla P_i - \left(1 - \frac{\bar{a}}{a_i} \nabla \left(\ln \left(\mu^\lambda\right)\right)\right) - \left(1 - \left(\frac{a_i}{\bar{a}}\right)^q\right) \nabla \left(\ln R_s\right) \right)$$

where

$$(5.4.14) \qquad D_i = -K\dot{\gamma}\bar{a}a_i\phi_\Sigma\phi_i$$

$$(5.4.15) \qquad P_i = \ln\left(\phi_i \dot{\gamma} \mu^\lambda R_s\right)$$

Finally the total flux of particle is

$$(5.4.16)$$

$$\mathbf{N} = \sum_i \mathbf{N}_i = -K\dot{\gamma}\bar{a}^2 \phi_\Sigma^2 \left(\nabla \left(\ln \left(\bar{a}\phi_\Sigma \dot{\gamma}\mu^\lambda R_s\right)\right) - \left(1 - \frac{\overline{a^{q+1}}}{\bar{a}^{q+1}}\right) \nabla \left(\ln R_s\right) \right)$$

where we have used the definition that

$$(5.4.17) \qquad \overline{a^q} = \frac{\sum_j a_j^q \phi_j}{\phi_\Sigma}$$

The first term in the total flux of a multicomponent slurry is reminiscent of that for a single particle slurry, (5.4.4). The second term, on the other hand may retard or enhance the total flux, depending on the distribution of particles in the suspension.

Let's now translate this result for the total flux to a continuous particle distribution. Firstly, we define $\phi(a) \, da$ the differential volume fraction of size a. Secondly we define the nondimensional moments of the distribution as

$$(5.4.18) \qquad \Phi^{(n)} = \int_0^\infty a^n d\phi = \int_0^\infty a^n \phi(a) \, da$$

Note that $\Phi^{(0)}$ is the total particle volume fraction. Also note that for a discrete distribution,

$$(5.4.19) \qquad \Phi^{(n)} = \sum_j a_j^n \phi_j$$

so that

$$(5.4.20) \qquad\qquad \Phi^{(0)} = \phi_\Sigma$$

We can obtain the fluxes of the moments by multiplying (5.4.13) by a^n and integrating over all particle sizes,

(5.4.21)

$$\mathbf{N}^{(n)} = \int_0^\infty a^n \mathbf{N}(a)\, da$$

$$= \int_0^\infty a^n D(a)\left(\nabla P(a) - \left(1 - \frac{\bar{a}}{a}\nabla\left(\ln\left(\mu^\lambda\right)\right)\right) - \left(1 - \left(\frac{a}{\bar{a}}\right)^q\right)\nabla\left(\ln R_s\right)\right) da$$

where now

$$(5.4.22) \qquad\qquad D(a) = -K\dot{\gamma}\bar{a}\phi(a)\,\Phi^{(1)}$$

$$(5.4.23) \qquad\qquad P(a) = \ln\left(\phi(a)\,\dot{\gamma}\mu^\lambda R_s\right)$$

If we can express the effective viscosity of the suspension in terms of the distribution of moments (see for example, Mwasame et al. (2016)), we can express the fluxes of moments as

$$(5.4.24) \quad \mathbf{N}^{(n)}$$

$$= -D^{(n)}\left(\nabla P^{(n)} - \left(1 - \frac{\Phi^{(1)}\Phi^{(n)}}{\Phi^{(0)}\Phi^{(n+1)}}\nabla\left(\ln\left(\mu^\lambda\right)\right)\right) - \left(1 - \frac{\Phi^{(n+q+1)}}{\Phi^{(1)}\Phi^{(n+1)}}\right)\nabla\left(\ln R_s\right)\right)$$

Where

$$(5.4.25) \qquad\qquad D^{(n)} = K\dot{\gamma}\Phi^{(1)}\Phi^{(n+1)}$$

$$(5.4.26) \qquad\qquad P^{(n)} = \ln\left(\Phi^{(n+1)}\dot{\gamma}\mu^\lambda R_s\right)$$

This allows us to track the evolution of the particle distribution with the flow as a function of the initial particle distribution, flow geometry, and various viscosity models.

Let's now turn back a discrete distribution of particle sizes and find the equilibrium (steady state) distribution of particles. We know that at steady state the net flux of each of the particle species is zero and the total flux of species is zero. Thus, from (5.4.13)

$$(5.4.27) \qquad \nabla P_i - \left(1 - \frac{\bar{a}}{a_i}\nabla\left(\ln\left(\mu^\lambda\right)\right)\right) - \left(1 - \left(\frac{a_i}{\bar{a}}\right)^q\right)\nabla\left(\ln R_s\right) = 0$$

and from (5.4.16)

$$(5.4.28) \qquad \nabla\left(\ln\left(\bar{a}\phi_\Sigma\dot{\gamma}\mu^\lambda R_s\right)\right) - \left(1 - \frac{\overline{a^{q+1}}}{\bar{a}^{q+1}}\right)\nabla\left(\ln R_s\right) = 0$$

Taking the difference between the zero fluxes of any two species j and k we obtain another equation,

$$(5.4.29) \qquad \nabla\left(\ln\left(\frac{\phi_j^{a_j}}{\phi_k^{a_k}}\dot{\gamma}^{a_j - a_k}\right)\right) + \frac{a_j^{q+1} - a_k^{q+1}}{\bar{a}^q}\nabla\left(\ln R_s\right) = 0$$

These equations are subject to the additional constraint that the total volume fraction remains conserved. Thus, the total volume fraction of each species integrated over the migration domain must remain unchanged. For pipe flow this means

$$(5.4.30) \qquad \int_A u_z \phi_i dA = Q_i$$

where we have taken the integral across the cross-sectional area of the pipe A; here u_z is the axial velocity in the pipe normalized by the mean velocity in the pipe; Q_i represents the total amount of species. To obtain a solution for a steady state volume fraction distributions in Poiseuille flow in a straight pipe we take advantage of the fact that in a pipe streamlines are not curved and so $\nabla (\ln R_s) = 0$. Applying this simplification to (5.4.28) and (5.4.29) and integrating each, we obtain

$$(5.4.31) \qquad \bar{a}\phi_\Sigma \gamma \mu^\lambda \dot{} = C$$

$$(5.4.32) \qquad \frac{\phi_j^{a_j}}{\phi_k^{a_k}}\gamma^{a_j - a_k}\dot{} = C_{jk}$$

where C and C_{jk} are constants found from applying boundary conditions. We can compare (5.4.31) and (5.4.32) to (5.2.15). Note that to calculate the volume fraction distribution, we need to calculate the shear rate distribution as we have done in Section 5.2. We also need a model for effective viscosity of a multicomponent slurry. Models for effective viscosity of a multicomponent slurries remain an active area of research (see for example Mwasame et al. (2016)). To get a sense for the effect of polydispersity on effective viscosity and so on particle distribution we will continue to use Krieger's model, (5.1.14), but with the maximum volume fraction determined by Probstein et al. (1994) model,

where

$$(5.4.33) \qquad \frac{\phi_m}{\phi_{m0}} = 1 + \frac{3}{2}|b|^{3/2}\left(\frac{\phi_1}{\phi_\Sigma}\right)^{3/2}\left(\frac{\phi_2}{\phi_\Sigma}\right)$$

where

$$(5.4.34) \qquad b = \frac{a_1 - a_2}{a_1 + a_2}$$

This model predicts that the maximum total volume fraction is higher for a bidisperse suspension than a monodisperse suspension. The reason for this is that the smaller particles can fit in the "nooks and crannies" between the larger particles that would otherwise be filled with the solvent. Shauly et al. report that this model agrees well with experimental data for maximum volume fraction.

While here we have developed an expression for the volume fraction distribution for Poiseuille flow in a straight pipe, Shauly et al. also provides similar relations for flow between parallel plates and for flow in a Couette concentric cylinder viscometer. For the viscometric flows Shauly et al. observe good agreement between model predictions and experimental results. Moreover, Shauly et al. observe that this model predicts particle segregation based on particle size, with the large particles preferentially segregating in the low shear rate region while the smaller particles accumulate in the higher shear rate regions. Such segregation is indeed experimentally observed. Finally, while we here considered the suspension of neutrally buoyant particles, Shauly et al. (2000) extended the model discussed here to non-neutrally buoyant particles.

5.5. Blood flow in capillaries

So far we have discussed viscosity and migration of particles in model slurries. One of the most important practical example of such slurry flow that combines many of the individual properties that we have discussed in previous sections is the flow of blood through capillaries. Blood is a slurry composed of a predominantly Newtonian aqueous solvent in which predominantly three types of particles are suspended. These particles are red blood cells (erythrocytes), white blood cells (leukocytes), and platelets (thrombocytes). These particles differ in size (the slurry is polydisperse), are deformable, and are non-neutrally buoyant. Erythrocytes, leukocytes, and thrombocytes have very different mechanical properties. Human erythrocytes have a shape of a biconcave disk, with a diameter of 6.2-8.2 µm, with a thickness at the thickest point of 2-2.5 µm, and with a thickness at the thinnest point of 0.8-1 µm (Turgeon (2005)). They have an average volume of 90 fL, a surface area of 136 µm^2 (McLaren et al. (1993)). Mature erythrocytes (the vast majority of them) do not have a nucleus. The typical erythrocyte volume fraction in human blood is 40-45% and they are by far the most common cell in the blood. In capillaries erythrocytes marginate to the center of the capillary.

Leukocytes are larger and much more spherical than erythrocytes. Their diameter rangers from 10-12 µm. Leukocytes is a class of cells made up of neutrophils (62%, diameter 10-12 µm), small (7-8 µm) and large (12-15 µm) lymphocytes (30%, density 1.073-1.077 g/ml), monocytes (5.3%, diameter 15-30 µm, density 1.067-1.077 g/ml), eosinophils (2.3%, diameter 10-12 µm), and basophils (62%, diameter 12-15 µm) (Daniels et al. (1979); Zipursky et al. (1976)). In healthy humans, leukocytes are a thousand times less abundant than red blood cells. Leukocytes are also substantially stiffer than erythrocytes (erythrocytes: Rand and Burton (1964); leukocytes: Schmid-Schonbein et al. (1981)). In capillaries leukocytes marginate to the walls of the capillary. This margination appears to decrease with increasing shear rate and is a weak function of hematoctrit for normal and lower hematocrit values (Kumar and Graham (2012)).

Platelets (thrombocytes) are anucleate cell fragments and are much smaller than both leukocytes and erythrocytes. They are discoid in shape with a diameter of 2 µm and thickness of 1 µm (Paulus (1975)). Platelets are about ten times less abundant than erythrocytes (AlMomani et al. (2008)). In capillaries platelets marginate to the walls of the capillary with their concentration near the wall can be nearly twice that at the centerline. This is quite advantageous as in case of injury to the vessel wall, platelets aggregate to plug the injury (Kumar and Graham (2012)).

Due to the presence of erythrocytes, leukocytes, and thrombocytes, blood has unusual rheology. Firstly, in glass (rigid) tubes with diameters under 300 µm, blood viscosity decreases with decreasing tube diameters! This is known as the Fahraeus-Lindqvist effect (Fahraeus and Lindqvist (1931)). This effect is attributed to the erythrocyte migration away from the vessel wall. Since shear rate scales inversely as the cube of the channel diameter (for the same flow rate), smaller tubes should have significantly higher shear induced lateral migration, pushing the cells out the higher shear zones. However, the decrease in viscosity understandably stops when the vessel diameter reaches the characteristic diameter of the red blood cells - around 6 µm. At this point, since the particles are commensurable with the tube diameter, much of the models that we discussed earlier cannot describe the flow, as they assume that the particle is much smaller than the channel. Pries et al.

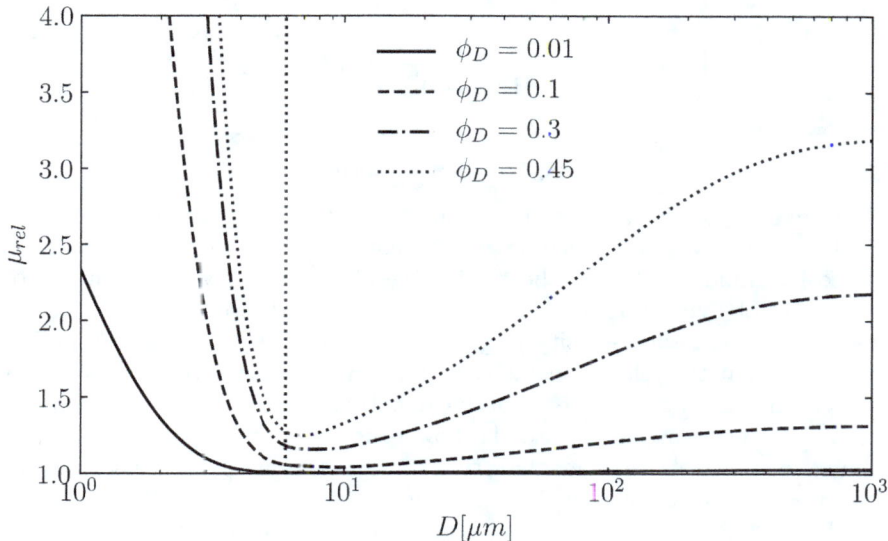

FIGURE 5.5.1. Relative viscosity of erythrocyte suspensions in rigid (glass) tubes as a function of tube diameter and discharge hematocrit according to the empirical fit of Pries et al. (1992). Vertical line is at 6 μm, roughly the point where cells have to travel single file through the tube and where the viscosity begins to rise again.

(1992) provide an empirical fit for the relative apparent viscosity as a function of erythrocyte volume fraction (hematocrit), and tube diameter (in micrometers) for flow in rigid glass tubes as,

$$(5.5.1) \qquad \mu_{rel} = \frac{\mu_{app}}{\mu_0} = 1 + (\mu_{45} - 1)\frac{(1 - \phi_D)^C - 1}{(1 - 0.45)^C - 1}$$

where μ_0 is the plasma viscosity and

$$(5.5.2) \qquad \mu_{45} = 220\exp{(-1.3D)} + 3.2 - 2.44\exp{(-0.06D^{0.645})}$$

is the relative apparent blood viscosity at discharge hematocrit of 0.45, and

$$(5.5.3) \qquad C = (0.8 + \exp{(-0.075D)})\left(\frac{1}{1 + 10^{-11}D^{12}} - 1\right) + \frac{1}{1 + 10^{-11}D^{12}}$$

We plot the relative apparent viscosity as a function of capillary size and hematocrit in Figure 5.5.1. For normal hematocrit of 0.45, relative viscosity in large vessels is about 3, while at the lowest point in a 6 μm diameter capillary the relative viscosity is only 1.2.

Pries et al. (1994) also obtained an empirical fit for relative viscosity of blood in living microvessels,

$$(5.5.4) \qquad \mu_{rel} = \left(1 + (\mu_{45} - 1)\frac{(1-\phi_D)^C - 1}{(1-0.45)^C - 1}\left(\frac{D}{D-1.1}\right)^2\right)\left(\frac{D}{D-1.1}\right)^2$$

where now

$$(5.5.5) \qquad \mu_{45} = 6\exp\left(-0.085D\right) + 3.2 - 2.44\exp\left(-0.06D^{0.645}\right)$$

and C is given by (5.5.3). Pries et al. found that for the flow regime relevant for flow in living microvessels (which have representative shear rates of order 100 s^{-1}) this viscosity is independent of shear rate. We plot the relative apparent viscosity as a function of living microvessel diameter and hematocrit in Figure 5.5.2 and comparison between the viscosity in glass tubes and living microvessels in Figure 5.5.3. We see that the relative viscosity behavior in living microvessels is markedly different from that in glass tubes. Specifically, for small channel diameters ($D <40$ µm) the relative viscosities and so the flow resistances are significantly higher in the living microvessels than in glass tubes. However, for large vessel diameters the two viscosities are roughly the same. Vogel et al. (2000) and Pries and Secomb (2005) attribute the increase in relative viscosity in microvessels over glass tubes to the presence of a glycocalyx on the walls of the microvessels. Glycocalyx, sometimes referred to as an endothelial surface layer (ESL), is a glycoprotein and glycolipoid covering the epithelial cells that make up the blood vessel wall. This layer is typically around 1 µm thick. Among the proteins embedded in this layer are enzymes and proteins that regulate leukocyte and thrombolytic adherence. Specifically in microvessels glycocalyx inhibits coagulation and leukocyte adhesion, and potentially controls the motion of fluid from capillaries into the interstitial space. (van den Berg et al. (2006); Drake-Holland and Noble (2009)).

While in blood flow in microvasculature investigated by Pries et al. did not exhibit strong shear dependence, blood viscosity is generally highly shear rate dependent and exhibits shear thinning behavior. There are many models for the shear rate dependence of blood (see Hund et al. (2017) and Formaggia et al. (2010)). Probably the simplest is to model blood as a power law (Ostwald–de Waele) fluid, and so express the viscosity as

$$(5.5.6) \qquad \mu = k\dot{\gamma}^{(n-1)}$$

where n is 0.828 and k is 0.00927 Pa s^n (Formaggia et al. (2010)). Blood viscosity ranges from order 10^{-1} Pa s at shear rates order 10^{-2} s^{-1} to 5×10^{-3}Pa s at shear rates 10^{-3} s^{-1}. The viscosity also is dependent on hematocrit and temperature and exhibits thixotropy (reduction of viscosity with time being subjected to a constant shear rate) and yield stress. Thixotropy and yield stress are primarily attributed to the aggregation of erythrocytes (Sousa et al. (2016)). Blood viscosity also depends on factors specific to the individual, such as age (Ajmani et al. (2000)).

Another consequence of margination of erythrocytes towards the center of the channel is the Fahraeus effect – where the channel hematocrit is less than the discharge hematocrit (hematocrit of the feed reservoir). This effect is due to the fact that since erythrocytes are marginated towards the center of the channel, they travel faster than the overall liquid (plasma) flow in the channel. From conservation of erythrocytes, we can obtain that

$$(5.5.7) \qquad \frac{\phi_T}{\phi_D} = \frac{U_l}{U_e}$$

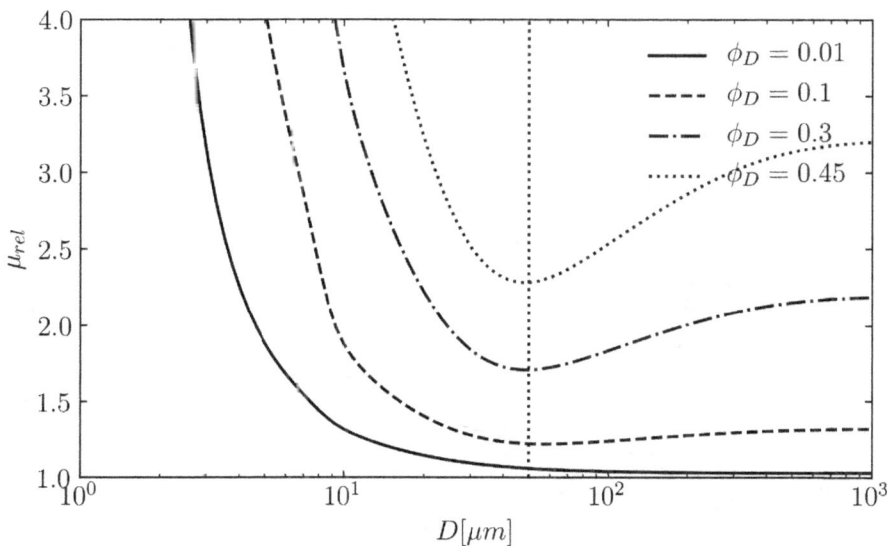

FIGURE 5.5.2. Relative viscosity of erythrocyte suspensions in rigid (glass) tubes as a function of tube diameter and discharge hematocrit according to the empirical fit of Pries et al. (1994). Vertical line is at 50 μm, roughly the point of minimum relative viscosity for hematocrit of 0.45.

where ϕ_T is the channel volume fraction of erythrocytes (channel hematocrit), ϕ_D is the discharge volume fraction of erythrocytes (discharge hematocrit), U_l is the cross-sectional area averaged velocity of the liquid, and U_e is the cross-sectional area averaged velocity of the erythrocytes (Sutera et al. (1970)). U_e is determined by both the erythrocyte volume fraction distribution, which is dependent on ϕ_D and channel diameter. Pries et al. (1992) found an empirical relation for the channel hematocrit as a function and channel diameter,

$$(5.5.8) \qquad \frac{\phi_T}{\phi_D} = \phi_D + (1 - \phi_D)(1 + 1.7 \exp(-0.35D) - 0.6 \exp(-0.01D))$$

We plot tube hematocrit in Figure 5.5.4. For very narrow channels the channel hematocrit is the same as the discharge hematocrit. Channel hematocrit has a minimum for a channel diameter of around 13 μm, where for discharge hematocrit of 0.45, the channel hematocrit becomes 0.32.

Erythrocyte margination also leads to the Zweifach-Fung effect ("bifurcation law"). Zweifach-Fung effect occurs when a vessel bifurcates into two, the hematocrit is reduced in the lower flow rate (small) branch, and increased in the higher flow rate (large) branch (Figure 5.5.5). Pries and Secomb (2005) give an empirical fit for this effect occurring in live microvessels:

$$(5.5.9) \qquad \mathrm{logit}(F_e) = A + B \mathrm{logit}\left(\frac{Q/Q_t - X_0}{1 - X_0}\right)$$

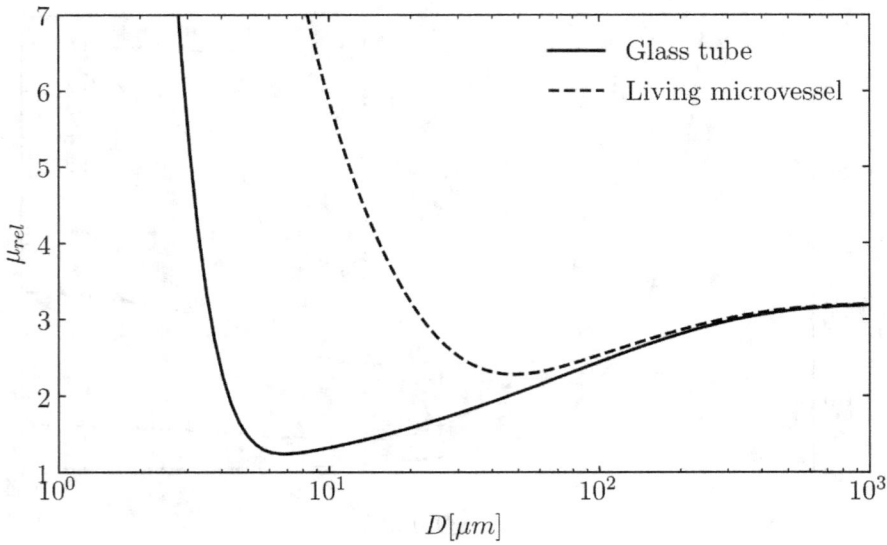

FIGURE 5.5.3. Comparison of relative viscosity in glass tube to that in a living micro vessel. For large channel diameters ($D >100$ µm) the relative viscosities in both types of channels are roughly the same. However, for small channel diameters ($D < 40$ µm) the relative viscosities and so the flow resistances are significantly higher in the living microvessels than in glass tubes.

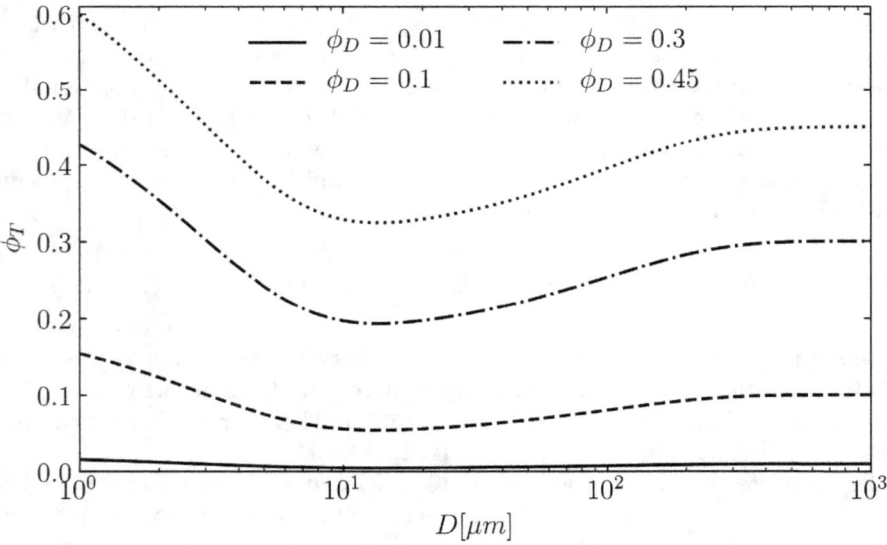

FIGURE 5.5.4. Tube hematocrit as a function of tube diameter and discharge (reservoir) hematocrit from the empirical fit of Pries et al. (1992).

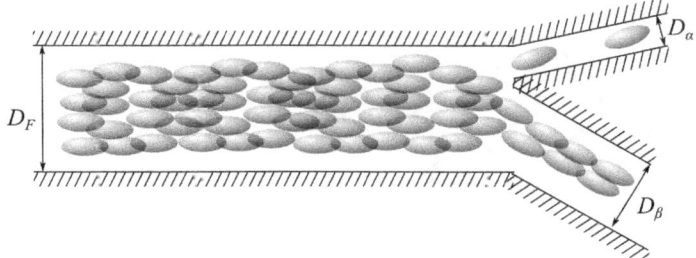

FIGURE 5.5.5. Sketch of uneven division of volume fraction of a slurry at a bifurcation. For blood flow, this is known as a Zweifach-Fung effect where the hematocrit is reduced in the lower flow rate (small) branch, and increased in the higher flow rate (large) branch. The division of hematocrit as well the cell free region near the wall is exaggerated for clarity.

$$(5.5.10) \qquad A = -13.29 \left(\frac{D_\alpha^2/D_\beta^2 - 1}{D_\alpha^2/D_\beta^2 + 1} \right) \frac{1 - \phi_D}{D_F}$$

$$(5.5.11) \qquad B = 1 + 6.98 \frac{1 - \phi_D}{D_F}$$

$$(5.5.12) \qquad X_0 = 0.964 \frac{1 - \phi_D}{D_F}$$

where F_e is the fraction of the total erythrocytes entering the branch of interest, Q/Q_t is the ratio of the flow rate in that branch to the total flow rate; D_α and D_β are the diameters of the two daughter branches, and D_F is the diameter of the parent (feeding) branch; logit (x) is the logistic function,

$$(5.5.13) \qquad \mathrm{logit}\,(x) = \ln \left(\frac{x}{1 - x} \right)$$

An extreme example of this effect occurs when the channel diameter is on the order of the cell diameter, and the flow rate in the high flow rate branch is 2.5 times that of the low flow rate branch. For this condition, the low flow rate branch becomes completely devoid of erythrocytes! (Tripathi et al. (2015)). This effect alone or in combination with other fluidic effects can be used for passive separation of plasma from the rest of the blood and concentration of blood cells (Tripathi et al. (2015)). Furthermore, Zweifach-Fung effect is also observed in suspensions of rigid spheres (Doyeux et al. (2011)) and likely can be used for depletion and enrichment of a wide variety of particles.

CHAPTER 6

Mixing in low and medium Reynolds number flows: dispersion

Dispersion is a process in which a non-uniform profile of the flow (along with transverse (e.g. radial) diffusion) helps spread the distribution of a compound of interest much further than ordinary axial diffusion, thus making it appear that axial (longitudinal) diffusion is stronger than it really is. Here we will be analyzing the Taylor-Aris dispersion due seminal papers by Taylor (1953) and Aris (1956). To understand this mechanism let's consider a slab of compound of interest in a channel with parabolic flow (Figure 6.0.1, t_0). We can create this situation experimentally, for example, by filling an entire channel with a photoactivatable fluorescent dye and then quickly pulsing a laser light sheet to activate the dye, generating a slab of fluorescent dye. We would then be able to track the bright dye, using fluorescent imaging. The parabolic flow would then stretch the dye slab into a parabolic shaped sheet (Figure 6.0.1, t_1). The surface area of the dyed slab would greatly increase. The dye then would diffuse transversely into the surrounding non-dyed regions (Figure 6.0.1, t_2), until that region is filled with dye (Figure 6.0.1, t_3). This thicker region is then stretched again, transverse diffusion will then spread the dye in the transverse direction again, and these steps will repeat over and over greatly increasing the axial distribution of the dye. Notice that even if the advection transport of the dye is greater than the true axial diffusion of the dye, the effective spreading of the dye can be large compared to the axial transport in this case. We term this axial spreading of the dye aided by hydrodynamic stretching dispersion. Also, note that to explain this process we discretized the process into several steps, however, of course in reality these steps are occurring simultaneously.

Now, let's consider the stretching and diffusing steps (Figure 6.0.1, t_1 - t_3) once again, considering that these steps are indeed occurring simultaneously. Notice that if the transverse diffusion was very fast, the dyed slab wouldn't get to stretch very far before the transverse diffusion filled in the concave region and the convex region of the parabola (t_2), and we would quickly obtain a broader slab (t_3). On the other hand if the transverse diffusion was slow the slab would have a lot of time to stretch into very long parabolic sheet, and so the dye would be transported very far from its original position. Now all that the transverse diffusion would have to do is to move the dye a fairly short distance - fill in the concave region and the convex region of the parabola. Thus, weaker transverse diffusion leads to greater dye dispersion! While it may seem paradoxical at first that smaller diffusion leads to greater dispersion, but since dispersion is driven by the hydrodynamic stretching of the dye profile, i.e., uneven hydrodynamic transport, giving more time to this transport helps disperse the dye.

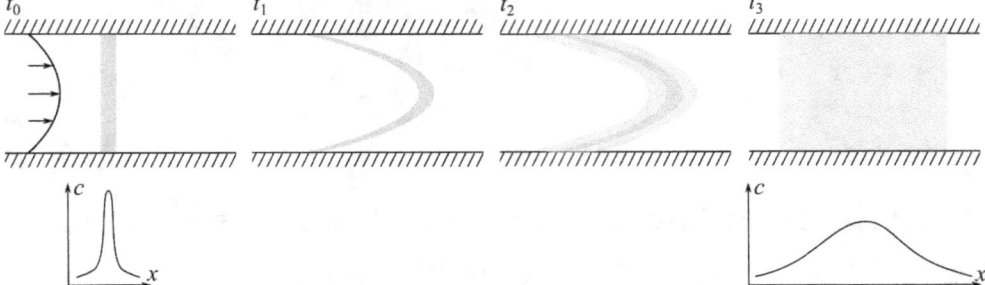

FIGURE 6.0.1. Schematic of the mechanism of dispersion. Consider a thin slab of compound of interest created in a channel with parabolic flow (t_0). Since the molecules of fluid (and therefore also our compound) travel faster near the center of the channel than the sides, the parabolic flow stretches the slab of compound into a parabolic shape (t_1). This greatly increases the surface area of the slab of compound. Radial diffusion then spreads the compound in the radial direction (t_2) and so the overall distribution broadens (t_3). While here we discretized the process into individual steps for clarity, in reality the process is continuous. Concentration profiles are exaggerated for clarity; in reality the integral of the c vs. x plot, i.e., the total amount of compound of interest, is constant. Notice that the combined advective stretching and radial diffusion moves the compound a much larger distance than ordinary axial diffusion could. Hence effectively the compound is diffused further in the axial direction via this mechanism.

Consider on the other hand, if the flow profile was completely uniform. In this case, the slab of compound would never be stretched. It would translate downstream, while axial diffusion would try to broaden the profile. Transverse (e.g., radial) diffusion would also be active of course, but because dye molecules would be diffusing into regions of virtually equal concentration of dye molecules, no net transport of dyed molecules in the transverse direction takes place. If in our flow the advective transport was greater than the axial diffusive transport, then the spreading of the compound would be minimal compared to its advection downstream in this case.

Let's now consider the dispersion process in a quantitative way. To illustrate dispersion, we will consider dispersion of solute in a pressure driven flow in a cylindrical pipe. Let $c = c(r, \theta, z, t)$ be the local concentration of a solute of interest (e.g., dye). Let the time evolution of the solute concentration is governed by the advection-diffusion equation,

$$(6.0.1) \qquad \frac{\partial c}{\partial t} = D\nabla^2 c - \mathbf{u} \cdot \nabla c$$

This equation states the change in concentration at a particular point with time is either due to the solute diffusing in or out of this point (first term on the right-hand side) or being carried (advected) by the fluid in or out of this point (second term on the right-hand side). This equation is analogous to the conduction-convection equation in heat transfer (with which the reader may be more familiar with) where conduction is really the diffusion of thermal energy, and convection is really the advection of thermal energy. In writing the advection-diffusion equation in this form we have also assumed that the diffusion coefficient D is isotropic and is spatially invariant, and thus we can pull it out of the derivative (like we did in the first term on the right-hand side). The diffusion coefficient is analogous to thermal diffusivity in heat transfer.

For steady, pressure driven flow in a cylindrical tube (Poiseuille flow), the velocity distribution is given by

$$(6.0.2) \qquad u_z(r) = 2\bar{u}\left(1 - \frac{r^2}{a^2}\right)$$

$$(6.0.3) \qquad u_r = u_\theta = 0$$

where a is the pipe radius and $\bar{u}$ is the average velocity over the pipe cross-section, defined as,

$$(6.0.4) \qquad \bar{u} = \frac{1}{\pi a^2}\int_0^{2\pi} d\theta \int_0^a r u_z dr = \frac{2}{a^2}\int_0^a r u_z dr$$

In this derivation, we shall denote the average of a quantity X over the cross-section of the tube by $\bar{X}$ (overbar). Note that in general $\overline{XY} \neq \bar{X}\bar{Y}$. We release an axisymmetric distribution of a solute (e.g., our dye), at $t = 0$. Since the distribution is axisymmetric (independent of θ), and that we only have a z-component of flow velocity, the advection-diffusion equation (6.0.1) in cylindrical coordinates simplifies to

$$(6.0.5) \qquad \frac{\partial c}{\partial t} + u_z\frac{\partial c}{\partial z} = D\left(\frac{\partial^2 c}{\partial z^2} + \frac{1}{r}\frac{\partial}{\partial r}\left(r\frac{\partial c}{\partial r}\right)\right)$$

For the advection-diffusion equation we have a no-penetration boundary condition on the tube wall (i.e., the dye does not pass through or react with the tube wall),

$$(6.0.6) \qquad \frac{\partial c}{\partial r} = 0 \text{ at } r = a$$

In order to solve this advection-diffusion equation (6.0.5), we separate the concentration into a cross-sectional average concentration and a perturbation to this concentration, and solve for each of these separately, and then combine the solution. We are able to do this because (6.0.5) is linear. Thus,

$$(6.0.7) \qquad c(r, z, t) = \bar{c}(z, t) + c'(r, z, t)$$

where the prime designates the perturbation-from-the-average term, and

$$(6.0.8) \qquad \bar{c} = \frac{2}{a^2}\int_0^a rc\,dr$$

Note that by this definition, $\overline{c'} = 0$. Substituting (6.0.7) into (6.0.5), we obtain

$$(6.0.9) \qquad \frac{\partial \bar{c}}{\partial t} + \frac{\partial c'}{\partial t} + u_z\frac{\partial \bar{c}}{\partial z} + u_z\frac{\partial c'}{\partial z} = D\left(\frac{\partial^2 \bar{c}}{\partial z^2} + \frac{\partial^2 c'}{\partial z^2} + \frac{1}{r}\frac{\partial}{\partial r}\left(r\frac{\partial c'}{\partial r}\right)\right)$$

Now we take the cross-sectional average of (6.0.9), exchanging the integration and differentiation order via the Leibniz integral rule, and expanding and obtain

(6.0.10)
$$\frac{\partial \bar{c}}{\partial t} + \bar{u}\frac{\partial \bar{c}}{\partial z} + \overline{u_z \frac{\partial c'}{\partial z}} = D\left(\frac{\partial^2 \bar{c}}{\partial z^2}\right)$$

where we have used the boundary condition

(6.0.11)
$$\frac{\partial c'}{\partial r} = 0 \text{ at } r = a$$

and that
(6.0.12)
$$\int_0^a u_z \frac{\partial \bar{c}}{\partial z} r dr = \int_0^a \frac{\partial (u_z \bar{c})}{\partial z} r dr = \frac{\partial}{\partial z}\left(\int_0^a (u_z \bar{c}) r dr\right) = \frac{\partial}{\partial z}\left(\bar{c}\int_0^a u_z r dr\right)$$

noting that from (6.0.2) we see that $\partial u_z / \partial z = 0$. Next, we subtract (6.0.10) from (6.0.9) and obtain the perturbation component

(6.0.13)
$$\frac{\partial c'}{\partial t} + (u_z - \bar{u})\frac{\partial \bar{c}}{\partial z} + u_z\frac{\partial c'}{\partial z} - \overline{u_z\frac{\partial c'}{\partial z}} = D\left(\frac{\partial^2 c'}{\partial z^2} + \frac{1}{r}\frac{\partial}{\partial r}\left(r\frac{\partial c'}{\partial r}\right)\right)$$

So far we have not introduced any additional assumptions. Now we need to introduce additional assumptions to solve these equations. Observe that on the time scale of order of diffusion time scale, a^2/D we expect the radial (transverse) diffusion to have almost smoothed out any concentration variation in the radial direction. In other words, on this time scale we expect the system to have gone from that in t_2 to that in t_3 in Figure 6.0.1 and the perturbations from the mean concentration to be small. Thus, for $t \sim O\left(a^2/D\right)$ we expect $\bar{c} \gg c'$. Furthermore, we expect the gradients in the radial direction to be greater than those in the axial direction, since the radial length scale is much smaller than the axial length scale. Hence with these additional assumptions we simplify (6.0.13) to

(6.0.14)
$$(u_z - \bar{u})\frac{\partial \bar{c}}{\partial z} = D\left(\frac{1}{r}\frac{\partial}{\partial r}\left(r\frac{\partial c'}{\partial r}\right)\right)$$

Now we substitute in the velocity profile (6.0.2) into (6.0.14) and obtain

(6.0.15)
$$\frac{\partial}{\partial r}\left(r\frac{\partial c'}{\partial r}\right) = \frac{\bar{u}}{D}\frac{\partial \bar{c}}{\partial z}\left(r - \frac{2r^3}{a^2}\right)$$

Note that by definition, $\bar{c}$ does not depend on r, so we can integrate (6.0.15) twice to find

(6.0.16)
$$c' = \frac{\bar{u}}{D}\frac{\partial \bar{c}}{\partial z}\left(\frac{1}{4}r^2 - \frac{1}{8a^2}r^4 + A + B\ln r\right)$$

Since c' cannot go to infinity at $r = 0$, (i.e. we want this point to be regular, not singular) this implies that $B = 0$. Furthermore, by definition

(6.0.17)
$$\int_0^a c' r dr = 0$$

which gives that $A = -a^2/12$, and so we obtain

(6.0.18)
$$c' = \frac{\bar{u}a^2}{24D}\frac{\partial \bar{c}}{\partial z}\left(6R^2 - 3R^4 - 2\right)$$

where

(6.0.19)
$$R = r/a$$

The cross-sectional average advection-diffusion equation, (6.0.10) requires a term $\overline{u_z \left(\partial c' / \partial z \right)}$ so let's calculate this term via (6.0.18):

(6.0.20)
$$\overline{u_z \left(\partial c' / \partial z \right)} = \frac{\bar{u}^2 a^2}{24D} \frac{\partial^2 \bar{c}}{\partial z^2} \int\limits_0^1 2 \left(1 - R^2 \right) \left(6R^2 - 3R^4 - 2 \right) 2RdR = -\frac{\bar{u}^2 a^2}{48D} \frac{\partial^2 \bar{c}}{\partial z^2}$$

Substituting this into (6.0.10) we obtain the cross-sectional average advection-diffusion equation to be

(6.0.21)
$$\frac{\partial \bar{c}}{\partial t} + \bar{u} \frac{\partial \bar{c}}{\partial z} = \frac{\partial^2 \bar{c}}{\partial z^2} \left(D + \frac{\bar{u}^2 a^2}{48D} \right)$$

Now we can define an effective diffusion coefficient (otherwise known as the dispersion coefficient) as

(6.0.22)
$$D_{eff} = D \left(1 + \frac{1}{48} \left(\frac{\bar{u}a}{D} \right)^2 \right) = D \left(1 + \frac{Pe^2}{48} \right)$$

where we have defined the Peclet number as $Pe = \bar{u}a/D$. Peclet number represents the ratio of transport via advection to transport via diffusion. We see that after a time $t \sim O \left(a^2/D \right)$, the concentration of solute will be fairly uniform across the pipe cross-section, meanwhile the solute will also have moved a distance $\bar{u}t$ downstream. These combined motions will have spread the solute in the axial direction a distance $O \left(\sqrt{D_{eff}t} \right)$. The effective diffusion coefficient has an interesting property that it has a minimum. This minimum is $D_{eff,min} = a\bar{u}/\sqrt{12}$ and occurs at $Pe = \sqrt{48}$. From this minimum value the effective diffusion coefficient increases as the diffusion coefficient decreases! As explained above, this is because we are giving the flow more time to stretch the solute disk into a longer parabolic sheet, and hence spread the solute over a longer distance.

Note that by obtaining the cross-sectional average advection-diffusion equation (6.0.21), we obtained most of information on how a solute spreads in a tube, as most of the time we only care about the cross-sectional average concentration of the solute, rather than the 3D solute distribution. Thus equation (6.0.21) is often more useful than equation (6.0.5). Similar cross-sectional average advection-diffusion equations, and effective diffusion coefficients can be obtained for other geometries as well. For example, for flow between parallel plates with spacing h, where the velocity profile is

(6.0.23)
$$u_z \left(y \right) = \frac{3}{2} \bar{u} \left(1 - \left(\frac{y}{h} \right)^2 \right)$$

and the cross-sectional average advection-diffusion equation is (Aris (1959))

(6.0.24)
$$\frac{\partial \bar{c}}{\partial t} + \bar{u} \frac{\partial \bar{c}}{\partial z} = \frac{\partial^2 \bar{c}}{\partial z^2} D_{eff}$$

(6.0.25)
$$D_{eff} = D \left(1 + \frac{2}{105} \left(\frac{\bar{u}h}{D} \right)^2 \right)$$

Doshi et al. (1978) give dispersion coefficients for open and closed rectangular conduits. Gill et al. (1969) give dispersion coefficients for diverging channels and concentric annuli. Ajdari et al. (2006) give dispersion coefficients for parabolic, D-shaped, and other cross-sections commonly obtained by lithography and thus of importance for microfluidic devices.

Part 2

Capillarity

Now that we discussed the fundamentals of (nominally) single phase fluid flow in microfluidic devices, we come to the second fundamental topic of this book: capillarity or two phase flows strongly influenced by surface tension. Capillarity problems arise quite often in microfluidics. They range from the simplest: "Now that we decided how liquid should flow through the chip, how do we now fill the chip with liquid?" to more advanced, such as moving and manipulating discrete portions of a particular liquid throughout the chip. Due the small dimensions in microfluidics, surface tension forces become of the same magnitude as the driving forces for the fluid, and so, as we will see in the next sections, surface tensions play a significant role. Hence, even seemingly simple problem of filling and voiding can be a challenge for microfluidic devices. (On the other hand, how many times have you worried about liquid completely filling or emptying out of a pot or kettle?). More advanced problems such as moving, manipulating and segregating discrete portions of liquids allow us to perform discrete reactions with controlled amount of sample and reagent and hence build mini reactors on chip. Capillarity thus adds powerful tools to the microfluidics toolbox.

We will start this section by discussing the fundamentals of surface tension and contact angle and deriving the fundamental equation of capillarity, the Young equation. We will then discuss static wetting, kinetics of wetting, interactions of capillary forces with gravitational and pressure forces; discuss surface tension driven flows; and lastly discuss droplet generation.

CHAPTER 7

Fundamentals of capillarity

In this section we will cover the three fundamental concepts of capillarity: surface tension, the Young-Laplace equation, and contact angle.

7.1. Surface tension

Liquids are held together by attractive molecular forces. Molecules near an interface have a different environment than those in the bulk (Figure 7.1.1a). A molecule in the bulk will experience forces in all directions due to surrounding molecules. The resultant force on such a molecule averaged over a time scale much larger than the molecular collision time scale will be zero. Molecules at the interface on the other hand, will partly experience forces from similar molecules from their own phase, and partly from forces of different molecules from the neighboring phase. For a liquid surface, molecules near the surface of the liquid will experience weaker attractive force from the gas phase than if the gas phase were replaced by a liquid phase of the same liquid, as the density of the gas is significantly smaller than that of the bulk liquid (typically order thousand times smaller). Thus the molecules in the liquid phase will, on average, experience a force directed toward the bulk of the liquid. This force will attempt to reduce the total area of the surface. We can conceptualize this force into a macroscopic surface force localized within about one molecular thickness of the surface. We mathematically consider this as a surface tension, (a force per unit length), as if it was in a membrane of negligible thickness. In our case this "membrane" is the surface of the liquid. The surface of a liquid will be in a state of uniform tension if (a) surface tension is perpendicular to any line drawn in the surface and have the same magnitude in all directions of the line; and (b) surface tension has the same value at all points on the surface. The surface of a liquid does not have to be in uniform tension - if the tension is not uniform, fluid motion will occur.

To imagine this conceptualized force better, we can imagine a rectangular frame (Figure 7.1.1b) with a height l in which there is a thin film of liquid. This liquid has two surfaces - a front and a back. For now, let's assume that the forces between the liquid and the material of the frame are quite strong, stronger than that between the molecules of the liquid. From our definition of surface tension above, when we pull on the movable slide AB to the right, we are pulling with a force

$$(7.1.1) \qquad\qquad\qquad F = 2\gamma l$$

where γ is the surface tension.

Let's also consider this situation from a thermodynamic standpoint. If the slide was initially at position x and we pulled it a small distance δx while maintaining the film temperature constant, the work done against the surface of the film is equal (a)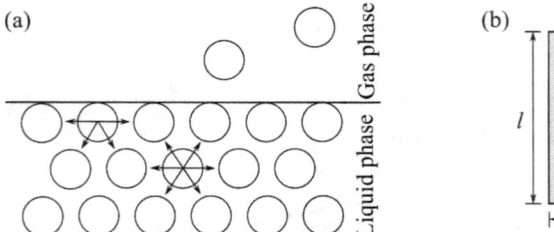

(b)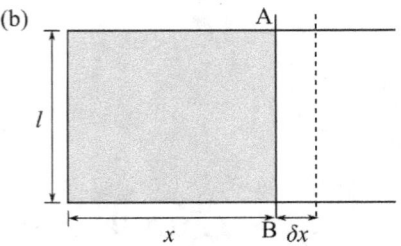

FIGURE 7.1.1. (a) Schematic of molecules in a liquid and gas phase near a surface. Molecules in the bulk of the fluid experience forces from their neighbors such that on average the net force on the molecule is zero. Right at the surface however, the net force on a molecule is directed towards the bulk of the liquid. This net force is the origin of surface tension. (b) Rectangular frame over which a liquid film is stretched for measuring the surface tension of the liquid. To move the movable slide AB to the right we must pull with a force proportional to the number of surfaces (here 2, front and back), the length of the surface (here l), and the surface tension γ.

to the product of the force and the distance moved,

$$(7.1.2) \qquad\qquad \delta W = 2\gamma l \cdot \delta x$$

The small change of area of both surfaces combined that resulted from our pulling is $2l \cdot \delta x = \delta A$ and so

$$(7.1.3) \qquad\qquad \frac{\delta W}{\delta A} = \gamma$$

Since this process occurred at constant temperature, pressure, and chemical composition, this work is equal to Gibbs free energy. Thus, at constant temperature, pressure, and chemical composition, surface tension equals Gibbs free energy per surface area:

$$(7.1.4) \qquad\qquad \gamma = \left(\frac{\partial G}{\partial A}\right)_{T,p,n}$$

Thus, since the surface tension is a property of the molecular forces between molecules in a liquid, we can see that any small change in the area of a surface of a liquid will produce a corresponding change it its free energy. Since spontaneous processes have a negative change in free energy, i.e., spontaneous processes tend to decrease the free energy of the system, liquids tend to decrease their area spontaneously.

We have defined surface tension as a force per unit length in a liquid-vapor interface. We can also extend this definition to two phases of different, immiscible fluids. In this case, the surface tension between two liquid phases would be called interfacial tension. Similarly the surface tension at an interface between a liquid and a solid is called adhesion tension. There is of course a surface tension also at a solid-gas interface.

Furthermore, in our derivations we have assumed that the liquid is at a constant temperature and in a thermodynamic equilibrium. Under these conditions the surface tension is called static surface tension. This is the most common surface tension tabulated. However the surface tension will differ from the static value if the liquid is not in thermodynamic equilibrium. For example, in a jet of water issuing out of an orifice the molecules at the surface of the jet are not in thermodynamic equilibrium, as a new surface is constantly being formed. Thus the surface tension will differ from that in the static case, as is called dynamic surface tension. The dynamic surface tension is situation dependent and also depends on the time since the formation of the new surface.

Surface tension of a liquid varies with temperature and this variation is different from associated and unassociated liquids. Associated liquids are liquids which have relatively stable groups of molecules, for example held together by hydrogen bonding. Associated liquids include water and formic acid as examples. Unassociated liquids do not have such groups. Examples of unassociated liquids include carbon tetrachloride, and benzene. Generally, surface tension decreases with the increase in temperature, reaching a value of zero at the critical temperature. The two most common empirical relations between surface tension and temperature are the Eotvos rule and the Guggenheim–Katayama relation. The Eotvos rule states

$$(7.1.5) \qquad \gamma = V^{-2/3} k \left(T_C - T \right)$$

where V is the molar volume of a substance, T_C is the critical temperature of the substance, and k is the Eotvos constant, equal to $2.1 \times 10^{-7} \mathrm{J}/(\mathrm{K}\,\mathrm{mol}^{2/3})$ (Adam (1941)). For water between 273K and 373K this rule gives

$$(7.1.6) \qquad \gamma = 0.07275 \, [\mathrm{N/m}] \left(1 - 0.002 \left(T - 291\mathrm{K} \right) \right)$$

while the Guggenheim–Katayama relation states

$$(7.1.7) \qquad \gamma = \gamma_0 \left(1 - \frac{T}{T_C} \right)^n$$

where γ_0 is a constant for each liquid, and n is an empirical factor, equal to about 1.2 for unassociated liquids (Adam (1941)).

Surface tension is also a function of solute type and solute concentration dissolved in the liquid. Most solutes decrease surface tension, with the slight exception of inorganic salts, which increase surface tension, but slightly. For example, 6 M aqueous solution of sodium chloride has a surface tension of 82.6 mN/m vs. that of pure water is 72.8 mN/m at 20°C. Alcohols tend to strongly decrease surface tension: 20% (w/w) ethanol in water at 25°C has a surface tension of 38.0 mN/m. Organic acids and small organic molecules also tend to decrease surface tension but less dramatically: 20% (w/w) acetic acid at 30°C has a surface tension of 43.3 mN/m; 20% (w/w) ethylene glycol at 20°C has a surface tension of 64.9 mN/m (Haynes (2014)). Finally, surfactants decrease the surface tension until a certain critical concentration. The CRC handbook of chemistry and physics is a good reference for tables of surface tensions of pure liquids as well as aqueous mixtures.

7.1.1. Interfacial tension. Analogously to surface tension, where a liquid is in contact with a gas, we can have interfacial tensions between any dissimilar phases, as once again molecules near an interface have a different environment than those in the bulk. Consider for example two immiscible liquids (a and b) in contact with one another (Figure 7.1.2). At the interface the dissimilar molecules in the adjacent layers have different potential energies than those in their respective bulks. The increased potential energy of the interface molecules of liquid a over those in

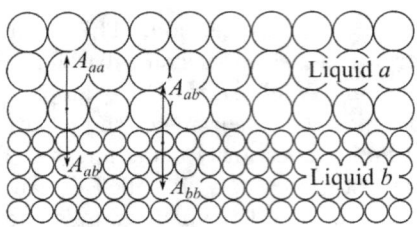

FIGURE 7.1.2. Schematic of an idealized interface between two immiscible liquids a and b. Only forces in the direction orthogonal to the interface that contribute to the different potential energies A in each phase are shown. Repeated subscript designates interaction within own phase. $A_{ba} = A_{ab}$ by symmetry.

the interior of liquid a is $A_{aa} - A_{ab}$ where A_{aa} is the molecular interaction energy between molecules of liquid a at the interface and the same molecules in the bulk of a; A_{ab} is the molecular interaction energy between the interfacial molecules of liquid a and molecules of liquid b across the interface. Analogously in liquid b we have an increased potential of interfacial molecules of liquid b over those in the bulk of b to be $A_{bb} - A_{ba}$. Note by symmetry, $A_{ba} = A_{ab}$. Thus, the increased potential energy of the interfacial molecules of both phases, i.e. the interfacial free energy is then

$$(7.1.8) \qquad \Delta G \sim (A_{aa} - A_{ab}) + (A_{bb} - A_{ab}) = A_{aa} + A_{bb} - 2A_{ab}$$

This is the minimum work required to create the interface. Thus from definition of surface tension (7.1.4), we can also define interfacial tension as the interfacial Gibbs free energy per unit area of the interface, as

$$(7.1.9) \qquad \gamma_I = \gamma_a + \gamma_b - 2\gamma_{ab}$$

where γ_a and γ_b are the surface Gibbs free energies per unit area (i.e. surface tensions) of the pure liquids (against a gas phase) a and b respectively. These are widely tabulated. γ_{ab} is the interaction energy per unit area across the interface. When the molecules of liquids a and b are similar to each other the interaction energy γ_{ab} is large (Rosen and Kunjappu (2012)), and so the interfacial surface tension is typically small. When the molecules of liquids a and b are very different from each other, γ_{ab} is small and can be even neglected. If for example one of the phases, b is not a liquid and is instead a gas, its molecules interact very weakly with each other and with another phase. Thus, in that situation both γ_b and γ_{ab} are negligible and the interfacial tension collapses to our familiar surface tension.

7.1.2. Reduction of surface and interfacial tension by surfactants.

Often we want to have precise control over the interfacial tension between two liquids or the surface tension of a liquid as to aid in their manipulation. This can be done by introducing surfactants to the liquids. Surfactants (surface active agents) are substances that are typically present in small amounts in a system, but adsorb onto the surface or interfaces of the system altering the interfacial energies (and so interfacial tensions) of that system. Surfactants typically have poor solubility in the bulk of the liquid and hence by diffusion are "squeezed out" by the liquid molecules to the interfaces of the liquid. Once at the interface, they alter the interaction energy between the liquid and the surrounding (air, another liquid, or a solid), and so reduce the minimum amount of work required to create that interface.

To achieve these properties, a typical surfactant molecule consist of a structural group that has little attraction to the liquid molecules (lyophobic group, or hydrophobic group when the liquid is water) together with a group that has a strong attraction to the liquid molecules (lyophilic group, or hydrophilic group when the liquid is water). Surfactants are classified into 4 classes based on their lyophilic group: anionic (e.g., sodium dodecyl sulfate, SDS), cationic (e.g., cetrimonium bromide, CTAB), zwitterionic (e.g., sodium lauroamphoacetate), nonionic (e.g., polysorbate). Cationic surfactants should be used with caution as many surfaces are natively negatively charged and will adsorb the cationic head group, making the surface more hydrophobic and changing the surface zeta potential (see Section 15.3.21). Surfactant adsorption to the surface of the fluid handling devices and the resulting change in contact angle and zeta potential should be considered whenever surfactants are used.

For the surfactant to do its job of lowering interfacial tension it needs to be located at the liquid-surrounding interface. Thus, we evaluate surfactant performance on (a) interfacial concentration of surfactant vs. bulk concentration, at equilibrium, (b) rate of transport of surfactant from the bulk to the interface, and (c) the surfactant energetic interactions at the interface. The equilibrium distribution of surfactant between the interface and the bulk and the associated reduction of surface tension can be modeled by standard interfacial distribution models, such as the Gibbs equation, the Langmuir equation, Szyszkowski equation, and the Frumkin equation. The Frumkin equation is particularly interesting as it relates the change in surface tension due to a surfactant as a function of saturation of the surface with surfactant,

$$(7.1.10) \qquad \gamma_0 - \gamma = -2.303 RT\Gamma_m \log\left(1 - \frac{\Gamma}{\Gamma_m}\right)$$

where γ_0 is the surface tension of the liquid without any surfactant (here in units of mN/m), Γ is the surface concentration of surfactant, and Γ_m is the maximum concentration of surfactant at the surface (here in units of mol/cm^2). Rosen and Kunjappu (2012) provide an extensive table of Γ_m. We may ask what is the surface tension reduction when the surface concentration of surfactant is near its maximum, e.g., $\Gamma/\Gamma_m = 0.999$. From tables in Rosen and Kunjappu we see that Γ_m is order 10^{-10} mol/cm^2. Substituting this into the Frumkin equation, we find that the surface tension reduction is order 20 mN/m. This has prompted the field of interfacial phenomena to define a concentration, C_{20}, the bulk concentration of surfactant that will produce a 20 mN/m reduction in surface tension. The effectiveness of a surfactant is often measured as pC_{20}, the negative log of the bulk phase concentration necessary to reduce the surface tension by 20 mN/m. These concentrations are tabulated in, for example, Rosen and Kunjappu. Increasing the concentration of surfactant past C_{20} does further decrease surface tension, until the critical micelle concentration (CMC, a point at which the surfactant concentration is high enough such that the surfactant begins to aggregate into surfactant clusters - micelles), but with significantly diminishing returns (see Figure 7.1.3). The reduction of surface tension between C_{20} and CMC is typically linear with the log of surfactant bulk concentration. However, if the CMC exceeds the solubility of a surfactant at a particular temperature, then we achieve the maximum reduction in surface tension at

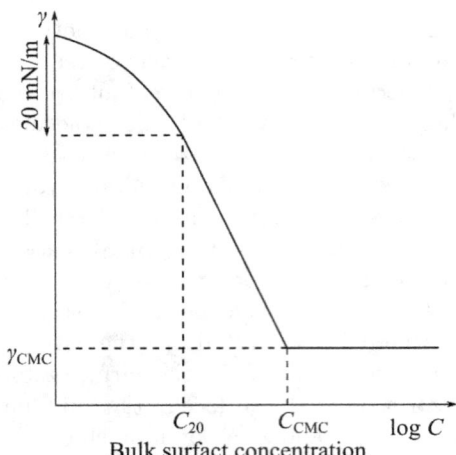

FIGURE 7.1.3. Schematic representing typical surface tension reduction as a function of the logarithm of the bulk surfactant concentration. Surfactant non-linearly reduces the surface tension by 20 mN/m up to C_{20}. From C_{20} till C_{CMC} the surface tension reduction is typically linear. Observe that since this is a linear-log plot, initially a small amount of surfactant causes a large reduction is surface tension, while later more and more surfactant has to be added to reduce the surface tension by a given amount.

the solubility limit. The temperature at which the CMC is equal to the maximum surfactant solubility is termed the Kraft point, and is tabulated in, for example, Rosen and Kunjappu.

7.2. Young-Laplace (capillarity) equation

In this section we derive the Young-Laplace equation, the governing equation for most surface tension driven situations. Fundamentally, it is a statement of normal stress balance for static fluids meeting at an interface, where the interface is considered to be a surface of zero thickness, but having a finite tension. As this equation is so important to the rest of the material, we will derive this equation via two methods: via infinitesimal elements, and via energy minimization.

7.2.1. Arc length and curvature.
Surface area (and its minimization) as well as the surface curvature play a major role in problems governed by capillarity. In this book, we will be dealing with these frequently, and so before we go on to the derivations of the capillary equation we would like to remind the reader of what these are.

Arc length is the length of the curve. If the curve were a road, it is the number of miles we would have to drive to get from one point to another. (Contrast this with another measure of distance between two points, the crow's flight between the two). For a planar curve defined in a parametric representation by a pair of functions $\gamma(t) = (x(t), y(t))$, the arc length between the point a and b is defined as

$$(7.2.1) \qquad s = \int_a^b \left| \frac{d\gamma}{dt} \right| dt$$

Of course, the functions $x(t)$ and $y(t)$ have to be appropriately differentiable. The length of a parametric curve is invariant under reparametrization and is therefore a property of the curve. This definition of arc length of course can be extended into more dimensions. For a planar curve defined as $y = f(x)$, where f is continuously differentiable, we can use this definition and obtain an even simpler expression for arc length

$$(7.2.2) \qquad s = \int_a^b \sqrt{1 + \left(\frac{dy}{dx} \right)^2} \, dx$$

The mean curvature for a surface will be defined below (7.2.24). Here we will consider a simple case of the curvature of a plane curve. Intuitively, we define the curvature of a straight line to be zero. Then the curvature of a circle with a radius R should be small (approaching zero) when R is large (where parts of it can be approximated by a straight line). Thus curvature should be large when R is small. Thus we define for a circle, the curvature

$$(7.2.3) \qquad \kappa = \frac{1}{R}$$

The radius of curvature is reciprocal of curvature. For a curve, at a given point we can always draw a circle that most closely approximates the curve near our point of interest. This circle is called the osculating (kissing) circle. Then we define the curvature of the curve at our point of interest as the curvature of this circle.

Having this definition, we can also consider curvature of the curve in another way. Imagine we are in a little vehicle moving along the curve at a unit speed. If we parameterize our trajectory as $\gamma(t) = (x(t), y(t))$, where t is time, because we are traveling at unit speed, we can also reparameterize the curve as $\gamma(t) = (x(s), y(s))$ where s is the distance traveled along the curve (arc length). As we go around each curve on our curvy road, we experience centripetal acceleration, even though we are traveling at a constant, unit speed along our road. Centripetal acceleration is inversely proportional to the radius of the circle, and so its directly proportional to curvature. The acceleration is also proportional to the change in the velocity vector of our vehicle, which in our case is tangent vector to the curve. Thus, curvature is the magnitude in the change of the tangent vector, or

$$(7.2.4) \qquad \kappa = \left| \frac{d\mathbf{T}}{ds} \right|$$

Using this definition, we find that for a curve parameterized as $\gamma(t) = (x(t), y(t))$, the curvature is

$$(7.2.5) \qquad \kappa = \frac{|x'y'' - y'x''|}{\left(x'^2 + y'^2 \right)^{3/2}}$$

and the signed curvature is

$$(7.2.6) \qquad k = \frac{x'y'' - y'x''}{\left(x'^2 + y'^2 \right)^{3/2}}$$

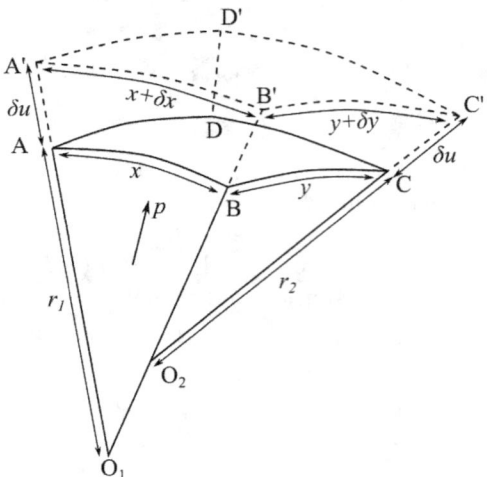

FIGURE 7.2.1. A curvilinear rectangular element $ABCD$ at the surface separating two fluids. When in the lower fluid there is a pressure p relative to the upper fluid is applied, the surface expands to move the rectangular element to $A'B'C'D'$.

where primes indicate derivatives with respect to t. Similarly, for a curve that we can express as $y = f(x)$ the curvature is then

(7.2.7)
$$\kappa = \frac{|y''|}{\left(1 + y'^2\right)^{3/2}}$$

while the signed curvature is

(7.2.8)
$$k = \frac{y''}{\left(1 + y'^2\right)^{3/2}}$$

where the primes indicate derivatives with respect to x. A very commonly used engineering assumption is that the slope of the curve y' is small compared to unity, and so its square is even smaller, and so

(7.2.9)
$$\kappa \approx \left|\frac{d^2 y}{dx^2}\right|$$

and

(7.2.10)
$$k \approx \frac{d^2 y}{dx^2}$$

7.2.2. Young-Laplace (capillarity) equation via infinitesimal elements. We begin our first derivation of Young-Laplace equation by considering a curvilinear rectangular element $ABCD$ separating two fluids, a lower fluid and an upper fluid, shown in Figure 7.2.1. The sides of this rectangle $ABCD$ have lengths x and y, and the rectangle has a surface area S. The radii of curvature of the sides of the rectangles are r_1 and r_2 as shown in Figure 7.2.1. r_1 is the radius of curvature lying in the plane ABO_1; the x-axis lies in this plane. r_2 is the radius of curvature lying in the plane BCO_2; the y-axis lies in this plane.

Now let's suddenly apply a pressure p to the lower fluid relative to the upper fluid. This causes the rectangular element to undergo a virtual displacement δu upwards, normal to the surface of the original rectangular element. A virtual displacement is an infinitesimal displacement that occurs in zero time. This moves and expands our rectangular element to a new position $A'B'C'D'$. This also increases the radii of curvature from r_1 to $r_1 + \delta u$ and r_2 to $r_2 + \delta u$ and the sides of the rectangular element from x to $x + \delta x$ and $y + \delta y$. We assume that our virtual displacement was done under isothermal conditions and so the work done by the pressure in the lower liquid, δW, will be equal to the increase in surface energy δG due to the change in the surface area of the element, δS. The infinitesimal work that is done by this pressure is then

$$(7.2.11) \qquad \delta W = pS\delta u$$

From definition of surface energy in equation (7.1.4),

$$(7.2.12) \qquad \delta G = \gamma \delta S$$

Expressing the change in surface area of the element in terms of the sides of the element,

$$(7.2.13) \qquad \delta G = \gamma \left((x + \delta x)(y + \delta y) - xy \right)$$

Next, we notice that ABO_1 and $A'B'O_1$ are similar triangles, thus,

$$(7.2.14) \qquad \frac{x + \delta x}{r_1 + \delta u} = \frac{x}{r_1}$$

which we rearrange to

$$(7.2.15) \qquad x + \delta x = x \left(1 + \frac{\delta u}{r_1} \right)$$

Similarly, we notice that BCO_2 and $B'C'O_2$ are also similar triangles, and so

$$(7.2.16) \qquad y + \delta y = y \left(1 + \frac{\delta u}{r_2} \right)$$

Substituting these two results into (7.2.13)

$$(7.2.17) \qquad \delta G = \gamma \left(xy \left(1 + \frac{\delta u}{r_1} \right) \left(1 + \frac{\delta u}{r_2} \right) - xy \right)$$

and expanding,

$$(7.2.18) \qquad \delta G = \gamma xy \delta u \left(\frac{1}{r_1} + \frac{1}{r_2} \right) + O\left(\delta u^2 \right)$$

Since, $S = xy$ and since the displacement is order δu and so $O\left(\delta u^2 \right)$ can be neglected,

$$(7.2.19) \qquad \delta G = \gamma S \delta u \left(\frac{1}{r_1} + \frac{1}{r_2} \right)$$

Since as we stated before the infinitesimal change in surface energy is equal to the infinitesimal work done by the pressure (7.2.11), we equate the two and obtain,

$$(7.2.20) \qquad \gamma S \delta u \left(\frac{1}{r_1} + \frac{1}{r_2} \right) = pS\delta u$$

or

(7.2.21) $$p = \gamma \left(\frac{1}{r_1} + \frac{1}{r_2} \right)$$

This result is valid for any orthogonal curvilinear coordinate system, in which x and y are measured in the interfacial surface. Thus we are free to orient our coordinates and use one where x and y are oriented in the planes of the principal, i.e. the maximum (R_1) and minimum (R_2), radii of curvature for a particular geometry. Thus, we can rewrite (7.2.21) as

(7.2.22) $$\Delta p = \gamma \left(\frac{1}{R_1} + \frac{1}{R_2} \right)$$

where we have also changed the notation from p to Δp to emphasize that it is the difference in pressure across the interface. Thus, we obtained the familiar Young-Laplace equation.

We can also express the Young-Laplace equation as a differential equation for the shape of the surface

$$z = f(x, y)$$

For any surface, the mean curvature $H(x, y)$ is defined as

(7.2.23) $$H(x, y) = \frac{1}{2} \left(\frac{1}{R_1(x, y)} + \frac{1}{R_2(x, y)} \right)$$

The mean curvature can also be defined in terms of a differential equation for the shape of the surface, (Geary et al. (1955)), as

(7.2.24) $$H = \frac{\left(1 + f_y^2\right) f_{xx} - 2 f_x f_y f_{xy} + \left(1 + f_x^2\right) f_{yy}}{2\left(1 + f_x^2 + f_y^2\right)^{3/2}}$$

where

(7.2.25) $$f_x = \frac{\partial f}{\partial x}, f_y = \frac{\partial f}{\partial y}, f_{xx} = \frac{\partial^2 f}{\partial x^2}, f_{yy} = \frac{\partial^2 f}{\partial y^2}, f_{xy} = \frac{\partial^2 f}{\partial x \partial y}$$

are the partial derivates. Combining, (7.2.24), (7.2.23) with (7.2.22), we see that

(7.2.26) $$\frac{\Delta p}{\gamma} = \frac{\left(1 + f_y^2\right) f_{xx} - 2 f_x f_y f_{xy} + \left(1 + f_x^2\right) f_{yy}}{\left(1 + f_x^2 + f_y^2\right)^{3/2}}$$

which is then the differential form of the Young-Laplace equation.

7.2.3. Young-Laplace (capillarity) equation via minimization of energy. Here we will take the minimization of energy approach to deriving the Young-Laplace equation. This approach is also equivalent to the one where we balance the forces on a fluid element, as we assume that no irreversible energy loss (e.g., viscous friction) occurs, and that all forces are conservative - that they have a potential. We constrain our minimization with the condition that the liquid volume is constant, and in the case of rotation, that the angular momentum of the liquid is also constant. Let's consider a general liquid surface (interface between the liquid and a gas), as shown in Figure 7.2.2.

We can express the position of this surface in Cartesian coordinates relative to an x, y plane as

(7.2.27) $$z = z(x, y)$$

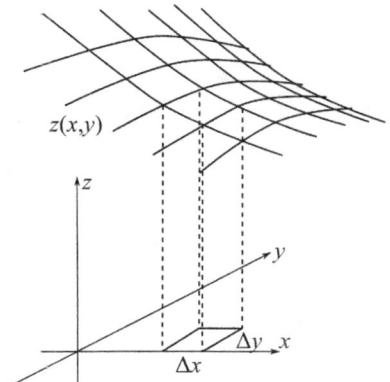

FIGURE 7.2.2. Schematic of an arbitrary liquid-gas interface surface in a Cartesian coordinate system.

The area of the arbitrary liquid surface element highlighted in Figure 7.2.2, is just the absolute value of the cross product

$$(7.2.28) \qquad \left| \left(\Delta x, 0, \frac{\partial z}{\partial x} \Delta x \right) \times \left(0, \Delta y, \frac{\partial z}{\partial y} \Delta y \right) \right| = \left| \left(-\frac{\partial z}{\partial x}, -\frac{\partial z}{\partial y}, 1 \right) \Delta x \Delta y \right| = S \Delta x \Delta y$$

where

$$(7.2.29) \qquad S = \sqrt{1 + \left(\frac{\partial z}{\partial x} \right)^2 + \left(\frac{\partial z}{\partial y} \right)^2}$$

Thus, if we want to find the area of some region of the liquid surface, we would sum such elements. Expressing this through integration,

$$(7.2.30) \qquad A = \iint S \, dx dy = \iint \sqrt{1 + \left(\frac{\partial z}{\partial x} \right)^2 + \left(\frac{\partial z}{\partial y} \right)^2} \, dx dy$$

The volume of the liquid under such surface is

$$(7.2.31) \qquad V = \iint z \, (x, y) \, dx dy$$

For gravity directed in the z-direction the gravitational potential energy for the liquid is

$$(7.2.32) \qquad E_{grav} = \frac{1}{2} \rho g \iint z^2 dx dy$$

If the liquid is on a spinning surface, such that the z-direction is normal to the axis of rotation, then the kinetic energy of rotation of the liquid is

$$(7.2.33) \qquad E_{rot} = -\frac{1}{2} \rho \omega^2 \iint z \left(x^2 + y^2 \right) dx dy$$

A relevant example of such rotating surface would be a capillary burst valve on a lab-on-a-CD type microfluidic device. The total energy of the liquid is then

$$(7.2.34) \qquad E = \gamma A + E_{grav} + E_{rot}$$

From calculus of variations, minimization of energy under the constraint of constant volume is equivalent to minimization of $E - pV$ where p is the pressure in the liquid, now acting as a Lagrange multiplier. Furthermore, minimization of a multidimensional integral

$$(7.2.35) \qquad F = \int f\left(z, \frac{\partial z}{\partial x_1}, \frac{\partial z}{\partial x_2}, ..., \frac{\partial z}{\partial x_N}\right) dx_1 dx_2 ... dx_N$$

with respect to a trial function $z = z(x_1, x_2, ..., x_N)$ is achieved by the relation

$$(7.2.36) \qquad \sum_{n=1}^{N} \frac{d}{dx_n} \frac{\partial f}{\partial(\partial z/\partial x_n)} = \frac{\partial f}{\partial z}$$

We can write this for a two dimensional function as

$$(7.2.37) \qquad \frac{d}{dx}\frac{\partial f}{\partial(\partial z/\partial x)} + \frac{d}{dy}\frac{\partial f}{\partial(\partial z/\partial y)} = \frac{\partial f}{\partial z}$$

where $\partial z/\partial x$ and $\partial z/\partial y$ are the variables of f as per

$$(7.2.38) \qquad f\left(z, \frac{\partial z}{\partial x}, \frac{\partial z}{\partial y}\right)$$

Substituting in $E - pV$ gathered from (7.2.30) through (7.2.34) into (7.2.35), we obtain,

$$(7.2.39) \qquad f\left(z, \frac{\partial z}{\partial x}, \frac{\partial z}{\partial y}\right) = \gamma S + \frac{1}{2}\rho g z^2 - \frac{1}{2}\rho \omega^2 z\left(x^2 + y^2\right) - pz$$

where the S term contains the variables $\partial z/\partial x$ and $\partial z/\partial y$. Applying (7.2.37) to (7.2.39) we obtain

$$(7.2.40) \qquad \gamma\left(\frac{dn_x}{dx} + \frac{dn_y}{dy}\right) = p - \rho g z + \frac{1}{2}\rho \omega^2\left(x^2 + y^2\right)$$

where n_x and n_y are the components of the surface normal

$$(7.2.41) \qquad \mathbf{n} = (n_x, n_y, n_z) = \frac{1}{S}\left(-\frac{\partial z}{\partial x}, -\frac{\partial z}{\partial y}, 1\right)$$

The term in the brackets on the left in (7.2.40) can be expended, and we write (7.2.40) as

$$(7.2.42) \qquad \gamma\left(\frac{1}{S^3}\left(\frac{\partial^2 z}{\partial x^2}\left(1 + \left(\frac{\partial z}{\partial y}\right)^2\right) - 2\frac{\partial z}{\partial x}\frac{\partial z}{\partial y}\frac{\partial^2 z}{\partial x \partial y} + \frac{\partial^2 z}{\partial y^2}\left(1 + \left(\frac{\partial z}{\partial x}\right)^2\right)\right)\right)$$
$$= p - \rho g z + \frac{1}{2}\rho \omega^2\left(x^2 + y^2\right)$$

Comparing (7.2.42) with (7.2.24), we obtain

$$(7.2.43) \qquad \gamma\left(\frac{1}{R_1} + \frac{1}{R_2}\right) = p - \rho g z + \frac{1}{2}\rho \omega^2\left(x^2 + y^2\right)$$

This is the more general Young-Laplace equation, including the contributions of gravity and rotation. While the equation in the form (7.2.43) is useful when simple predictions of the radii of curvature are known (or to give an estimate of the pressure from a known, measured radii of curvature), the form in (7.2.42) is far more useful. The form in (7.2.42) is the fundamental PDE of capillarity and combined with appropriate boundary conditions is useful for finding the complete shape of the liquid surface.

7.2.4. Young-Laplace (capillary) equation in cylindrical coordinates.
Here as a reference we provide capillary equation in cylindrical coordinates; for more coordinate systems see Langbein (2002).

$$(7.2.44) \qquad \mathbf{n} = \frac{1}{S}\left(-\frac{\partial z}{\partial r}, -\frac{1}{r}\frac{\partial z}{\partial \phi}, 1\right)$$

$$(7.2.45) \qquad S = \sqrt{1 + \left(\frac{\partial z}{\partial r}\right)^2 + \left(\frac{1}{r}\frac{\partial z}{\partial \phi}\right)^2}$$

$$(7.2.46)$$
$$\frac{p}{\gamma}\frac{1}{S^3} = \frac{1}{r}\frac{\partial}{\partial r}r\frac{\partial z}{\partial r}\left(1 + \left(\frac{1}{r}\frac{\partial z}{\partial \phi}\right)^2\right) + \frac{1}{r}\frac{\partial z}{\partial r}\left(\left(\frac{\partial z}{\partial r}\right)^2 - 2\frac{1}{r}\frac{\partial z}{\partial \phi}\frac{\partial^2 z}{\partial \phi \partial r} + \left(\frac{1}{r}\frac{\partial z}{\partial \phi}\right)^2\right)$$
$$+ \frac{1}{r^2}\frac{\partial^2 z}{\partial \phi^2}\left(1 + \left(\frac{\partial z}{\partial r}\right)^2\right)$$

$$(7.2.47) \qquad A = \iint S r\, dr\, d\phi$$

$$(7.2.48) \qquad V = \iint z\,(r,\phi)\, r\, dr\, d\phi$$

$$(7.2.49) \qquad E_{grav} = \frac{1}{2}\rho g \iint z r^2 d\phi dz$$

$$(7.2.50) \qquad E_{rot} = -\frac{1}{4}\rho \omega^2 \iint r^4 d\phi dz$$

7.2.5. Young-Laplace equation for axisymmetric situations.
The most fruitful application of the Young-Laplace PDE is to axisymmetric situations as for these the equation collapses to an ODE and therefore can be easily (and often analytically) solved. Axisymmetric situations also arise quite frequently - consider a drop or a jet of water. Let's orient our coordinate system such that the first plane of principle curvature, $\kappa_1 = 1/R_1$ is the plane through the symmetry axis, while the second plane of principle curvature, $\kappa_2 = 1/R_2$ extends in the azimuthal direction. This produces in cylindrical coordinates a surface $r\,(z)$ such that

$$(7.2.51) \qquad \kappa_1 = -\left(1 + (dr/dz)^2\right)^{-3/2}\frac{d^2 r}{dz^2}$$

$$(7.2.52) \qquad \kappa_2 = \frac{1}{r}\left(1 + (dz/dr)^2\right)^{-1/2}$$

where, as usual,

$$(7.2.53) \qquad \frac{\Delta p}{\gamma} = \left(\frac{1}{R_1} + \frac{1}{R_2}\right) = \kappa_1 + \kappa_2$$

If we invert the coordinate system such that the liquid surface is defined by $z\,(r)$, then

$$(7.2.54) \qquad \kappa_1 = \left(1 + (dz/dr)^2\right)^{-3/2}\frac{d^2 z}{dr^2}$$

$$(7.2.55) \qquad \kappa_2 = \frac{1}{r}\left(1 + (dz/dr)^2\right)^{-1/2}\frac{dz}{dr}$$

where

$$(7.2.56) \qquad \frac{\Delta p}{\gamma} = \left(\frac{1}{R_1} + \frac{1}{R_2}\right) = \kappa_1 + \kappa_2 = \frac{1}{r}\frac{d}{dr}\left(\frac{r}{\sqrt{1 + (dz/dr)^2}}\frac{dz}{dr}\right)$$

For the general case, accounting for gravity and rotation, the capillary equation can be written as

$$(7.2.57) \qquad \kappa_1 + \kappa_2 = \frac{1}{r}\frac{d}{dr}\left(\frac{r}{\sqrt{1 + (dz/dr)^2}}\frac{dz}{dr}\right) = \frac{\Delta p}{\gamma} - \frac{g\Delta\rho z}{\gamma} + \frac{\Delta\rho\omega^2 r^2}{2\gamma}$$

where $\Delta\rho$ is the density difference between the liquid and the surrounding gas. Typically $\Delta\rho \approx \rho$ where ρ is the density of the liquid. When numerically integrating this to avoid problems with vertical or horizontal slopes, $dr/dz = 0$ or $dz/dr = 0$ it is convenient to change into a new set of variables using arc length, where

$$(7.2.58) \qquad ds = \sqrt{(dr)^2 + (dz)^2}$$

$$(7.2.59) \qquad dr = ds\cos\phi$$

$$(7.2.60) \qquad dz = ds\sin\phi$$

$$(7.2.61) \qquad \frac{\Delta p}{\gamma} = \frac{d\phi}{ds} + \frac{\sin\phi}{r}$$

$$(7.2.62) \qquad \frac{d^2 r}{ds^2} = -\frac{dz}{ds}\left(\frac{\Delta p}{\gamma} - \frac{1}{r}\frac{dz}{ds}\right)$$

$$(7.2.63) \qquad \frac{d^2 z}{ds^2} = \frac{dr}{ds}\left(\frac{\Delta p}{\gamma} - \frac{1}{r}\frac{dz}{ds}\right)$$

7.2.6. Axisymmetric solutions to capillary equation: Delaunay surfaces. For axisymmetric situations, the capillary equation has special family of solutions called the Delaunay surfaces. Notice that for situations where gravity and rotation can be neglected, pressure is equal everywhere in the fluid and is constant (when the motion of the fluid is negligible). Thus, by capillary equation, (7.2.22) and (7.2.23), mean curvature should be the same everywhere on the surface of the liquid. Thus, we are interested in constant mean curvature surfaces, and Delaunay surfaces are just these. The family of Delaunay surfaces consists of unduloids, periodic nodoids, and catenoinds. The sphere and the right circular cylinder are very special cases of these (Ciric (2009)).

An unduloid is an undulary rotated around an axis of rotation, here the x-axis. An undulary (see Figure 7.2.3) is the locus of an ellipse as the point of contact rolls along a flat surface without slip. Right circular cylinders are unduloids made by rolling a circle. A sphere is created from rolling a degenerate ellipse of zero eccentricity. A parameterization for an undulary (Zeleny (2014)) can be written, with parameter u, as

$$(7.2.64) \qquad f(u) = (x(u), y(u))$$

(7.2.65)
$$x(u) = \int_0^u \sqrt{a^2\sin^2\phi + b^2\cos^2\phi}\,d\phi + \frac{\sqrt{a^2 - b^2}\sin(u)\left(\sqrt{a^2 - b^2}\cos(u) + a\right)}{\sqrt{a^2\sin^2 u + b^2\cos^2 u}}$$

(7.2.66)
$$y(u) = \frac{b\left(\sqrt{a^2 - b^2}\cos(u) - a\right)}{\sqrt{a^2\sin^2 u + b^2\cos^2 u}}$$

where a and b are the parameters of an ellipse, $x^2/a^2 + y^2/b^2 = 1$.

A nodoid is a nodary rotated around an axis of rotation, here the x-axis. A nodary (see Figure 7.2.3) is the locus of a focus of a hyperbola as the point of contact rolls along a flat surface without slip. Periodic nonoids intersect themselves and so only sections of them may be realized in practice as real liquid surfaces. A parameterization for a nodary (Ciric (2009)) can be written as

(7.2.67)
$$f(u) = (x(u), y(u))$$

(7.2.68)
$$x(u) = a\left(1 - \cos u + \int_0^u \frac{\sin^2\phi}{\sqrt{\sin^2\phi + b^2/a^2}}\,d\phi\right)$$

(7.2.69)
$$y(u) = a\left(\sin u + \sqrt{\sin^2 u + b^2/a^2}\right)$$

where a and b are the parameters of a hyperbola, $x^2/a^2 - y^2/b^2 = 1$.

A catenoid is obtained by rotating a catenary, (a hyperbolic cosine), around an axis of rotation. A catenoid surface is special in that it has zero mean curvature, implying that the pressure under such surface is zero.

For both unduloid and nodoid surfaces it has been proven that they are stable only if their radius (maximum y-coordinate of the corresponding undulary or nodary) does not exceed a single period (Figure 7.2.3) . If their radius exceeds the period, they break antimetrically in the middle, that it is one half widens and the other half narrows. This instability is called the Plateau-Rayleigh instability, and we will discuss this in a later section. This instability is quite useful for making liquid drops.

Let's now consider the solutions to an axisymmetric capillary equation (7.2.57) in more detail. We write the equation here again for convenience,

(7.2.70)
$$\frac{1}{r}\frac{d}{dr}\left(\frac{r}{\sqrt{1 + (dz/dr)^2}}\frac{dz}{dr}\right) = \frac{\Delta p}{\gamma} - \frac{g\Delta\rho z}{\gamma} + \frac{\Delta\rho\omega^2 r^2}{2\gamma}$$

The expression

(7.2.71)
$$\frac{1}{\sqrt{1 + (dz/dr)^2}}\frac{dz}{dr} = \sin\phi$$

is the radial component of the surface normal (see (7.2.44)) and so is equal to the sine of the surface's slope (see Figure 7.2.4). Neglecting the gravity term in (7.2.70)

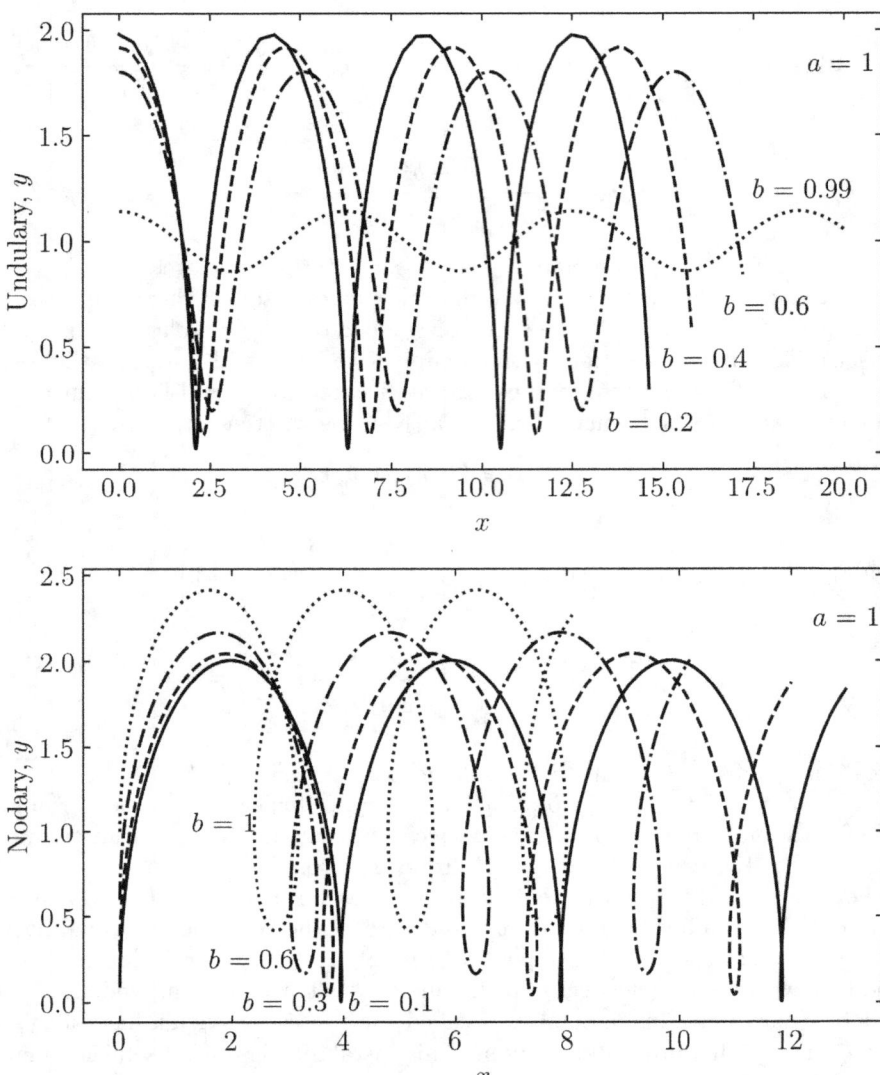

FIGURE 7.2.3. Udualary and nodary curves. When rotated around the x-axis these make the unduloid and nodoid surfaces - solutions to the Young-Laplace capillary equation. Unduloid with $a = 1$, $b = 1$, is a right circular cylinder. The transition between unduloids and nodoids is a periodic series of coaxial spheres. Unlike udualaries, nodaries self intersect. Self intersection cannot exist in a real liquid surface, so only portions of nodary (or rather nodoid) surfaces are taken as solutions to the capillary equation.

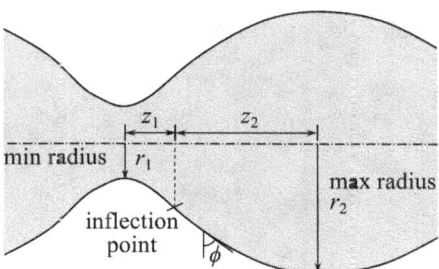

FIGURE 7.2.4. Section of a Delaunay surface, showing the minimum and the maximum radii of the liquid tube as well as the inflection point in the surface.

(for simplicity of solving), multiplying both sides of (7.2.70) by r, substituting in (7.2.71), and then integrating with respect to r, we obtain

$$(7.2.72) \qquad \frac{r}{\sqrt{1 + (dz/dr)^2}} \frac{dz}{dr} = r \sin \phi = r_0 + \frac{\Delta p}{\gamma} \frac{r^2}{2} + \frac{\Delta \rho \omega^2 r^4}{8\gamma}$$

where r_0 is the constant of integration to be determined. If the surface is not rotating, i.e., $\omega = 0$ then the right-hand side of (7.2.72) is a quadratic equation. If both r_0 and $\Delta p/\gamma$ have the same sign, then the right-hand side of (7.2.72) has no zero, and so $\sin \phi$ is positive, and so we have unduloids, which have no radial tangent. If on the other hand, r_0 and $\Delta p/\gamma$ differ in sign, $\sin \phi$ may equal to zero and we obtain nodoids, which intersect themselves and have a tangent in the radial direction. Finally, if $\Delta p/\gamma$ is equal to zero, we have

$$\frac{r}{\sqrt{1 + (dz/dr)^2}} \frac{dz}{dr} = r_0$$

for which there is a solution,

$$(7.2.73) \qquad \frac{r}{r_0} = \cosh \left(\frac{z - z_0}{z_0} \right)$$

where z_0 is another constant of integration. This is the familiar catenary equation that leads to a catenoid.

Following Langbein, let's analyze the Delaunay surfaces, such as that in Figure 7.2.4, further. We define minimum and maximum radii of the surface, r_1 and r_2. Langbein found the following relations for Delaunay surfaces:

$$(7.2.74) \qquad r \sin \phi = \frac{r^2 + r_1 r_2}{r_1 + r_2}$$

$$(7.2.75) \qquad \frac{r_1 r_2}{r_1 + r_2} = r_0$$

$$(7.2.76) \qquad \frac{2}{r_1 + r_2} = \frac{\Delta p}{\gamma}$$

The radii for an unduloid are as shown in Figure 7.2.4. For a catenoid, r_2 is equal to infinity and for a nodoid, r_2 is equal to the maximum negative radius. For an unduloid the inflection point lies at the geometric mean of the two radii,

$$(7.2.77) \qquad r_{inf} = \sqrt{r_1 r_2}$$

The angle ϕ is minimum at the inflection point, where we label $\phi_{inf} = \psi$. This angle is given by

$$(7.2.78) \qquad \sin\psi = \frac{2\sqrt{r_1 r_2}}{r_1 + r_2}$$

$$(7.2.79) \qquad \cos\psi = \frac{r_2 - r_1}{r_1 + r_2}$$

Finally, the distances z are given by

$$(7.2.80) \qquad z_{1,2} = \frac{r_1 + r_2}{2}\left(E\left(\cos^2\psi\right) \mp \cos\psi\right)$$

where $E(k)$ is the complete elliptic integral of the second kind.

7.3. Vapor pressure at a curved interface: Kelvin equation

We have seen that the curvature of a surface produces a pressure jump across the surface. However, curvature produces another effect - it alters the vapor pressure of the liquid! To get a qualitative picture of this, consider the liquid-gas interface in Figure 7.1.1a. If we curve that interface just a bit downward, i.e., towards the shape of a drop or balloon (so that the pressure in the liquid is higher than in the gas), the forces on the surface molecules will change their directions, and the overall force on the liquid molecules in the direction toward the liquid decreases. Thus it is easier for these molecules to escape the liquid and so the vapor pressure is higher. Conversely, if we curve the interface upward, the forces on the surface molecules align such that for these molecules the net force directed towards the liquid is higher. Thus it is more difficult for the liquid molecules to escape and the vapor pressure is lower.

In fact, based on this qualitative argument we might start to believe that the surface tension would also be affected by curvature, and indeed it is. The surface tension dependence on curvature is described by the Tolman equation,

$$(7.3.1) \qquad \gamma(R) = \gamma_\infty (1 - 2\delta/R)$$

where γ_∞ is the surface tension of a flat surface and δ is the first order Tolman correction or Tolman length. One estimate for the Tolman length for water δ is equal -0.056 nm (Joswiak et al. (2013)), and so we can see that the surface tension dependence on curvature is important only for very small curvatures. Such curvatures are important in the early phase of nucleation, such as during boiling, condensation, or cavitation. However, we will not consider such small curvatures in this book and will neglect this correction. Furthermore, surface tension dependence on curvature is still a debated topic and is not fully understood. For more details, see Blokhuis and Kuipers (2006).

Let's now consider vapor pressure dependence on curvature from a quantitative point of view. For our derivation of the relationship between vapor pressure and curvature - the Kelvin equation we will roughly follow the derivation given by Elliott (2001). For purposes of the derivation, we will consider the device shown in Figure 7.3.1, but the results obtained can be generalized to any two phase system. Our device consists of a cylinder with a movable piston and a capillary. This device is

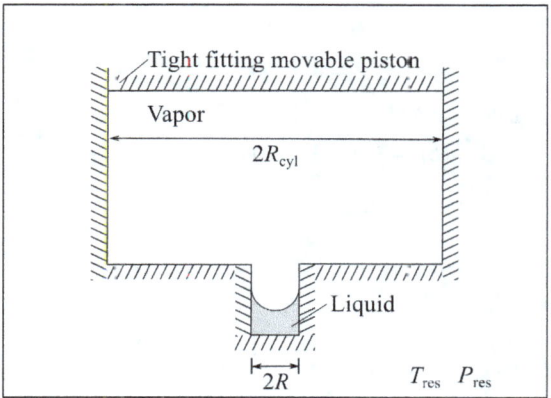

FIGURE 7.3.1. Schematic of an isolated system under consideration: a movable, tight fitting piston with a capillary inside a constant pressure, constant temperature enclosure. The radius of the
capillary is much smaller than the radius of the cylinder and is not
drawn to scale.

surrounded by a constant temperature and pressure reservoir. The reservoir and
the device comprise our isolated thermodynamic system. The movable piston allows
the pressure inside the device to be equal to that of the reservoir at equilibrium, but
does not allow mass transfer from inside the cylinder to the reservoir. Furthermore,
the walls of the device allow heat transfer between the cylinder and the reservoir,
such that at equilibrium the temperature inside the device and the reservoir are
the same. A portion of the capillary is filled with liquid and the rest of the device
is filled with the vapor of that liquid. The capillary is small enough such that
the gravitational forces on the liquid can be neglected (more on this in the later
sections). We are interested in finding the vapor pressure, i.e. the pressure of the
vapor in the cylinder at equilibrium and compare that to the vapor pressure of a
liquid interface with infinite radius of curvature - a flat surface.

At equilibrium the total variation in the entropy of an isolated system approaches zero, which for this system is

(7.3.2) $$dS_l + dS_v + dS_s + dS_{lv} + dS_{sv} + dS_{sl} + dS_{res} = 0$$

where the subscripts l, v, s indicate the entropies of the liquid, vapor, and solid
phases respectively. Note that the solid here is the container walls, and is not made
of the same material as the liquid and vapor (which are of the same molecular
composition). The subscripts lv, sv, sl refer to the entropy of the associated interphases; the subscript res refers to the reservoir. Recall the fundamental equation
of thermodynamics,

(7.3.3) $$dU = TdS - PdV + \sum_{i=1}^{n} \mu_i dN_i$$

where U is the internal energy, P is pressure, V is volume, μ_i is the chemical
potential of a particular species, and N_i is the particle number of the species. In
our system, we are only interested a single chemical species - the molecules that
comprise the liquid and the gas, for example, water molecules. For this species

(e.g., water molecules), we can rewrite the above equation as

(7.3.4) $$dS = (dU + PdV - \mu dN)/T$$

and so for the pure phases, we write

(7.3.5) $$dS_l = (dU_l + P_l dV_l - \mu_l dN_l)/T_l$$

(7.3.6) $$dS_v = (dU_v + P_v dV_v - \mu_v dN_v)/T_v$$

(7.3.7) $$dS_s = (dU_s + P_s dV_s - \mu_s dN_s)/T_s$$

For the interphases, the PdV term is replaced by the $-\gamma dA$ term (see Section 7.1)

(7.3.8) $$dS_{lv} = (dU_{lv} - \gamma_{lv} dA_{lv} - \mu_{lv} dN_{lv})/T_{lv}$$

(7.3.9) $$dS_{sv} = (dU_{sv} - \gamma_{sv} dA_{sv} - \mu_{sv} dN_{sv} - \mu_{sv,wall} dN_{sv,wall})/T_{sv}$$

(7.3.10) $$dS_{sl} = (dU_{sl} - \gamma_{sl} dA_{sl} - \mu_{sl} dN_{sl} - \mu_{sl,wall} dN_{sl,wall})/T_{sv}$$

Solid-vapor and solid-liquid interphases contain both the molecules of the fluid we are interested in (for which we omit the second subscript on the chemical terms) and the molecules of the solid wall material for which we call out the second subscript "wall". In using the surface tensions as we did, we assumed that the surface tension has no dependence on curvature. Finally, for the reservoir

(7.3.11) $$dS_{res} = (dU_{res} + P_{res} dV_{res} - \mu_{res,res} dN_{res,res})/T_{res}$$

where the second subscript on the chemical terms indicate the chemical species of the material of the reservoir, for example some other fluid that helps us keep the reservoir a constant temperature and pressure reservoir. As we will see later, the nature of this material does not matter.

Our system also has a number of constraints: Firstly, the internal energy of an isolated system is constant,

(7.3.12) $$dU_l + dU_v + dU_s + dU_{lv} + dU_{sv} + dU_{sl} + dU_{res} = 0$$

Second, the number of molecules of the fluid of interest, of the solid, and of the reservoir material must remain constant,

(7.3.13) $$dN_l + dN_v + dN_{lv} + dN_{sv} + dU_{sl} = 0$$

(7.3.14) $$dN_{s,wall} + dN_{sv,wall} + dU_{sl,wall} = 0$$

(7.3.15) $$dU_{res,res} = 0$$

Third, the total volume of an isolated system is constant, so

(7.3.16) $$dV_l + dV_v + dV_{res} = 0$$

and the volume of the walls of the device is constant,

(7.3.17) $$dV_s = 0$$

Forth, the liquid vapor interface is spherical in shape since the capillary is round and of constant cross-section. We assume that capillary is long enough so that when more liquid is added, it does not over flow into the cylinder. Only the height of the liquid in the capillary changes. Thus the liquid vapor interface area remains unchanged,

(7.3.18) $$dA_{lv} = 0$$

and the change in the area of the solid liquid interface from geometry is,

$$(7.3.19) \qquad dA_{sl} = 2dV_l/R$$

To obtain the area of the solid vapor interface we simplify subtract the area of the solid liquid interface from the total internal surface area of the device. Since the change in the amount of liquid only changes the height of the liquid in the capillary

$$(7.3.20) \qquad dA_{sv} = 2\,(dV_v - dV_l)/R_{cyl} - dA_{sl}$$

Combining equations (7.3.5) through (7.3.20) with (7.3.2) we obtain

(7.3.21)

$$\left(\frac{1}{T_l} - \frac{1}{T_{res}}\right) dU_l + \left(\frac{1}{T_v} - \frac{1}{T_{res}}\right) dU_v + \left(\frac{1}{T_s} - \frac{1}{T_{res}}\right) dU_s$$

$$+ \left(\frac{1}{T_{lv}} - \frac{1}{T_{res}}\right) dU_{lv} + \left(\frac{1}{T_{sv}} - \frac{1}{T_{res}}\right) dU_{sv} + \left(\frac{1}{T_{sl}} - \frac{1}{T_{res}}\right) dU_{sl}$$

$$+ \left(\frac{\mu_{s,wall}}{T_s} - \frac{\mu_{sv,wall}}{T_{sv}}\right) dN_{sv,wall} + \left(\frac{\mu_{s,wall}}{T_s} - \frac{\mu_{sl,wall}}{T_{sl}}\right) dN_{sl,wall} + \left(\frac{\mu_l}{T_l} - \frac{\mu_v}{T_v}\right) dN_v$$

$$+ \left(\frac{\mu_l}{T_l} - \frac{\mu_{lv}}{T_{lv}}\right) dN_{lv} + \left(\frac{\mu_l}{T_l} - \frac{\mu_{sv}}{T_{sv}}\right) dN_{sv} + \left(\frac{\mu_l}{T_l} - \frac{\mu_{sl}}{T_{sl}}\right) dN_{sl}$$

$$+ \left(\frac{p_l}{T_l} + \frac{2\gamma_{sv}}{T_{sv}}\left(\frac{1}{R} - \frac{1}{R_{cyl}}\right) - \frac{2\gamma_{sl}}{T_{sl}}\frac{1}{R} - \frac{p_{res}}{T_{res}}\right) dV_l$$

$$+ \left(\frac{p_v}{T_l} + \frac{2\gamma_{sv}}{T_{sv}}\frac{1}{R_{cyl}} - \frac{p_{res}}{T_{res}}\right) dV_l = 0$$

This equation must be valid for arbitrary variations about the equilibrium state, and so all the terms in parentheses must be zero. Furthermore, $R_{cyl} \gg R$ and so $1/R - 1/R_{cyl} \approx 1/R$. This leads to the following relations for the system at equilibrium. Firstly, and quite unsurprising, the temperature throughout the system must be the same,

$$(7.3.22) \qquad T_l = T_v = T_s = T_{lv} = T_{sv} = T_{sl} = T_{res}$$

Secondly, the chemical potentials of the wall must be the same throughout the system,

$$(7.3.23) \qquad \mu_{s,wall} = \mu_{sv,wall} = \mu_{sl,wall}$$

and the chemical potentials of the fluid of interest must the same throughout the system,

$$(7.3.24) \qquad \mu_l = \mu_v = \mu_{lv} = \mu_{sv} = \mu_{sl}$$

Unsurprisingly again, the pressure of the vapor is the same as the pressure of the reservoir, as our piston specifically is there to allow this,

$$(7.3.25) \qquad P_v = P_{res}$$

Lastly, we also have a relation between the pressure in the liquid and surface tensions:

$$(7.3.26) \qquad p_v - p_l = \frac{2}{R}\,(\gamma_{sv} - \gamma_{sl})$$

To use these relations we just obtained, we make a simplifying assumption that the vapor is an ideal gas, and recall the ideal gas law

$$(7.3.27) \qquad N = PV/R_g T$$

where R_g is the ideal gas constant. Combining this with the fundamental equation of thermodynamics, (7.3.3), we obtain that for an ideal gas (or in our case our vapor),

$$(7.3.28) \qquad \mu_v\left(T, P_v\right) = \mu_v\left(T, P_\infty\right) + R_g T \ln\left(P_v/P_\infty\right)$$

where P_∞ is any reference pressure, defining the chemical potential of the reference state, the first term on the right-hand side. Here we take it as the vapor pressure of a flat interface, i.e., the saturated vapor pressure at temperature T. Furthermore, we assume our liquid is incompressible. For an incompressible liquid,

$$(7.3.29) \qquad \mu_l\left(T, P_l\right) = \mu_l\left(T, P_\infty\right) + \bar{v}_M\left(P_l - P_\infty\right)$$

where $\bar{v}_M$ is the molar volume of the liquid and again P_∞ is any reference pressure, here taken as the vapor pressure of the liquid with a flat interface. Using the fact that the chemical potential of the liquid and the vapor phases are equal, (7.3.24),

$$(7.3.30) \qquad \mu_v\left(T, P_v\right) = \mu_l\left(T, P_l\right)$$

and

$$(7.3.31) \qquad \mu_v\left(T, P_\infty\right) = \mu_l\left(T, P_\infty\right)$$

and so we combine (7.3.28) and (7.3.29) and obtain,

$$(7.3.32) \qquad R_g T \ln\left(P_\nu/P_\infty\right) = \bar{v}_M\left(P_l - P_\infty\right)$$

We can rearrange this as

$$(7.3.33) \qquad P_\infty - P_l = \frac{R_g T}{\bar{v}_M} \ln\left(\frac{P_\infty}{P_v}\right)$$

This is known as the Kelvin equation, but in an unfamiliar form. To put it in the more familiar form, we need to use a relation between the pressure in the liquid phase and the vapor phase. We have such a relation - the Young-Laplace equation (7.2.22),

$$(7.3.34) \qquad P_l - P_v = 2\gamma_{lv}/R$$

which after substituting into (7.3.33), we obtain

$$(7.3.35) \qquad \frac{R_g T}{\bar{v}_M} \ln\left(\frac{P_\infty}{P_v}\right) = \frac{2\gamma_{lv}}{R} + \left(P_v - P_\infty\right)$$

This is the more familiar form of the Kelvin equation. Often the second term on the right-hand side is neglected, and we have the simplified Kelvin equation,

$$(7.3.36) \qquad \ln\left(\frac{P_v}{P_\infty}\right) = \frac{2\gamma_{lv}}{R} \frac{\bar{v}_M}{R_g T}$$

Now having derived the familiar form of the Kelvin equation, let's look at some implications. Substituting in material properties of water, γ_{lv} of 72 mN/m, and $\bar{v}_M$ of 18 ml/mol, we see that the change in vapor pressure is only significant for very small radii of curvature. For example, for a 1 μm radius of curvature, P_v/P_∞ is only 1.001. Only for a radius of curvature of 10 nm is change in the vapor pressure significant, P_v/P_∞ of 1.11, about 10% change. At smaller radii, such as that of 1 nm, the change in vapor pressure is very significant, with P_v/P_∞ of 2.8. The conclusion we can draw from this is that we don't have to worry about a significant change in liquid evaporation in typical (10 - 100 μm) microfluidic channels. However, the change in vapor pressure as a function of curvature has a big impact on condensation of liquids in small corners and crevices, and as a result, as we shall see later, strong attachment of particles to walls. Additionally, by making a connection between vapor pressure and solubility, another equation can be derived that shows that small particles dissolve at a much faster rate than large ones - a "Kelvin equation for solids" so to speak.

One major implication of the Kelvin equation is that in small crevices (small radius of curvature) vapor condensation occurs significantly below the saturation vapor pressure. This is sometimes known as Kelvin condensation or capillary condensation. The physical nature of this result is because the fluid molecules in the crevice experience a larger number of fluid-solid interactions. A classical example of this is a hydrophilic particle sitting on top of a hydrophilic flat surface (e.g., silicon oxide). If the particle is approximately spherical, near the contact with the flat surface, the radius of curvature approaches zero! Hence, at almost any percent humidity water will condense in this crevice. How much of the crevice and how quickly this crevice is filled depends of course on the relative humidity of the room. As we will see later, this little meniscus that forms can create a significant attractive force between the particle and the surface, preventing the removal of the particle from the surface. Since in this case the meniscus causes vapor pressure depression, significantly higher temperatures (than expected for drying the liquid on a flat surface) is required to drive the liquid from such crevices. This example is relevant for maintaining appropriate particle cleanliness level in clean rooms (one of the reasons for stringent humidity requirements); for avoiding condensation bridges in contact mode atomic force microscopy (AFM); and for avoiding stiction in microelectromechanical (MEMS) devices. Kelvin condensation also impacts particle sintering (currently of major importance in 3D printing and in powder metallurgy), and is important in nucleation theories, and in measurement of properties of porous media.

7.4. Contact angle

Consider the commonly encountered three phase system: a liquid drop on a surface. The three phases are the solid, the liquid, and the surrounding gas. For this system there are three interfaces: solid-gas, solid-liquid, and the interface we have already considered in detail: liquid-gas. Each interface has a respective surface tension. Consider a cutaway of a drop on a solid surface as shown in Figure 7.4.1. In this plane, the surface tensions are directed parallel to the interfaces and are as shown. In an equilibrium situation, the surface tensions must balance each other such that the net force on the point of contact (in our plane) of the three phases is zero. (If the net force on the line of contact was not zero, the line would move and accelerate). Thus taking the force balance in the horizontal direction we obtain,

(7.4.1) $$\gamma_{sl} + \gamma_{lg} \cos\theta - \gamma_{sg} = 0$$

or,

(7.4.2) $$\cos\theta = \frac{\gamma_{sg} - \gamma_{sl}}{\gamma_{lg}}$$

Here the angle θ is known as the equilibrium contact angle. As we can see, the contact angle depends on the properties of the solid, the liquid, and also the gas atmosphere surrounding the two phases. Notice that whether the contact angle is greater or less than 90° (whether $\cos\theta$ is negative or positive) is determined

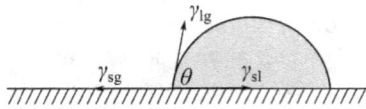

FIGURE 7.4.1. Interfacial tensions at the three-phase (gas, liquid, solid) contact line.

by whether the solid-liquid surface tension is greater than that of solid-gas. If the attraction forces of the surface to the gas phase are greater than those of the surface and the liquid phase than the surface-gas surface tension is smaller than the surface-liquid surface tension, and the contact angle is greater than 90°. As a reminder, the higher the attractive forces of one phase to another phase, the smaller the difference between the attractive forces for the other phase and itself (as for the phase to keep together it has to have higher attractive forces for itself than to another phase) and the lower the surface tension between the phases. On the other hand, if the attraction forces of the surface to the gas phase are less than those of the surface and the liquid phase, the contact angle is less than 90°.

If the contact angle is zero degrees, (the droplet spreads on the surface), the liquid perfectly wets the surface. For contact angle between zero and 90° the liquid is considered to have high wettability towards the surface. If the liquid is water, the surface is considered hydrophilic. For contact angle between 90° and 180° the liquid is considered to have low wettability towards the surface. In this case if the liquid is water, the surface is considered hydrophobic. For contact angles above 150° the surface is considered superhydrophobic.

In predicting the behavior of a liquid in contact with a particular material (and so potentially selecting the correct material for a desired behavior), it is desired to understand whether the liquid will spread over the surface or will it have a finite contact angle. For this, there is an approximate rule, the Zisman rule (after William A. Zisman). This rule assigns surface-liquid interaction a critical surface tension γ_c (see Table 1). For non-polar liquids γ_c is independent of the liquid and is just a property of the solid. If the surface tension of the liquid is less than the critical surface tension, the liquid totally wets the surface (spreads). Critical surface tension is usually obtained by studying the wetting properties of a series of liquids, commonly n-alkanes (with varying n), plotting the resulting cosine of contact angles as a function of the surface tension of the liquids (Zisman plot), and then extrapolating to where the cosine of the contact angle is equal to unity. Alkanes are chosen because they are non-polar and are easily available. This rule gives most accurate predictions for non-polar surfaces.

Another useful rule for determining if a liquid will spread is to consider if the surface is a "high energy" surface or a "low energy" surface. High energy surfaces are those for which the chemical binding energy is order 1 eV. These typically are ionic, covalent, and metallic surfaces. These surfaces typically have solid-gas interfacial tensions γ_{sg} of order 500-5000 mN/m. On these surfaces most liquids spread, as they have surface tensions much less than 500 mN/m. Low energy surfaces are for which chemical binding energy is order kT. These include polymeric surfaces (plastics) and molecular crystals. These surfaces have γ_{sg} of order 10-50 mN/m. Since many liquids have surface tensions higher than 10 mN/m, these surfaces are not typically wettable (de Gennes et al. (2004)).

Table 1: Contact angle of water and critical surface tension on common microfluidic substrates

Substrate	Contact angle (deg.)	Critical surface tension (mN/m)	Ref.
Polyvinyl alcohol (PVOH)	51	37	Accudynetest (2017)
Polyvinyl acetate (PVA)	60.6	35.3	Accudynetest (2017)
Nylon 6	62.6	43.9	Accudynetest (2017)
Polyethylene oxide (PEO, PEG)	63	43	Accudynetest (2017)
Polysulfone (PSU)	70.5	42.1	Accudynetest (2017)
Polymethyl methacrylate (PMMA, acrylic)	70.9	37.5	Accudynetest (2017)
Polyethylene terephthalate (PET)	72.5	39	Accudynetest (2017)
Polyoxymethylene (POM, polyacetal, Delrin)	76.8	37	Accudynetest (2017)
Polyvinylidene chloride (PVDC, Saran, old style Saran wrap)	80	40.2	Accudynetest (2017)
Polyphenylene sulfide (PPS)	80.3	38	Accudynetest (2017)
Acrylonitrile butadiene styrene (ABS)	80.9	38.6	Accudynetest (2017)
Polycarbonate (PC)	82	44	Accudynetest (2017)
Polyvinyl fluoride (PVF)	84.5	32.7	Accudynetest (2017)
Polyvinyl chloride (PVC)	85.6	37.9	Accudynetest (2017)
Polystyrene (PS)	87.4	34	Accudynetest (2017)
Cyclic olefin copolymer (COC)	88		Cidraprecisionservices (2017)
Polyvinylidene fluoride (PVDF)	89	31.6	Accudynetest (2017)
Polybutadiene	96	29.3	Accudynetest (2017)
Polyethylene (PE)	96	31.6	Accudynetest (2017)
Polypropylene (PP)	102.1	30.5	Accudynetest (2017)

Polydimethylsiloxane (PDMS)	107.2	20.1	Accudynetest (2017)
Fluorinated ethylene propylene (FEP)	108.5	19.1	Accudynetest (2017)
Paraffin	108.9	24.8	Accudynetest (2017)
Polytetrafluoroethylene (PTFE, Teflon)	109.2	19.4	Accudynetest (2017)
Polyisobutylene (PIB, butyl rubber)	112.1	27	Accudynetest (2017)
Glass (soda-lime, clean)	<15	30 (wet), 47 (dry)	Arkles (2011)
Quartz	29		Ethington (1990)
Silicon (typical) **	70		Yang et al. (2014)
Silicon (etched)	86-88		Arkles (2011)
Silicon nitride	28-30		Arkles (2011)
SU-8 *	79		Walther et al. (2007)
Gold (typical)	66		Arkles (2011)
Gold (clean)	<10		Arkles (2011)
Platinum	40		Arkles (2011)
Steel	70-75		Arkles (2011)
Graphite	86		Accudynetest (2017)Arkles (2011)
Diamond	87		Arkles (2011)

We assume here that the water as well as the substrate is clean (free of both residues and particles). * SU-8 can be made temporarily more hydrophilic using oxygen and other plasmas; see Walther et al. (2007) and Walther et al. (2010); ** Angle depends on cleaning method and changes with time; see Yang et al. (2014)

7.5. Detailed look into contact angle

In the last section we took a simplified look at what contact angle is and obtained the so called "equilibrium contact angle". From this we learned that the contact angle is the angle between the tangent of the liquid-gas interface and the tangent to the solid liquid interface. This definition is correct, and we will keep using this definition for other contact angles as well. However, as we will discover shortly, the picture of the droplet on a solid surface in Figure 7.4.1 is overly simplistic. Firstly, the picture of a sharp jump at the three-phase contact line as shown in Figure 7.4.1 is unrealistic. Secondly, typically the equilibrium conditions required for (7.4.2) to hold (based on the assumptions of the derivations), are almost impossible to achieve in practice. Thirdly, a single contact angle does not uniquely define the interaction between a liquid and a solid surface, especially if one is in motion with respect to the other; in practice there are multiple contact angles.

The sharp jump (discontinuity) from liquid to solid at the three-phase contact line as often drawn and as we have drawn Figure 7.4.1 is impossible. A sharp jump in a curve can be only approximated by a kissing circle of infinitesimal radius and so the curvature at such a jump would approach infinity. Thus, from the Young-Laplace equation, the capillary pressure at this location would be infinite. This is certainly not physical. To get a clearer picture, we must understand that if the molecules that comprise the liquid have any affinity to the molecules comprising solid, at true equilibrium, the solid surface will be covered by a thin film of liquid. For clarity, consider a perfectly dry quartz plate placed into an environment containing water vapor. We know that quartz is hydrophilic and so water molecules have an attraction towards silicon oxide of quartz. Thus, over time, some of the water molecules will condense on the surface of the quartz. Furthermore, these molecules will diffuse around. At equilibrium, there will be a very thin film of liquid water on the surface. As another example, imagine placing a drop of water on again an initially dry quartz substrate. The water molecules at the edges of the drop are held by attraction of other water molecules inside the drop, but are also attracted by the silicon oxide of the quartz. Over time, some of the water molecules will break the bond with their water neighbors and jump on the quartz. Here is where diffusion is balanced against the forces of attraction, and once in a while (depending on the attractive forces in the liquid) diffusion wins. Once on the quartz, the water molecules can jump further and further from the parent drop - diffuse over the quartz surface. They will do this until a very thin film of water molecules is covering the surface. In fact, since many of the surfaces we deal with in daily life are hydrophilic and since typically are in an environment of non-negligible humidity, most of those surfaces are normally covered by a very thin layer of water. Hydrophobic (oleophilic) surfaces, such as the surfaces of some plastics, can become coated with a thin layer of oil from our hands if we touch them, even if we touch them in one spot, by diffusion of oil molecules.

For the sake of pedagogical clarity, we will assume we are dealing with clean surfaces. Thus if a drop is placed on a surface, due to evaporation, surface diffusion, and attractive forces between the solid and the molecules of the liquid, a thin film of liquid will form around the drop, which for our purposes we assume extends to infinity. Thus the three-phase contact line appears more like that in Figure 7.5.1. But now there are no three phases in contact! Therefore, we refer to this object as the "apparent three-phase contact line" as we can observe it and its location macroscopically, but we know that at that location three phases do not actually contact. The shape of the liquid-gas interface in the region of the apparent three-phase contact line must have an inflection point as we transition from a region of negative curvature (in the main drop) to the flat infinitesimal film. Thus the radius of curvature changes and so there must be a force accounting for that change. As we have hinted at above, this force is the attractive force between the liquid and the solid, and we will model it as disjoining pressure. We will explain disjoining pressure in detail below.

The next detail we will consider is the assumption of equilibrium in deriving (7.4.2). Equilibrium here means that (a) the liquid in the droplet is in equilibrium with its own vapor (the gas phase); (b) the liquid in the droplet is in equilibrium with the solid substrate; and (c) the vapor is also in equilibrium with the solid substrate. We have examined what it means to be in perfect equilibrium in deriving the Kelvin equation, (7.3.36). This equation would imply that for certain properties of a fluid and surrounding vapor pressure there is a unique droplet radius. But in fact from everyday experience we observe a wide variety of droplet radii. Thus,

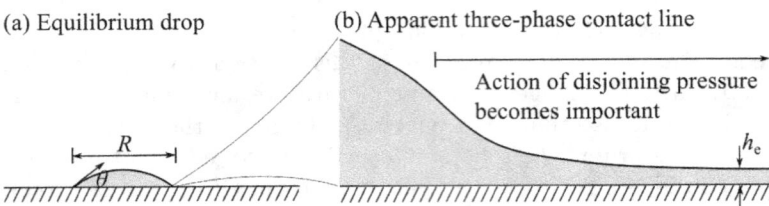

FIGURE 7.5.1. Geometry of a realistic apparent three-phase contact line. It is much more realistic to model the three-phase contact line as a transition from a large volume of liquid (shape governed by capillary forces - Young-Laplace equation) to a thin film of liquid (with shape governed by the disjoining pressure). The thin film surrounding the droplet is sometimes referred to as the precursor film.

what we observe are droplets which are not truly at equilibrium. Furthermore, this implies that the equilibrium contact angles that we simplify measure by placing a droplet on a surface and observing the drop shape are not truly in equilibrium. Furthermore, when $\gamma_{sv} > \gamma_{sl}$ the molecules in the vapor phase are more attracted to the surface than the molecules in the liquid phase. This would imply that at equilibrium, the surface would get coated with a thin film of condensed vapor. This in turn changes the initial surface tension as the droplet of liquid actually would sit on a thin film of (potentially another) fluid. This is sometimes accounted for by replacing γ_{sv} with γ_{hv} where h represents the thin fluid film, thickness h adsorbed onto the surface of the solid, as it is this film that is truly in contact with the vapor phase.

Finally, let's examine the last problem with the simple expression for the contact angle, the absence of a unique contact angle. To observe this, we can perform an experiment shown in Figure 7.5.2. We place a drop on a clean, smooth plate (e.g, an atomically smooth silicon wafer) with a small orifice that allows us to pump in liquid into the drop. Before pumping liquid in, we measure the drop's contact angle with the material (Figure 7.5.2a). This is what we call the equilibrium contact angle. Next, very carefully and slowly we pump in liquid into the drop. As we watch the drop something unexpected happens, (at least to a first time observer) - the drop does not change its radius at the wafer! Instead, the apparent three-phase contact line remains fixed, but the drops contact angle increases! (Figure 7.5.2b). Eventually, however, the contact angle becomes large enough, and the apparent three-phase contact line moves. The contact angle at which this occurs is termed the advancing contact angle.

We can repeat this experiment again with a slight alteration. Starting with a drop at equilibrium, we instead of pumping liquid into the liquid, carefully withdraw the liquid. Once again, while the droplet shrinks, and the contact angle decreases, the apparent three-phase contact line does not move until a certain angle is reached! This angle is termed the receding contact angle. For these angles, $\theta_r < \theta_e < \theta_a$ where the subscripts indicate receding, equilibrium, and advancing respectively.

The "sticking" of the contact line in both the advancing and receding case is often blamed on particles, contamination, or imperfections in the surface. While these definitely are observed to pin contact lines, we would like to strongly stress

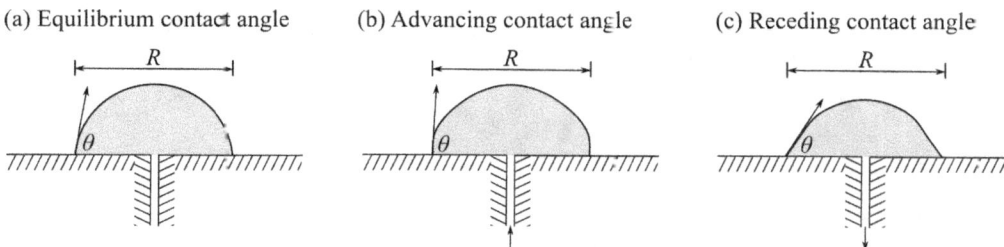

FIGURE 7.5.2. Advancing and receding contact angles of a liquid drop on a solid surface with an orifice. In (a) we have a drop that has reached equilibrium. In (b) we pump liquid into the drop and the common observation is that the radius of the drop does not change, while the contact angle increases (until a point). The contact angle at the point at which the three-phase contact line begins to move (the radius increase) is termed the advancing contact angle. In (c) we pump the liquid out of the drop, and again, the common observation is that the radius of the drop does not change, while the contact angle decreases (again, until a point). The point at which the radius of the drop decreases and the three-phase contact line begins to move is called the receding contact angle.

that the advancing and receding contact angle phenomena occurs even on smooth surfaces, carefully cleaned as to be contamination and particle free (Extrand and Kumagai (1997)). Furthermore, if we plot the contact angle as a function of velocity for the advancing and receding contact angles together, we get a jump at zero velocity (where we switch directions of pumping in vs. out of the drop). This discontinuity can be removed if we imagine that at the contact line, the droplet is in a state of very slow, "microscopic" motion, as liquid is transferred from the bulk of the drop (left portion of Figure 7.5.1b) to the thin film (right of Figure 7.5.1b) (Starov et al. (2007)). We list the equilibrium, advancing and receding contact angles for a few common solvents on a quartz substrate in Table 2.

7.6. Contact angle on rough surfaces

Up till now we have examined contact angle on ideal, smooth surfaces, and we have already managed to run into considerable trouble. Real surfaces are often rough and present even more trouble. We here will examine two most common models for contact angle on real rough, heterogeneous surfaces: the Wenzel model and the Cassie-Baxter model. These models are commonly used to model both common real naturally rough surfaces as well as specially made nanostructured surfaces. While these two models have been extensively experimentally verified and agree with experiments rather well, there are also many cases where experimental behavior deviates from that predicted by these models. Thus the validity of these models is (of this writing) still hotly debated and improvements to these models are still being sought (for further discussion see Bormashenko (2008)). Nevertheless, because of their practical usefulness and widespread use, we will provide a simple conceptual derivation of these as an extension of Young's equation.

TABLE 2. Equilibrium, advancing, and receding contact angles for various common solvents on quartz.

Solvent	Surface tension (mN/m)	Equilibrium contact angle (deg.)	Advancing contact angle (deg.)	Receding contact angle (deg.)
Acetic acid	27.1	6	12	3
Acetone	22.7	6	-	-
Benzene	28.2	12	18	6
Carbon tetrachloride	26.2	9	12	8
Chloroform	26.7	8	11	8
Cyclohexane	26.2	9	12	8
Ethanol	21.9	0	0	0
Ethylene glycol	48.0	35	36	4
Mercury	485.5	136	-	-
Methanol	22.2	0	0	0
Toluene	27.9	9	11	7
Water	72.1	29	39	8

Contact angles from Ethington (1990), surface tension from Haynes (2017) at 25°C.

We begin by defining the work of adhesion per unit area, $w_{a,sl}$, as the amount of energy per unit area required to separate a liquid of interest and the solid of interest from each other in the vapor media. When the liquid comes in contact with the solid as in Figure 7.4.1, then

$$(7.6.1) \qquad w_{a,sl} = \gamma_{sg} + \gamma_{lg} - \gamma_{sl}$$

When this value is positive, the liquid is attracted to the solid, and when it is negative, the two are repelled from each other. On the other hand, when the liquid is just a drop "floating" in the gas phase, it's total interfacial energy, from definition of surface tension, is just

$$(7.6.2) \qquad W_{tot,l} = \gamma_{lg} A_{sphere}$$

where A_{sphere} is the surface area of the spherical droplet. When the liquid is in contact with a smooth ideal solid, (Figure 7.4.1) then the total area of the liquid drop is just the sum of the area in contact with the gas A_{lg} and the area in contact with the solid A_{sl}. However, if we somehow first deform the drop into this shape, but not place it onto the solid surface, its total interfacial energy would just be

$$(7.6.3) \qquad W_{tot,l} = \gamma_{lg} \left(A_{lg} + A_{sl} \right)$$

When, after this deformation we do finally place it onto the solid, energy of adhesion comes into play and so now the total interfacial energy of the drop is

$$(7.6.4) \qquad W_{tot,l} = \gamma_{lg} \left(A_{lg} + A_{sl} \right) - w_{a,sl} A_{sl}$$

Substituting in the definition of energy of adhesion, (7.6.1),

$$(7.6.5) \qquad W_{tot,l} = \gamma_{lg} A_{lg} - \left(\gamma_{sg} - \gamma_{sl} \right) A_{sl}$$

Furthermore, recall that we model the liquid drop (Figure 7.4.1) as a section of a sphere. For a section of a sphere, it can be shown that $dA_{lg}/dA_{sl} = \cos\theta$, with θ

as defined in Figure 7.4.1. Thus taking a variation of (7.6.5),

$$(7.6.6) \qquad dW_{tot,l} = \gamma_{lg} dA_{sl} \cos\theta - (\gamma_{sg} - \gamma_{sl}) \, dA_{sl}$$

At equilibrium, the interfacial energy is minimized, so the differential $dW_{tot.l} = 0$. Applying this to (7.6.6), we arrive at

$$(7.6.7) \qquad \cos\theta = \frac{\gamma_{sg} - \gamma_{sl}}{\gamma_{lg}}$$

the classical Young's equation for the equilibrium contact angle. Now, let's use the same approach to derive an expression for contact angle where both the surface is rough and the liquid wets all the nooks and crannies of this roughness, a behavior of how a liquid interacts with a rough surface as indicated by Wenzel (Figure 7.6.1a). (The question of whether the liquid actually does wet all or some of the nooks and crannies and how it does this is currently a subject of debate in the field and this question is what leads some to question the validity of Wenzel's model; nevertheless here we will assume that the liquid indeed does). In this case, we can define a roughness factor, $r = A_{r,sl}/A_{g,sl}$ where $A_{r,sl}$ is the true area of the solid liquid interface, taking into account the area of nooks and crannies; and $A_{g,sl}$ is the geometric area or the projection of the rough area vertically onto the drop. $A_{g,sl}$ is the area which we observe macroscopically. Substituting in this factor into (7.6.5)

$$(7.6.8) \qquad W_{tot,l} = \gamma_{lg} A_{lg} - (\gamma_{sg} - \gamma_{sl}) \, r A_{g,sl}$$

and proceeding as before, where now $dA_{lg}/dA_{g,sl} = \cos\theta$, we obtain

$$(7.6.9) \qquad \cos\theta_W = r \frac{\gamma_{sg} - \gamma_{sl}}{\gamma_{lg}}$$

where θ_W is still the observed contact angle as defined Figure 7.4.1. We add the subscript W to emphasize that it comes from the Wenzel's model. We see that in Wenzel's model the roughness acts to increase the wetting property of the substrate. If the substrate was initially hydrophobic, roughness makes it even more hydrophobic. If the substrate is hydrophilic, the roughness in turn makes it even more hydrophilic.

We can make another generalization to the derivation of Young's equation by considering the case where the solid-liquid interfacial tension is a weighted mean of solid-liquid interfacial tensions of various materials comprising the real surface; and solid-gas interfacial tension is also a weighted mean of solid-gas interfacial tensions of various materials comprising the real surface (Figure 7.6.1b). We can use this to model heterogeneous surfaces made up of materials with different solid-liquid interfacial tensions. This model is called the Cassie-Baxter model. Substituting this into (7.6.6) we obtain

$$(7.6.10) \qquad \cos\theta_{CB} = \frac{\sum_i w_i \left(\gamma_{i,sg} - \gamma_{i,sl} \right)}{\gamma_{lg}}$$

where w_i are the weights that are assumed to be proportional to the composition of the surface so that

$$(7.6.11) \qquad \sum_i w_i = 1$$

FIGURE 7.6.1. Schematic of liquid behavior as modeled by the Wenzel (a) and Cassie-Baxter (b) models. In the Wenzel model, the surface has finite roughness and the liquid wets all of the rough walls. In the Cassie-Baxter model the liquid sits atop of a smooth surface made up of different, but uniformly distributed materials. The model assumes that the droplet experiences a weighted average of the properties of these solid materials.

Expanding (7.6.10) in terms of the contact angle of each of the individual material θ_i comprising the composite surface, we can write

$$(7.6.12) \qquad \cos\theta_{CB} = \sum_i w_i \cos\theta_i$$

Cassie-Baxter model is often used to model nanostructured surfaces (natural, e.g., lotus leaf, or man-made) where a drop is thought to sit atop of asperities (spikes) of the surface. In this case there are two materials, the first being the material of the spikes, and the second being air that is trapped between the spikes. The spike material is assumed to have some contact angle (typically assumed to be that of the bulk material), while the cosine of the contact angle of air is set to be -1.

7.7. Disjoining pressure

When we bring two objects together, the molecules at the surface of one object apply forces on the molecules at the surface of the other object. The macroscopic observation and treatment of net action of these forces is termed the disjoining pressure. The term disjoining is used because this was studied from a point of view of the two surfaces repelling each other; it is termed pressure because it is a force per unit of the surface area. Out the several components comprising disjoining pressure, we will consider three that are most relevant to liquid-solid interactions: (a) the molecular component, (b) the electrostatic component, and (c) the structural component.

The molecular component of disjoining pressure arises from the molecular (van der Waals or dispersion) forces between molecules. These attractive forces are frequently modeled by a Lennard-Jones potential. At large distances (compared to typical molecular spacing in a solid), this potential scales as $1/r^6$, where r is the distance between two molecules. We here are interested in a sandwich of three phases, where a thin phase 3 with thickness h, is sandwiched between two semi-infinite phases 1 and 2. For this problem, one may estimate the molecular component of the disjoining pressure by either summing individual London-van der Waals interactions, or by considering the fluctuating electromagnetic field at the surfaces. From the later analysis (Deryaguin et al. (1987)) we obtain the following scaling for the disjoining pressure:

$$(7.7.1) \qquad \Pi_m(h) = \begin{cases} A/h^3 \text{ for } h < \lambda \\ B/h^4 \text{ for } h > \lambda \end{cases}$$

Here λ is a characteristic wavelength of IR absorption spectra of the three materials, typically order 2 - 10 μm (Bergstrom (1997)); A is the net scaled Hamaker constant, and B is another constant. Positive A indicates repulsion of the surfaces. For thin films we are interested in, we will be mostly using the top expression. The scaled Hamaker constant is

$$(7.7.2) \qquad A = -A_H/6\pi$$

where A_H is the net Hamaker constant, which for our sandwich is

$$A_H = A_{33} + A_{12} - A_{13} - A_{23}$$

where A_{ij} are the Hamaker constants for the individual materials or their interactions and can be sometimes found tabulated for common material (quartz, mica, water); net Hamaker constants can also sometimes be found tabulated; for a large table, see Bergstrom (1997). For many materials (e.g., quartz) for a material-water-gas sandwich A is order 10^{-21} J (Bergstrom (1997)). Hence for water layers thickness h of 1 nm and 100 nm, the disjoining pressure is 10^6 Pa ($\sim$10 atm) and 1 Pa respectively. We are mostly interested in films of order 1 nm thickness as those are the ones we expect to coat substrates ahead of drops (Figure 7.5.1b). For water drops, capillary (Young-Laplace) pressure reaches the magnitude of 10^6 Pa only for radius of curvature of 100 nm! And is of course much smaller for larger radii of curvature. For example, for a drop with a radius of curvature of 1 mm, the capillary pressure is only 10^2 Pa. Hence we see that disjoining pressure can substantially distort droplet shape in the vicinity of the apparent three-phase contact line.

Electrostatic component of disjoining pressure arises due to the presence of an electrical double layer at material interfaces. We will discuss electrical double layers in considerable detail in the electrokinetics section of the book. Here, briefly, for a liquid touching a solid surface, the solid becomes charged when in contact with the liquid. Thus the surface carries a potential, which roughly can be approximated by a ζ (zeta) potential (see Section 15.3.21). This charge brings counterions (ions of opposite charge) to the liquid-solid interface. However, the attractive force of the electrical field attracting the ions to the solid-liquid interface is balanced by the diffusion of ions, hence the counterion layer is diffuse. This counterion cloud has a characteristic length known as the Debye length (κ^{-1}). Thus, when two surfaces are separated by a water film and they try to interact electrostatically, some of this interaction is shielded by the presence of these diffuse ion clouds. For the case of low zeta potentials ($|\zeta| < R_g T/F$), the electrostatic component of zeta potential is given by

$$(7.7.3) \qquad \Pi_e(h) = \frac{\varepsilon\varepsilon_0 \kappa^2}{8\pi} \frac{2|\zeta_1||\zeta_2|\cosh(\kappa h) - (\zeta_1^2 + \zeta_2^2)}{\sinh^2(\kappa h)}$$

where ε is the dielectric constant of water (Deryaguin et al. (1987)). This relation can be further simplified assuming the two surfaces are oppositely charged and $h \gg \kappa^{-1}$ to

$$(7.7.4) \qquad \Pi_e(h) = -\frac{\varepsilon\varepsilon_0}{8\pi} \frac{(|\zeta_1| - |\zeta_2|)^2}{h^2}$$

Here we have an inverse square scaling in h for the disjoining pressure, and as a result its effects are felt at larger distances than the molecular disjoining pressure. Also notice that for oppositely charged surfaces, unsurprisingly the pressure between them is attractive (Deryaguin et al. (1987)). Structural component of disjoining pressure arises due to the orientation of polar liquid molecules (e.g., water) near solid-liquid and liquid gas interfaces. For example, near a negatively charged surface water molecules orient themselves such that the positive part of the dipole is facing the charged surface. The layer of water molecules above this also align themselves according to this orientation. This alignment would continue forever if it was not opposed by thermal motion of the molecules. Thus the balance between the electric field and the thermal motion of the molecules sets a certain characteristic length from the surface where this alignment is considerable. This layer is sometimes referred to as the hydration layer. If we have two surfaces with such hydration layers, and these are brought in close to each other such that hydration layers overlap, this results in either attraction or repulsion of the two surfaces. The structural component of disjoining pressure scales as

$$(7.7.5) \qquad \Pi_s(h) = K \exp(-h/\lambda_K)$$

where K is an empirically determined constant and λ_K is the characteristic thickness of the hydration layer and is roughly the correlation length of the water molecules in an aqueous solution. Typically λ_K is order 1 nm (Deryaguin et al. (1987)).

The total disjoining pressure is then the sum of the molecular, electrostatic, and structural components, and so has the form of

$$(7.7.6) \qquad \Pi(h) = \frac{A}{h^3} + \frac{Z}{h^2} + K \exp(-h/\lambda_K)$$

where Z represents the contribution of the zeta potentials, see (7.7.4). The values of the constants determine which components dominate but in general three shapes of $\Pi(h)$ arise. These are termed disjoining pressure isotherms because they represent disjoining pressure at constant temperature and we show these in Figure 7.7.1. The complete wetting curve arises when the molecular component (first term) dominates. The partial wetting curve arises due to the addition of the other components. Complete wetting is prominent in the case of oils on common materials (e.g., glass, quartz, metal) and partial wetting is prominent in the case of aqueous solutions on the same materials.

7.8. Balance of forces in capillary phenomena

A number of forces including surface tension, viscous, inertial, gravitational, and others are important in capillary phenomena. However, considering them all at once in all situations is of course laborious and unnecessarily difficult. Thus, we consider their relative magnitude and so consider the appropriate dimensionless numbers. We are already familiar with the Reynolds number, the ratio of inertial to viscous forces. We here are typically concerned with low Reynolds number flows, where viscous forces dominate. Here we will also consider the capillary number,

$$(7.8.1) \qquad Ca = \frac{\mu U}{\gamma}$$

where U is the characteristic fluid velocity. The capillary number represents the ratio of viscous forces to surface tension forces. We will be typically concerned with

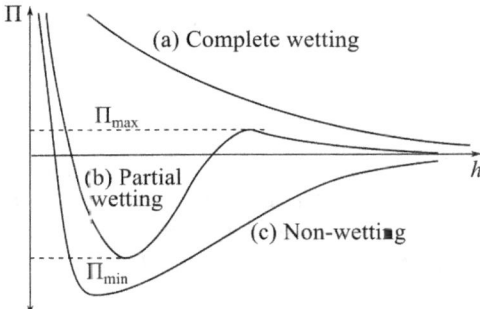

FIGURE 7.7.1. Three types of disjoining pressure isotherms: (a) case of complete wetting (e.g., oil films on glass quartz, or metal), (b) partial wetting (aqueous solutions on glass, quartz, or metal) (c) non-wetting.

low capillary numbers. A related dimensionless number is the Marangoni number, which is often written as

$$(7.8.2) \qquad Ma = -\frac{d\gamma}{dT}\frac{L\Delta T}{\mu\alpha}$$

where L is the characteristic length scale, and here α is the thermal diffusivity. Marangoni number is a ratio of surface tension forces specifically caused by surface tension gradient to the viscous forces. The surface tension gradient is usually caused either by a temperature gradient (as we have written in (7.8.2)) or by a chemical gradient. The Marangoni number is not be confused with the Mach number, which often has the same abbreviation. The next number of interest is the Bond number (sometimes called the Eotvos number),

$$(7.8.3) \qquad Bo = \frac{\Delta\rho g L^2}{\gamma}$$

where $\Delta\rho$ is the density difference between two phases, and g is the gravitational acceleration. Bond number is the ratio of gravitational forces to surface tension forces. It is also related to the Deryagin number, De (sometimes called the Goucher number), $Bo = 2De^2$. Related to the Bond number is the concept of capillary length. It is the characteristic length scale for an interface between two fluids that is subject to both gravity and surface tension and is defined as

$$(7.8.4) \qquad \lambda_c = \sqrt{\frac{\gamma}{\rho g}}$$

This length scale is equivalent to the length scale in the definition of the Bond number, when the Bond number is equal to unity. When length scales in the problem (such as droplet size or orifice size) are smaller than the capillary length, gravitational effects can be typically neglected. Typically for clean water at room temperature, capillary length is about 2.7 mm. For mercury (surface tension, 486.5 mN/m, density, 13.5 g/cm^3), capillary length is 1.9 mm. For soap bubbles (filled with air) the capillary length is about 4 m!

The next number of interest is the Weber number, which represents the ratio of inertial forces to surface tension forces,

$$(7.8.5) \qquad\qquad We = \frac{\rho U^2 L}{\gamma}$$

Weber number is also the product of capillary number and Reynolds number.

Another number of interest is the Laplace number (also called Suratman number), which represents the ratio between surface tension and momentum transport (and momentum dissociation) in a fluid. It is written as

$$(7.8.6) \qquad\qquad La = \frac{\gamma \rho L}{\mu^2} = \frac{Re^2}{We}$$

Laplace number is related to the Ohnesorge number, where

$$(7.8.7) \qquad\qquad Oh = \frac{1}{\sqrt{La}} = \frac{\mu}{\sqrt{\rho \lambda L}}$$

Laplace (or Ohnesorge) number is important in droplet ejection out of nozzles, such as inkjet printing. For example, in inkjet printing liquids that have a Laplace number between 1 and 100 are considered jettable with a nozzle of the characteristic length scale (Derby (2010)).

CHAPTER 8

Static wetting

In this section we consider the behavior of liquids interacting with a solid (either a flat substrate or in a microfluidic channel) which are governed purely by surface tension and disjoining pressure. We will neglect the effects of gravity, effects of inertia, and of user imposed pressure. We will first consider stability of thin liquid films on flat solid substrates as a function of the surrounding vapor pressure. Next we will consider the shape of a droplet in the vicinity of the apparent three-phase contact line sitting on a flat surface and that of a meniscus. We then will discuss the shape of a drop between two plates and the forces involved, as well as the forces on a spherical particle on a plate surrounded by a meniscus. Lastly we will discuss the wetting of corners.

8.1. Stability of thin films as a function of vapor pressure

As we have discussed in previous sections real solid substrates are normally covered by a thin liquid film. The free energy of such thin film coated substrate is lower than that of a bare substrate. Hence there is no true three-phase contact line at equilibrium - the three-phase contact line is only apparent. Here we will discuss the stability of such films.

The excess free energy per unit area of a flat equilibrium film of thickness h_e on a solid and in equilibrium with the vapor is

$$(8.1.1) \qquad \frac{\Delta G}{S} = \gamma + h_e \Delta p + f_D(h_e) + \gamma_{sl} - \gamma_{sv}$$

where S is the surface area of the region covered by the film; $\Delta p = p_a - p_l$ is the express pressure (difference between the pressure in the air and that in the liquid film); and f_D is the disjoining pressure potential defined as

$$(8.1.2) \qquad \Pi(h) = -\frac{df(h)}{dh}$$

or

$$(8.1.3) \qquad f_D(h) = \int_h^\infty \Pi(h)dh$$

For a process to occur spontaneously, ΔG should be negative. Thus for spontaneous creation of the liquid film and so for a liquid film to stably exist ΔG should be negative. Furthermore, since we take the liquid film to be in equilibrium with vapor, excess pressure is already determined from the Kelvin equation

$$(8.1.4) \qquad \Delta p = \frac{R_g T}{\bar{v}_M} \ln\left(\frac{P_\infty}{P_v}\right)$$

At equilibrium, excess free energy should be at a minimum, which implies

$$(8.1.5) \qquad \frac{d\Delta G}{dh_e} = 0$$

which determines that the excess free energy is at an extremum, and

$$(8.1.6) \qquad \frac{d^2\Delta G}{dh_e^2} > 0$$

which determines that the extrema is a minimum. The first requirement (8.1.5) applied to (8.1.1) produces

$$(8.1.7) \qquad \Delta p = \Pi(h_e)$$

and the second requirement results in

$$(8.1.8) \qquad \frac{d\Pi(h_e)}{dh_e} < 0$$

Let's now use these to consider the disjoining pressure curves in Figure 7.7.1. The complete wetting curve always monotonically decreases and so (8.1.8) is always satisfied. This means that for a complete wetting case the liquid film is always stable. This is unsurprising because this is how we define "complete wetting" - a case where the liquid film on the solid is always stable. For the partial wetting case the situation is more complicated. For films in the range between zero and when $\Pi_{\min}$ is reached, (8.1.8) is satisfied and so the films are always (absolutely) stable. These films are known as α-films. For films of greater thickness, (8.1.8) is not satisfied and these films are metastable. These films are known as β-films.

Now let's consider the stability of a droplet on a liquid film. Since normally solid substrates are covered by liquid films, we are interested in a surface tension between the solid, the thin liquid film and the vapor to replace the surface tension between the solid and the vapor in Young's equation, (7.4.2). We can rewrite (8.1.1) as

$$(8.1.9) \qquad \frac{\Delta G}{S} = \gamma_{svh} - \gamma_{sv}$$

where

$$(8.1.10) \qquad \gamma_{svh} = \gamma + h_e\Delta p + \int_h^\infty \Pi(h)dh + \gamma_{sl}$$

is the new solid-film-vapor surface tension that we seek. Substituting this into Young's equation (7.4.2) we obtain

$$(8.1.11) \qquad \cos\theta_e = \frac{\gamma + h_e\Delta p + \int_h^\infty \Pi(h)dh}{\gamma} = 1 + \frac{h_e\Delta p}{\gamma} + \frac{1}{\gamma}\int_h^\infty \Pi(h)dh$$

This result is known as the Deryaguin-Frumkin equation (Deryaguin et al. (1987)). For a drop to exist on the substrate (here covered by a liquid film), cosine term has to be bounded between -1 and 1. From Young-Laplace equation, $\Delta p/\gamma \sim -1/R$, where R is the radius of curvature of the drop. Thus, the second right-hand side term scales as h_e/R. Typically h_e is order 1 nm (and certainly < 100 nm), while R typically scales as the radius of the drop, typically 1 mm. Hence h_e/R is typically negligible. Thus, to satisfy that cosine term be bounded between -1 and 1, the integral term in (8 1.11) must be negative and sufficiently comparable with unity in magnitude, i.e.,

$$(8.1.12) \qquad \int\limits_{h}^{\infty} \Pi(h)dh < 0$$

When we apply this condition to the complete wetting case (Figure 7.7.1), we see that the integral of the disjoining pressure is never negative and so in this situation equilibrium droplets do not exist. Instead such a droplet placed on such a surface would spread out - hence the name complete wetting. Furthermore, when the vapor above the droplet is oversaturated, according to the Kelvin equation Δp is negative. However, from the stability of films condition (8.1.7), excess pressure must equal to the disjoining pressure for a stable film. From Figure 7.7.1 we see that the disjoining pressure is never negative and so the film that results in this case is not stable! The droplet spreads and then evaporates! Under oversaturation conditions, the liquid droplets are not on the surface, but are in the form of mist in the air. On the other hand when the excess pressure is positive (8.1.7) is satisfied and a film exists.

When we apply the (8.1.12) condition to the partial wetting case we see a different picture. The integral of disjoining pressure can be negative, and depending on the shape of the curve, there can be a stable droplet on the surface all the way from a thickness h where the disjoining pressure first crosses zero to a large h. For the oversaturated condition, $\Delta p < 0$, a constant value of Δp intersects the disjoining pressure curve in two places. The intersection at the lower thickness represents the stable film thickness, and intersection at the higher thickness represents an unstable film thickness. So what would be a radius of curvature of such a stable drop on a stable film? Since $\Delta p/\gamma \sim -1/R$ and from (8.1.7) $\Delta p = \Pi(h_e)$

$$(8.1.13) \qquad R \sim \gamma/|\Pi(h_e)|$$

For water with surface tension of 72 mN/m and for surfaces having a $|\Pi(h_e)|$ of 10^5 Pa ($\sim$1 atm) and 10^6 Pa ($\sim$10 atm), the equilibrium radius of curvature is 720 nm and 72 nm respectively. These are a fairly small drops, and for the 72 nm, the drop would likely not be spherical as the disjoining pressure would begin to affect the droplet shape.

For the partial wetting case where $\Delta p > 0$ (a meniscus rather than a droplet) the situation splits into two different subcases, depending if $\Delta p > \Pi_{max}$ or $\Delta p < \Pi_{max}$ (see Figure 7.7.1). For $\Delta p > \Pi_{max}$ there is only one crossing where Δp crosses the disjoining pressure curve. At this point the slope of the disjoining pressure curve is negative and (8.1.8) is also satisfied. This subclass results in a stable α-film with a small thickness. For $\Delta p < \Pi_{max}$ there are three crossings of Δp and the disjoining pressure curve and hence three solutions. The smallest thickness solution, the first crossing, has a negative slope of the disjoining pressure curve and hence is stable by (8.1.8). This thickness (and solution) corresponds to the stable α-film. The next crossing has a positive slope of the disjoining pressure curve and so is unstable by (8.1.8). The third crossing has a negative slope of the disjoining pressure curve and so corresponds to a stable solution again. This film is referred to as a β-film. The excess free energy of the β-films is higher than that of the α-films and so β-films are less stable. Furthermore, if we take a completely dry surface and place it in an atmosphere of vapor, the molecules of the vapor will adsorb to the surface until an α-film is created. The thickness of the film does not increase past this value and we cannot obtain a β-film with this method. We can only obtain a β-film by coating a surface with a puddle of liquid (or a thick film) and allowing the liquid to evaporate and the film to thin until a stable β-film is formed. For this reason α-films are sometimes referred to as adsorption films, whereas β-film are referred to as wetting films (Deryaguin et al. (1987)).

If we imagine a meniscus, for example in a microfluidic chamber approximated by two parallel flat plates, we may ask is there any special chamber height at which the behavior of the equilibrium meniscus changes. In fact, there indeed is such a height, as a result of existence of $\Pi_{\max}$. Using (8.1.13) we can define a critical radius of curvature,

$$(8.1.14) \qquad\qquad R_{\max} \sim \gamma/\Pi_{\max}$$

For water with surface tension of 72 mN/m and for surfaces having $\Pi_{\max}$ of 1 kPa, $R_{\max}$ is 72 μm - very much in the range of common microfluidic chamber dimensions. If the height of the chamber is significantly smaller than $R_{\max}$ then there can be only a single crossing between the excess pressure and the disjoining pressure and only α-films can form in such a chamber. On the other hand, if the chamber is significantly higher than $R_{\max}$ both α-films and the thicker β-films can exist in such a chamber.

One implication of the thickness of the film ahead of the meniscus is that the amount of diffusive transport through the film. For example, we may be concerned about species dissolved in the liquid forming the meniscus from reaching some area of the device prematurely. For example, they maybe antigens dissolved in the liquid from reaching antibodies or aptamers immobilized on the surface in some part of the microfluidic device. If the film is thicker, there will be higher flux of molecules (product of diffusive flux and cross-sectional area) and so more antigens will prematurely reach the antibodies. However, depending on the application, such flux might be negligible to be concerned about.

8.2. Liquid shape in the transition from bulk to film

Having discussed the presence and stability of liquid films ahead of drops and menisci, let's now discuss the transition region between this film and the bulk liquid inside the drop, and how disjoining pressure affects this shape. The shape of this region is what we macroscopically observe and describe as the equilibrium contact angle. Let's first consider a meniscus between two flat plates. The excess free energy of the surface at equilibrium, similarly to (8.1.1) can be written as

$$(8.2.1) \qquad \Delta G = \int \left(\gamma \left(\sqrt{1 + h'} - 1 \right) + \Delta p \left(h - h_e \right) + f_D \left(h, h' \right) - f_D \left(h_e \right) \right) dx$$

where $h' = dh/dx$. Note that

$$(8.2.2) \qquad\qquad \int \sqrt{1 + h'}\, dx$$

represents the arc length, and for this geometry is proportional to S in (8.1.1). In writing this, two assumptions are made: (1) that the surface force energy density, f_D depends only on the film thickness; (2) that the surface tension retains it bulk value even in the interface region. Additionally, the flat film in front of the meniscus must be stable, and so we must also satisfy (8.1.5). Thus, taking the derivative with respect to h of (8.2.1) we obtain,

$$(8.2.3) \qquad \gamma \frac{h''}{\left(1 + h'^2\right)^{3/2}} - \frac{\partial f_D}{\partial h} + \frac{\partial^2 f_D}{\partial h'^2} h'' + \frac{\partial^2 f_D}{\partial h \partial h'} h' = \Delta p$$

The solution for this is given by Starov et al. (2007) and here we will just present and discuss the result. For a chamber height $2H$ (see Figure 8.2.1), the liquid profile is determined by

$$(8.2.4) \qquad \frac{1}{\sqrt{1 + h'}} = \frac{1}{\gamma} \left[\Delta p \left(H - h \right) - \int_h^H \Pi \left(h \right) dh \right]$$

The liquid profile can then be determined numerically from this equation. However, for the purposes of analytical estimation, we consider here three special regions of the meniscus. In the first region, one at the center of the chamber (far from the walls), disjoining pressure can be neglected and (8.2.3) approaches the Young-Laplace equation,

$$(8.2.5) \qquad \Delta p = \frac{h''}{\left(1 + h'^2\right)^{3/2}}$$

For this geometry, this has a solution of

$$(8.2.6) \qquad \left(H - h \right)^2 + \left(\frac{\gamma}{\Delta p} - x \right)^2 = \left(\frac{\gamma}{\Delta p} \right)^2$$

or since $\gamma / \Delta p = R$

$$\left(H - h \right)^2 + \left(R - x \right)^2 = R^2$$

The second region, is where the film is completely flat, $h = h_e$ so $h' \approx 0$ and $h'' \approx 0$. In this region, as expected,

$$(8.2.7) \qquad \Delta p = \Pi \left(h \right)$$

Now, let's consider the third region - the transition between the first two. To estimate the effective observed equilibrium contact angle, which is defined by this region, we construct that for our geometry (see Figure 8.2.1),

$$(8.2.8) \qquad \Delta p = \frac{\gamma \cos \theta_e}{H}$$

In this region, we can assume, $h \approx h_e$ and $h' \approx 0$, and so (8.2.4) becomes,

$$(8.2.9) \qquad \sim = \left[\frac{\gamma \cos \theta_e}{H} \left(H - h_e \right) - \int_h^H \Pi \left(h \right) dh \right]$$

or

$$(8.2.10) \qquad \cos \theta_e = \frac{1 + \frac{1}{\gamma} \int_h^H \Pi \left(h \right) dh}{1 - h_e / H} \approx 1 + \frac{1}{\gamma} \int_h^\infty \Pi \left(h \right) dh$$

Once again we obtained the Deryaguin-Frumkin equation (8.1.11) with the assumption of h_e / H is small or equivalently of h_e / R is small. This equation tells us that it is the disjoining pressure that determines the equilibrium contact angle.

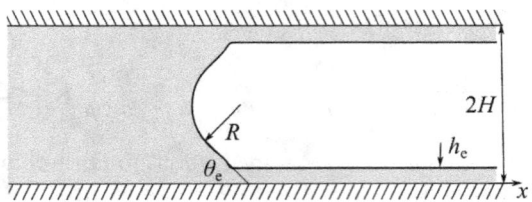

FIGURE 8.2.1. Meniscus shape of a liquid constrained by two flat plates. The thin film ahead of the meniscus is greatly exaggerated for clarity. Typically $R \approx H$ and $h_e \ll R$.

8.3. Shape of a drop between two plates and associated forces

A drop sandwiched between two parallel plates is a situation that often is encountered in microfluidic devices. We see this geometry in droplet microfluidics, as a cuvette in certain low volume spectrometers, and even in things as mundane as holding a cover slip on a glass slide. Here we are interested in finding the shape of the drop at the liquid-gas interface as this determines the forces on the plates. To do this, we perform a geometrical construction shown in Figure 8.3.1. We consider two plates with, for generality, different contact angles for the liquid. This is quite realistic, as microfluidic devices are often fabricated with different materials. Depending on the sum of the contact angles on the two plates, the interface can be either convex (Figure 8.3.1a), concave (Figure 8.3.1b), or if the angles add to exactly 180° the interface would be straight and the radius of curvature infinite. As we have seen in Section 7.2, every surface has two radii of curvature. We have drawn only one in Figure 8.3.1, R_1. The second radius of curvature, R_2, corresponds to the radius of the drop and is not shown. From the geometric construction for the convex interface,

$$(8.3.1) \qquad R_1 \cos \theta_1 = H - \delta$$

where H is the distance from the center of the circle whose arc makes the liquid-gas interface (curvature center) and the top plate (convex interface) or bottom plate (concave interface). δ is the distance between the plates; θ_1 is the contact angle with the bottom plate. We also observe that

$$(8.3.2) \qquad R_1 \cos (\pi - \theta_2) = -R_1 \cos \theta_2 = H$$

where θ_2 is the contact angle with the upper plate. Combining (8.3.1) and (8.3.2) we obtain for the convex interface,

$$(8.3.3) \qquad R_1 = -\frac{\delta}{\cos \theta_1 + \cos \theta_2}$$

For the concave interface,

$$(8.3.4) \qquad R_1 \cos (\pi - \theta_2) = -R_1 \cos \theta_2 = H - \delta$$

and

$$(8.3.5) \qquad R_1 \cos \theta_1 = H$$

and so

$$(8.3.6) \qquad R_1 = \frac{\delta}{\cos \theta_1 + \cos \theta_2}$$

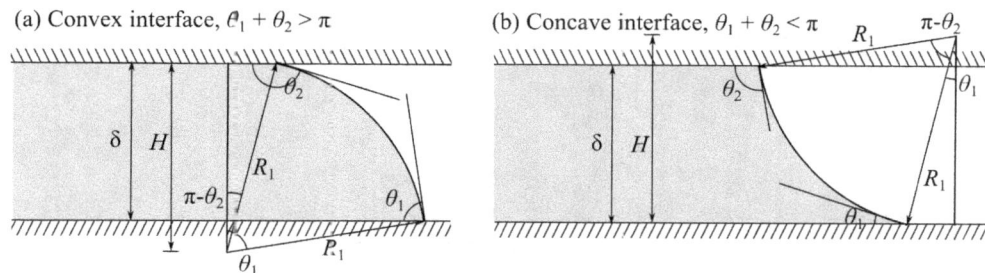

FIGURE 8.3.1. Schematic of the geometry of a drop sandwiched between two parallel plates of different contact angles to the liquid.

The force on the plates is due to the Young-Laplace pressure in the liquid,

$$(8.3.7) \qquad \Delta p = \gamma \left(\frac{1}{R_1} + \frac{1}{R_2} \right)$$

If the second radius of curvature is large, the pressure becomes,

$$(8.3.8) \qquad \Delta p = \frac{\gamma}{R_1}$$

And the force on the plate

$$(8.3.9) \qquad f = \pi R_2^2 \frac{\gamma}{R_1}$$

assuming that the area occupied by the curved area near the interface is negligible compared to the overall area of the drop. When the interface is concave, the pressure in the liquid is lower than that in the gas. Typically the gas is air and has the same pressure on the surfaces of the plates not contacting the liquid (think cover slip on a glass slide). We can see then in this case the liquid is pulling the surfaces together.

Let's calculate the force between a cover slip and a glass slide. For the case of pure water and clean surfaces, δ practically has no limitation. However, in practice, either our solution is filled with particles of interest (e.g., cells) or the water or the surfaces have particle impurities. Let's assume our cells force δ to be 10 µm. Furthermore, let's assume both contact angles are about 30° (see Table 1); surface tension of our solution is 72 mN/m; and that the size of the cover slip is 20 mm in diameter. We obtain that R_1 is -5.77 µm, Δp equals -12.5 kPa, and so the solution pulls the cover slip and the glass slide together with a force of 3.93 N. The volume of the solution sandwiched between the glass slide and the cover slip is approximately 3 µl. For comparison, an object weighing 1 lb presses on its support with a force of 4.45 N. There is more than enough force to keep the cover slip from falling off!

Another practical application of this is temporarily closing microfluidic chips made from two pieces of glass. We would clean the glass thoroughly in a clean room environment, and use water passed through a 0.45 µm syringe filter. Let's assume that this sets a δ to be 0.5 µm. Assuming similar material properties as in the last paragraph, the two parts of the microfluidic chip would then be held together with a pressure of 249 kPa - more than twice the atmospheric pressure! This is often more than enough to perform quick experiments.

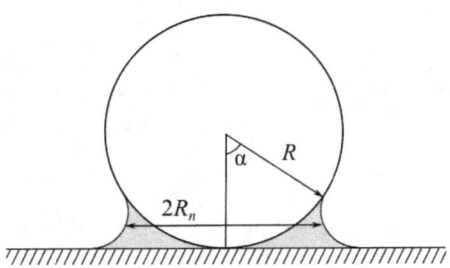

FIGURE 8.4.1. Schematic of a sphere on a flat surface with a liquid (capillary) bridge connecting the two.

8.4. Force between a spherical particle and a plate

Here we will give an approximate result for the shape of the meniscus and the forces between a sphere and a plane (Figure 8.4.1). The complete analysis is quite involved and the final results, while analytical are complicated. These can be found in Orr et al. (1975). We here use a simpler, experimentally verified approximation of Clark et al. (1968). These results are useful for estimating the sticking of particles to surfaces, as in contamination of wafers in clean rooms, and the sticking of particles in microfluidic chips after the liquid has been evacuated.

For a perfectly wetting liquid, Clark et al. (1968) presents that the volume of the liquid in the bridge is

$$(8.4.1) \qquad \frac{V_{bridge}}{2R^3} = \pi \left(\frac{1 - \cos \alpha}{1 + \cos \alpha} \right)^2 \left(2 - \frac{\sin \alpha \, (\pi - \alpha)}{1 + \cos \alpha} \right)$$

the neck radius

$$(8.4.2) \qquad \frac{R_n}{R} = \frac{2 \sin \alpha + \cos \alpha - 1}{1 + \cos \alpha}$$

and the pressure drop across the liquid-gas interface,

$$(8.4.3) \qquad \frac{\Delta p}{\gamma / R} = 2 \left(\frac{1 + \cos \alpha}{1 - \cos \alpha} \right) \left(\frac{\sin \alpha + \cos \alpha - 1}{2 \sin \alpha + \cos \alpha - 1} \right)$$

and finally the attractive capillary force between the sphere and the surface

$$(8.4.4) \qquad \frac{f}{2\pi R\gamma} = \frac{2 \sin \alpha + \cos \alpha - 1}{\sin \alpha}$$

When the situation involving a sphere and a flat surface is highly experimentally controlled, we can either control the bridge volume, or measure the resulting neck radius. Once one of these is known, we can obtain α and then obtain the interface pressure drop and the attractive force.

8.5. Wetting in corners

In microfluidic chips we often encounter cross-sections of fairly unusual shape (see Figure 8.5.1), with corners of different angles. Moreover, microfluidic chips are often fabricated of several different materials, such that the fluid sees multiple surfaces with different contact angles. Often it is desired that the working fluids fill the chip by capillary action, without the need for action of external or internal pump. Sometimes capillary filling (priming) is even absolutely necessary for the

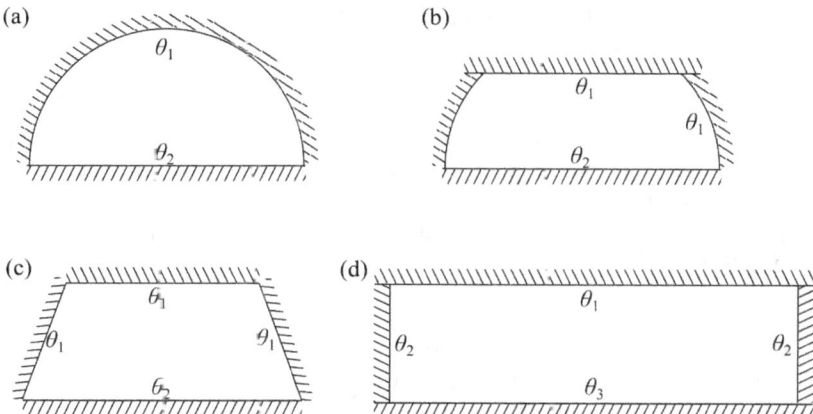

FIGURE 8.5.1. Common microfluidic chip cross-sections. (a) Simple D shape cross-section formed by anisotropic etching of substrate 1 (contact angle θ_1) and bonding it to a flat cover substrate 2 (contact angle θ_2). (b) Another D shape cross-section, again formed by anisotropic etch from a rectangular pattern The "wings" of the D are sections (sometimes 90° sections) of a circle. (c) A trapezoidal cross-section often found on hot embossed chips. Typically the trapezoidal section is embossed in a substrate 1 (contact angle θ_1) and is then covered by another material, substrate 2 (contact angle θ_2). The trapezoidal cross-section results from the necessary draft angle of the embossing tool. (d) A rectangular section formed from bonding three substrates. Substrate 2 is etched through (e.g., via reactive ion etch) and is sandwiched between substrates 1 and 3. Depending on the etch or patterning of substrate 2, sometimes trapezoidal (c) or D shaped (b) cross-sections also result. Cross-section (c) is probably the most common cross-section encountered.

device to work as many microfluidic integrated pumps used are not self priming (e.g., electroosmotic, thermal inertial, piezo inertial pumps). Thus, two questions arise: (a) how the liquid fills a particular cross-section (especially with respect to corners), and (b) how quickly (if at all) does the liquid penetrate into the channel network. The second question is especially important as designers need to know what variables (including corner geometry and material wetting properties) to tune to allow effective capillary filling (priming), and will be answered in later sections. Here we will focus on the first question, a necessary foundation to answering the second question.

8.5.1. Wetting of an ideal wedge. Let's consider a corner of a microfluidic channel, redrawn as an ideal wedge in Figure 8.5.2. At first, let's consider that the wedge is formed by two substrates of the same material, and so the same contact angle. Depending on the wetting of the liquid on this substrate, we can have a situation shown in Figure 8.5.2a, b or c. In Figure 8.5.2a the liquid is clearly wetting the material of the wedge. Due to the curvature formed (concave interface), according to the Young-Laplace equation the pressure in the liquid is less than that in the atmosphere (gas) above the liquid. Thus, the liquid is pushed down into the wedge, and this situation is stable. In Figure 8.5.2c the liquid clearly does not wet the material of the wedge. Due to the curvature formed (concave interface), the pressure inside the liquid is higher than that in the gas. Thus the liquid is pushed out of the wedge. Thus, in this situation the liquid will not fill the wedge. The situation in Figure 8.5.2b is the intermediate case between that in Figure 8.5.2a and Figure 8.5.2c. Here the liquid-gas interface is flat, and so the pressure in the liquid equals the pressure in the gas. Since the contact angle of the liquid with the substrate is θ, for this case

$$(8.5.1) \qquad\qquad \theta + \theta + 2\alpha = \pi$$

Notice if θ was larger, we would shift to the situation in Figure 8.5.2c (the unstable state), and if θ was smaller we would shift to the stable situation in Figure 8.5.2a. Thus, the criteria for stability is

$$(8.5.2) \qquad\qquad \theta + \theta < \pi - 2\alpha$$

We can generalize this to a situation where the substrates forming the wedge are not the same. The intermediate state for this is shown in Figure 8.5.2d, and the criteria for a stable wetting of the wedge is

$$(8.5.3) \qquad\qquad \theta_1 + \theta_2 < \pi - 2\alpha$$

This is sometimes known as the Concus-Finn relation (Concus and Finn (1969)).

Note, that for a microfluidic channel in Figure 8.5.1c, we can have a situation where the lower corners are wetted, while the top corners are dry. This means, that the lower corners can be filled with liquid all the way to the end of the channel, while the top corners would remain dry. This creates a slanted overall meniscus in the microfluidic channel. When pressure is applied to fill the channel with liquid, depending on how the pressure is applied, liquid might run ahead in the lower section, and close off a bubble of air above it, especially in corners, thus introducing a bubble in the channel. This bubble may be harmless, but more often, it might be highly unwanted, as for example, it may significantly alter the hydrodynamic field in the channel, or it may alter electric field in the channel, or it may even detach and block a sensing surface downstream. Care should be taken to avoid such situations.

8.5.2. Wetting of an arbitrary polygon. A cross-section of a microfluidic channel can often be treated as polygon, where more than one corner is wetted at the same time. Thus it is interesting to know what is the capillary pressure (and so the pressure in the liquid) when a certain area of the channel is filled with the liquid. Conversely, if a certain pressure is applied to the liquid, how does this change the area of the channel filled with the liquid.

Following Langbein (2002) we derive the integral theorem for long channels using the following arguments (for more details see Langbein (2002)). Consider a long channel with a polygonal cross-section (Figure 8.5.3) filled with liquid in the shaded area all the way form end to end of the channel, and with an endless supply of liquid at one end. From the Young-Laplace equation, the curvature of the liquid surface is equal to $\Delta p/\gamma$. Furthermore, the curvature of the liquid surface is equal to the divergence of its normal (e.g., see (7.2.40)). The divergence theorem gives us a relation between the area integral of the curvature of the surface and a boundary integral of the surface normal, in this case, the 2D scalar product of the normal to

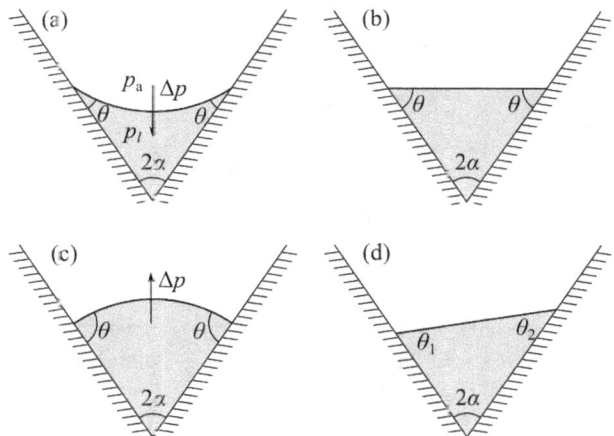

FIGURE 8.5.2. Wetting of a wedge, with dihedral angle 2α that extends infinitely into the plane. In (a-c) the wedge is formed by identical substrates, while in (d) the wedge is formed by two different substrates with different contact angles. In (a) the liquid wets the solid substrate, the liquid surface is concave, the capillary pressure is directed into the wedge, and the liquid interface is stable. In (c) the liquid does not wet the substrate, the liquid surface is convex, the capillary pressure is directed out of the wedge, and the liquid leaves the wedge. Situation in (b) is the transition from (a) to (c). (d) shows the same transition but for a wedge formed by different substrates.

the liquid surface and the normal to the substrate. However, notice that the scalar product of these two normals is equal to the cosine of the contact angle. Hence we have,

$$\int_\Omega \nabla \cdot \mathbf{n}\, df = \int_\Sigma \cos\theta\, ds$$

where Ω is the cross-sectional area of the polygon, and Σ is its perimeter. To use this result we must know which sides of the polygon (and which corners) are covered by the liquid. Using this result together with the stability condition (8.5.3), we can derive that

(8.5.4)
$$\left(\frac{\Delta p}{\gamma}\right)^2 \Omega - \frac{\Delta p}{\gamma} \sum_i L_i \cos\theta_i$$
$$+\frac{1}{2}\frac{\gamma}{\Delta p} \sum_{[i,i+1]} (\Sigma_{i,i}\cos\theta_i + \Sigma_{i,i+1}\cos\theta_{i+1} - \Sigma_{i+1}) = 0$$

where $[i, i+1]$ represents summation over all the wetted wedges and, $\Sigma_{i,j}$ the lengths of the edges of the wedges. For full derivation see Langbein (2002). Using this result and measurements of the position of the liquid interface in the channel (e.g., using confocal microscopy) we can determine the pressure in the liquid. Additionally, knowing (and then applying) a pressure to liquid we can predict how we can alter which sides (and corners) of the channels will be wetted.

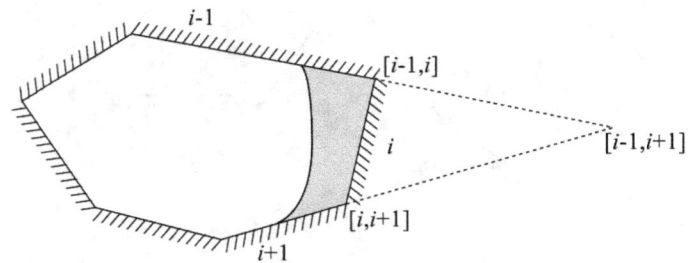

FIGURE 8.5.3. Channel with a polygonal cross-section, with faces i (1 to N) and wedges $[i, i+1]$. When the face i is wetted completely, the liquid behaves as if the faces $i-1$ and $i+1$ neighbor each other and forgets about the existence of face i.

CHAPTER 9

Kinetics of wetting

We begin our study of the kinetics of wetting by studying the simple case of spreading of a liquid drop over dry and pre-wetted surfaces to understand the fundamentals of kinetics of wetting, spreading of liquids on surfaces and penetration of liquids into channels. This will lead us then to consider more practical problems of meniscus advance between flat plates and inside a circular capillary - phenomena frequently encountered wherever microfluidic chips are filled.

The most interesting parts of the kinetics of wetting occur at the apparent three-phase contact line. This region can be subdivided into four regions (Figure 9.0.1). Region (1) represents the bulk, spherical part of the droplet or the liquid in a meniscus. The slope of this region interfacing with the surface underneath represents the dynamic contact angle. This angle depends on time from the start of the spreading and we can measure this angle as the droplet is spreading. In region (2) the profile of the liquid-gas interface is distorted from the spherical by the outflow of liquid from the bulk of the drop (region (1)); disjoining pressure is still negligible in this region. In region (3) however, disjoining pressure becomes the dominant driver of fluid flow. In region (4), continuum description of fluid flow breaks down as the characteristic film thickness becomes on the order of molecular size. Fluid motion in this region is represented by surface diffusion.

We begin our dive into the kinetics of wetting by considering the Navier-Stokes equations in cylindrical coordinates for an incompressible fluid (the liquid in our

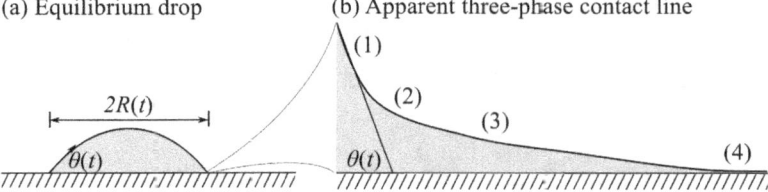

FIGURE 9.0.1. Schematic of spreading of a liquid drop. $R(t)$ represents the position of the apparent three-phase contact line (from the center, origin) of the drop. $\theta(t)$ is the dynamic contact angle. The region around the apparent three-phase contact line can be divided into four regions. Region (1) is the spherical part of the drop; in region (2) the hydrodynamics of spreading distort the drop shape; in region (3) the liquid film becomes thin enough so that disjoining pressure begins to be important; in region (4) continuum description breakdown and surface diffusion describes the motion of molecules comprising the drop.

drop). Using these, we will first investigate the situation around the apparent three-phase contact line and then investigate the situation inside the bulk of the drop. For both cases, the continuity equation is

$$(9.0.1) \qquad \frac{1}{r}\frac{\partial}{\partial r}(ru_r) + \frac{1}{r}\frac{\partial u_\phi}{\partial \phi} + \frac{\partial u_z}{\partial z} = 0$$

The radial component of the momentum equation is

$$(9.0.2)$$
$$\rho\left(\frac{\partial u_r}{\partial t} + u_r\frac{\partial u_r}{\partial r} + \frac{u_\phi}{r}\frac{\partial u_r}{\partial \phi} + u_z\frac{\partial u_r}{\partial z} - \frac{u_\phi^2}{r}\right) =$$
$$-\frac{\partial p}{\partial r} + \mu\left(\frac{1}{r}\frac{\partial}{\partial r}\left(r\frac{\partial u_r}{\partial r}\right) + \frac{1}{r^2}\frac{\partial^2 u_r}{\partial \phi^2} + \frac{\partial^2 u_r}{\partial z^2} - \frac{u_r}{r^2} - \frac{2}{r^2}\frac{\partial u_\phi}{\partial \phi}\right) + \rho g_r$$

the tangential component of the momentum equation is

$$(9.0.3)$$
$$\rho\left(\frac{\partial u_\phi}{\partial t} + u_r\frac{\partial u_\phi}{\partial r} + \frac{u_\phi}{r}\frac{\partial u_\phi}{\partial \phi} + u_z\frac{\partial u_\phi}{\partial z} - \frac{u_r u_\phi}{r}\right) =$$
$$-\frac{1}{r}\frac{\partial p}{\partial \phi} + \mu\left(\frac{1}{r}\frac{\partial}{\partial r}\left(r\frac{\partial u_\phi}{\partial r}\right) + \frac{1}{r^2}\frac{\partial^2 u_\phi}{\partial \phi^2} + \frac{\partial^2 u_\phi}{\partial z^2} - \frac{u_\phi}{r^2} + \frac{2}{r^2}\frac{\partial u_r}{\partial \phi}\right) + \rho g_\phi$$

and the axial component of the momentum equation is

$$(9.0.4)$$
$$\rho\left(\frac{\partial u_z}{\partial t} + u_r\frac{\partial u_z}{\partial r} + \frac{u_\phi}{r}\frac{\partial u_z}{\partial \phi} + u_z\frac{\partial u_z}{\partial z}\right) =$$
$$-\frac{\partial p}{\partial z} + \mu\left(\frac{1}{r}\frac{\partial}{\partial r}\left(r\frac{\partial u_z}{\partial r}\right) + \frac{1}{r^2}\frac{\partial^2 u_z}{\partial \phi^2} + \frac{\partial^2 u_z}{\partial z^2}\right) + \rho g_z$$

Next we assume there is no motion in the tangential direction in the drop, so $u_\phi = 0$. Furthermore, we assume that the spreading is axisymmetric and there is no dependence on ϕ. We lastly assume, for now, that the Bond number is low, and gravitational effects can be neglected. This simplifies the Navier-Stokes equations to:

$$(9.0.5) \qquad \frac{1}{r}\frac{\partial}{\partial r}(ru_r) + \frac{\partial u_z}{\partial z} = 0$$

$$(9.0.6) \quad \rho\left(\frac{\partial u_r}{\partial t} + u_r\frac{\partial u_r}{\partial r} + u_z\frac{\partial u_r}{\partial z}\right) = -\frac{\partial p}{\partial r} + \mu\left(\frac{1}{r}\frac{\partial}{\partial r}\left(r\frac{\partial u_r}{\partial r}\right) + \frac{\partial^2 u_r}{\partial z^2} - \frac{u_r}{r^2}\right)$$

$$(9.0.7) \quad \rho\left(\frac{\partial u_z}{\partial t} + u_r\frac{\partial u_z}{\partial r} + u_z\frac{\partial u_z}{\partial z}\right) = -\frac{\partial p}{\partial z} + \mu\left(\frac{1}{r}\frac{\partial}{\partial r}\left(r\frac{\partial u_z}{\partial r}\right) + \frac{\partial^2 u_z}{\partial z^2}\right)$$

where the first two assumptions have completely eliminated the equation for the tangential component of the momentum equation. Finally, let's assume that we are interested in the spreading behavior when the spreading is already steady (or at least quasi steady), and so neglect the unsteady terms:

$$(9.0.8) \quad \rho\left(u_r\frac{\partial u_r}{\partial r} + u_z\frac{\partial u_r}{\partial z}\right) = -\frac{\partial p}{\partial r} + \mu\left(\frac{1}{r}\frac{\partial}{\partial r}\left(r\frac{\partial u_r}{\partial r}\right) + \frac{\partial^2 u_r}{\partial z^2} - \frac{u_r}{r^2}\right)$$

$$(9.0.9) \quad \rho\left(u_r\frac{\partial u_z}{\partial r} + u_z\frac{\partial u_z}{\partial z}\right) = -\frac{\partial p}{\partial z} + \mu\left(\frac{1}{r}\frac{\partial}{\partial r}\left(r\frac{\partial u_z}{\partial r}\right) + \frac{\partial^2 u_z}{\partial z^2}\right)$$

Next, to simplify the equations further, we define the characteristic length, and velocity scales and nondimensionalize the equations. We let U and v be the characteristic velocity scales in the radial and axial direction respectively. We let R^* and h^* be the characteristic length scales in the radial and axial directions respectively. R^* represents the typical radius of the drop, while h^* represents the height of the thin film ahead of the drop. Let's first consider the case near the apparent three-phase contact line. In this case $h^* \ll R^*$. Thus we define a smallness parameter,

$$(9.0.10) \qquad \varepsilon = h^*/R^*$$

Thus, we obtain the dimensionless variables:

$$(9.0.11) \qquad \begin{aligned} r^* &= r/R^* \\ z^* &= z/h^* \\ u_r^* &= u_r/U \\ u_z^* &= u_z/v \end{aligned}$$

We then use these to nondimensionalize the continuity equation and obtain

$$(9.0.12) \qquad \frac{U}{R^*}\frac{h^*}{v}\frac{1}{r^*}\frac{\partial}{\partial r^*}\left(r^* u_r^*\right) + \frac{\partial u_z^*}{\partial z^*} = 0$$

Since we anticipate both terms of the continuity equation are important in this problem, we conclude

$$(9.0.13) \qquad \frac{U}{R^*}\frac{h^*}{v} = O(1)$$

where $O(1)$ indicates order unity. This in turn implies that,

$$(9.0.14) \qquad v/U = \varepsilon$$

Or that the velocity scale in the axial (vertical) direction is much smaller than the velocity scale in the radial direction. Now let's nondimensionalize the momentum equations:

$(9.0.15)$

$$\rho \frac{U^2}{R^*}\frac{h^{*2}}{\mu U}\left(u_r^*\frac{\partial u_r^*}{\partial r^*} + u_z^*\frac{\partial u_r^*}{\partial z^*}\right) = -\frac{\partial p^*}{\partial r^*} + \left(\frac{h^{*2}}{R^{*2}}\frac{1}{r^*}\frac{\partial}{\partial r^*}\left(r^*\frac{\partial u_r^*}{\partial r^*}\right) + \frac{\partial^2 u_r^*}{\partial z^{*2}} - \frac{h^{*2}}{R^{*2}}\frac{u_r^*}{r^{*2}}\right)$$

$$(9.0.16) \quad \rho\frac{vU}{R^*}\frac{h^{*2}}{\mu v}\left(u_r^*\frac{\partial u_z^*}{\partial r^*} + u_z^*\frac{\partial u_z^*}{\partial z^*}\right) = -\frac{\partial p^*}{\partial z^*} + \left(\frac{h^{*2}}{R^{*2}}\frac{1}{r^*}\frac{\partial}{\partial r^*}\left(r^*\frac{\partial u_z^*}{\partial r^*}\right) + \frac{\partial^2 u_z^*}{\partial z^{*2}}\right)$$

where we have used (9.0.13) and have nondimensionalized the pressure with the vicious pressure scale

$$(9.0.17) \qquad p^* = \frac{p}{\mu U/R^*}$$

We see that in both equations in front of the inertial terms we have a Reynolds number,

$$(9.0.18) \qquad Re = \rho\frac{U}{R^*}\frac{h^{*2}}{\mu} = \varepsilon^2\frac{\rho U R}{\mu}$$

where we have used (9.0.10) to obtain the second equality. We see that even if the traditional Reynolds number $\rho U R/\mu$ is not small, the governing Reynolds number here is small due to the presence of the smallness parameter, and so the inertial terms are negligible. Neglecting all terms order ε^2 we obtain our set of governing equations, including the continuity equation and the two simplified momentum equations:

$$(9.0.19) \qquad \frac{U}{R^*} \frac{h^*}{v} \frac{1}{r^*} \frac{\partial}{\partial r^*} (r^* u_r^*) + \frac{\partial u_z^*}{\partial z^*} = 0$$

$$(9.0.20) \qquad \frac{\partial p^*}{\partial r^*} = \frac{\partial^2 u_r^*}{\partial z^{*2}}$$

$$(9.0.21) \qquad \frac{\partial p^*}{\partial z^*} = \frac{\partial^2 u_z^*}{\partial z^{*2}}$$

We see that the Navier-Stokes equations have collapsed down to the Stokes equations. Furthermore, near the apparent three-phase contact line, since we expect the slope of the liquid-gas interface to be low, i.e., $dz/dr \ll 1$ or $h^* \ll R^*$, we expect the interface to be flat. Hence, the radius of curvature of the interface is very large and so the Young-Laplace pressure is negligible. Thus, we can conclude that the stress tangential to the free liquid interface is zero.

Next, let's consider the situation in the bulk of the drop. Here the relevant length and velocity scales are different and so we must re-nondimensionalize our equations. Here we let U be the characteristic velocity scale for both the radial and axial direction. Similarly, we let R^* be the characteristic length scales in both the radial and axial directions, as we expect the drop to be spherical. Thus,

$$(9.0.22) \qquad \begin{aligned} r^* &= r/R^* \\ z^* &= z/R^* \\ u_r^* &= u_r/U \\ u_z^* &= u_z/U \end{aligned}$$

Furthermore, here we scale the pressure with the Laplace pressure scale,

$$(9.0.23) \qquad p^* = \frac{p}{\gamma/R^*}$$

After this nondimensionalization we obtain that the continuity equation is

$$(9.0.24) \qquad \frac{1}{r^*} \frac{\partial}{\partial r^*} (r^* u_r^*) + \frac{\partial u_z^*}{\partial z^*} = 0$$

and the momentum equations are
(9.0.25)
$$\frac{\rho U R^*}{\mu} \frac{\mu U}{\gamma} \left(u_r^* \frac{\partial u_r^*}{\partial r^*} + u_z^* \frac{\partial u_r^*}{\partial z^*} \right) = -\frac{\partial p^*}{\partial r^*} + \frac{\mu U}{\gamma} \left(\frac{1}{r^*} \frac{\partial}{\partial r^*} \left(r^* \frac{\partial u_r^*}{\partial r^*} \right) + \frac{\partial^2 u_r^*}{\partial z^{*2}} - \frac{u_r^*}{r^{*2}} \right)$$
(9.0.26)
$$\frac{\rho U R^*}{\mu} \frac{\mu U}{\gamma} \left(u_r^* \frac{\partial u_z^*}{\partial r^*} + u_z^* \frac{\partial u_z^*}{\partial z^*} \right) = -\frac{\partial p^*}{\partial z^*} + \frac{\mu U}{\gamma} \left(\frac{1}{r^*} \frac{\partial}{\partial r^*} \left(r^* \frac{\partial u_z^*}{\partial r^*} \right) + \frac{\partial^2 u_z^*}{\partial z^{*2}} \right)$$

We see that the inertia terms in the momentum equation are multiplied by the product of a standard Reynolds number and a capillary number and the viscous terms are multiplied by the capillary number. In cases where both the Reynolds number is small and the capillary number is small the inertial term is negligible. In the case where Reynolds number is large, but capillary number is small, we keep the inertial term but drop the viscous term. To get a feeling for these numbers, let's consider a typical spreading situation. Let's consider a water droplet with a radius 1 mm, surface tension of 72 mN/m, viscosity of 1 mPa s, and density of 1000 kg/m^3, spreading a considerably high spreading velocity of 1 radius per second. This gives a Reynolds number of 1, and a capillary number of 10^{-5}. Note, that for a water droplet, for Ca to be close to unity, the surface would have to be moving with a speed of 72 m/s! We can see that for the situations we are interested in both the inertial and the viscous terms are negligible. Thus, the momentum equations for the bulk of the drop reduce to,

$$(9.0.27) \qquad \frac{\partial p^*}{\partial r^*} = 0$$

$$(9.0.28) \qquad \frac{\partial p^*}{\partial z^*} = 0$$

This implies that the pressure remains constant inside the main part of the spreading droplet. This in turn implies that the bulk of the droplet retains its spherical shape. While we have assumed that the droplet radius changes over time, we also assumed that the equations are quasi steady, and so the droplet radius of curvature must change somewhat slowly.

9.1. Spreading of a liquid drop over dry surface

Now we are ready to consider the spreading behavior of a drop over a dry, completely wetting surface. We begin with the simplified momentum equation for the velocity in the radial direction near the apparent three-phase contact line, as the velocity in the radial direction is much larger than that in the axial (vertical) direction. This equation, (9.0.20), back in dimensional form is

$$(9.1.1) \qquad \frac{\partial p}{\partial r} = \mu \frac{\partial^2 u_r}{\partial z^2}$$

Additionally, we also have two boundary conditions, the no-slip boundary condition at the solid surface,

$$(9.1.2) \qquad u_r = 0 \text{ at } z = 0$$

and that the tangential stress at the liquid-gas interface is zero,

$$(9.1.3) \qquad \mu \frac{\partial u_r}{\partial z} = 0 \text{ at } z = h$$

Integrating (9.1.1), subject to these boundary conditions, we obtain,

$$(9.1.4) \qquad u_r = -\frac{1}{\mu} \frac{\partial p}{\partial r} \left(hz - \frac{z^2}{2} \right)$$

Next, we integrate over the thickness of the film in this region to find that the total flow rate out of the bulk of the droplet is,

$$(9.1.5) \qquad Q = 2\pi \int_0^h r u_r dz = -\frac{2\pi}{3\mu} r h^3 \frac{\partial p}{\partial r}$$

However, when the liquid is leaving the bulk of the droplet, it must go into the thin film. Thus the change of the thin films thickness h at a given r and t must be related to this flow rate though,

$$(9.1.6) \qquad 2\pi r \frac{\partial h}{\partial t} + \frac{\partial Q}{\partial r} = 0$$

Combining this with (9.1.5), we obtain the equation of spreading,

$$(9.1.7) \qquad \frac{\partial h}{\partial t} = \frac{1}{3\mu r} \frac{\partial}{\partial r} \left(r h^3 \frac{\partial p}{\partial r} \right)$$

To solve it, we need to incorporate two more conditions. The first is the conservation of the total liquid volume, and the second is an expression for pressure in the liquid. We assume that the liquid is relatively non-volatile, and so does not evaporate on the time scale of the spreading, thus its total volume is conserved:

$$(9.1.8) \qquad 2\pi \int_0^{R(t)} rh\,dr = V_{liq}$$

Secondly, in the general case, the pressure in the liquid in this region is due to the ambient (atmospheric) pressure, due to curvature of the interface, due to disjoining forces, and due to gravity. We will deal with the relative magnitudes of each of these contributions later - under certain conditions each of these can be negligible. For now we write,

$$(9.1.9) \qquad p = p_a - \gamma K - \Pi(h) + \rho g h$$

where p_a is the ambient (gas) pressure, Π is the disjoining pressure, and K is the interface curvature (see (7.2.26)). For the low slope approximation K simplifies to

$$(9.1.10) \qquad K = \frac{1}{r} \frac{\partial}{\partial r} \left(h \frac{\partial h}{\partial r} \right)$$

And so,

$$(9.1.11) \qquad p = p_a - \gamma \frac{1}{r} \frac{\partial}{\partial r} \left(h \frac{\partial h}{\partial r} \right) - \Pi(h) + \rho g h$$

For different thickness of the spreading film, different set of pressure terms in equation (9.1.11) are applicable. For example, when the thickness of the film $h > 100\,\text{nm}$, the disjoining pressure term is negligible. Furthermore, if the main droplet size is greater than the capillary length, the droplet spreads via the gravitational spreading regime (the main driving force pushing the liquid out of the main drop is gravity); versus if the main droplet size is less than the capillary length, the droplet spreads via the capillary spreading regime (the main driving force pushing the liquid out of the main drop is Laplace pressure). We will discuss both below. Thus, under these conditions

$$(9.1.12) \qquad p = p_a - \gamma \frac{1}{r} \frac{\partial}{\partial r} \left(h \frac{\partial h}{\partial r} \right) + \rho g h$$

If the film thickness $10\,\text{nm} < h < 100\,\text{nm}$ then both the capillary pressure and disjoining pressure are important, but the gravitational contribution is likely negligible. Thus,

$$(9.1.13) \qquad p = p_a - \gamma \frac{1}{r} \frac{\partial}{\partial r} \left(h \frac{\partial h}{\partial r} \right) - \Pi(h)$$

Finally, if the film thickness $h < 10\,\text{nm}$, then only the disjoining pressure is important,

$$p = p_a - \Pi(h)$$

9.1.1. Capillary spreading regime: Tanner's law. Let's first consider spreading in the capillary regime. Here the main droplet size is less than the capillary length and we assume that the pressure inside the drop is governed by Laplace pressure,

$$(9.1.14) \qquad p = p_a - \gamma \frac{1}{r}\frac{\partial}{\partial r}\left(h\frac{\partial h}{\partial r}\right)$$

Combining this with the spreading equation (9.1.7), we obtain

$$(9.1.15) \qquad \frac{\partial h}{\partial t} = -\frac{\gamma}{3\mu r}\frac{\partial}{\partial r}\left(rh^3\frac{\partial}{\partial r}\left(\frac{1}{r}\frac{\partial}{\partial r}\left(r\frac{\partial h}{\partial r}\right)\right)\right)$$

We solve this by applying a symmetry boundary condition

$$(9.1.16) \qquad \left.\frac{\partial h}{\partial r}\right|_{r=0} = \left.\frac{\partial^3 h}{\partial r^3}\right|_{r=0} = 0$$

as well as the conservation of liquid volume condition, (9.1.8), using a similarity solution approach. The full solution is lengthy and is given in full in Starov et al. (2007) and we here just provide the results. We find that the radius of the droplet as a function of time is

$$(9.1.17) \qquad R\left(t\right) = 0.65\lambda_C\left(\frac{\gamma V_{liq}^3}{\mu}\right)^{1/10}(t + t_{in})^{1/10}$$

where λ_C is an unknown constant of order unity and t_{in} is the duration of the inertial spreading of the drop, which we assume precedes the capillary spreading. We did not consider or evaluate inertial spreading here and assume that the time for inertial spreading is short. The height of the bulk of the drop as a function of time scales as

$$(9.1.18) \qquad H\left(t\right) \sim \left(\frac{\mu V_{liq}^2}{\gamma}\right)^{1/5}(t + t_{in})^{-1/5}$$

and the dynamic contact angle scales as

$$(9.1.19) \qquad \theta(t) \sim \left(\frac{\mu^3 V_{liq}}{\gamma^3}\right)^{1/10}(t + t_{in})^{-3/10}$$

where we define the dynamic contact angle as $\theta\left(t\right) = H\left(t\right)/R\left(t\right)$. The velocity of spreading,

$$(9.1.20) \qquad U = \frac{dR\left(t\right)}{dt} = 0.065\lambda_C\left(\frac{\gamma V_{liq}^3}{\mu}\right)^{1/10}(t - t_{in})^{-9/10}$$

We can also express this in terms of the dynamic contact angle,

$$(9.1.21) \qquad U \sim 0.065\lambda_C\frac{\gamma}{\mu}\theta^3$$

What we have derived here is well known as Tanner's law (Tanner (1979)), a well experimentally verified scaling for spreading of completely wetting liquids. An example of complete wetting is the spreading of silicone oils (including lower molecular weight polydimethylsiloxane (PDMS) oils) on microscope slides (borosilicate and soda lime glasses).

9.1.2. Gravitational spreading regime. Let's now consider the gravitational spreading regime. Here the main droplet size is greater than the capillary length and we assume that the pressure inside the drop is governed mainly by gravity,

$$(9.1.22) \qquad\qquad p = p_a + \rho g h$$

Combining this with the spreading equation (9.1.7), we obtain

$$(9.1.23) \qquad\qquad \frac{\partial h}{\partial t} = \frac{\rho g}{3\mu r}\frac{\partial}{\partial r}\left(rh^3\frac{\partial h}{\partial r}\right)$$

We solve this equation once again by combining this with a symmetry boundary condition (9.1.16), as well as the conservation of liquid volume condition, (9.1.8), once again, using a similarity solution approach. Once again, the full solution is lengthy and is given in full in Starov et al. (2007) and we here just write the results. We find that the radius of the droplet as a function of time is

$$(9.1.24) \qquad\qquad R\left(t\right) = 0.78\left(\frac{\rho g V_{liq}^3}{\mu}\right)^{1/8}\left(t + t_0\right)^{1/8}$$

where t_0 is the duration of another type of spreading (such as capillary spreading) that may have preceded gravitational regime spreading. Notice that now the exponent is 1/8 rather than 1/10. Gravitational spreading occurs faster than capillary spreading.

9.1.3. Disjoining pressure spreading regime. When the droplets are really thin, disjoining pressure dominates the pressure inside the drop,

$$(9.1.25) \qquad\qquad p = p_a - \Pi\left(h\right)$$

which results in a spreading equation of

$$(9.1.26) \qquad\qquad \frac{\partial h}{\partial t} = \frac{1}{3\mu r}\frac{\partial}{\partial r}\left(rh^3\frac{d\Pi}{dh}\frac{\partial h}{\partial r}\right)$$

We solve this equation also by combining this with a symmetry boundary condition (9.1.16), as well as the conservation of liquid volume condition, (9.1.8), once again, using a similarity solution approach. Before proceeding with the solution, we have to choose an expression for the disjoining pressure. We here choose to investigate two such form. The first,

$$(9.1.27) \qquad\qquad \Pi_1\left(h\right) = \begin{cases} b\left(t_s - h\right) \text{ for } 0 \leqslant h \leqslant t_s \\ 0 \text{ for } h > t_s \end{cases}$$

where t_s is a critical thickness of the film at which disjoining pressure becomes negligible, for example, 100 nm. And the second,

$$(9.1.28) \qquad\qquad \Pi_2\left(h\right) = A/h^n$$

where we investigate two subcases where n is 2 and 3. For the first case, (9.1.27), the spreading equation for $0 \leqslant h \leqslant t_s$ becomes

$$(9.1.29) \qquad\qquad \frac{\partial h}{\partial t} = \frac{b}{3\mu r}\frac{\partial}{\partial r}\left(rh^3\frac{\partial h}{\partial r}\right)$$

Notice, that this has the exact same form as (9.1.23), with ρg replaced by b. Since it is subject to the same boundary conditions, it has the same solution, with just ρg replaced by b. Thus for this case,

$$(9.1.30) \qquad R(t) = 0.78 \left(\frac{b V_{liq}^3}{\mu} \right)^{1/8} (t + t_0)^{1/8}$$

For the second case, the spreading equation simplifies to

$$(9.1.31) \qquad \frac{\partial h}{\partial t} = \frac{nA}{3\mu r} \frac{\partial}{\partial r} \left(r h^{2-n} \frac{\partial h}{\partial r} \right)$$

For these we find the similarity solution as before (see Starov et al. (2007) for details). For $n = 2$

$$(9.1.32) \qquad R(t) = 3.04 \left(\frac{2A}{3\mu} \right)^{1/2} (t + t_2)^{1/2}$$

where A is the disjoining pressure constant and for $n = 2$ has units of the product of pressure and length squared; t_2 is the duration of another type of spreading that may have preceded this regime. For $n = 3$ we find that the radius of the droplet scales exponentially with time,

$$(9.1.33) \qquad R(t) \sim \exp \left(\frac{2\pi A}{\mu V} (t + t_3) \right)$$

where A is the disjoining pressure constant and for $n = 3$ has units of the product of pressure and length cubed; t_3 is the duration of another type of spreading that may have preceded this regime. Notice that these two past relations for spreading predict a much faster spreading than gravitational or capillary driven spreading. The exponential spreading in (9.1.33) is especially fast.

9.1.4. Capillary and disjoining pressure spreading regime. Having predicted rates of spreading when spreading was governed by individual causes, we now turn to the case where spreading is caused by more than one cause. Namely, where spreading is caused by both the capillary and disjoining pressure together. As before, we will assume that the capillary number remains small, and so the main part of the droplet remains spherical. Secondly, as before, we will also assume a low slope in the vicinity of the apparent three-phase contact line. From the first assumption by geometrical considerations (see Figure 9.1.1) we find that in the main part of the droplet,

$$(9.1.34) \qquad h(r) = \sqrt{R_1^2 - r^2} - (R_1 - H)$$

where R_1 is the radius of curvature of the droplet as shown in Figure 9.1.1. We also find that

$$(9.1.35) \qquad R_1 = R/\sin\theta$$

and

$$(9.1.36) \qquad H = \frac{R(1 - \cos\theta)}{\sin\theta}$$

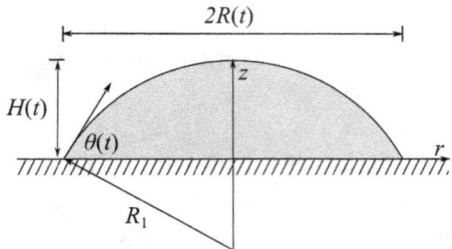

FIGURE 9.1.1. Geometry of an axisymmetrically spreading droplet in a situation where capillary number remains low and the main part of the droplet remains spherical.

We assume that for most of the time the majority of the liquid is located in the spherical part of the droplet (as opposed to the thin film) and so volume of the liquid is

$$(9.1.37) \qquad V_{liq} = \frac{\pi}{6} R^3 \tan\left(\frac{\theta}{2}\right) \left(3 + \tan^2\left(\frac{\theta}{2}\right)\right)$$

or upon inverting this,

$$(9.1.38) \qquad R(t) = \left(\frac{6V_{liq}}{\pi}\right)^{1/3} \left(\tan\left(\frac{\theta(t)}{2}\right)\left(3 + \tan^2\left(\frac{\theta(t)}{2}\right)\right)\right)^{-1/3}$$

For small angles, $0 < \theta < \pi/4$, we can approximate (9.1.38) as

$$(9.1.39) \qquad R(t) = \left(\frac{4V_{liq}}{\pi\theta}\right)^{1/3}$$

We will use these relations in combination with the spreading equation to obtain the whole profile of the droplet. Namely, we will solve for the entire profile using matched asymptotic expansions where we will match the droplet profile obtained by the spreading equation in the region near the apparent three-phase contact line to the spherical profile of (9.1.34). In this case we want to consider that both the capillary pressure and disjoining pressure are important in the vicinity of the three-phase contact line. Thus, there,

$$(9.1.40) \qquad p = p_a - \gamma\frac{1}{r}\frac{\partial}{\partial r}\left(h\frac{\partial h}{\partial r}\right) - \Pi(h)$$

where we will consider disjoining pressures in the form of

$$(9.1.41) \qquad \Pi(h) = A/h^n$$

where n is 2 or 3. Substituting this into the general spreading equation, we obtain the following spreading equation,

$$(9.1.42) \qquad \frac{\partial h}{\partial t} = -\frac{\gamma}{3\mu r}\frac{\partial}{\partial r}\left(rh^3\frac{\partial}{\partial r}\left(\frac{1}{r}\frac{\partial}{\partial r}\left(r\frac{\partial h}{\partial r}\right)\right) - \frac{nA}{\gamma h^{n+1}}\frac{\partial h}{\partial r}\right)$$

where the first term on the right-hand side is due to the capillary pressure and the second is due to disjoining pressure. We solve this equation, once again, by combining this with a symmetry boundary condition (9.1.16), as well as the conservation of liquid volume condition, (9.1.8). The full solution to this problem is given in

Starov et al. (2007). Here we will just summarize the approach and present the results.

To obtain the droplet profile, we will split the drop into three regions as we show in Figure 9.0.1, solve the profile in each region and combine the solutions using a matched asymptotic solution approach. The first region (1) is the spherical region, the region from the center of the drop to the point where the film thickness is still large. In this region both viscous and disjoining pressure effects are negligible compared to the capillary effects. For the purposes of the solution we refer to this region as the outer region. Further out radially is region (2). Here both viscous and capillary effects dominate. Lastly, even further out radially is region (3). Here the film is thin and disjoining force is comparable in magnitude to capillary and viscous forces. For the purposes of the solution, we refer to these two regions as the inner regions.

We obtain that, interestingly for both n of 2 or 3 that the drop radius evolves with time as

$$(9.1.43) \qquad R\left(t\right) = 0.89 \left(\frac{\gamma V_{liq}^3}{\mu}\right)^{1/10} (t + t_{in})^{1/10}$$

where here

$$(9.1.44) \qquad t_{in} = 0.145 \frac{\mu R_0^{10} B^3}{\gamma V_{liq}^3}$$

where R_0 is the initial drop radius and B is equal to 2.68 for $n = 2$, and 2.82 for $n = 3$. We also obtain that the dynamic contact angle in radians

$$(9.1.45) \qquad \theta\left(t\right) = B(2Ca)^{1/3}$$

and

$$(9.1.46) \qquad H\left(t\right) = \left(2B^3\right)^{1/5} \left(\frac{3}{40\pi^2} \frac{\mu V_{liq}^3}{\gamma}\right)^{1/5} (t + t_{in})^{-1/5}$$

Here we have now derived a more complete version of Tanner's law (Tanner (1979)). The 1/10 power scaling in the equation for the drop radius is of course the same as we have obtained before in Section 9.1.1. This relation has been extensively experimentally verified, (see, for example, Chen (1988) or Tanner (1979)) and the theory, specifically the scaling. agrees well with experimental observations.

9.2. Spreading of a liquid drop over a pre-wetted surface

Having considered spreading of drops on dry surfaces, let's consider spreading of drops on a surface covered with a thin film of the same liquid. This situation is more interesting from a practical point of view, as we have seen in the previous sections that at equilibrium most surfaces are covered with a thin layer of liquid, typically water absorbed from the environment. Thus, surfaces in microfluidic chips, especially those that have been in storage for a while, unless special precautions are taken otherwise, will be covered with a thin layer of water.

We will consider a drop of a viscous liquid on a planar horizontal surface covered with a layer, thickness h_0 of the same liquid. We assume that the thickness h_0 is thick enough so that the liquid is not affected by disjoining pressure. We will also limit our analysis to droplets with characteristic dimensions smaller than the capillary length, so that the action of gravity on them is negligible. Once again, we will assume that the droplet will spread axisymmetrically, and so we again have a symmetry condition, (9.1.16). We once again, also assume that the liquid is non-volatile, at least on the time scale of spreading, and so the total volume of liquid is conserved. Due to the presence of the preexisting thin film we now write this as,

$$(9.2.1) \qquad 2\pi \int_0^{R(t)} r\,(h - h_0)\,dr = 2\pi \int_0^\infty r\,(h - h_0)\,dr = V_{liq}$$

Lastly, far from the drop the liquid profile tends to approach the film thickness, so

$$(9.2.2) \qquad h\,(r \to \infty, t) = h_0$$

We assume here that the dominant driving force of spreading is the capillary force, and so

$$(9.2.3) \qquad p = p_a - \gamma \frac{1}{r}\frac{\partial}{\partial r}\left(h\frac{\partial h}{\partial r}\right)$$

and so once again the spreading equation becomes

$$(9.2.4) \qquad \frac{\partial h}{\partial t} = -\frac{\gamma}{3\mu r}\frac{\partial}{\partial r}\left(rh^3\frac{\partial}{\partial r}\left(\frac{1}{r}\frac{\partial}{\partial r}\left(r\frac{\partial h}{\partial r}\right)\right)\right)$$

The problem is again solved using matched asymptotic expansions (see Starov et al. (2007) for details). We obtain that the droplet radius evolves with time as

$$(9.2.5) \qquad R\,(t) = \left(\frac{5120}{24\pi^3 B^3}\right)^{1/10}\left(\frac{\gamma V_{liq}^3}{\mu}\right)^{1/10}\left(t + \frac{24\pi^3}{5120}\frac{\mu R_0^{10}}{\gamma V_{liq}^3}B^3\right)^{1/10}$$

where R_0 is the initial drop radius and B is the solution to

$$(9.2.6) \qquad 2B^3 \exp\left(B^3\right) = \left(\frac{\pi}{2}\frac{h_0 R_0^2}{V}\right)^3$$

Lastly, the dynamic contact angle evolves with time as

$$(9.2.7) \qquad \tan\left(\theta\,(t)\right) = B(3Ca)^{1/3}$$

Note that for small angles, $\tan\theta \approx \theta$. Note the similar scaling between this case, the spreading of drops on a surface covered with a thin film of the same liquid, to the previous case, the spreading of drops on dry surface. The Tanner's law $1/10$ power scaling in the equation for the drop radius still holds! The experimental success of this scaling (Tanner's law) might be attributed to the fact that it is so insensitive to the surface conditions - we have seen this same scaling for dry or wetted surfaces, and spreading due to capillary pressure alone or capillary pressure with disjoining forces. Thus for any smooth flat surface, we would be quite safe to assume that the drop radius will evolve roughly as

$$R\,(t) \sim \left(\frac{\gamma V_{liq}^3}{\mu}\right)^{1/10}(t + t_0)^{1/10}$$

(where t_0 is some constant) and so the velocity of the spreading will be,

$$\frac{dR}{dt} \sim \left(\frac{\gamma V_{liq}^3}{\mu}\right)^{1/10}t^{-9/10}$$

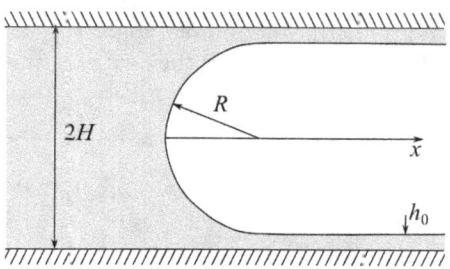

FIGURE 9.3.1. Schematic of a meniscus between two plates. Ahead of the meniscus there is a liquid film, thickness h_0 (exaggerated for clarity).

9.3. Meniscus between flat plates and in circular capillaries

Having thoroughly investigated the spreading of a drop on a single flat plate, lets now turn to the case of a moving meniscus between flat plates. This situation occurs frequently in microfluidic devices, when, for example, liquid is filling a large flat chamber under capillary action (rather than pressure, gravity, or centrifugal forces).

In this section we will consider a meniscus spreading between flat plates spaced apart by a distance $2H$ (see Figure 9.3.1). The meniscus is spreading onto a surface already wetted with an equilibrium film with thickness h_0. Similarly to the previous section, we will consider both the capillary pressure and the disjoining pressure important. Here we will consider a meniscus that is infinitely long into the page. In other words, considering that major meniscus surface has two radii of curvature (Section 7.2), the first radius of curvature scales as the half height between the plates, H while the second is infinite. Thus for our situation we use two dimensional Cartesian coordinates. Hence we can write our spreading relation, similarly to the one in the previous section, (9.1.42), but now in Cartesian coordinates as,

$$(9.3.1) \qquad \frac{\partial h}{\partial t} = -\frac{\gamma}{3\mu} \frac{\partial}{\partial x} \left(h^3 \frac{\partial h}{\partial x} - \frac{nA}{\gamma h^{n-2}} \frac{\partial h}{\partial x} \right)$$

with the boundary condition of

$$h(x \to \infty, t) = h_0$$

To make it easier to solve the equation (eliminate the time variable), we transform this equation into one moving with the meniscus at a steady state meniscus velocity, $y = x - Ut$, and thus convert the PDE (9.3.1) into an ODE,

$$(9.3.2) \qquad U \frac{dh}{dy} = \frac{\gamma}{3\mu} \frac{d}{dy} \left(h^3 \frac{dh}{dy} - \frac{nA}{\gamma h^{n-2}} \frac{dh}{dy} \right)$$

with the boundary condition of

$$(9.3.3) \qquad h(y \to \infty) = h_0$$

Taking the integral of both sides (with respect to y) of (9.3.2) subject to (9.3.3) we obtain,

$$(9.3.4) \qquad U(h - h_0) = \frac{\gamma h^3}{3\mu} \left(\frac{d^3 h}{dy^3} - \frac{nA}{\gamma h^{n+1}} \frac{dh}{dy} \right)$$

Unfortunately, (9.3.4) does not have a simple analytical solution, and so we have to integrate it numerically. To help us with this, we nondimensionalize it. We scale the meniscus height with the height of the film,

(9.3.5) $\xi = h/h_0$

the horizontal dimension with a combination of capillary number and film thickness:

(9.3.6) $z = \dfrac{y}{h_0}(3Ca)^{1/3}$

We have undertaken the last scaling, (9.3.6), using the intuition we gathered from the previous sections that the relationship between the horizontal and vertical dimensions (which scales as the dynamic contact angle), scales with capillary number to the 1/3 power (e.g., see (9.1.45)). Finally, we have a scaling for the disjoining pressure constant A,

(9.3.7) $\alpha = \dfrac{nA^{1/n}\Delta p^{(n-1)/n}}{\gamma^{1/3}(3\mu U)^{2/3}}$

where Δp is the pressure drop across the liquid-gas interface (as in Section (8.2)) Substituting this into (9.3.4), we obtain the dimensionless equation,

(9.3.8) $\xi - 1 = \xi^3\dfrac{d^3\xi}{dz^3} - \alpha\dfrac{1}{\xi^2}\dfrac{d\xi}{dz}$

This equation can then be solved numerically to obtain the meniscus profile. However, let's attempt to analyze it some more analytically. We can transform (9.3.8), introducing $v = \xi - 1$ and linearizing it to

(9.3.9) $\dfrac{d^3v}{dz^3} - \alpha\dfrac{dv}{dz} = 0$

For this equation, we seek a solution in the form $v = \exp(\lambda z)$. Substituting this into (9.3.9), we obtain

(9.3.10) $\lambda^3 - \alpha\lambda - 1 = 0$

We find the discriminant of (9.3.10) to be

(9.3.11) $Q = \dfrac{1}{4} - \left(\dfrac{\alpha}{3}\right)^3$

Thus, we see that the discriminant can be either positive or negative. When the discriminant is positive, (9.3.10) has one real positive root and two complex conjugate roots, whose real parts are negative. This implies that there are damped oscillations (waves) in the meniscus profile near the three-phase contact line. The film ahead of the meniscus is wavy. The discriminant is positive when $\alpha < 1.89$. This occurs when the velocity or viscosity is high, or when the disjoining pressure constant is low. When the discriminant is negative ($\alpha > 1.89$) (9.3.10) has two negative real roots and positive real root. This hints that $\xi(z)$ has only a single minimum ahead of the main meniscus. Here the film ahead of the meniscus has one depression but generally is not wavy. This analysis implies that if the disjoining pressure is neglected or is very low, i.e., $\alpha \to 0$, we would predict that the film ahead of the meniscus is wavy. Disjoining pressure acts to dampen these oscillations.

While we do not present the numerical solution of (9.3.8) here, as this can be solved using standard numerical ODE solvers, we present some salient results.

(For more details on the solution, see Starov et al. (2007)). We define the dynamic contact angle θ for this situation as

$$(9.3.12) \qquad \cos \theta = \frac{H}{R} = \frac{H}{\gamma} \frac{d^2 h}{dx^2}$$

where this is valid for $H/r < 1$. We find that for thicker microfluidic chambers, $H > 10$ µm, the solution for the dynamic contact angle collapses fairly well to

$$(9.3.13) \qquad \tan \theta = 2.36 \sqrt{Ca}$$

a result from a different analysis by Friz (1965), which is independent of the disjoining pressure constant. For thin chambers, $H < 10$ µm disjoining pressure becomes important and (9.3.13), for a given capillary number, underpredicts $\cos \theta$. Kalliadasis and Chang (1994) obtained, using the method of matched asymptotic expansions, that for circular capillaries,

$$(9.3.14) \qquad \tan \theta = 7.48 Ca^{1/3}$$

This relation shows good agreement with experimental data for capillary numbers between 10^{-4} and 10^{-1}.

9.4. Penetration of liquid into capillaries and wedges

A commonly encountered problem in microfluidic device design is the so called "passive" or capillary driven motion of liquid in microfluidic chips. It is often desired that a chip will fill itself with liquid either to prime the channels and the pumping regions so that the pumps can take over or so that the entire motion of the fluid in the device is done passively. In this section we will look into problems central to this question, namely the penetration of liquid in circular and non-circular channels.

9.4.1. Liquid penetration into circular channels (Washburn equation).
To derive the rate of liquid penetration into circular channels we follow the derivation of Washburn (1921). To do this, we consider the situation shown in Figure 9.4.1. Here we have a chamber (e.g., a well or a reagent storage reservoir in a microfluidic chip) that is filled with liquid with a volume much larger than can ever fit into the attached capillary. Above this liquid is there is gas pressure p_0 (i.e., the well can be pressurized by an outside pressure source, e.g., a pressurized tank). The height of liquid in this chamber is large enough so that the hydrostatic head is non-negligible. Similarly, the attached capillary goes up at an angle ψ as shown and so hydrostatic head inside the capillary is non-negligible. However, despite this, we assume that the radius of the capillary is small enough, such that the Bond number is small, and so we can neglect the gravitational effects on the meniscus as we have been doing before. In this section we will briefly jump ahead and consider the effects of both external pressure, gravity (but not at the surface), and pure capillarity together on the filling of the circular capillary.

We consider the cases where the flow in the circular capillaries is laminar ($Re < 2000$) and so Poiseuille flow applies. Furthermore, we assume that the capillary is filled with a fluid of negligible viscosity. Note that at room temperature, the dynamic viscosity of water is about 500 times that of air! Therefore, this is quite a good assumption for most situations. Thus, we can immediately write that the

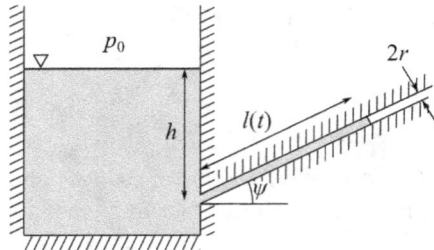

FIGURE 9.4.1. Large reservoir connected to a small circular capillary. Gravity is directed downward. While the hydrostatic pressure provided by gravity is non-negligible, here we neglect the effect of gravity on the meniscus both in the large reservoir and in the small capillary. We assume that the liquid-gas interface in the reservoir is flat, and so neglect the effects of surface tension in the reservoir.

flow rate in the capillary is

$$(9.4.1) \qquad \frac{dV}{dt} = \frac{\pi \sum P}{8\mu l} \left(r^4 - 4\varepsilon r^3 \right)$$

where $\sum P$ is the sum of all the pressure contributions. The first term on the right-hand side is the familiar Poiseuille flow term. The second term comes from a solution of Poiseuille's problem when slip is present in the capillary. Here ε is the coefficient of slip. While slip is a rare phenomena in microfluidic devices, we include this term, just as Washburn did firstly for generality and secondly as a way to account for the effects of disjoining pressure as a fudge factor, i.e., without directly accounting for them. In our case, the cross-section of the capillary is constant and so

$$(9.4.2) \qquad dV = \pi r^2 dl$$

giving us

$$(9.4.3) \qquad \frac{dl}{dt} = \frac{\sum P}{8r^2 \mu l} \left(r^4 - 4\varepsilon r^3 \right)$$

The pressure contribution consists of hydrostatic pressure,

$$(9.4.4) \qquad P_h = \rho g h - \rho g l \sin \psi$$

capillary (Young-Laplace) pressure,

$$(9.4.5) \qquad P_c = \frac{2\gamma}{r} \cos \theta$$

and the pressure of the pressurized well, p_0. In fact a more general hydrostatic term can be defined for a more complicated network of channels, since we would know the vertical position of the meniscus as a function of the distance from the entrance of the channel, and can write this as a function of l (where now we would treat l as the distance from the entrance). Here, however, we will just continue considering the simple geometry in Figure 9.4.1 for simplicity. Here the contact angle θ is the dynamic contact angle since the meniscus is moving. Substituting these pressures into (9.4.3) we obtain

$$(9.4.6) \qquad \frac{dl}{dt} = \frac{\left(p_0 + \rho g \left(h - l \sin \psi \right) + \frac{2\gamma}{r} \cos \theta \right)}{8\mu l} \left(r^2 - 4\varepsilon r \right)$$

The good news is that this gives us an instantaneous velocity of the meniscus in the capillary. The bad news is we often want to know the position of the meniscus directly as a function of time. We do not know an analytical solution to (9.4.6) but we can solve it numerically. However, when the capillary is horizontal, i.e., $\psi = 0$, (the most common and useful case for us), Washburn integrated (9.4.6) analytically to obtain,

$$(9.4.7) \qquad l^2(t) = \frac{\left(p_0 + \rho g h + \frac{2\gamma}{r}\cos\theta\right)\left(r^2 - 4\varepsilon r\right)}{4\mu} t$$

This result is often referred to as Washburn's equation. When the external pressurization and hydrostatic pressures are negligible and slip is not important, this equation simplifies to

$$(9.4.8) \qquad l(t) = \left(\frac{\gamma}{\mu}\frac{r\cos\theta}{2}t\right)^{1/2}$$

Notice from (9.4.7) that the square root dependence of distance traveled on time is not a result of the fact that the flow is driven by capillary pressure. This dependence is typical of any laminar flow of a Newtonian liquid subject to a constant pressure difference when the length of the liquid column is increased by the flow. Differentiating (9.4.8) we obtain the velocity of the meniscus,

$$(9.4.9) \qquad \frac{dl}{dt} = \frac{1}{2}\left(\frac{\gamma}{\mu}\frac{r\cos\theta}{2}\right)^{1/2}t^{-1/2}$$

or in terms of the distance the meniscus traveled along the capillary,

$$(9.4.10) \qquad \frac{dl}{dt} = \frac{\gamma}{\mu}\frac{r}{4l}\cos\theta$$

We see from (9.4.8) to (9.4.10) that an important quantity characterizing the properties of the liquid emerges,

$$(9.4.11) \qquad u_p = \frac{\gamma}{\mu}\cos\theta$$

This quantity is sometimes referred to as the coefficient of penetration or penetrativity. Notice it is only a property of the liquid and the material of the channel, and has a unit of velocity. It characterizes how quickly a given liquid is going to capillarily penetrate into any channel of a certain material. To make penetrativity independent of the channel material, we can define maximum penetrativity

$$(9.4.12) \qquad u_{p,\max} = \frac{\gamma}{\mu}$$

The maximum penetrativity for water is 7.9 m/s at room temperature. Interestingly, the maximum penetrativity for the common household penetrant WD-40 is only 0.6 m/s (TSI301.com (2017)). Mercury on the other hand, has a maximum penetrativity of 32 m/s, mostly due to its high surface tension. For a quick and dirty estimate of penetrativity we can use the equilibrium contact angles from Table 1 instead of the dynamic contact angles. For water in very hydrophilic channels (e.g., glass, <15°, see Table 1) penetrativity is 7.6 m/s, whereas for water is more hydrophobic materials (e.g., SU8, 79°), the penetrativity is only 1.5 m/s. Note that the liquid never travels this fast in most of the channel - for a typical glass microfluidic channel several centimeters in length and 50 μm radius, at 1 cm mark the liquid will be travel at only 9.5 mm/s (Re of 0.05). This is still fast for a microfluidic channel (and so capillary filling is effective) but the flow is very much laminar. Even 1 radius away from the entrance the Re is 42 and so is laminar, justifying the use of Poiseuille's solution.

Historical note: The result $l(t) \sim t^{1/2}$ was obtained Washburn (1921), and prior to him, Lucas (1918), and even prior by Bell and Cameron (1906). Thus, this result is often referred to as Washburn equation, Lucas-Washburn equation, Washburn-Lucas equation, Bell-Cameron-Lucas-Washburn equation, etc. This sometimes leads to confusion among students. Here for conciseness we will refer to it as the Washburn equation.

9.4.2. Penetration into a vertical circular capillary. Washburn also found an analytical (but more involved) solution to (9.4.6) for the case where the capillary is pointed vertically (i.e., $\psi = \pi/2$). For the case where external pressurization and hydrostatic pressures are negligible and slip is not important, we can obtain a solution in dimensionless form

$$(9.4.13) \qquad T = -R^2 \left(\frac{L}{R} + \ln \left(1 - \frac{L}{R} \right) \right)$$

$$(9.4.14) \qquad T = t/\tau$$

$$(9.4.15) \qquad L = 4l/r$$

$$(9.4.16) \qquad R = 4h^*/r$$

$$(9.4.17) \qquad \tau = \frac{\mu}{\gamma \cos \theta} \frac{r}{4}$$

$$(9.4.18) \qquad h^* = \frac{2\gamma \cos \theta}{\rho g R}$$

Here τ is the characteristic time scale of wicking, which happens to be the penetrativity scaled by the radius of the capillary. h^* is the equilibrium (maximum) capillary rise. Similar equations were also found by Rideal (1922) and independently Lucas (1918). Hence this equation is often referred to as Washburn-Rideal-Lucas (WRL) equation. We plot this solution in Figure 9.4.2. We see that at very short times the slope is small - the velocity of the meniscus is low. As time progresses the capillary force accelerates the meniscus and the curve's slope and so the meniscus velocity increases. At later times, the curve's slope begins to decrease - gravity is decelerating the meniscus, and eventually the meniscus settles to the equilibrium (maximum) height.

There is however a challenge with using the capillary penetration equations derived in this section and in the previous one - the equations contain a finicky quantity, the dynamic contact angle. As we saw in the previous sections, the dynamic contact angle is a function of the speed of the meniscus itself, depends on the geometry, and to some extent depends also on the relevance of characteristic forces (e.g., disjoining pressure vs. capillary pressure). Furthermore, a number of different relations for dynamic contact angle are available. Examples include an analytically obtained one of Kalliadasis and Chang (1994), our (9.3.14), or an experimentally obtained one of Joos et al. (1990),

$$(9.4.19) \qquad \cos \theta_d = \cos \theta_0 - 2 \left(1 - \cos \theta_0 \right) Ca^{1/2}$$

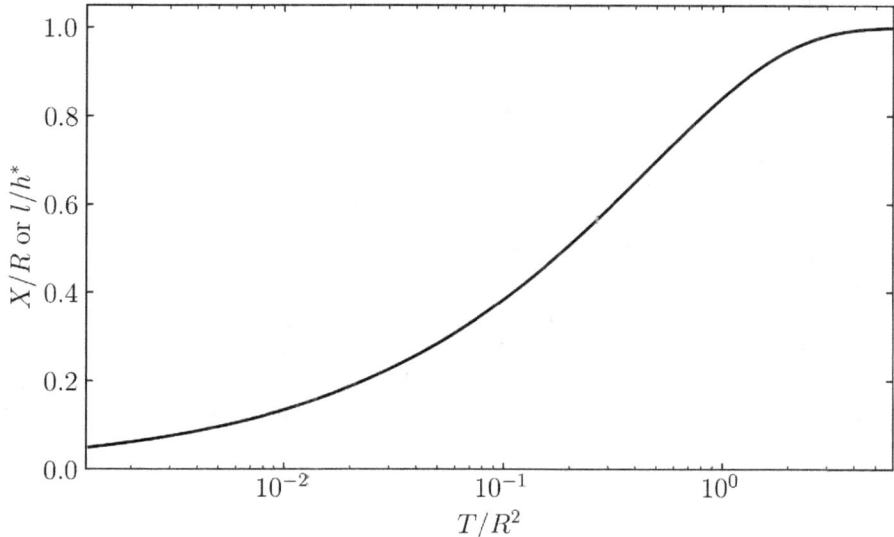

FIGURE 9.4 2. Capillary rise l scaled by the maximum (equilibrium) capillary rise h^* as a function of time scaled by the wicking time scale and the square of the dimensionless channel size, R.

where θ_0 is the static advancing contact angle. However, even assuming a constant dynamic contact angle, as was originally done by Washburn, gives a fairly reasonable results, correct to better than an order of magnitude, and often better than within 20%. Even quick and dirty estimates using the equilibrium contact angle as a static dynamic contact angle can give a reasonable estimate.

9.4.3. Penetration into a circular capillary with inertia (Bosanquet equation).
In deriving the Washburn equation, by directly assuming Poiseuille flow, we assumed that the flow is quasi steady, as that is assumed when deriving Poiseuille's equation. Furthermore, we assumed that the inertia of the fluid was negligible. These assumptions are in general good for most liquids for small channels (< 1 mm) - dimensions we are usually concerned with in microfluidics. However, sometimes we are also concerned with larger channels (e.g., mesofluidics, mm scale channels). For these situations it is useful to consider the non-steady case. For this, we follow the derivation of Bosanquet (1923) and obtain the Bosanquet equation. Consider again a capillary in Figure 9.4.1, with $\psi = 0$, i.e. a horizontal capillary. Let P be the total pressure acting on the column of liquid in the capillary at the entrance of the capillary, i.e., where the capillary connects to the reservoir. Let T be the capillary line force (force per length) acting on the meniscus, i.e., $T = \gamma \cos \theta$. Furthermore, the liquid is viscous, and so has a viscous shear force, a retarding force, proportional to the length of the liquid column, and the velocity of the column's travel,

(9.4.20)
$$8\pi \mu l \frac{dl}{dt}$$

The column of liquid has a momentum of

$$(9.4.21) \qquad\qquad \rho \pi r^2 x \frac{dl}{dt}$$

Using Newton's second law, i.e., change in an object's momentum is equal to the force applied to it, and balancing the three forces, we obtain

$$(9.4.22) \qquad \frac{d}{dt}\left(\rho \pi r^2 x \frac{dl}{dt}\right) = P\pi r^2 + 2\pi r T - 8\pi \mu l \frac{dl}{dt}$$

which is a second order differential equation. Notice that we captured both the unsteady nature of the problem and the inertia of the fluid, as well as the viscous forces. The only (potentially) relevant force not explicitly captured is the disjoining pressure, but that can enter through θ. Note that since the column of liquid is moving, θ is the dynamic contact angle. We have assumed that the column is fairly long and its volume can be described without explicitly taking account of the volume of the meniscus. For convenience, let's define

$$(9.4.23) \qquad\qquad \tau = \frac{r^2 \rho}{8\mu}$$

$$(9.4.24) \qquad\qquad b = \frac{rP + 2T}{r\rho} = \frac{rP + 2\gamma \cos\theta}{r\rho}$$

with which (9.4.22) becomes

$$(9.4.25) \qquad\qquad \frac{d}{dt}\left(x\frac{dl}{dt}\right) + \frac{1}{\tau}\left(x\frac{dl}{dt}\right) = b$$

Integrating (9.4.25) twice, we obtain,

$$(9.4.26) \qquad l(t)^2 - l(t=0)^2 = 2b\tau \left[t - \tau\left(1 - \exp\left(-\frac{t}{\tau}\right)\right)\right]$$

This result is known as the Bosanquet equation. Notice that when τ is small (compared to t) the exponential decays quite quickly, and the Bosanquet equation collapses to the Washburn equation. Thus, we see that the inertia and the non-steadiness are only important at early times, when $t < \tau$. Notice from (9.4.23) that τ scales at the square of the tube radius, and so the smaller the channel the less we have to worry about early times (start up) behavior. In fact, for long times (compared to τ) the liquid forgets this initial stage, and behaves completely according to the Washburn equation. For water in a 50 µm radius capillary, τ is 34 µs - a time certainly negligible for most situations.

9.4.4. Liquid penetration into wedges or corners. As we have discussed above, cross-sections of typical microfluidic channels often have "sharp" corners - wedges. We call these corners "sharp" (in quotes) because while in principle they should be mathematically sharp - in contrast to a round circular capillary, in practice due to the different practical constraints of different fabrication methods their internal shape might be rounded or this internal shape might be non-uniform. A closer SEM inspection is usually warranted. However, sometimes (e.g., glass or silicon bonded via anodic bonding) these corners are sharp down to 10-100 nm level, for example, for a channel with characteristic width of e.g., 500 µm. These corners present an interesting problem. For such corners, a meniscus will develop in a corner stretching out a liquid finger far into the corner (Figure 9.4.3). If the

(a) (b)

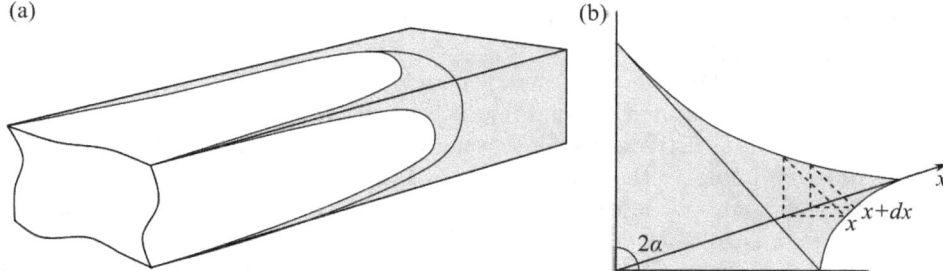

FIGURE 9.4.3. Penetration of liquid into corners of channels. (a) Fingers of liquid from the main meniscus extend far into corners, well ahead of the main meniscus. It has been experimentally observed for square glass capillaries for these fingers to rise (against gravity) all the way to the end of the capillary, far above the maximum rise height of the main meniscus. When filter paper was placed on top of such capillary, the fingers were able to transport enough liquid to saturate the filter, without the main meniscus rising (Bico and Quere (2002)). (b) Simplified view of one of such fingers in a wedge.

corners were mathematically sharp, the radius of curvature of this meniscus at its tip would tend to zero, and so the capillary pressure driving the meniscus would tend to infinity. While real corners will not allow a zero radius of curvature, the real radii of curvature can be quite small (e.g. O(10 nm)) and so capillary pressures quite high (O(10 MPa)) and so the fingers can penetrate quite far, or even to the end of such capillaries. Of course for radii of curvature of O(10 nm) disjoining pressure would become important and may supersede capillary pressure as the main driving force of the spreading. However, it has been observed, for example, by Bico and Quere (2002) that such fingers can penetrate into the capillary far ahead of the main meniscus even under force of gravity working against penetration. These fingers can also have significant liquid carrying capacity: Bico and Quere (2002) found that placing a filter paper on top of such capillary would saturate the filter paper. This means that if you have a meniscus (including a meniscus stopped by some types of capillary valves, especially 2D capillary valves) in a tube with cornered channels, the liquid in the corners may penetrate much farther than this main meniscus and begin to interact with "dry areas" such as lyophilized reagents or sensor surfaces far earlier than intended. One has to be very careful with channels with corners.

The problem of liquid penetrating into a corner has been studied for quite a while: even Taylor (of Taylor expansion fame) has studied this problem and published on this as early as 1712 (Taylor (1712)). This problem has been approached in various ways. For example see, Ponomarenko et al. (2011) - for a series of small tubes ("organ") approach, or Higuera et al. (2008) - small angle approach. However, at the time of this writing no simple analytical solutions exist, and here we will follow what we believe is the current simplest approach, that of Dong and Chatzis (1995). While this approach requires us to use numerically calculated parameters, it gives us thorough appreciation of the dynamics of the liquid in a corner. Moreover, one major advantage of Dong and Chatzis (1995) approach over many others is that it can account for realistic corners with roundedness rather than only perfectly sharp, mathematical ones.

Consider a liquid imbibing a corner of a horizontal tube, a liquid finger, as shown in Figure 9.4.3b. We assume as usual that the gas phase exerts a negligible stress on the liquid-gas interface. We consider a situation where the Bond number is low, so gravity does not influence the liquid interface. Lastly, we neglect the components of the fluid velocity components orthogonal to the surfaces of the walls of the corner. Let's consider an arbitrary position along the liquid finger, for example as outlined by a dashed triangles in Figure 9.4.3b. The change in the flow rate, q between this position, x and that slightly upstream, $x + dx$ is equal to a change in the cross-section area (S vs. $S - dS$) and change in area averaged velocity (u vs. $u + du$). Thus we can write,

$$(9.4.27) \qquad \frac{dq}{dx} = -\frac{dS}{dx}\frac{du}{dx}$$

Since $u = dx/dt$,

$$(9.4.28) \qquad \frac{dq}{dx} = -\frac{dS}{dt}$$

From geometry, for a square (90°) corner, the relationship between the liquid cross-section S, the contact angle θ, and the radius of curvature of the liquid-gas interface r is

$$(9.4.29) \qquad S = 4r^2 \left(\frac{\cos\theta \cos\left(\frac{\pi}{4} + \theta\right)}{\sin(\pi/4)} - (\pi/4 - \theta) \right)$$

Similar relations can be found for other corner angles. Note that of course the contact angle has to be such that it satisfies the Concus-Finn condition of wetting, (8.5.3), i.e., for a 90° corner of the same material, $\theta < \pi/4$. For calculation of area averaged velocity in a wedge we employ a dimensionless flow resistance β, proposed by Ransohoff and Radke (1988), that is defined by

$$(9.4.30) \qquad u(x,t) = \frac{r^2(x,t)}{\beta\mu}\frac{\partial P_c}{\partial x}$$

where P_c is the Young-Laplace capillary pressure,

$$(9.4.31) \qquad P_c(x,t) = \frac{\gamma}{r(x,t)}$$

Here we have assumed that the radius of curvature in the (y, z) plane, r, is much more significant than the orthogonal radius of curvature, i.e., the curvature along the stretch of the liquid finger. As calculated by Ransohoff and Radke, the dimensionless resistance β has a constant value for a given corner geometry and contact angle. The corner geometry is determined by the dihedral angle (2α, Figure 9.4.3b), and the degree of roundness of a corner. (Here, for simplicity, we will only consider, the mathematically sharp corner case for which roundness is zero. For rounded corners see Dong and Chatzis (1995) and Ransohoff and Radke (1988)). The flow rate in the liquid finger is then the product of the area averaged velocity and the area, i.e.,

$$(9.4.32) \qquad q(x,t) = S(x,t)\frac{r^2(x,t)}{\beta\mu}\frac{\partial P_c}{\partial x}$$

Substituting in (9.4 31), we obtain

(9.4.33)
$$q = -S\frac{\gamma}{\beta\mu}\frac{\partial r}{\partial x}$$

and substituting this result into (9.4.28), we obtain a partial differential equation for the cross-section of the liquid finger,

(9.4.34)
$$\frac{\partial S}{\partial t} = B\frac{\partial}{\partial x}\left(S^{1/2}\frac{\partial S}{\partial x}\right)$$

where

(9.4.35)
$$B = \frac{1}{2\beta}\frac{\gamma}{\mu}C^{-1/2}$$

(9.4.36)
$$C = 4\left(\frac{\cos\theta\cos\left(\frac{\pi}{4}+\theta\right)}{\sin(\pi/4)} - (2.334)(\tau/4-\theta)\right)$$

Let's consider the motion of the finger in a reference frame moving with the main meniscus. The partial differential equation we have derived does not change, since if we subtract a constant velocity from (9.4.30) dq/dt does not change, nor the condition (9.4.28) is affected. In this reference frame it is as if the reservoir of liquid for the liquid finger is traveling slowly and at a constant speed with it. In this moving coordinate system, we can create a reference point, some small distance away from the point of the main meniscus and call this point $x = 0$. (Note, this is x in the moving coordinate system; while we could have used primes to differentiate the two, we drop these for convenience). This point is sufficiently far away from the tip of the advancing finger so that here the liquid reached its equilibrium profile and remains at this profile. Thus at this point,

(9.4.37)
$$S = S_0 \text{ at } x = 0$$

Furthermore, far away from the advancing finger tip, the liquid cross-section area is zero, so

(9.4.38)
$$S = 0 \text{ at } x \to \infty$$

Let's start the clock at $t = 0$, such that the tip of the liquid finger is at x_0, some position ahead of $x = 0$. Thus,

(9.4.39)
$$S = 0 \text{ at } t = 0, \ x > x_0$$

Finally, from the first boundary condition we have

(9.4.40)
$$S = S_0 \text{ at } t = 0, \ x = 0$$

What we have here is a non-linear diffusion equation. Dong and Chatzis (1995) found an approximation to the exact solution for this equation and these boundary and initial conditions in the reference frame moving with the main meniscus to be

(9.4.41)
$$S = S_0\left(1 - \frac{ax}{x_0\sqrt{1+2vt}} + \frac{(a-1)x^2}{x_0^2(1+2vt)}\right)^2$$

where v is a time constant related to the original shape of the meniscus,

(9.4.42)
$$v = K\frac{\gamma}{\mu\beta}\frac{R}{x_0^2}$$

Here R is the radius of the liquid meniscus in the corners of the capillary under equilibrium at $x = 0$. In other words, it is the radius of curvature of the liquid measured in the plane of S_0. The parameters a and K are fitting parameters, fitting this approximate solution to the exact solution for this PDE (obtained numerically). For mathematically sharp 90° corners, Dong and Chatzis (1995) calculate a and K to be 0.59 and 1.447 respectively, and β to be 93.5. For water at room temperature, in a 100 x 100 µm channel, assuming R to be 50 µm and x_0 to be 500 µm, v equals $10\,\text{s}^{-1}$. For x_0 to be 100 µm, v equals $250\,\text{s}^{-1}$. Thus we expect the initial travel of the finger to take place on millisecond to hundreds of milliseconds time scale. In the reference frame moving with the main meniscus, the tip of the finger moves according to

$$(9.4.43) \qquad x_f = x_0(1 + 2vt)^{1/2}$$

and the finger cross-sectional area along its length is given by

$$(9.4.44) \qquad S = S_0\left(1 - a\frac{x}{x_f} + (a-1)\left(\frac{x}{x_f}\right)^2\right)^2$$

For long times, $t \gg 1/v$, we can simplify (9.4.43) and combine it with (9.4.42), eliminating the pesky typically a priori unknown x_0, and obtain

$$(9.4.45) \qquad x_f = \left(\frac{2K}{\beta}\frac{\gamma}{\mu}Rt\right)^{1/2}$$

By differentiating (9.4.44) we find the saturation gradient,

$$\left(\frac{\partial S}{\partial x}\right)_{x=0} = -\frac{2aS_0}{x_f}$$

which we then combine with (9.4.36) and (9.4.29) and substitute into (9.4.33) to obtain the flow rate to be

$$(9.4.46) \qquad q_{x=0} = aC(2K\beta)^{-1/2}\left(\frac{\gamma}{\mu}\right)^{1/2} R^{5/2}t^{-1/2}$$

and integrating this, gives the volume of the finger penetrated at a particular time,

$$(9.4.47) \qquad V = 2aC(K\beta)^{-1/2}\left(\frac{\gamma}{\mu}\right)^{1/2} R^{5/2}t^{1/2}$$

By differentiating (9.4.43) we find that the velocity of the liquid finger tip (in the reference frame moving with the main meniscus),

$$(9.4.48) \qquad u_f = K(2\beta)^{-1/2}\left(\frac{\gamma}{\mu}\right)^{1/2} R^{1/2}t^{-1/2}$$

We can observe that penetration into corners follows the $t^{1/2}$ scaling ($t^{-1/2}$ for velocity), just like in the Washburn penetration case. Secondly, notice that the dependence of penetration velocity on penetrativity is also the same as in the Washburn case, both scale as one-half power. This should not be surprising because here, just as in the Washburn case we have balanced the viscous forces (Hagen-Poiseuille eqn. in Washburn, eqn. (9.4.30) here) against Young-Laplace capillary pressure.

9.4.5. Penetration into rectangular channels (generalized Bosanquet equation).

Rectangular channels are one of the most common channels encountered in microfluidic devices due to their ease of fabrication. Approximately rectangular channels can be manufactured by injection molding, hot embossing, and SU-8 and even PDMS lithography. To investigate penetration in these channels we generalize the Bosanquet approach, following Ouali et al. (2013). As we shall see the Bosanquet approach is very powerful and can be easily generalized to other shapes as well, e.g., triangular channels or channels of custom unusual cross-section.

Just as before, we will balance the change in momentum of the fluid with the capillary (surface tension driven) forces and viscous forces; for generality we also consider the effect of gravitational force. As before we treat the liquid as a column (or slab), and for the purposes of calculating its mass and momentum ignore the fine features of the meniscus at the liquid-gas interface. As before, we consider a column of liquid of constant density ρ traveling in a constant cross-section area A_c pipe. Thus, we can write that the rate of change of momentum of the liquid column when it is at position x in the channel is equal to

$$(9.4.49) \qquad \frac{d}{dt}\left(\rho A_c x \frac{dx}{dt}\right) = \rho A_c x \frac{d^2 x}{dt^2} + \rho A_c \left(\frac{dx}{dt}\right)^2$$

For generality, the capillary may be pointed vertically at some angle (see Figure 9.4.1). Thus, the force of gravity on the column of liquid in the capillary is,

$$(9.4.50) \qquad f_g = -\rho g \sin\psi A_c x$$

Capillary forces arise in the fluid due to creation of new solid-vapor, solid-liquid, and/or liquid-vapor interfaces at the location of the main meniscus. The small change in surface free energy, ΔG due to a small movement of meniscus forward Δx is then

$$(9.4.51) \qquad \Delta G = \Delta x \left[\sum_i (\gamma_{SL,i} - \gamma_{SV,i}) L_i^{SV \to SL} + \sum_i \gamma_{LV,i} L_i^{LV}\right]$$

where the subscript i refers to the ith wall (i.e., bottom, or top, or left, or right wall in a rectangular channel). As a reminder, $\gamma_{LV} = \gamma$. $L_i^{SV \to SL}$ is the perimeter length of the ith solid wall on which the contact with the vapor is replaced with contact with the liquid. L_i^{LV} is the perimeter of the new liquid-vapor interface created. The last term is important in open channels, where one of the walls is "missing"; in this model this solid wall is replaced by a vapor wall, and as the fluid slug extends in the channel, it extends the area of contact with the vapor. This way, this generalized Bosanquet model can even capture open channel geometries! Since the interfacial energies $\gamma_{SL,i}$ and $\gamma_{SV,i}$ are not always known, and it is more convenient to use a contact angle instead, we employ Young's equation, (7.4.2), to write (9.4.51) in terms of contact angle.

$$(9.4.52) \qquad \frac{-\Delta G}{\Delta x} = \left[\sum_i L_i^{SV \to SL} \cos\theta_i + \sum_i L_i^{LV} \cos\pi\right]$$

where we use $\cos\pi = -1$ to emphasize the similarity between the liquid contacting solid walls, and liquid contacting the vapor (gas) surface (open wall) in this model. Recall from Section 7.2.2 that capillary force is equal to $-\Delta G/\Delta x$ as $\Delta x \to 0$.

Thus we can write (9.4.52) as

$$(9.4.53) \qquad f_{cap} = \left[\sum_i L_i^{SV \to SL} \cos \theta_i + \sum_i L_i^{LV} \cos \pi \right]$$

We arrived at a general equation for the capillary force in an arbitrarily shaped (and even open!) channel. Here we will apply this to a circular geometry (to rederive the Bosanquet equation), to completely enclosed rectangular geometry as well as that for an open wall. As you can imagine, this equation can be just as easily applied to triangular, octagonal, ellipsoidal, and even limacon shapes! Interestingly, the solution for viscous force, which we will consider next, is available analytically for many of these shapes, even for the limacon! (See Berker (1963), or White (1991)). Fortunately, limacon cross-sections are rarely encountered in practice. For the rectangular geometries we will make a simplifying assumption that the geometric dimension of the wall is equal to the wetting perimeter, i.e., $L_i^{SV \to SL}$ for that wall. From the previous section and Figure 9.4.3, we see that in reality this is not the case - the actual wetting perimeter is larger due to presence of liquid fingers in corners. Thus, with this approach we are slightly underpredicting the capillary force on the column of liquid. However, by neglecting the presence of the liquid fingers in the corners, we also neglect the viscous forces in these small sized fluid columns, for which the viscous resistance is large. However, as Ouali et al. (2013) found neglecting liquid fingers in corners underpredicts the total motion slightly, typically less than by 20%. For the case of a circular cross-section capillary, (9.4.53) becomes

$$(9.4.54) \qquad f_{cap,circ} = 2\pi R \gamma \cos \theta$$

just like in our original Bosanquet derivation. For a closed rectangular channel,

$$(9.4.55) \qquad f_{cap,r} = \gamma W \left(\cos \theta_B + \cos \theta_T + \varepsilon \left(\cos \theta_L + \cos \theta_R \right) \right)$$

where $\varepsilon = H/W$ and where H and W are the channel height and width respectively. Here the subscripts B, T, L, and R refer to bottom, top, left, and right walls of the channel. We have left the equation in this form to be able to specify the contact angle on each wall independently as often times in microfluidic devices different walls are made from different materials. For an open rectangular channel, $\theta_T = \pi$. For the first approximation, (and as was done by Ouali et al. (2013)) we can use the equilibrium contact angle as θ. Strictly speaking, this is not correct as the meniscus is moving. Thus, for a better approximation, we may use a dynamic contact angle, for example, based on the derivations for the dynamic contact angle in above sections. For channels in which one of the sections is rough or nanostructured, we can of course use the equivalent contact angle predicted by the Cassie-Baxter or Wenzel models.

Viscous force arises from shearing of the viscous fluid as it flows. To obtain it, we assume as usual that the fluid is Newtonian, incompressible, flows in a laminar manner, and that there is no-slip at the walls. As before, for a circular cross-section channel the viscous force is

$$(9.4.56) \qquad f_{visc,circ} = -8\pi \mu x u_{ave}$$

where u_{ave} is the area averaged velocity. For other geometries this can be found by taking partial derivatives of the velocity profile to obtain the shear stress, and then integrating the shear stress over the cross-sectional area of the channel. Analytic expressions for the velocity profile for various cross-sections can be found in Berker (1963) and in White (1991). Ouali et al. (2013) found that the viscous force for closed rectangular channel to be

$$(9.4.57) \qquad f_{visc,RC} = \frac{-12\mu x u_{ave}}{\varepsilon \zeta_c(\varepsilon)}$$

$$(9.4.58) \qquad \zeta_c^{-1}(\varepsilon) \approx 1 + 0.362374\varepsilon + 1.020980\varepsilon^2$$

and open rectangular channel to be,

$$(9.4.59) \qquad f_{visc,RC} = \frac{-3\mu x u_{ave}}{\varepsilon \zeta_o(\varepsilon)}$$

$$(9.4.60) \qquad \zeta_o^{-1}(\varepsilon) \approx 1 + 0.671004\varepsilon + 4.169711\varepsilon^2$$

where here ζ are the fits for the sum of infinite terms that arise from those sums found in the expressions for the velocity profiles. The exact ζ functions are given in Ouali et al. (2013) as well. These approximate fits are valid for aspect ratios $0 \leqslant \varepsilon \leqslant 2$.

Combining the expressions for the viscous, capillary, and gravitational forces with Newton's second law - that the change in momentum of a body is equal to the sum of the forces on it, we obtain

$$(9.4.61) \qquad \frac{a}{dt}\left(x\frac{dx}{dt}\right) = x\frac{d^2x}{dt^2} + \left(\frac{dx}{dt}\right)^2 = b - g\sin\psi x - ax\frac{dx}{dt}$$

where for the circular cross-section, rectangular closed cross-section, and rectangular open cross-section,

$$(9.4.62) \qquad a = \begin{cases} 8\mu/\rho R^2 \\ 12\mu/\rho H^2 \zeta_c(\varepsilon) \\ 3\mu/\rho H^2 \zeta_c(\varepsilon) \end{cases}$$

$$(9.4.63) \qquad b = \begin{cases} 2\gamma\cos\theta/\rho R \\ \gamma(\cos\theta_B + \cos\theta_T + \varepsilon(\cos\theta_L + \cos\theta_R))/\rho H \\ \gamma(\cos\theta_B + \cos\theta_T + \varepsilon(\cos\theta_L + \cos\theta_R))/\rho H \end{cases}$$

As a reminder, for the open cross-section, $\theta_T = \pi$. a has dimensions of inverse time (e.g., s^{-1}) and is the inverse of the time constant in our original derivation of the Bosanquet solution. b has dimensions of speed square (e.g., m^2/s^2). For ease of solution we nondimensionalize (9.4.61), and obtain dimensionless time $T = at$, and dimensionless meniscus position $X = ax/\sqrt{2b}$. Thus, nondimensionalized (9.4.61) becomes

$$(9.4.64) \qquad 2\frac{d}{dT}\left(X\frac{dX}{dT}\right) = 2\left(X\frac{d^2X}{dT^2} + \left(\frac{dX}{dT}\right)^2\right) = 1 - G\sin\psi X - 2X\frac{dX}{dT}$$

where

$$(9.4.65) \qquad G = \frac{g\sqrt{2}}{a\sqrt{b}}$$

Compare this equation to the original Bosanquet equation (9.4.25). We can see that if G was small or $\sin\psi$ was small or even zero (i.e. a horizontally orientated capillary, the most common case) than the $G\sin\psi X$ term may be neglected and so our equation would be identical to the Bosanquet equation (9.4.25). When this is the case, we already have a solution, the Bosanquet solution,

$$(9.4.66) \qquad X^2(T) = T - (1 - \exp(-T))$$

As we have seen before, for long time scales, i.e., $T \gg 1$ ($t \gg 1/a$), which are frequently encountered, this equation simplifies to the Washburn solution,

$$(9.4.67) \qquad X(T) \approx T^{1/2}$$

At short time scales, i.e., $T \ll 1$ ($t \ll 1/a$), this simplifies to (Quere (1997)),

$$(9.4.68) \qquad X(T) \approx T\big/\sqrt{2}$$

Having obtained these very useful extensions of the Bosanquet solution for horizontal capillaries of rectangular cross-section, let's examine cases where the gravitational term cannot be neglected. Firstly, for convenience, let's rewrite (9.4.64) using derivative identities, as

$$(9.4.69) \qquad \frac{d^2 X^2}{dT^2} = 1 - G\sin\psi X - \frac{dX^2}{dT}$$

The term on the left represents the inertial term. When the Reynolds number is particularly small, the inertial term can become negligible, and so

$$(9.4.70) \qquad 0 = 1 - G\sin\psi X - \frac{dX^2}{dT}$$

This ODE can be integrated using the initial condition,

$$(9.4.71) \qquad X = 0 \text{ at } T = 0$$

and we obtain,

$$(9.4.72) \qquad T = \frac{-2}{(G\sin\psi)^2}\,(G\sin\psi X + \log(1 - G\sin\psi X))$$

Unfortunately, this equation is difficult to invert and write in the form of $X(T)$. For more on the solution of this equation see Fries and Dreyer (2008). However, we have no problem in graphing $X(T)$ and producing the appropriate predictions with the aid of a computer.

The last term on the right-hand side of (9.4.69) represents the viscous forces. When capillary filling is dominated by inertia and gravitational forces (possibly in the very early stages of filling), we may neglect the influence of viscous forces, and obtain,

$$(9.4.73) \qquad \frac{d^2 X^2}{dT^2} = 1 - G\sin\psi X$$

While there is no closed form solution to this non-linear ODE, it can be solved via perturbation methods (Quere (1997)). This solution is

$$(9.4.74) \qquad X(T) = \frac{T}{\sqrt{2}}\left(1 - \frac{\sqrt{2}G\sin\psi T}{12}\right)$$

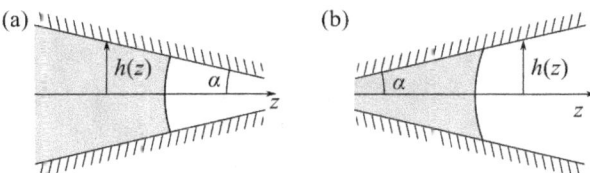

FIGURE 9.4.4. Schematic of liquid penetration into a diverging geometry (a) and converging geometry (b). The fluid is supplied by a large (endless) reservoir. While here we draw a linearly tapered geometry, in the text we consider other functions $h(z)$ as well. The liquid has a contact angle θ with the walls. The position of the meniscus is at $l(t)$.

9.4.6. Penetration into diverging and converging capillaries and plates.

Very often in microfluidic devices channels and chambers are not of constant cross-section - they either diverge or converge. When we design these channels and chambers we can leverage divergence or convergence to control the rate of liquid penetration (imbibition) into a chamber. Here to model penetration of liquid into diverging or converging channels we will employ the Washburn approach once again - balancing the viscous forces with the capillary forces.

Let's consider first a converging geometry as shown in Figure 9.4.4a. Converging channels are interesting in that as the liquid penetrates deeper into the channel, the effective radius of curvature of the meniscus decreases, so increasing the capillary force on the liquid. However, the hydraulic diameter of the capillary shrinks as well, increasing the resistance. Thus if we correctly design the converging channel we can actually accelerate the fluid in the channel. We will follow the approach of Gorce et al. (2016) in our analysis. Here we will consider only cases where the half angle of the cone α is small. We do this so that we can employ the lubrication approximation, i.e., $dh/dz| \ll 1$, so that we can assume that the velocity in the channel is predominantly in the z direction. As a result of this assumption, the pressure within the liquid depends only on z, and the Navier-Stokes momentum equations reduce to

$$(9.4.75) \qquad \frac{1}{r}\frac{\partial}{\partial r}\left(r\frac{\partial u_z}{\partial r}\right) = \frac{1}{\mu}\frac{\partial p}{\partial z}$$

Here u_z is the velocity component in the z-direction. Employing the no-slip condition at the walls and integrating this, yields

$$(9.4.76) \qquad u_z = \frac{1}{4}\frac{dp}{dz}\left(r^2 - h^2(z)\right)$$

This is quite similar to the classical Hagen-Poiseuille profile, with the exception for the dependence on the wall geometry. Fluid flux (volumetric flow rate) through a cross-section is

$$(9.4.77) \qquad Q(z) = 2\pi \int\limits_{0}^{h(z)} r u_z dr = -\frac{\pi h^4(z)}{8\mu}\frac{dp}{dz}$$

However, since the liquid is incompressible, mass conservation (continuity) tells us that the fluid flux should be independent of z. Hence we can rearrange (9.4.77) to obtain

$$(9.4.78) \qquad p(z) = -\frac{8\mu Q}{\pi} \int_0^z (h(z))^{-4} dz$$

where we set $p(z = 0) = 0$. We are interested in position of the meniscus l and the velocity of the meniscus dl/dt. By mass conservation,

$$(9.4.79) \qquad Q = \pi h^2(l) \frac{dl}{dt}$$

Our driving force comes from the capillary pressure at the meniscus, and accounting for the convergence of the channel and the contact angle, this pressure is

$$(9.4.80) \qquad p_{cap} = -\frac{2\gamma \cos(\theta - \alpha)}{h(l)}$$

Since we have taken the lubrication theory assumption, $\alpha \ll 1$ and so

$$(9.4.81) \qquad p_{cap} \approx -\frac{2\gamma \cos\theta}{h(l)}$$

Combining this result with (9.4.78) and (9.4.79), we obtain,

$$(9.4.82) \qquad \frac{dl}{dt} = \frac{\cos\theta}{4} \frac{\gamma}{\mu} \frac{(h(l))^{-3}}{\int_0^l (h(l))^{-4} dz}$$

Now we need an explicit form for the shape of the capillary walls, $h(z)$. Let's consider a cone of the form

$$(9.4.83) \qquad h(z) = h_0 \left(\frac{z_0 - z}{z_0} \right)^n$$

for $n \geqslant 0$. This cone has a width h_0 at its base ($z = 0$) and decays to a zero radius at $z = z_0$. For practical purposes of course the cone has a hole at $z = z_0$ so that air can escape.

Now, for convenience, we nondimensionalize the equations, such that the dimensionless meniscus position is

$$(9.4.84) \qquad L = \frac{l}{z_0}$$

and dimensionless time is

$$(9.4.85) \qquad T = t \frac{\gamma}{\mu} \frac{h_0 \cos\theta}{z_0^2}$$

Substituting (9.4.83) into (9.4.82), and nondimensionalizing the result, we obtain

$$(9.4.86) \qquad \frac{dL}{dT} = \left(n - \frac{1}{4} \right) \left((1 - L)^{1-n} - (1 - L)^{3n} \right)^{-1}$$

Integrating this subject to the initial condition $L(T = 0) = 0$, we obtain

$$(9.4.87) \qquad T = T_{end} + \frac{4}{4n - 1} \left(\frac{(1 - L)^{3n+1}}{3n + 1} - \frac{(1 - L)^{2-n}}{2 - n} \right)$$

where

$$(9.4.88) \qquad T_{end} = \frac{4}{(2 - n)(3n + 1)}$$

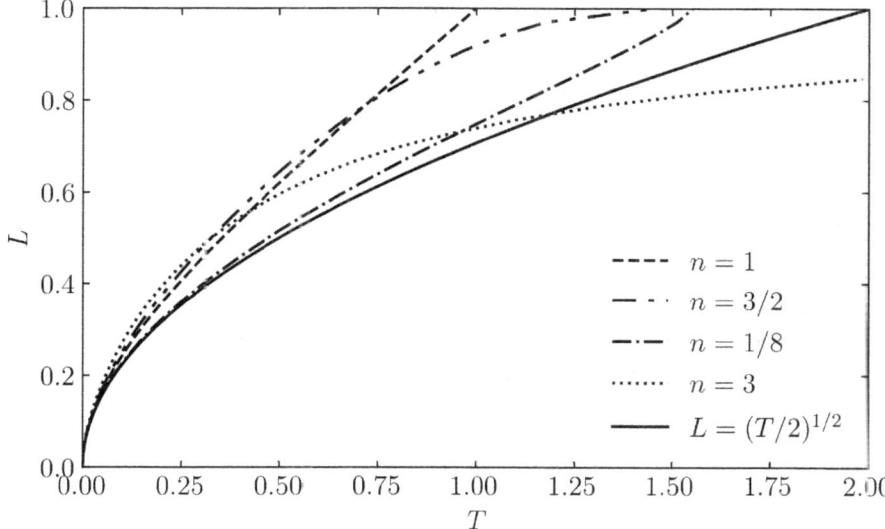

FIGURE 9.4.5. Location of meniscus (nondimensionalized) as a function of nondimensionalized time for various converging channels. The linearly sloping capillary, $n = 1$, has the fastest liquid penetration for the shapes that are practical to fabricate, and is only about 2% slower than the optimal shape.

is the time for the meniscus to reach the theoretical tip of the cone (where in practice we have a small opening), i.e., the time to reach $z = z_0$. We plot the meniscus location as a function of time for various capillary profiles in Figure 9.4.5. We also plot the time to reach the end of the converging capillary as a function of capillary geometry in Figure 9.4.6. We observe that linearly shaped capillary ($n = 1$) has the fastest penetration of the practical to fabricate shapes. In fact if the fabrication does not produce a perfectly linear capillary, we should not worry too much - the minimum of the T_{end} plot is fairly flat (Figure 9.4.6). Furthermore, it is interesting to note that for $0 \leqslant n \leqslant 5/6$, making the sloping more aggressive (raising n) causes the capillary forces due to a decreasing radius of curvature to more and more dominate the viscous resistance. After $n = 5/6$ viscous resistance begins to grow faster than the capillary forces with increasing n. After $n = 5/3$, the penetration becomes even slower than the Washburn penetration.

Having considered converging geometry, let's now consider diverging geometry as shown in Figure 9.4.4b. We will simultaneously, side-by-side consider a geometry as that formed by infinite plates (i.e., a 2D case, which we will call a wedge), or that formed by revolving the profile around the z-axis (i.e., a 3D case, which we will call a cone). In our derivation we will follow the approach of Reyssat et al. (1922). Once again, here we will only consider cases where the half angle of the cone α is small. We do this so that we can employ the lubrication approximation, i.e., $|dh/dz| \ll 1$, so that we can assume that the velocity in the channel is predominantly in the z-direction. As a result of this assumption, the pressure within the liquid depends

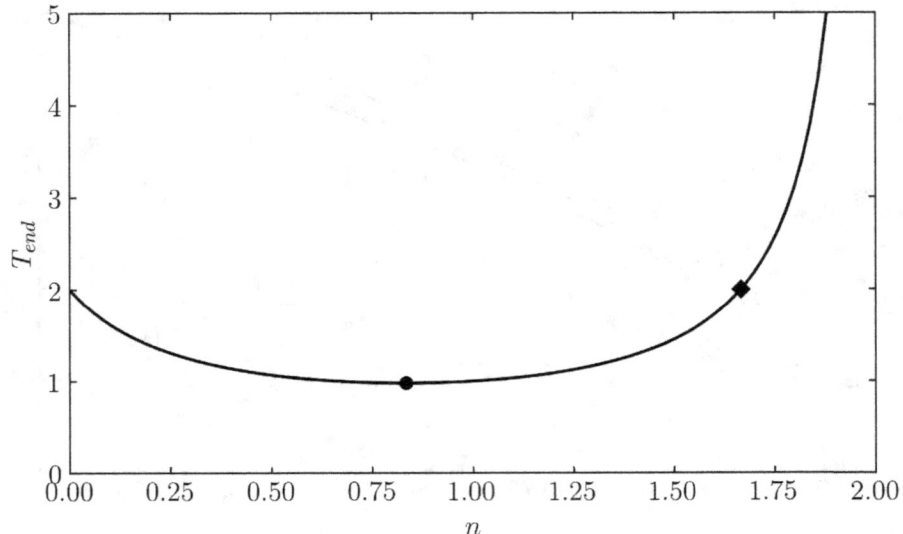

FIGURE 9.4.6. Time to reach the tip of a converging capillary (nondimensionalized) as a function of the power for a power law capillary shape (see (9.4.83)). The linearly sloping capillary, $n = 1$, has the fastest liquid penetration for the shapes that are practical to fabricate, and is only about 2% slower than the optimal shape, $n = 5/6$ (circle). For $0 \leqslant n \leqslant 5/3$ the penetration is faster than the Washburn penetration (penetration in a cylindrical capillary). For $n > 5/3$ the penetration is slower than the Washburn penetration as the viscous resistance grows faster than the capillary force for these tapered geometries.

only on z, and the Navier-Stokes momentum equations reduce to

$$(9.4.89) \qquad \frac{\partial^2 u_z}{\partial y^2} = \frac{1}{\mu} \frac{\partial p}{\partial z}$$

for the infinite plates, wedge case, and as before

$$(9.4.90) \qquad \frac{1}{r} \frac{\partial}{\partial r} \left(r \frac{\partial u_z}{\partial r} \right) = \frac{1}{\mu} \frac{\partial p}{\partial z}$$

for the cylindrical cone case. As we saw before, generally, the solution for the average velocity, u has a form of

$$(9.4.91) \qquad u = -\frac{k}{\mu} \frac{\partial p}{\partial z}$$

where $k(z)$ is a factor dependent on the $h(z)$ profile of the walls. This form is reminiscent of the Darcy equation in porous media. Reysat et al. found that for our cases $k = h^2/\lambda$, where $\lambda = 3$ for the wedge and 8 for the cone case. In addition to the momentum equation, we also have the mass conservation equation. Mass conservation states that for the wedge case, the flow rate per unit width $q(t) = 2uh$ and for the cone case, the volumetric flow rate $Q(t) = \pi u h^2$. Once again, the position of the meniscus is at $l(t)$ and we are interested in finding this function. At the meniscus the capillary pressure has the form of

(9.4.92)
$$p = p_0 - \gamma\kappa = p_0 - c\gamma/h$$

for both cases, where c is a function of geometry and the contact angle.

Combining the result of the momentum equation and the mass conservation, we obtain for the wedge case,

(9.4.93)
$$\frac{\partial p}{\partial z} = -\frac{3}{2}\frac{\mu q(t)}{h^3(z)}$$

and for the cone case,

(9.4.94)
$$\frac{\partial p}{\partial z} = -\frac{8}{\pi}\frac{\mu Q(t)}{h^4(z)}$$

Integrating these between the starting position ($z = 0$) and the location of the meniscus and substituting in the expression for the capillary force, (9.4.92), we obtain for the wedge case,

(9.4.95)
$$q(t) = \frac{2}{3}\frac{\gamma}{\mu}\frac{c}{h(l(t))\int_0^{l(t)}(h(z))^{-3}dz}$$

and for the cone case,

(9.4.96)
$$Q(t) = \frac{\pi}{8}\frac{\gamma}{\mu}\frac{c}{h(l(t))\int_0^{l(t)}(h(z))^{-4}dz}$$

Since the speed of the meniscus is just dl/dt, and by mass conservation for the wedge case it is just

(9.4.97)
$$\frac{dl}{dt} = \frac{q(t)}{2h(l(t))}$$

and for the cone case,

(9.4.98)
$$\frac{dl}{dt} = \frac{Q(t)}{\pi h(l(t))^2}$$

Combining these results with (9.4.95) and (9.4.96), we obtain for the wedge case,

(9.4.99)
$$\frac{dl}{dt} = \frac{1}{3}\frac{\gamma}{\mu}\frac{c}{h(l(t))^2\int_0^{l(t)}(h(z))^{-3}dz}$$

and for the cone case,

(9.4.100)
$$\frac{dl}{dt} = \frac{1}{8}\frac{\gamma}{\mu}\frac{c}{h(l(t))^3\int_0^{l(t)}(h(z))^{-4}dz}$$

Let's now consider a linear tapered channel as shown in Figure 9.4.4a. Since α is small (as was necessitated by the lubrication approximation) we can write the equation for the channel profile as

(9.4.101)
$$h(z) = h_0 + \alpha z$$

From geometry, we can obtain that for a fluid having contact angle θ with the wall the curvature of the meniscus for the wedge case,

$$(9.4.102) \qquad \kappa = \frac{\cos(\theta + \alpha)}{h}$$

for the cone case

$$(9.4.103) \qquad \kappa = \frac{2\cos(\theta + \alpha)}{h}$$

Substituting these profiles into (9.4.99) and (9.4.100) we obtain the following ODEs for the wedge and cone cases respectively,

$$(9.4.104) \qquad \left(\left(1 + \frac{\alpha l}{h_0}\right)^2 - 1 \right) \frac{dl}{dt} = \frac{2}{3}\frac{\gamma}{\mu}\alpha\cos(\theta + \alpha)$$

$$(9.4.105) \qquad \left(\left(1 + \frac{\alpha l}{h_0}\right)^3 - 1 \right) \frac{dl}{dt} = \frac{3}{4}\frac{\gamma}{\mu}\alpha\cos(\theta + \alpha)$$

For convenience, we nondimensionalize the variables. We nondimensionalize the position of the meniscus to be

$$(9.4.106) \qquad L = \frac{\alpha l}{h_0}$$

and nondimensionalize time to be

$$(9.4.107) \qquad T = t\frac{\gamma}{\mu}\frac{\alpha^2}{h_0}\cos(\theta + \alpha)$$

With these, the ODEs (9.4.104) and (9.4.105) become,

$$(9.4.108) \qquad \left((1 + L)^2 - 1\right)\frac{dL}{dT} = \frac{2}{3}$$

$$(9.4.109) \qquad \left((1 + L)^3 - 1\right)\frac{dL}{dT} = \frac{3}{4}$$

We then integrate these with an initial condition $L(T = 0) = 0$ and obtain for the wedge case,

$$(9.4.110) \qquad 2T = L^3 + 3L^2$$

and for the cone case,

$$(9.4.111) \qquad 3T = L^4 + 4L^3 + 6L^2$$

We plot these in Figure 9.4.7, as well as the Washburn scaling $L \sim T^{1/2}$. At short times for both the wedge and the cone case, $L \sim T^{1/2}$. For longer times, $L \sim T^{1/3}$ for the wedge and $L \sim T^{1/4}$ for the cone. As the liquid continues to penetrate into the diverging channel, the penetration slows down. The crossover between the fast regime and the slower regime occurs at $L \approx 1$ or $l \approx h_0/\alpha$.

Let's now consider a more general case, a power-law-shaped diverging channel, with a profile defined as

$$(9.4.112) \qquad h(z) = h_0 + \alpha z^n$$

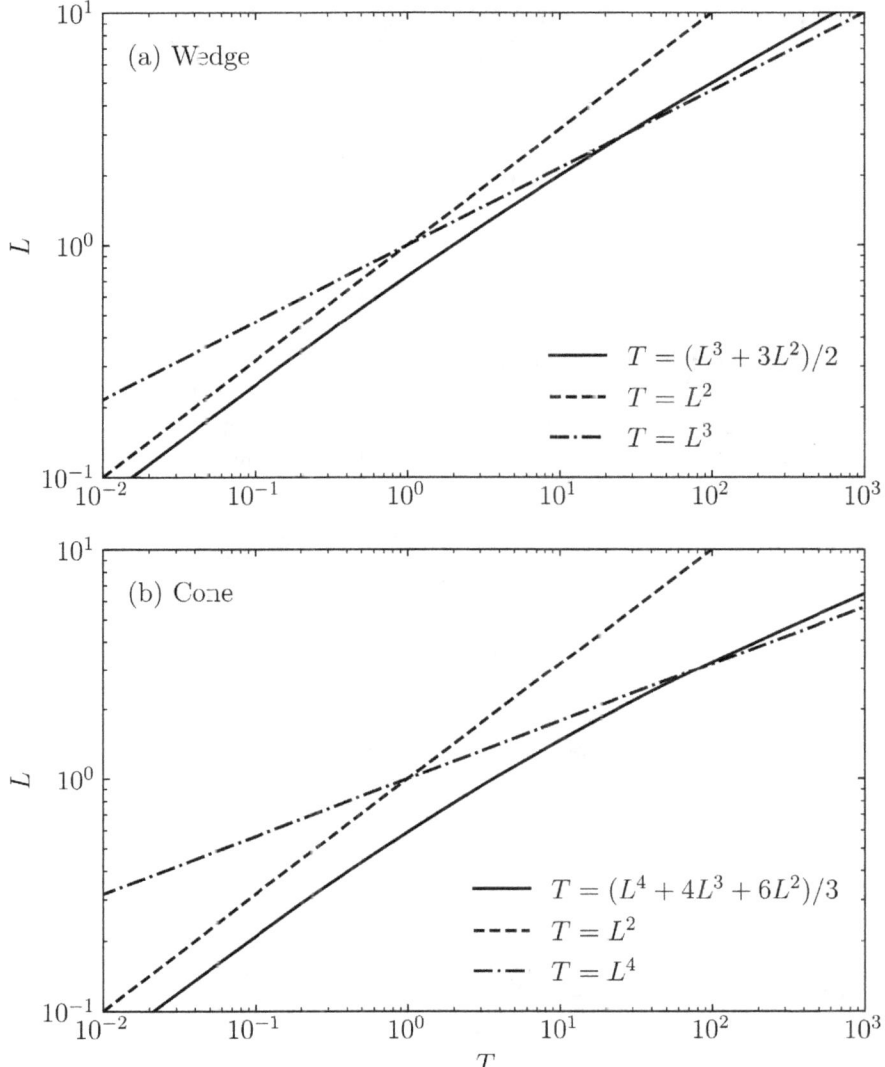

FIGURE 9.4.7. Dimensionless position of the meniscus as a function of dimensionless time for liquid penetrating into a wedge formed by infinite plates (a), and a cylindrical cone (b). For comparison we also plot the Washburn scaling $L \sim T^{1/2}$ or $T = L^2$ which is approximately valid for both cases at early times and the appropriate approximate scaling for the late times.

where $n \geqslant 1$. Substituting this into (9.4.99) and (9.4.100), we obtain the following integro-differential equations, for the wedge and the cone respectively,

$$(9.4.113) \qquad \left(1 + \frac{\alpha}{h_0}l^n\right)^2 \frac{dl}{dt} \int\limits_0^{l(t)} \left(1 + \frac{\alpha}{h_0}z^n\right)^{-3} dz = \frac{c\,(l)\,h_0}{3}\frac{\gamma}{\mu}$$

$$(9.4.114) \qquad \left(1 + \frac{\alpha}{h_0} l^n\right)^3 \frac{dl}{dt} \int_0^{l(t)} \left(1 + \frac{\alpha}{h_0} z^n\right)^{-4} dz = \frac{c(l) h_0}{8} \frac{\gamma}{\mu}$$

where

$$(9.4.115) \qquad c(l) = \cos\left(\theta + \arctan\left(\frac{dh}{dz}(l)\right)\right)$$

$$(9.4.116) \qquad c(l) = 2\cos\left(\theta + \arctan\left(\frac{dh}{dz}(l)\right)\right)$$

for the wedge and the cone respectively. Since for most realistic channels c is of $O(1)$ as cosine is bounded and as we are interested in a wetting fluid, this cosine is should be $O(1)$ for successful penetration. Thus, as an approximation, we can set $c(l)$ to be a constant, $c(l) = c_0$. We nondimensionalize the equations once again, such that

$$(9.4.117) \qquad L = \left(\frac{\alpha}{h_0}\right)^{1/n} l$$

$$(9.4.118) \qquad T = t \frac{\gamma}{\mu} \frac{c_0}{h_0} \left(\frac{\alpha}{h_0}\right)^{2/n}$$

which simplify our integro-differential equations to

$$(9.4.119) \qquad \frac{dL}{dT}(1 + L^n)^2 \int_0^{L(T)} (1 + Z^n)^{-3} dZ = \frac{1}{3}$$

$$(9.4.120) \qquad \frac{dL}{dT}(1 + L^n)^3 \int_0^{L(T)} (1 + Z^n)^{-4} dZ = \frac{1}{3}$$

These can now be solved, for example by a computer algebra system, analytically to give the penetration behavior in a power-law channel. Here quickly, we use this result to derive the penetration between flat plates ("zero angle wedge" case). The analogous case for the "zero angle cone" is the Washburn solution for penetration in a circular capillary. For the penetration between flat plates, $n = 0$, simplifying (9.4.119)

$$(9.4.121) \qquad \frac{3}{2} L \frac{dL}{dT} = 1$$

Integrating this subject to the usual initial condition $L(T = 0) = 0$, we obtain

$$(9.4.122) \qquad L = (4T/3)^{1/2}$$

where now $c_0 = \cos\theta$. As expected, we see the familiar Washburn scaling, $L \sim T^{1/2}$.

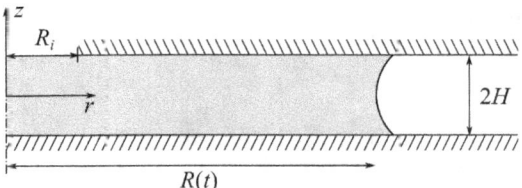

FIGURE 9.4.8. cross-section of a cylindrical geometry with the z-axis as the axis of rotation. This geometry serves as a model geometry for capillary penetration into a chamber (chamber filling) from a single small opening as may occur is an sample inlet port.

9.4.7. Penetration into a chamber from a single opening (chamber filling).

At the end of Section (9.4.6) we briefly considered the Washburn type solution for penetration of a liquid through a narrow planar space as created by two parallel plates. In that case the liquid was fed by an infinite plane. That situation can serve a model for capillary filling of a rectangular chamber from a gently sloping manifold. Here we consider a chamber with a hole roughly in the center (far from the walls), such as might be found in a lab-on-a-chip sample input chamber, and investigate the capillary penetration of liquid in this space. We assume that such a chamber is horizontal, or fairly small and so the effect of gravity on the meniscus and so on the filling such a chamber is minimal. We show a cross-section of this model chamber in Figure 9.4.8.

Once again we follow the Washburn approach were we assume a quasi-steady flow and equate viscous forces from a Hagen-Poiseuille type solution to capillary forces. The Hagen-Poiseuille type solution for velocity profile for the flow between infinite flat plates emanating from a point at the center (Middleman (1995)) is

$$(9.4.123) \qquad u_r(r,z) = \left(\frac{3Q}{8\pi H}\right)\frac{1}{r}\left(1 - \left(\frac{z}{H}\right)^2\right)$$

For this geometry, the relationship between flow rate and pressure drop between the entrance and the outlet is

$$(9.4.124) \qquad p(R_i) - p(R) = -\frac{3\mu Q}{4\pi H^3}\ln\left(\frac{R_i}{R}\right)$$

where R is the position of the liquid-gas interface. By mass conservation, the flow rate is related to the speed of the interface by

$$(9.4.125) \qquad Q = 4\pi H R \frac{dR}{dt}$$

The expression for capillary pressure across the liquid-gas interface in this geometry is particularly interesting as, unlike in previous solutions, both radii of curvature are important. In the rz plane, the radius of curvature R_{rz} is given by

$$(9.4.126) \qquad \frac{1}{R_{rz}} = -\frac{\cos\theta}{H}$$

while in the $r\phi$ plane the radius of curvature is given by

$$(9.4.127) \qquad \frac{1}{R_{r\phi}} = \frac{1}{R}$$

Thus the total capillary pressure across the liquid-gas interface is

$$(9.4.128) \qquad \Delta p = \gamma \left(-\frac{\cos\theta}{H} + \frac{1}{R} \right)$$

Combining this with (9.4.124) and (9.4.125) we obtain a non-linear ODE for the meniscus position,

$$(9.4.129) \qquad -\frac{3\mu R}{H^2}\frac{dR}{dt}\ln\left(\frac{R_i}{R}\right) = \gamma\left(\frac{\cos\theta}{H} - \frac{1}{R}\right)$$

Next, we nondimensionalize the equations for convenience, introducing dimensionless meniscus position as

$$(9.4.130) \qquad X = R/H$$

and dimensionless time as

$$(9.4.131) \qquad T = \frac{\gamma}{\mu}\frac{\cos\theta}{3H}t$$

This is of course similar to the nondimensionalization we have performed for the other Washburn approaches. Rewriting (9.4.129) in the nondimensionalized form and integrating,

$$(9.4.132) \qquad \int_{X_i}^{X} \frac{\ln\left(X/X_i\right)X}{1-(1/(X\cos\theta))}dX = T$$

To solve this analytically, let's consider the cases where the liquid has penetrated into the chamber several times the distance of the half separation between the plates. In other words, let's consider cases where

$$(9.4.133) \qquad \frac{1}{X\cos\theta} \ll 1$$

For these cases we obtain a solution,

$$(9.4.134) \qquad \left(\frac{X}{X_i}\right)^2\left(\ln\left[\frac{X}{X_i}\right]\right)^2 - \left(\left(\frac{X}{X_i}\right)^2 - 1\right) = \frac{4T}{X_i^2}$$

We plot this result in Figure 9.4.9. It is interesting to note that unlike the other Washburn type solutions we have considered the velocity of the meniscus accelerates with time, rather than decelerates.

9.4.8. Penetration into a circular capillary fed by a drop. In designing various microfluidic devices, and especially lab-on-a-chip devices, we often desire to make the reagent well or storage reservoir feeding the microfluidic network as small as possible (and as practical). We desire this to both conserve (often expensive) reagents and to minimize the overall foot print of the device (thus reducing fabrication cost per device). Thus, penetration of liquids into channels fed by finite reservoirs is of great interest. Since in general reservoir shapes and fluidic networks can be quite complicated, there is no simple solution for all of these cases, and so we often need to obtain the solution (e.g., by numerical simulation) on case by case basis. Here however, to get the feel for the problem, we present a simple case of liquid penetrating a circular capillary fed by a drop. We will follow the approach of Middleman (1995) and that of Marmur (1988). This example, for example, can serve as a rough model for sampling capillary blood using a capillary collection

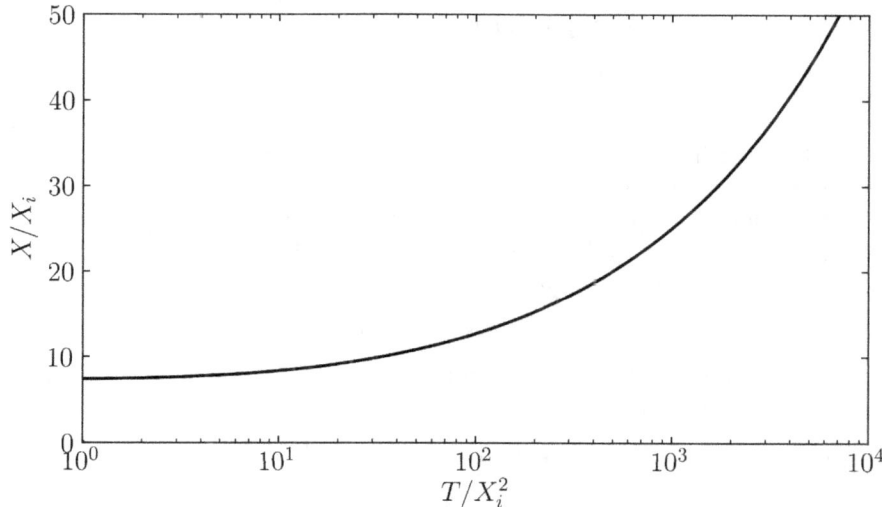

FIGURE 9.4.9. Dimensionless position of the meniscus penetrating into a space between flat plates from a small central opening as a function of dimensionless time. Unlike other Washburn type penetrations, this one accelerates with time.

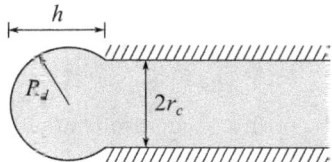

FIGURE 9.4.10. Geometry of a finite drop pinned at a circular opening feeding a circular capillary. We here consider situations where the drop is small enough such that gravity has no effect on the shape of the drop (small Bond number). Positive pressure associated with its curvature aids the penetration.

tube. (Here "capillary blood" refers to blood that was in the capillary blood vessels just prior to sampling).

Consider the geometry in Figure 9.4.10. A spherical drop is pinned to an entrance of our capillary, with the capillary having a radius of r_c. First from the geometry of the situation we can obtain a relationship between the volume of the drop and the radius of the drop,

$$(9.4.135) \qquad \frac{R_d}{r_c} = \frac{1}{\pi}\left(\frac{r_c}{h}\right)^2 \phi + \frac{h}{3r_c}$$

$$(9.4.136) \qquad \frac{h}{r_c} = \left(\frac{3\phi}{\pi} + \sqrt{\left(\frac{3\phi}{\pi}\right)^2 + 1}\right)^{1/3} + \left(\frac{3\phi}{\pi} - \sqrt{\left(\frac{3\phi}{\pi}\right)^2 + 1}\right)^{1/3}$$

where $\phi = V/r_c^3$. These equations are somewhat involved and a bit clunky to use. Luckily, for large ϕ, we can approximate

$$(9.4.137) \qquad \frac{R_d}{r_c} \approx \left(\frac{3}{4\pi}\phi\right)^{1/3} \approx 0.620\phi^{1/3}$$

We are interested in the radius of the drop as it gives rise to one of the driving forces, driving the liquid into the capillary. We consider here a case where a drop is brought to an empty capillary and the drop if what fills the capillary. Thus, the volume of the drop is related to the position of the meniscus in the capillary (neglecting the shape of the meniscus) as,

$$(9.4.138) \qquad V = V_0 - \pi r_c^2 x(t)$$

where V_0 is the initial volume of the drop. Combining this with (9.4.137) and the definition of ϕ, we obtain

$$(9.4.139) \qquad R_d = 0.620\big(V_0 - \pi r_c^2 x(t)\big)^{1/3}$$

Now that we have the radius of the drop as a function of the meniscus position, we can obtain the capillary pressure driving the liquid into the capillary from the shrinking drop. As usual we follow the Washburn approach and assume a quasi steady penetration and so equate the viscous resistance given by the Hagen-Poiseuille equation to capillary driving force. In this case, however, there are two capillary driving forces: one from our drop, and one from the meniscus in the tube. From Poiseuille equation, the average velocity of the fluid in the tube (and so the meniscus velocity), is

$$(9.4.140) \qquad \frac{dx}{dt} = \frac{r_c^2 \Delta p}{8\mu x}$$

The capillary pressure due to both the meniscus in the capillary and the drop, is

$$(9.4.141) \qquad \Delta p = \frac{2\gamma \cos\theta}{r_c} + \frac{2\gamma}{R_d}$$

Combining this with (9.4.140) and (9.4.138), we obtain

$$(9.4.142) \qquad \frac{dx}{dt} = \frac{r_c^2}{8\mu x}\left(\frac{2\gamma\cos\theta}{r_c} + \frac{2\gamma}{0.620(V_0 - \pi r_c^2 x)^{1/3}}\right)$$

To obtain the exact solution for the penetration dynamics we would have to solve this equation numerically. However, to get the feel for the penetration dynamics, we would like to have an analytical solution. We have already limited ourselves to situations where the drop diameter is large, i.e., large ϕ. We can utilize this approximation further, i.e., consider cases where penetration is small enough, and approximate

$$(9.4.143) \qquad \frac{1}{(V_0 - \pi r_c^2 x)^{1/3}} \approx \frac{1 + \pi r_c^2 x/3V_0}{V_0^{1/3}}$$

Substituting this into (9.4.142),

$$(9.4.144) \qquad \frac{dx}{dt} = \frac{r_c^2}{4\mu x}\frac{\gamma}{\mu}\left(\frac{\cos\theta}{r_c} + \frac{1}{0.620}\frac{1 + \pi r_c^2 x/3V_0}{V_0^{1/3}}\right)$$

We now nondimensionalize the equations, such that

(9.4.145)
$$X = 4x/r_c$$

(9.4.146)
$$T = t\frac{4\cos\theta}{r_c}\frac{\gamma}{\mu}$$

and obtain

(9.4.147)
$$X\frac{dX}{dT} = A + BX$$

where

(9.4.148)
$$A = 1 + \frac{1.612\phi_0^{-1/3}}{\cos\theta}$$

(9.4.149)
$$B = \frac{0.422\phi_0^{-4/3}}{\cos\theta}$$

where $\phi_0 = V_0/r_c^3$. Integrating this equation, we obtain

(9.4.150)
$$T = \frac{X}{B} - \frac{A}{B^2}\ln\left(1 + \frac{B^2}{A}\frac{X}{B}\right)$$

The simplification (9.4.143) however, puts a limitation on validity of the solution to

(9.4.151)
$$\pi r_c^2 x/V_0 \ll 1$$

or

(9.4.152)
$$X \ll 4\phi_0/\pi$$

Hence the analytical solution is really useful only for large values of ϕ_0. We plot this result is Figure 9.4.11 with A/B^2 as a parameter. Notice that the velocity of the meniscus increases with time. This is the second case where we have seen this for a Washburn type solution. Here this is due to the drop's radius of curvature decreasing with time and so increasing the driving force, pushing the liquid into the capillary.

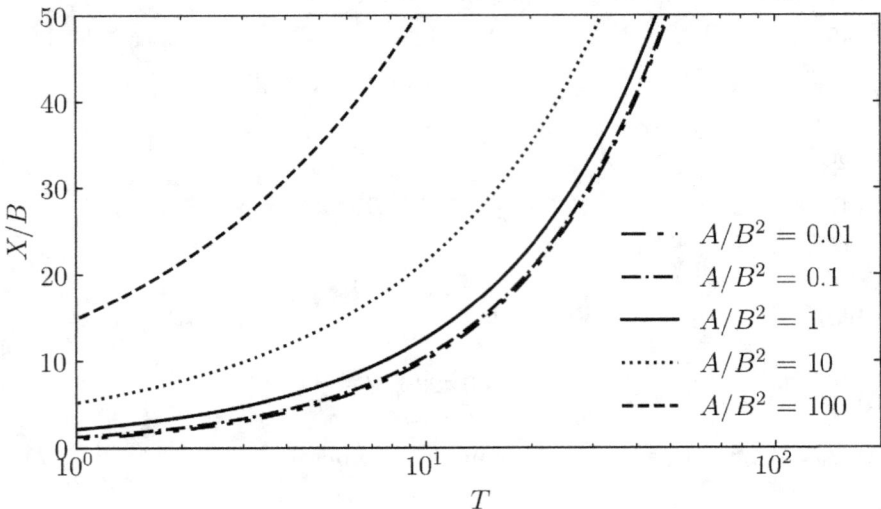

FIGURE 9.4.11. Scaled dimensionless meniscus position as a function of dimensionless time for penetration of a liquid into a capillary fed by a drop. The penetration velocity increases as a function of time. For $A/B^2 \leqslant 0.1$ the scaled meniscus position vs. time practically collapses onto a single curve.

CHAPTER 10

Capillarity with gravity

Up till now we have considered situations where the effects of gravity were either completely negligible or played minimal roles. Such situations are quite typical in microfluidics mostly due to small dimensions of microfluidic devices. However, there are some interesting situations in microfluidics where gravity is the defining force - either shaping the resulting liquid-gas interfaces or defining the fluid motion or equilibrium positions. We will consider such situations in this section.

10.1. Drops on a plate

10.1.1. Shape of a puddle. In the past few chapters we have considered simple drops sitting on top of a rigid surface. Previously, we have considered only small drops - for which their characteristic dimensions were smaller than the capillary length. Here let's consider a drop whose dimensions are larger than the capillary length - a puddle. A defining characteristic of a puddle is its height, and this is what we are after here.

Consider a large drop (a puddle) shown in Figure 10.1.1. At the center of the drop (designated by Q) and at the top the two principle radii of curvature will be equal by symmetry. Let these radii be equal to R_Q. Below this point, the capillary pressure across the liquid-gas interface will be

(10.1.1)
$$\Delta p_Q = 2\gamma/R_Q$$

For a liquid of density ρ_l surrounded by a gas of density ρ_g, the pressure drop a distance y below the center of the drop, for example at a location designated P in Figure 10.1.1 is then

(10.1.2)
$$2\gamma/R_Q + gy(\rho_l - \rho_g)$$

At point P the surface has two principle radii of curvature R_1 and R_1 and so the total balance of forces at this point is

(10.1.3)
$$\frac{2\gamma}{R_Q} + gy(\rho_l - \rho_g) = \gamma\left(\frac{1}{R_1} + \frac{1}{R_2}\right)$$

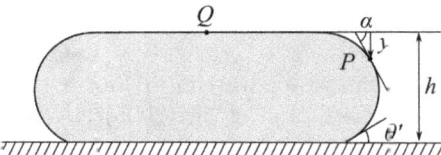

FIGURE 10.1 1. Schematic of puddle geometry. The puddle's height scales as the product of capillary length and sine of half of the contact angle.

237

Let R_2 be the radius of curvature lying in the plane parallel to the plane of the substrate the drop is sitting on; and R_1 be in the plane perpendicular to this plane. Typically for large drops, $R_2 \gg R_1$, so $1/R_2 \ll 1/R_1$. Furthermore, for large drops the top of the puddle is pretty much flat, and so the radius of curvature there is infinite. Hence, (10.1.3) reduces to,

$$(10.1.4) \qquad\qquad gy\left(\rho_l - \rho_g\right) = \gamma/R_1$$

Let's introduce the angle α between the tangent at point P and the horizontal surface through Q, and let $\rho_l - \rho_g = \Delta\rho$ so that

$$(10.1.5) \qquad\qquad gy\Delta\rho = \gamma \frac{d\alpha}{ds}$$

where s is the coordinate running along the liquid-gas interface; From geometry $dy/ds = \sin\alpha$. Thus, we can write (10.1.5) as

$$(10.1.6) \qquad\qquad gy\Delta\rho = \gamma \sin\alpha \frac{d\alpha}{dy}$$

We can integrate this ODE, subject to the initial condition that at point Q, $y = 0$ and $\alpha = 0$, to obtain

$$(10.1.7) \qquad\qquad y^2 = \frac{2\gamma}{g\Delta\rho}\left(1 - \cos\alpha\right)$$

or

$$(10.1.8) \qquad\qquad y = 2\sqrt{\frac{\gamma}{g\Delta\rho}} \sin\frac{\alpha}{2}$$

From geometry, at $y = h$, $\alpha = \pi - \theta'$, where θ' is π - the contact angle (as shown in Figure 10.1.1). Thus the contact angle θ is equal to α. Substituting this in,

$$(10.1.9) \qquad\qquad h = 2\sqrt{\frac{\gamma}{g\Delta\rho}} \sin\frac{\theta}{2}$$

Since typically $\rho_l \gg \rho_g$, we can write,

$$(10.1.10) \qquad\qquad h = 2\sqrt{\frac{\gamma}{g\rho}} \sin\frac{\theta}{2}$$

where θ is the contact angle. Notice that the height of the puddle is solely determined by the capillary length (see (7.8.4)) and contact angle.

10.1.2. Sliding of drops and roll off (or sliding) angle. Sliding of drops is widely encountered, from drops sliding on windshields and windows (and so in design of window surfaces) to drops sliding off leaves during pesticide spraying (and so in design of pesticide formulations). In lab-on-chip applications droplet sliding occurs in the field of digital microfluidics. While there sliding motion is not due to gravity but due to electrical forces, these forces can be modeled as a body force, just like gravity.

Sliding motion of drops can be separated into four regimes (Figure 10.1.2): oval, corner, cusp, and pearling (Le Grand et al. (2005)). When a surface is tilted, so that the drop just begins to flow, the drop has an oval shape, and the flow is said to be in an oval regime (Figure 10.1.2a). The curvature (for the radius of curvature lying in the plane parallel to the substrate) for the front and rear of the drop is low and practically constant for increasing capillary numbers. In the beginning of the oval regime (when the velocity of the droplet is small) the front and rear

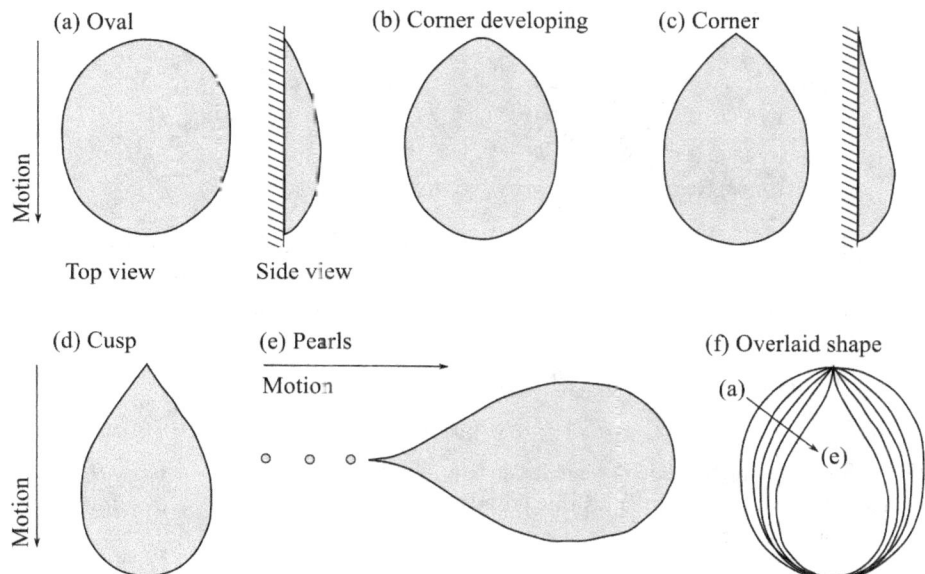

FIGURE 10.1.2. Schematics of drop geometries over different sliding regimes (a) oval, (b) and (c) corner, (d) cusp, and (e) pearls. For the oval and corner regime we also provide a schematic of the side view of the drop. In (f) we overlay the shapes of the drops to emphasize the change in the rear curvature of the droplet across the regimes. These sketches were drawn based on photographs in Le Grand et al. (2005).

curvatures are equal. As the drop velocity (and so capillary number) increases the drop transitions into the corner regime (Figure 10.1.2b). In this regime the rear curvature increases rapidly with increasing capillary number (drop velocity) while the front curvature remains roughly constant (Figure 10.1.2c). The rear curvature effectively becomes infinite (diverges) as we transition into the cusp regime (Figure 10.1.2d). This regime is characterized by a sharp cusp, or tail at the rear of the droplet. As capillary number increases further this tail grows in length, becomes unstable (likely by a form of Plateau-Rayleigh instability, to be discussed later) and forms satellite droplets behind the main droplet. The first droplet (and one that winds up being furthest from the parent droplet) is the largest, while the subsequent droplets are smaller and smaller (Figure 10.1.2e). Much of this behavior and droplet shape is attributed to contact angle hysteresis - the difference between the advancing and receding contact angles.

One of the most interesting part of this motion, is the critical angle at which the droplet starts to move on an inclined surface. This is the roll off angle (sometimes called sliding angle) and it is a measure of contact angle hysteresis - the difference between advancing and receding contact angles. Let's begin with a simple model for the sliding angle following the approach of Furmidge (1962).

Consider an idealized droplet shown in Figure 10.1.3. The gravitational force on the drop is $mg \sin \alpha$ where m is the mass of the drop. The work done by the

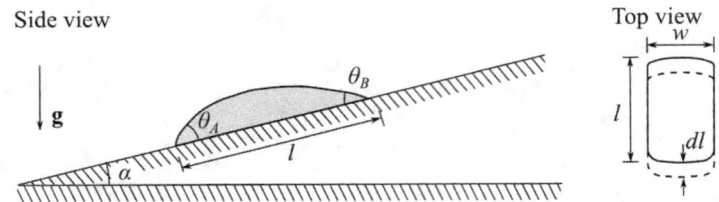

FIGURE 10.1.3. Side view and top view of an idealized droplet on an inclined plane. For simplicity of calculations we approximate that the droplet has a rectangular plan as shown in the top view.

drop in sliding a small distance dl is then

$$(10.1.11) \qquad W_g = mg \sin \alpha dl$$

When the drop moves down a surface, the leading edge wets an area wdl (for our idealized rectangular drop) and dewets also an area wdl. The work done in wetting a unit area of a surface is

$$(10.1.12) \qquad \gamma \left(1 + \cos \theta_A\right)$$

and dewetting a unit area is

$$(10.1.13) \qquad \gamma \left(1 + \cos \theta_R\right)$$

Thus the energy spent in wetting and dewetting is

$$(10.1.14) \qquad W_c = \gamma wdl \left(\cos \theta_R - \cos \theta_A\right)$$

We assume that the energy to wet and dewet the surface is supplied by the gravitational potential energy and that there are no energy losses in this system. Thus, putting together

$$(10.1.15) \qquad mg \sin \alpha dl = \gamma wdl \left(\cos \theta_R - \cos \theta_A\right)$$

or

$$(10.1.16) \qquad \sin \alpha = \gamma \frac{w}{mg} \left(\cos \theta_R - \cos \theta_A\right)$$

Since we assumed that the droplet has a rectangular plan, this relation is not completely accurate at predicting the sliding angle α. Nevertheless it gives the correct scaling. This scaling has been extensively verified by experiment (Furmidge (1962)). We can see that the sine of the sliding angle scales with the surface tension and the difference between the cosines of the advancing and receding contact angles. It is inversely related to the density of the fluid and scales as the $V^{-2/3}$ of the original volume of the drop. Thus, if we desire drops to slide (or slide off) easily we would like $\cos \theta_R - \cos \theta_A$ to be as small as possible. Note, from this model, we don't care what the actual contact angle is! Secondly, if we can't change the contact angle hysteresis of the surface, we can decrease the surface tension of the liquid by simply adding surfactants. In some situations, such as electrowetting displays, you get to tune both. In electrowetting displays while you want surface to have fairly small resistance to movement (so that little electrical energy is consumed in moving drops), you don't want the surface to have too little resistance to movement, otherwise your screen will be sensitive to how you tilt it! Thus, there you might, for example, want to have small drops, but low hysteresis and surface tension.

Another application of the sliding angle is the discrimination of whether a nanostructured (or roughened) surface should be described by the Wenzel model or the Cassie-Baxter model. McHale et al. (2004) found that Wenzel surfaces should have more contact angle hysteresis and thus are more "sticky" whereas Cassie-Baxter surfaces have less contact angle hysteresis and are more "slippy". McHale found this by considering the amplification of hydrophobicity or hydrophlicity based on increase in the roughness factor.

A more complicated relation for the sliding angle was found Dussan (1985)

$$(10.1.17) \qquad \sin\alpha = \gamma\frac{V^{-2/3}}{\rho g}\left(\cos\theta_R - \cos\theta_A\right)\frac{(24/\pi)^{1/3}(1 + \cos\theta_A)^{1/2}}{(2 + \cos\theta_A)^{1/3}(1 - \cos\theta_A)^{1/6}}$$

This relation is valid for $\theta_R - \theta_A < 10°$. Notice that the scaling in this relation agrees well with the much simpler one in (10.1.16).

10.2. Equilibrium capillary rise and jumping menisci

Another problem that tightly combines capillarity and gravity is the problem of equilibrium capillary rise. Like many problems in capillarity, the work on this began long ago, with Jurin publishing on this as early as 1719 (Jurin (1719)). While this problem appears to be a simple one at first glance, even it can produce surprising results. Consider a round capillary (or two plates) immersed in a large reservoir of liquid as shown in Figure 10.2.1. For simplicity, let's first consider that the walls of the capillary or the plates are parallel to one another (i.e., the angle $\alpha = 0$). We know that at equilibrium, the capillary pressure across the liquid-gas interface must balance the hydrostatic pressure of the liquid (i.e., the weight of the liquid). Thus we can write,

$$(10.2.1) \qquad p_0 - c\frac{\gamma}{r}\cos\theta = p_0 - \rho g h$$

where p_0 is the atmospheric (ambient) pressure. Here c is a constant accounting for the geometry of liquid-gas interface, with $c = 2$ for a cylindrical capillary ($1/R_0 + 1/R_0$), and $c = 1$ for flat plates ($1/R_0 + 1/\infty$). Thus, rearranging this, we obtain the capillary rise

$$(10.2.2) \qquad h = c\frac{\gamma}{\rho g}\frac{\cos\theta}{R_0} = c\lambda_c^2\frac{\cos\theta}{R_0}$$

where λ_c is the capillary length, $\lambda_c = \sqrt{\gamma/\rho g}$. This result is known as Jurin's law and has been known since 1700s. For room temperature water and a completely wetting circular capillary, Jaurin's law predicts a rise of 14.7 mm, 147 mm, 1.47 m, 14.7 m, 147 m, for 1 mm, 100 µm, 10 µm, 1 µm, 100 nm radius capillaries respectively. As an aside, the xylem and phloem, the "capillaries" that transport liquid in trees, are typically on the order of 25 µm in radius. Jurin's law would imply that a tree can never be taller than about half a meter! This is of course not correct, because it is not correct to assume that simple capillary forces balancing gravitational forces account for transport of liquid in trees. In fact this transport is driven by a complicated chemical gradients, including the gradient created, for example, by sugars produced in the leaves as they are transported to the roots. For details on these processes, including theories of electrokinetic transport in plants, see for example the work of Zimmermann and Milburn (2012). For close to atomic level dimensions assumption leading to derivation of Jurin's law are invalid, but

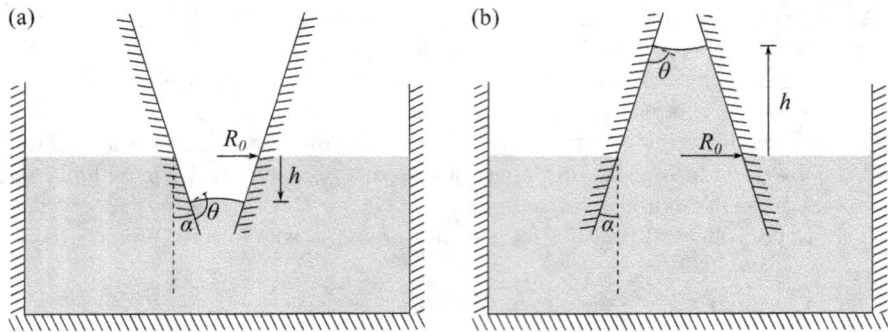

FIGURE 10.2.1. Schematic of a tapered circular capillary or inclined plates immersed into a large liquid reservoir. The angle of taper α is defined from the surface of the liquid in the reservoir.

capillary rise can be obtained by other means, and capillary rise on the order of tens of kilometers is predicted! (see Caupin et al. (2008))

Now let's consider the cases where the walls are not parallel, following the derivations of Tsori (2006) and Tsori (2007). Firstly relation between capillary pressure and hydrostatic pressure at equilibrium, (10.2.1) becomes

$$(10.2.3) \qquad p_0 - c\frac{\gamma}{r\left(h\right)}\cos\left(\theta + \alpha\right) = p_0 - \rho g h$$

where

$$(10.2.4) \qquad r\left(h\right) = R_0 + h\tan\alpha$$

Where we have now accounted for the varying radius of curvature and the angle of the meniscus with the wall. Now for convenience we nondimensionalize the equations, scaling both the horizontal and vertical distances with capillary length, such that,

$$(10.2.5) \qquad h^* = h/\lambda_c$$

$$(10.2.6) \qquad R_0^* = R_0/\lambda_c$$

Implementing this nondimensionalization in (10.2.3) and rearranging, we obtain

$$(10.2.7) \qquad \cos\left(\theta + \alpha\right) = f\left(h^*\right)$$

$$(10.2.8) \qquad f\left(h^*\right) = \frac{1}{c}h^*\left(R_0^* + h^*\tan\alpha\right)$$

Or written in another form and dropping the asterisks for clarity,

$$(10.2.9) \qquad h^2\frac{\tan\alpha}{c} + h\frac{R_0}{c} - \cos\left(\theta + \alpha\right) = 0$$

Observe that the resulting equation is a quadratic equation for h. This right-of-the-bat means that there are sometimes two solutions for the equilibrium height, and so hints that the behavior of capillary rise in tapered channels can be quite a bit more interesting than in straight channels. This equation has thus the following solutions,

$$(10.2.10) \qquad h = \frac{-R_0 \pm \sqrt{R_0^2 + 4c\tan\alpha\cos\left(\theta + \alpha\right)}}{2\tan\alpha}$$

When the discriminant $R_0^2 + 4c \tan \alpha \cos(\theta + \alpha)$ is positive, then there are two real roots; when it is zero, then there is a single real root, and when it is negative there are no real roots. The physical implications of this are that when there are two real roots, the liquid will travel to the position closest to the level of the large reservoir and stop there as it has found a stable equilibrium position. If we perturb its location gently, the forces on it will become unbalanced and it will be accelerated back to this stable position. However if we perturb it heavily so that is near the second (distant) stable position, it may then also stabilize there.

When the discriminant is zero, the meniscus stabilizes itself at a single position given by

$$(10.2.11) \qquad\qquad h_c = -R_0 / 2 \tan \theta$$

The conditions for this (including the critical value of the angle α, α_c can be found by numerically solving the transcendental equation,

$$(10.2.12) \qquad\qquad R_0^2 + 4c \tan \alpha \cos(\theta + \alpha) = 0$$

When the discriminant is negative the meniscus does not have a stable position inside the capillary - one of the forces dominates the other in the tug of war. If gravity where to win then the meniscus would be at the same level as the rest of the fluid in the reservoir and so $h = 0$. But $h = 0$ is a finite real solution, so gravity could not have won the tug of war. Thus, capillarity wins and the meniscus is propelled towards the narrow end of the capillary (or tapered plates). The meniscus in a tapered capillary or between tapered plates can have very peculiar behavior - for example, as the angle α is gradually changed, the meniscus gradually changes its position for a while; then at a critical angle, it jumps to the narrow end of the capillary!

Let's consider the four cases that arise here in more detail: (a) hydrophobic walls tapered downward (Figure 10.2.1a); (b) hydrophilic walls tapered downward; (c) hydrophilic walls tapered upward (Figure 10.2.1b); (d) hydrophobic walls tapered upward. When the walls are tapered downward, we define α to be positive, and when the walls are tapered upward, we define α to be negative.

In the first case, α is positive (tapered downward) and the surface is hydrophobic, i.e., $\cos \theta < 0$. Thus for small enough α, $\cos(\theta + \alpha)$ is also negative. This means that the two solutions h_1 and h_2 are also negative, which physically means that the meniscus level is below level of the reservoir. Physically the two solutions mean that if left unperturbed the meniscus will travel to h_1 (closest to the level of the reservoir) and if perturbed about this position, will stabilize at h_1. However, if perturbed all the way to h_2, it might settle there as well. However perturbations about this position, if they are large enough, may land the meniscus back at h_1. Hence h_2 is known as the unstable solution. If the parameters of the discriminant of (10.2.10) are changed so that the discriminant is negative, then there is no real solution, and the meniscus jumps to the bottom of the capillary. To induce this jump we can alter α (for example by altering the angle of the plates), by altering R_0 (by moving the plates together), or by altering the contact angle θ. Solving the discriminant equation, (10.2.12) for the critical contact angle at which this can occurs, we obtain

$$(10.2.13) \qquad\qquad \theta_c = \arccos\left(\frac{R_0^2}{4c \tan \alpha}\right) - \alpha$$

For $\theta > \theta_c$ the meniscus jumps to the bottom of the capillary.

In the second case, α is positive (tapered downward) and the surface is hydrophilic, i.e., $\theta < \pi/2$. When both α and θ are small, $\cos(\theta + \alpha)$ is positive and there is always a positive solution for h. Thus, physically the meniscus is found above the level of the liquid in the reservoir. However, if $\theta + \alpha > \pi/2$ the capillary effectively behaves as a hydrophobic capillary and the solution for h is negative. For this case, the jump to the bottom of the capillary is possible, just like in case (a).

For the tapered upward (negative α) cases let's define new variables to see how the behavior of (10.2.9) changes. Let's define $\alpha' = -\alpha$, $\theta' = \pi - \theta$, and $h' = -h$. Substituting these into (10.2.9),

$$(10.2.14) \qquad -\left(h'^2 \frac{\tan \alpha'}{c} + h' \frac{R_0}{c} - \cos(\theta' + \alpha') \right) = 0$$

Thus, qualitatively the behavior is the same for this transformation. This means that the third case, where α is negative (tapered upward) and the surface is hydrophilic, i.e., $\theta < \pi/2$, maps directly onto the first case, where α is positive, (α' is negative), and $\theta > \pi/2$ ($\theta' < \pi/2$). Thus, when we saw the two solutions h_1 and h_2 were negative for the first case, we have two positive solutions for the third case. Physically this means the meniscus rises above the level of the reservoir, and again if left unperturbed the meniscus will travel to h_1 (closest to the level of the reservoir) and if perturbed about this position, will stabilize at h_1. However, if perturbed all the way to h_2, it might settle there as well. However, perturbations about this position, if they are large enough, may land the meniscus back at h_1. Once again, if the parameters of the discriminant of (10.2.10) are changed so that the discriminant is negative, then there is no real solution, and the meniscus jumps to the top (tapered end) of the capillary.

Using this symmetry, for the fourth case, α is negative (tapered upward) and the surface is hydrophobic, when the value of α' and θ' are small, $\cos(\theta' + \alpha')$ is positive and there is always a positive solution for h' and so a negative solution for h. Physically this means that there is a meniscus found below the level of the liquid in the reservoir. However, if $\theta' + \alpha' > \pi/2$ the capillary effectively behaves as a hydrophilic capillary and the solution for h' is negative (h is positive), and so the meniscus is found above the liquid level in the reservoir. For this case, the jump to the top of the capillary is possible, just like in the third case.

To make the behavior of a meniscus in a tapered capillary even more interesting, and to add another practical knob to control the meniscus position we can make the gas pressure inside the capillary be different from that the surrounding capillary (by connecting a pressure controller to the capillary). This just replaces p_0 in (10.2.3) for a single pressure difference Δp. Implementing this change, we obtain in dimensionless form

$$(10.2.15) \qquad h^2 \tan \alpha + h (R_0 + \Delta p \tan \alpha) + R_0 \Delta p - c \cos(\theta + \alpha) = 0$$

where we have nondimensionalized there pressure to $\Delta p^* = \lambda_c \Delta p / \gamma$, where we have dropped the asterisk in (10.2.15) and going forth, for convenience. The solution to this equation is

$(10.2.16)$

$$h = \frac{-(R_0 + \Delta p \tan \alpha) \pm \sqrt{(R_0 + \Delta p \tan \alpha)^2 + 4 \tan \alpha (R_0 \Delta p + c \cos(\theta + \alpha))}}{2 \tan \alpha}$$

When the descriminant is equal to zero,

$$(R_0 + \Delta p \tan \alpha)^2 + 4 \tan \alpha \left(R_0 \Delta p + c \cos(\theta + \alpha) \right) = 0 \tag{10.2.17}$$

we have a single solution, the critical height,

$$h = \frac{-(R_0 + \Delta p \tan \alpha)}{2 \tan \alpha} \tag{10.2.18}$$

For a positive α and a hydrophobic surface, $\theta > \pi/2$ and the meniscus is found below the surface of the reservoir when

$$-\frac{(R_0 - \Delta p \tan \alpha)^2}{4c \tan \alpha} < \cos(\theta - \alpha) \tag{10.2.19}$$

i.e., when the discriminant is positive. When the discriminant is negative,

$$-\frac{(R_0 - \Delta p \tan \alpha)^2}{4c \tan \alpha} > \cos(\theta + \alpha) \tag{10.2.20}$$

the meniscus jumps to the bottom of the capillary. We can also solve the discriminant equation (10.2.17) for the critical contact angle at which the jump occurs,

$$\theta_c = \arccos\left(-\frac{(R_0 - \Delta p \tan \alpha)^2}{4c \tan \alpha} \right) - \alpha \tag{10.2.21}$$

and similarly find the critical pressure at which the jump occurs,

$$\Delta p_c = \frac{\left(R_0 - \sqrt{-4c \tan \alpha \cos(\theta + \alpha)} \right)}{\tan \alpha} \tag{10.2.22}$$

Raising the contact angle past the critical angle induces the jump and increasing the pressure past the critical pressure also causes the jump of the meniscus to the tapered end of the capillary.

Having observed that external pressure can also cause the meniscus to have this interesting non-linear, jumping behavior, we can ask what happens when in a tapered channel or capillary when the capillary pressure is just balanced by pressure applied manually to the liquid, and gravity is absent, i.e., where (10.2.3) collapses to

$$c \frac{\gamma}{r(h)} \cos(\theta + \alpha) = \Delta p \tag{10.2.23}$$

where Δp is the pressure applied across the interface. Expanding this dimensional equation,

$$h \Delta p \tan \alpha - R_0 \Delta p + c \gamma \cos(\theta + \alpha) = 0 \tag{10.2.24}$$

we see that the equation is linear and so we can conclude that for a linearly tapered channel or capillaries under these conditions no jumping behavior occurs.

In deriving the above behavior for a meniscus in a tapered capillary, we assumed that the walls taper linearly. In practice this does not have to be the case - it is fairly easy to make the walls of any desired shape (including sinusoidaly varying). This might be of particular interest in designing opto-microfluidic devices where the liquid-gas interface can be used as an adjustable lens. By changing the position of the interface in the capillary of an appropriate shape, we can tune the position of the lens there as well as its radius of curvature, and hence focal point.

10.2.1. Capillary rise on a single inclined plate. Another interesting example of capillary rise is that about a single inclined flat plate (Figure 10.2.2). When such a plate is immersed in a reservoir of liquid, the liquid rises around the plate. We consider a reservoir that is large enough such that the liquid-gas interface far away from the plate is horizontal. The density of the gas is ρ_1 and that of the liquid is ρ_2 so that their difference is $\rho = \rho_2 - \rho_1$. While the density of the gas is almost always so much smaller than that of the liquid so that it is negligible, we consider it here in this form, so that we can easily generalize this problem to that of a slanted plate at the interface of two liquids, whose densities are comparable.

We consider a plate slanted at an angle β as shown in Figure 10.2.2. The Young-Laplace capillary equation for the pressure difference across the liquid-gas interface in this cases collapses to

$$(10.2.25) \qquad \Delta p = \gamma / R_1$$

since the second radius of curvature tends to zero, since we are considering a plate that is very long (infinite) into the page. Let's consider a small element of the liquid-gas interface PQ. In the liquid phase, there pressure under this element is equal to the external atmospheric pressure minus the pressure due to the height y of the liquid between the vertical location of PQ and the surface of the liquid in the reservoir. At this point, the capillary pressure balances this hydrostatic pressure and so,

$$(10.2.26) \qquad \rho g y = \gamma / R_1$$

We now have to express R_1 in terms of easy to use coordinates. We do this via a geometric construction shown in Figure 10.2.2. From this construction we see

$$(10.2.27) \qquad R_1 = \delta s / \delta \theta$$

and

$$(10.2.28) \qquad \delta y = \delta s \sin \theta$$

Combining (10.2.26) through (10.2.28),

$$(10.2.29) \qquad \rho g y \, dy = \gamma \sin \theta \, d\theta$$

Integrating this equation with the boundary condition $\theta = 0$ at $y = 0$, we obtain

$$(10.2.30) \qquad y = 2 \left(\frac{\gamma}{\rho g} \right)^{1/2} \left(\sin \frac{\theta}{2} \right)^2 = 2 \lambda_c \left(\sin \frac{\theta}{2} \right)^2$$

This gives us the location of the interface on the left-hand side of the inclined plate. Applying the same derivation to the right-hand side of the plate, we notice it is convenient to have θ be defined as negative as shown in Figure 10.2.2. The derivation follows similarly, and we find that again

$$(10.2.31) \qquad y = 2 \left(\frac{\gamma}{\rho g} \right)^{1/2} \left(\sin \frac{\theta}{2} \right)^2 = 2 \lambda_c \left(\sin \frac{\theta}{2} \right)^2$$

From observation, the maximum height of the liquid on the left side of the plate occurs at $\theta = \beta - \alpha$ and so

$$(10.2.32) \qquad y_l = 2 \lambda_c \left(\sin \frac{\beta - \alpha}{2} \right)^2$$

(a) 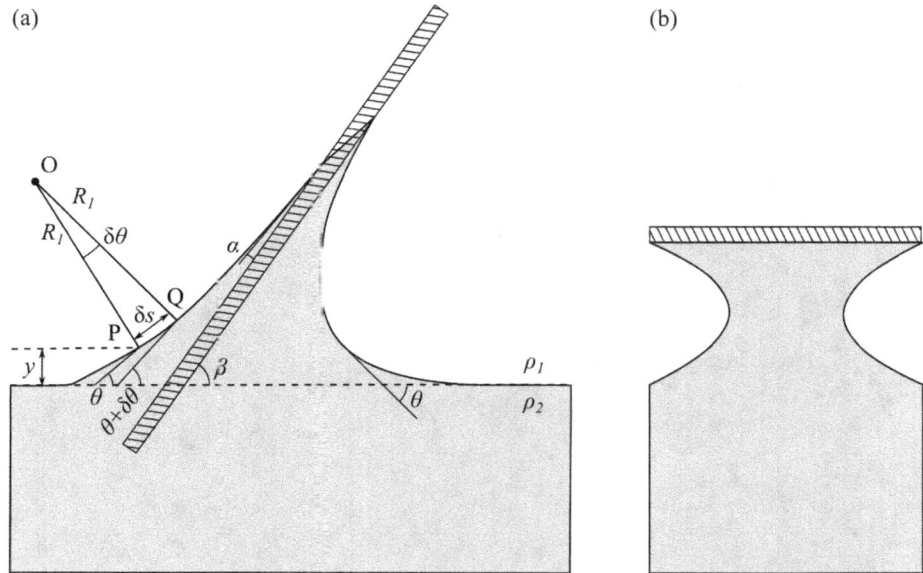 (b)

FIGURE 10.2.2. Schematic of the geometry of an inclined flat plate for an arbitrary inclination angle β (a), and when $\beta = 0$ (b). α is the contact angle of the liquid to the plate. The small element of the liquid-gas interface PQ is greatly exaggerated for clarity. Segments OP and OQ are normal to points P and Q respectively and have a length of R_1. The tangents of points P and Q meet the horizontal surface representing the distant surface of the reservoir at angles θ and $\theta + \delta\theta$. The point P is located a distance y above this horizontal surface.

where α is the contact angle of the liquid to the plate. Similarly, the maximum height of the liquid on the right side of the plate occurs at $\theta = \pi - \beta - \alpha$ and so

(10.2.33)
$$y_r = 2\lambda_c \left(\sin \frac{\beta + \alpha}{2} \right)^2$$

If we turn the plate very carefully so it is horizontal (Figure 10.2.2b, $\beta = 0$) and so we obtain that for the liquid to be stably located between the reservoir surface and our plate, the plate should be no higher than

$$y_0 = 2\lambda_c \left(\sin \frac{\alpha}{2} \right)^2$$

above the surface of the reservoir.

10.3. Pendent drops: Tate's law and corrections

Here we are interested in the problem of liquid passively dripping from a capillary due to gravity. This problem has many applications, including the measurement of surface tension of a liquid. We consider cases where the liquid flows slowly through the capillary such that the drop grows under its own weight and eventually the weight becomes so large that the force of surface tension holding it to the tube

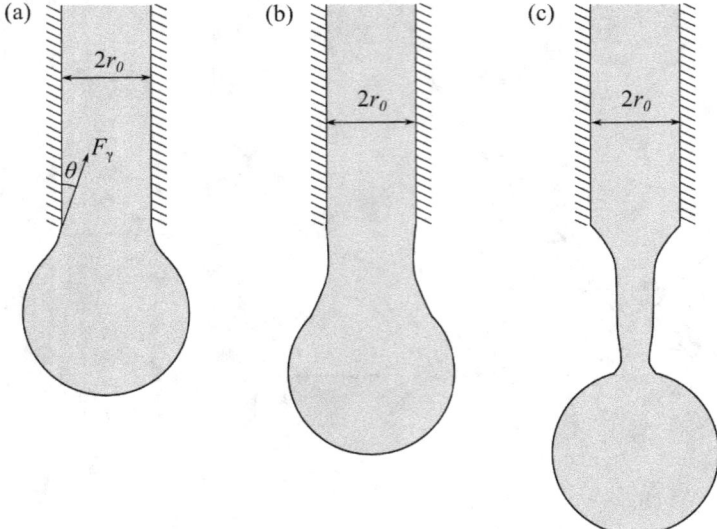

FIGURE 10.3.1. Breakaway of a pendant drop. Some features of the drop and drop neck are exaggerated for clarity. At first the drop fills under gravity to the point just below which the surface tension cannot support the drop (a). Then a liquid neck forms between the drop and the parent capillary (b). The neck elongates and thins at a particular point (c). Such thinning (necking) is unstable and the neck breaks at this point releasing the main drop. Often the neck itself disintegrates into a number of smaller (satellite) drops due to Plateau–Rayleigh instability (to be discussed later).

is not sufficient to hold it and the drop falls off. Consider such a capillary as shown in Figure 10.3.1.

In our case the droplet is supported on the inner surface of the capillary, radius r_0. The drop is held in place by surface tension, where the vertical component of the surface tension force is equal to

$$(10.3.1) \qquad\qquad \gamma 2\pi r_0 \cos\theta$$

This force is balanced by the force of the weight of the drop, $\rho g V$ and so

$$(10.3.2) \qquad\qquad \gamma 2\pi r_0 \cos\theta = \rho g V$$

As the drop fills with more liquid and its volume increases the force holding it to the capillary must increase, otherwise the drop breaks off. The force can change by changing the angle θ and so the shape of the neck. Thus, the neck contorts until $\theta = 0$ at which point any additional liquid in the drop causes the drop to break off. Hence the maximum volume that the capillary can support is,

$$(10.3.3) \qquad\qquad V_{\max} = \frac{2\gamma\pi r_0}{\rho g}$$

As the drop is flying through the air, it can regain a spherical shape. The radius of that drop, as predicated by this simple model is then

$$(10.3.4) \qquad r_d = \left(\frac{3}{2}\frac{\gamma r_0}{\rho g}\right)^{1/3}$$

or

$$(10.3.5) \qquad \frac{r_d}{r_0} = \left(\frac{3}{2}\frac{\gamma r_0^2}{\rho g}\right)^{1/3}$$

If we define the Bond number based on the diameter of the capillary,

$$(10.3.6) \qquad Bo_D = 4r_0^2\rho g/\gamma$$

then

$$(10.3.7) \qquad r_d/r_0 = (6/Bo_D)^{1/3}$$

or

$$(10.3.8) \qquad r_d/r_0 = 1.82 Bo_D^{-1/3}$$

Another way to express this is in terms of drop volume and capillary length $\lambda_c = \sqrt{\gamma/\rho g}$. Rewriting (10.3.3) in these terms, we obtain

$$(10.3.9) \qquad \frac{V_{\max}}{\lambda_c^3} = \frac{2\pi r_0}{\lambda_c} = 2\pi Bo^{1/2}$$

where here for convenience we defined the Bond number in terms of the capillary radius, $Bo = r_0^2\rho g/\gamma$. The result (10.3.9) is often referred to as Tate's law, after Tate who studied this in 1864 (Tate (1864)). When we measure the volume of the drop right before break off, this measured volume agrees very well with that predicted by Tate's law. However, the agreement with the volume of the broken droplet is less good. This is because we assumed here that all the liquid that was weighing down the drop to make it break off goes into the final drop that is flying through the air. Closer observations of drop break off behavior (see for example Shi et al. (1994)) show that during drop break off, a neck forms (Figure 10.3.1b), the neck necks again, usually very close to the drop surface (Figure 10.3.1c), that necking narrows greatly, and only then the drop is free. This causes some of the liquid that was weighing down the drop to go back into the capillary. In fact, Middleman (1995) found that for Weber numbers between 0.001 and 1, (10.3.8) agrees very well with experiment when the prefactor 1.82 is replaced with 1.6. This reduction in the prefactor is to account for the some of the liquid weighing the drop reproducibly returning to the capillary. For this case the Weber number was defined as

$$(10.3.10) \qquad We = \frac{\rho\left(4Q/\pi D_{ci}^2\right)^2 D_{co}}{\gamma}$$

where Q was the flow rate through the tube, D_{ci} was the tube's inner diameter, and D_{co} was the tube's outer diameter. Thus, the actual volume of the drop dripping out of a capillary with $We < 1$ is closer to

$$(10.3.11) \qquad V_{detach} = 1.37\left(\frac{\pi\gamma r_0}{\rho g}\right)$$

For details on the approach to the correction between the drop volume predicted by Tate's law and the actual volume of the detached drop see Harkins and Brown

(1919). Using Tate's law (or even more accurate predictions of pendant drop shape) and using a camera to determine to volume of the hanging drop, we can obtain the surface tension of the liquid. We can also use relations such as (10.3.11) or similar and obtain the surface tension of the liquid by measuring the resulting detached drops. For a detailed discussion on this method as well as details of pendant drop profiles see Boucher and Evans (1975).

CHAPTER 11

Capillarity and pressure

Pressure driven flows with menisci are commonly encountered in microfluidic applications and include the pressure driven filling and emptying of reservoirs and the pressure driven of transport of bubbles and drops. We will focus on these topics in this section.

11.1. Saffman-Taylor instability

Filling and emptying of flat reservoirs is a very common occurrence in microfluidic devices, especially in lab-on-a-chip devices. Such reservoirs include reagent storage reservoirs that need to be emptied of their reagents, as well as reaction, separation, and sensing reservoirs. Almost always these reservoirs are filled (or are to be filled) with aqueous solutions (with liquid properties very similar to water), and are emptied by replacing the contents with air or nitrogen gas. We would ideally like these to be filled completely without air bubbles and emptied completely as well. For example, we would ideally like the reagent storage reservoirs to be emptied without leaving precious reagents behind, we would like to move all of the material from the reaction chamber into a downstream sensing chamber, and we would like not to leave any of the aqueous wash solution in the sensing chamber. As we will see shortly, filling an air-filled chamber with water should be fairly easy and stable, under ideal conditions. (Here by ideal conditions, we mean, parallel, smooth, flat, particle free chamber walls; when chamber walls are not perfectly parallel, filling can also become unstable, and we will consider that in the next section). On the other hand, even under these ideal conditions emptying water filled chambers with air can be difficult and lead to interesting phenomena. Hence, we will consider emptying of flat reservoirs first.

A flat microfluidic reservoir can be modeled as a Hele-Shaw cell. The problem of displacing a more viscous fluid (e.g., water) with a less viscous fluids (e.g., air) in a Hele-Shaw cell has been intensely studied since 1950s and is known to result in a famous instability know as a Saffman-Taylor instability. (Note that at room temperature, the dynamic viscosity of water is about 500 times that of air.) This instability has major importance in crude oil extraction (for which it was originally studied, see Bensimon et al. (1986)), in the performance and health of lung airways (Huh et al. (2007)), in the performance and health of the stomach (Bhaskar et al. (1992)), and of course in microfluidic devices. Hence, let's consider this problem where a less viscous fluid displaces a more viscous fluid as the two fluids move in the narrow space between flat plates. Here, we let the two plates be perfectly parallel. Let's first consider a single fluid between two plates (in a Hele-Shaw cell, Figure 11.1.1). We place the origin on one of the plates, with the z-axis perpendicular to the plates. We assume, as usual, that the velocity component in the x-direction, u_x is only a function of z and is not a function of x or y. We also assume that the y

and z components of the velocity are zero. Since, for a Newtonian fluid shear stress parallel to x-axis on any plane perpendicular to the z-axis is equal to

$$(11.1.1) \qquad \tau = \mu \frac{\partial u_x}{\partial z}$$

the difference between the shear stresses on a two faces of a parallelepiped (lamina or stratum) of a unit area and thickness δz is

$$(11.1.2) \qquad \mu \left(\frac{\partial u_x}{\partial z} + \frac{\partial}{\partial z}\left(\frac{\partial u_x}{\partial z} \right)\delta z - \frac{\partial u_x}{\partial z} \right) = \mu \frac{\partial^2 u_x}{\partial z^2}\delta z$$

This must be balanced by the pressure difference in the parallelepiped, which is equal to $-\partial p/\partial x$ per unit volume of the parallelepiped. Thus,

$$(11.1.3) \qquad \mu \frac{\partial^2 u_x}{\partial z^2} = \frac{\partial p}{\partial x}$$

We could have obtained the same result from simplifying the Navier-Stokes momentum equation (rather than effectively re-deriving it for this particular case) assuming velocity component in the x-direction, u_x is only a function of z and that the y and z components of the velocity are zero. Furthermore, since the z component of the velocity is zero, the Navier-Stokes momentum equation implies that $\partial p/\partial z = 0$. Thus, $\partial p/\partial x$ in (11.1.3) is a constant. Integrating (11.1.3),

$$(11.1.4) \qquad u_x = A + Bz + \frac{1}{2\mu}z^2 \frac{\partial p}{\partial x}$$

Applying the boundary conditions, $u_x = 0$ at $z = 0$ and $z = b$, we find

$$(11.1.5) \qquad u_x = -\frac{1}{2\mu}z\left(b - z \right)\frac{\partial p}{\partial x}$$

To find the gap averaged velocity, we integrate the velocity in (11.1.5) from 0 to b and divide by the gap height, b,

$$(11.1.6) \qquad u_{ave} = -\frac{b^2}{12\mu}\frac{\partial p}{\partial x}$$

Now let's consider the more general problem with two fluids and the fluid moving not only in the x-direction but also with some motion in the y-direction. Considering that the gap averaged velocity parallel to the plates is proportional to the local force, equation (11.1.6) can be generalized to,

$$(11.1.7) \qquad \mathbf{u}_i(x, y) = -\frac{b^2}{12\mu_i}\nabla p(x, y)$$

where the subscript i refers to each of the two fluids and $u_i(x, y)$ is the velocity in each of the fluids. This gap averaging of the velocity is sometimes known as the Darcy approximation, because it resembles the Darcy equation where the average velocity in a porous medium is proportional to the pressure gradient and inversely related to the viscosity. Next, we assume that both fluids are incompressible (air can be considered incompressible for Mach number < 0.3). From the continuity equation we find that the divergence of the velocity vanishes,

$$(11.1.8) \qquad \nabla \cdot (\mathbf{u}_i) = 0$$

This implies that the pressure has to obey Laplace's equation,

$$(11.1.9) \qquad \nabla^2 p = 0$$

(a)   (b)

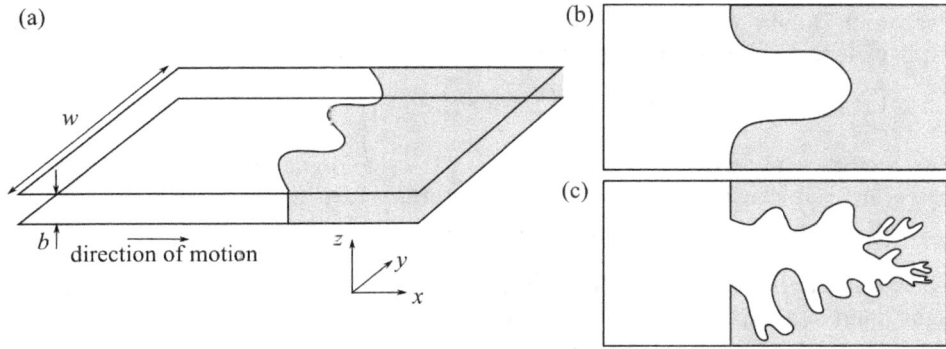

(c)

FIGURE 11.1.1. Schematic of less viscous fluid (e.g., air) displacing a more viscous fluid (e.g., water) between two flat plates with a small gap b (a Hele-Shaw cell). (a) Initially the interface between the two fluids may be perturbed, for example by encountering a pinning point (particle or defect on the surfaces of the plates). Here we greatly exaggerated the amplitude of the perturbation. At moderate velocities (or capillary numbers) greater than that allowed by the stability condition a single large finger order one-half width of the channel, w forms (b). At much higher velocities (or capillary numbers) more viscous fingers emerge at the interface (only one is shown here) and the fingers appear fractal in nature (c). The branching ("fingers on fingers") is especially prominent at points furthest from the main interface.

in each fluid. Furthermore continuity implies that the velocity component normal to the interface must be the same in both fluids. Next, we need an expression for the pressure jump across the liquid-gas interface. One option is to use the Young-Laplace equation,

$$(11.1.10) \qquad \Delta p = \gamma \left(\frac{1}{R_1} + \frac{1}{R_2} \right)$$

where R_1 is the radius of curvature in the xy plane and R_2 is the radius of curvature in the xz plane. Interestingly, most researchers choose to simplify this equation to

$$(11.1.11) \qquad \Delta p = \gamma / R_1$$

At first glance, especially from the experimental point of view, this does not make sense. One would expect that for a small spacing between the plates, $R_2 \ll R_1$ and so R_2 should be dominant in (11.1.10). Equation (11.1.11) only makes sense if the flow was truly two dimensional (i.e., had infinite span in the z-direction). Equation (11.1.11) would also make sense from an experimental perspective if the contact angle was exactly 90°. A more reasonable equation for pressure drop across the liquid gas interface would be

$$(11.1.12) \qquad \Delta p = \frac{2\gamma}{b} \cos\theta + \frac{\gamma}{R_1}$$

Park and Homsy (1984) suggested another relation for this pressure drop with explicit dependence on capillary number,

$$(11.1.13) \qquad \Delta p = \frac{\gamma}{b/2}\left(1 + 3.80 Ca^{2/3}\right) + \frac{\pi}{4}\frac{\gamma}{R_1}$$

Now we are ready to look at the stability of the liquid-gas interface in more detail. Let's begin by considering an almost flat interface for which we can write

$$(11.1.14) \qquad x(y) = Ut + A(t)\cos qy$$

where the interface moves with an average velocity U and the second term on the right-hand side represents a small sinusoidal perturbation from flatness. Here q is the wave vector of the perturbation. We consider $A(t)$ to be small. If A vanished, i.e., the interface was flat, then according to (11.1.7), U would be produced by a pressure gradient of

$$(11.1.15) \qquad \nabla p = -\frac{U}{b^2/12\mu}$$

where μ is the viscosity of the more viscous liquid (here water). If A did not vanish, but just is small, then the velocity of the interface would approximately be (using the first term of Taylor expansion of A)

$$(11.1.16) \qquad U_n = U + A(t)\cos qy$$

and so the pressure gradient would have the form of

$$(11.1.17) \qquad \nabla p = -\frac{U + A\cos qy}{b^2/12\mu}$$

Integrating (11.1.17), we find

$$(11.1.18) \qquad p(x,y) = p_0 - \frac{U}{b^2/12\mu}(x - Ut) + B(x,t)\cos qy$$

where p_0 is a constant and B is to be found. Recall that the pressure has to obey Laplace's equation (11.1.9). If we substitute (11.1.18) into Laplace's equation, we find that $B(x,t)$ must vary as a linear combination of $\exp(qx)$ and $\exp(-qx)$. If the pressure varied as $\exp(q)$ then as x tends to infinity, pressure must also, and this is not physical. Thus,

$$(11.1.19) \qquad p(x,y) = p_0 - \frac{U}{b^2/12\mu}(x - Ut) + B(t)\exp(-qx)\cos qy$$

From substituting the result into Laplace's equation we also learn that the no-slip condition on the side walls requires the wave vector to be

$$(11.1.20) \qquad q = \frac{2\pi n}{W}$$

Substituting (11.1.19) into (11.1.17), and neglecting terms of order A^2 and AB, we obtain

$$(11.1.21) \qquad U_n = U + \frac{b^2}{12\mu}qB(t)\cos qy$$

or

$$(11.1.22) \qquad \frac{dA}{dt} = B(t)\frac{b^2}{12\mu}q$$

Now we need another relationship between A and B. So far we have considered what happens in the water phase. Now we consider in what happens in the air phase. Recall that because of the continuity at the liquid-gas interface, the air must move at the same velocity as the water. Since air is about 500 times less viscous than water, the pressure gradient in the air is 500 times less than that in the water! So we can assume that the pressure gradient in the air is negligible - pressure in the air is constant. Thus for air, (11.1.17) collapses to

$$(11.1.23) \qquad 0 = -\frac{U + A\cos qy}{b^2/12\mu}$$

and so combining this result with (11.1.18) and (11.1 14), we obtain that the pressure jump across the liquid gas interface should be

$$(11.1.24) \qquad \Delta p = \left(B - \frac{U}{b^2/12\mu}A \right)\cos qy$$

However, we also have a relation for the pressure jump across the liquid gas interface from the simplified two dimensional Young-Laplace equation, (11.1.11),

$$(11.1.25) \qquad \Delta p = \gamma/R_1 \approx \gamma \frac{d^2}{dy^2}x\,(y)$$

Combining (11.1.22) with (11.1.24) and (11.1.25) we obtain,

$$(11.1.26) \qquad \frac{dA}{dt} = A\left(\frac{U}{b^2/12\mu} - \gamma q^2 \right)\frac{b^2}{12\mu}q$$

This is the equation for the growth of the amplitude of the perturbation in time. We can see by observation that the solution to (11.1.26) is an exponential function of time, with the terms in the parenthesis and to the right of the parenthesis multiplying time. If these terms are positive, than the amplitude of perturbation grows over time - the interface is not stable. If they are negative, the amplitude decays over time, and so the perturbation decays, and so the interface is stable to perturbations. We see that the interface is stable if the velocity is negative - that is if it is the water that is displacing air and not the other way around. Thus, under ideal conditions, the interface of water displacing air, i.e., the filling of a reservoir, the air - water interface should be smooth and the filling should be simple. In fact the interface is more stable, the faster the water displaces the air! However, if we need to empty a reservoir, i.e., displace the water with air, we see the interface is stable if

$$(11.1.27) \qquad \frac{U}{b^2/12\mu} < \gamma q^2$$

for the smallest wave vector q. From (11.1.20) this is

$$(11.1.28) \qquad 1 < \frac{\pi^2}{3}\frac{b^2}{W^2}\frac{\gamma}{\mu U}$$

We define the dimensionless group on the right-hand side as d_0, the surface tension parameter (Bensimon et al. (1986)),

$$(11.1.29) \qquad d_0 = \frac{\pi^2}{3}\frac{b^2}{W^2}\frac{\gamma}{\mu U} = \frac{\pi^2}{3}\frac{b^2}{W^2}\frac{1}{Ca}$$

Another way to look at this is that the interface is stable when

$$(11.1.30) \qquad \frac{3}{\pi^2} \frac{W^2}{b^2} \frac{\mu U}{\gamma} = \frac{3}{\pi^2} \frac{W^2}{b^2} Ca < 1$$

The interface is stable when the product of the capillary number and the square of the ratio of width to spacing between the plates is less than 1. Since the spacing between the plates is small, the square of the ratio of width to spacing between the plates is always very large. Thus, for the interface to be stable the capillary number must be small. It is interesting to observe that while in many instabilities viscosity acts to dampen out the instability, here viscosity amplifies the instability instead.

The behavior we predict is indeed observed experimentally (see Bensimon et al. (1986) for a review). When U is negative or positive but small (such that it satisfies (11.1.30)) if the initial interface has a few bumps in it, it straightens out over time and a straight boundary exists between the two fluids. If U is still small but exceeds that allowed by (11.1.30) any small bump present in the interface grows and forms a stable finger in the channel! (see Figure 11.1.1). These fingers give the instability its name - the viscous fingering instability. When U is large (or d_0 is small) a chaotic behavior is observed where several fingers form and then branch and split. The tallest fingers appear to get ahead and leave the shorter ones behind. There is a cascade of different sized fingers with the largest fingers being limited by the width of the Hele-Shaw cell. The interface is understood to be chaotic (in the mathematical sense of the word). For emptying of reservoirs it is best to stay in the stable high d_0 (ultra low capillary number) regime. However, for creating a large area gas-liquid interface, for example for mixing of the aqueous solution or for enhancing an interfacial reaction or extracting components of the solutions (e.g., proteins) to the liquid gas interface, this may be beneficial.

Note that the stability criteria, (11.1.30) that we have derived depends our equation for amplitude growth, which in turn depends on the relation for the pressure drop across the liquid-gas interface used (here we used (11.1.25)). If we used a more realistic pressure drop equation, (11.1.13), our amplitude growth equation would be

$$(11.1.31) \qquad \frac{dA}{dt} = \frac{Uq - \frac{\pi}{4}\gamma q^3 b^2/12\mu}{1 + 0.42qb(\gamma/\mu U)^{1/3}} A$$

and so our stability relation would be

$$(11.1.32) \qquad U < \frac{\pi}{4} \frac{\gamma q^2 b^2}{12\mu}$$

or using (11.1.20),

$$(11.1.33) \qquad 1 < \frac{\pi^3}{12} \frac{\gamma}{\mu U} \frac{b^2}{W^2} = \frac{\pi^3}{12} \frac{b^2}{W^2} Ca$$

Notice that both stability expressions have exactly the same scaling, despite one completely ignoring the radius of curvature in the xz direction and one not only keeping this radius of curvature, but also accounting for dependence of contact angle on meniscus travel speed. Furthermore, the prefactor, $\pi^2/3 = 3.29$ in our simple relation, agrees not too poorly with the prefactor $\pi^3/12 = 2.58$ in our second equation.

11.2. Filling and emptying of tapered Hele-Shaw reservoirs

Having established the criteria for flat plates, we might ask about what happens to this stability criteria when the plates are tapered either in the direction of the flow or against it. Does this tapering make the interface more or less stable? To answer this question we follow the analysis of Al-Housseiny and Stone (2013).

Let's consider a modified Hele-Shaw cell where the top plate is titled such that the height of the cell is specified as

$$(11.2.1) \qquad\qquad h = h_0 + \alpha x$$

Hence, if α is positive, the less viscous fluid (e.g., air) is chasing the more viscous fluid (e.g., water) into an expanding channel, whereas if α is negative, the channel contracts in the direction of the motion of the fluids. We assume that the small taper of the plates does not significantly affect the velocity behavior inside the channel and that the velocity still has the same dependence on pressure gradient as in (11.1.7), thus

$$(11.2.2) \qquad\qquad \mathbf{u}_j\left(x,y\right) = -\frac{h(x)^2}{12\mu_j}\nabla p_j\left(x,y\right)$$

where j designates the phase. We designate $j = 2$ as the displacing phase (e.g., air) and $j = 1$ as the displaced phase (e.g., water). Furthermore, we have the continuity equation, again assuming that both fluids are incompressible,

$$(11.2.3) \qquad\qquad \nabla \cdot \left(h\mathbf{u}_j\right) = 0$$

Substituting (11.2.2) into (11.2.3) we obtain,

$$(11.2.4) \qquad\qquad \nabla^2 p_j + \frac{3\alpha}{h\left(x\right)}\frac{\partial p_j}{\partial x} = 0$$

While this equation has an analytical solution using separation of variables, the result is fairly difficult to use to predict the growth of the amplitude of the oscillations of the interface. Hence if we drop the terms of $O\left(\alpha^2\right)$, we get a constant coefficient PDE,

$$(11.2.5) \qquad\qquad \frac{\partial^2 p_j}{\partial x^2} + \frac{\partial^2 p_j}{\partial y^2} + \frac{3\alpha}{h_0}\frac{\partial p_j}{\partial x} = 0$$

Now that we have an equation for the pressure in the bulk of the phases, we need to see how the pressure and so interface position behaves at the interface between the two fluids. As before, we let the position of the interface to be at

$$(11.2.6) \qquad\qquad x = x_0 + \varepsilon\left(y,t\right)$$

where ε is a function representing the perturbation,

$$(11.2.7) \qquad\qquad \varepsilon\left(y,t\right) \sim \exp\left(iky + \sigma t\right)$$

where k is the wave number of the perturbation, and σ is the so called dispersion coefficient, which just represents the growth of the amplitude of the perturbation with time. Note, the velocity of the interface $U = dx_0/dt$. Since the governing PDE (11.2.5) is shift invariant in x, we can just let the interface to be initially at zero, and so let $x_0 = 0$. Thus we can write the pressure in each of the phases near the interface to be

$$(11.2.8) \qquad\qquad p_j\left(x,y,t\right) = f_j\left(x\right) + g_{jk}\left(x\right)\varepsilon\left(y,t\right)$$

where f_j represents the base pressure, i.e. the pressure if the interface was propagating without any perturbations with velocity U, g_{jk} is a weighting factor representing the magnitude of the contribution of a particular wave number k, and ε is our sinusoidal perturbation. The component f_j must of course satisfy (11.2.2). g_{jk} must of course vanish in both directions from the interface as there is no perturbation far from the interface. Taking these into account and substituting (11.2.8) into (11.2.5) we obtain the following equations for each of the pressure terms:

$$(11.2.9) \qquad \frac{d^2 f_j}{dx^2} + \frac{3\alpha}{h_0} \frac{df_j}{dx} = 0$$

$$(11.2.10) \qquad \frac{df_j}{dx}\bigg|_{x=0} = -\frac{12\mu_j U}{h_0^2}$$

$$(11.2.11) \qquad \frac{d^2 g_{jk}}{dx^2} + \frac{3\alpha}{h_0} \frac{dg_{jk}}{dx} - k^2 g_{jk} = 0$$

$$(11.2.12) \qquad \begin{aligned} \lim_{x \to \infty} g_{1k}(x) &= 0 \\ \lim_{x \to -\infty} g_{2k}(x) &= 0 \end{aligned}$$

Solving these and substituting the result back into (11.2.8) we obtain

$$(11.2.13) \qquad p_j(x, y, t) = \frac{4\mu_j U}{\alpha h_0} \exp\left(-\frac{3\alpha}{h_0} x\right) + \sum_k b_{jk} \exp\left(m_{jk} \varepsilon(y, t)\right)$$

where the eigenvalue for each phase m_{jk} is

$$(11.2.14) \qquad m_{jk} = -\frac{3\alpha}{2h_0}\left(1 - (-1)^j \frac{\alpha}{|\alpha|}\sqrt{1 + \frac{4k^2 h_0^2}{9\alpha^2}}\right)$$

If we linearize the equation around the position of the interface, $x = 0$, then we obtain,

$$(11.2.15) \qquad p_j|_{x=\varepsilon} = -\frac{12\mu_j U}{h_0^2}\varepsilon + \sum_k b_{jk}\varepsilon + O\left(\varepsilon^2\right)$$

We determine the constants b_{jk} from the fact that the velocity of the interface in one phase must be the same as that in the other phase (the so called kinematic boundary condition). This condition can be written as

$$(11.2.16) \qquad U + \frac{\partial \varepsilon}{\partial t} = u_{jx}|_{x=\varepsilon} + \frac{\partial \varepsilon}{\partial y} u_{jy}|_{x=\varepsilon}$$

where from (11.2.2) we have the velocities in each of the two phases,

$$(11.2.17) \qquad u_{jx} = -\frac{h^2}{12\mu_j} \frac{\partial p_j}{\partial x}$$

$$(11.2.18) \qquad u_{jy} = -\frac{h^2}{12\mu_j} \frac{\partial p_j}{\partial y}$$

Using orthogonality of eigenfunctions we find

$$(11.2.19) \qquad b_{jk} = \frac{12\mu_j U}{m_{jk} h_0^2}\left(\frac{\sigma}{U} + \frac{\alpha}{h_0}\right)$$

As in the previous parallel plate Hele-Shaw cell case, all that we need now is an expression for the pressure jump across the interface between the two fluids. Here Al-Housseiny and Stone (2013) suggest accounting for both the lateral curvature of any of the fingers that develop, as well as the curvature due to the depth of the Hele-Shaw cell (which scales as $1/h$). Thus they suggest a pressure jump of

$$(11.2.20) \qquad p_2 - p_1 = \gamma \left(\frac{2\cos\theta}{h_0 + \alpha\varepsilon} - \frac{\partial^2\varepsilon/\partial y^2}{\left(1 + (\partial\varepsilon/\partial y)^2\right)^{3/2}} \right)$$

Here the first term accounts for the curvature due to the depth of the Hele Shaw cell and the second term accounts for the curvature of the fingers. To simplify this relation, we linearize in ε,

$$(11.2.21) \qquad p_2 - p_1 = \gamma \left(\frac{2\cos\theta}{h_0} + \varepsilon \left(k^2 - \frac{2\alpha\cos\theta}{h_0^2} \right) \right) + O\left(\varepsilon^2\right)$$

To obtain a relation for σ, the growth of the amplitude of the instability with time, we substitute the expression for pressure (11.2.15) with (11.2.19) into our pressure jump relation (11.2.21); we then apply orthogonality of eigenfunctions and the approximation that depth varies negligibly on the scale of the perturbation size scale (or $|kh_0/\alpha| \gg 1$). We find that

$$(11.2.22) \qquad \sigma = \frac{U}{1 + \mu_2/\mu_1} \left(\left(1 - \frac{\mu_2}{\mu_1} + \frac{2\alpha\cos\theta}{12Ca} \right) k - \frac{h_0^2 k^3}{12Ca} \right)$$

Where the capillary number is defined with the viscosity of the displaced phase ($j = 1$). For the interface to be stable we would like the disturbance to decay with time, and so we would like $\sigma < 0$. Let's compare this result with the one we obtained previously in the limit where $\alpha \to 0$ and $\mu_1 \gg \mu_2$ (the case of air displacing water). Using our previous result for the wave number (wave vector), (11.1.20),

$$(11.2.23) \qquad k = q = \frac{2\pi n}{W}$$

for $n = 1$,

$$\frac{\pi^2}{3} \frac{1}{Ca} \frac{h_0^2}{W^2} > 1$$

This is exactly the relation (11.1.30) we have obtained previously. Note that this relation came solely from the second term in (11.2.22). When the Hele-Shaw cell is slightly tapered, i.e., $\alpha \neq 0$, we see from (11.2.22) that the stability criteria is

$$(11.2.24) \qquad \left(1 - \frac{\mu_2}{\mu_1} + \frac{2\alpha\cos\theta}{12Ca} \right) k < \frac{h_0^2 k^3}{12Ca}$$

In the usual case where $\mu_1 \gg \mu_2$, we can simplify this further to

$$(11.2.25) \qquad \frac{12Ca + 2\alpha\cos\theta}{h_0^2 k^2} < 1$$

We see that the interface is stable for both negative values of α (channel converging in the direction of travel) and even for small but positive values of α. A more conservative stability estimate can be made by limiting

$$(11.2.26) \qquad 1 - \frac{\mu_2}{\mu_1} + \frac{2\alpha\cos\theta}{12Ca} < 0$$

Here we have a critical capillary number, above which this transition threshold is crossed,

(11.2.27) $$Ca < -\frac{1}{6}\frac{\alpha \cos \theta}{(1 - \mu_2/\mu_1)}$$

So far here we have considered the stability for $U > 0$, i.e., the less viscous fluid displacing the more viscous fluid. Now let's consider the case where we reverse the flow direction, i.e., set $U < 0$. For parallel plate Hele-Shaw cell, we saw a negative velocity (a more viscous fluid displacing a less viscous fluid) always leads to the interface being stable. But, this is not so in the tapered Hele-Shaw cell. For the tapered Hele-Shaw cell in this case, the interface is stable when

(11.2.28) $$\left(1 - \frac{\mu_2}{\mu_1} + \frac{2\alpha \cos \theta}{12Ca}\right)k - \frac{h_0^2 k^3}{12Ca} > 0$$

Or for the usual case where $\mu_1 \gg \mu_2$, since we know $U < 0$, and so $Ca < 1$

(11.2.29) $$\frac{-12|Ca| + 2\alpha \cos \theta}{h_0^2 k^2} < 1$$

Thus, the conditions that made the less viscous fluid displacing a more viscous fluid stable, will make the more viscous fluid displacing the less viscous fluid unstable! Stable filling and emptying a tapered chambers needs to happen at different capillary numbers. Letting,

(11.2.30) $$k = q = \frac{2\pi n}{W}$$

and taking $n = 1$, (11.2.29) can be rewritten as

(11.2.31) $$\frac{-12|Ca| + 2\alpha \cos \theta}{4\pi^2}\frac{W^2}{h_0^2} < 1$$

We see that increasing the velocity, i.e., increasing $|Ca|$, especially to the point where the angle term is negligible, makes the interface stable. This is also what we expect from our analysis of the parallel plate Hele-Shaw case. To have a feel for the conditions that cause instabilities when filling or emptying such reservoir, we substitute in some realistic values for the parameters. For example, for capillary number based on water with an interface moving at a brisk 1 cm/s, $Ca = 10^{-4}$. If the material of the chamber is fairly water wet, $\cos \theta \to 1$, then the angle of the taper dictates whether the interface is stable or not. The typically very large prefactor W^2/h_0^2, typically order 10^4, just amplifies the effect. Consider that a change in the height by 20 µm over 1 cm gives an angle of 0.002 rad or 0.1°. Let's consider the case of filling the chamber with water, $Ca = 10^{-4}$. (We see from (11.2.31) that if the taper angle α is even as small as 0.1° but positive, the filling interface is unstable! (Here the air-water interface is moving in the direction of the contraction of the chamber). Unintuitively, stability can be regained by increasing the interface speed (capillary number) to cancel the effect of the angle. Now let's consider emptying this chamber. If the taper angle α is even as small as 0.1° but again positive, by (11.2.25) the emptying interface is unstable. (Here the air-water interface is now moving in the direction of the expansion of the chamber). Increasing interface velocity (increasing Ca) just makes the interface even more unstable. While we knew that the emptying interface (i.e., air displacing water) has the propensity to be unstable, now we see that even the filling interfaces (water displacing air, a more viscous fluid displacing a less viscous fluid) can be unstable as well. We also see that a chamber that supports both stable filling and emptying is one where the interface moves in the direction of expansion of the chamber upon filling and in the direction of contraction of the chamber upon emptying.

Here we have observed the very common and well studied interface instability of a less viscous fluid displacing a more viscous fluid, and a much less studied instability (due to chamber angle) of a more viscous fluid displacing a less viscous fluid, both in closed geometries. In open geometries, such that on a single flat plane, the instability of a more viscous fluid displacing a less viscous fluid, for example driven by a body force, is much more commonly encountered and studied. An everyday example of this is the viscous fingering of oil on a highly tilted frying pan (displacing air). For more on this instability see Troian et al. (1989).

11.3. Using phase guides to control filling of reservoirs

In the above section we saw that filling and emptying large flat (Hele-Shaw) reservoirs can be tricky. This suggests that it is sometimes beneficial to have smaller flat chambers. But what if you need to have a large chamber? The answer to this is to segment your large chamber into smaller chambers using pinning barriers or phase guides. Phase guides are material or geometry changes to the chamber floor or ceiling which pin the liquid gas boundary and so control its overall shape. By controlling the shape of the liquid-gas boundary in this way, various, highly controlled filling geometries of large chambers can be achieved that we would not be able to be achieve otherwise. Such unusual filling geometries can be for example, used to control the operation of surface based sensors in the chamber. In this section we will focus on the broad characteristics of phase guided chambers and general guidelines for their design. We refer the reader to Vulto et al. (2011) for details. We will not go into detail into the design of pinning barriers and calculating their burst pressure. Currently, design of effective pinning barriers is more an art then a science, specifically due to the difficulties in precisely controlling the geometries of the micro and nanostructures that form the pinning barriers; the contact angle of these materials; and their roughness - factors that play a major role in how pinning barriers behave. We refer the reader to Kalinin et al. (2009) for details on pinning barriers.

To understand what phase guides can do for us, let's examine specifically difficult to fill butterfly shaped chamber from Vulto et al. (2011) in Figure 11.3.1. While such chamber, with extra sharp corners that make filling and emptying extra difficult, is unlikely to arise in practice, let's examine it as the worst case scenario. If such phase chamber were to be filled without phase guides, the liquid would enter the chamber, expand in a circular arc until it reached the outlet, enter the outlet, and trap air in the remaining sections of the chamber. The chamber will fill as shown in Figure 11.3.1b. Since the majority of the chamber is not filled with liquid, such filling would typically be undesirable.

Let's see if we can do a better job of filling the chamber with phase guides. A phase guide, as the name implies, is a pinning barrier that guides the location of the air liquid interface. This pinning barrier can be as an evaporated or sputtered metal stripe (e.g., order 100 nm in thickness), a bump made from resist (e.g., SU8, and typically the more hydrophobic the better) or even same material (for embossed plastic chambers), or a grove embossed or etched into the material. For each of

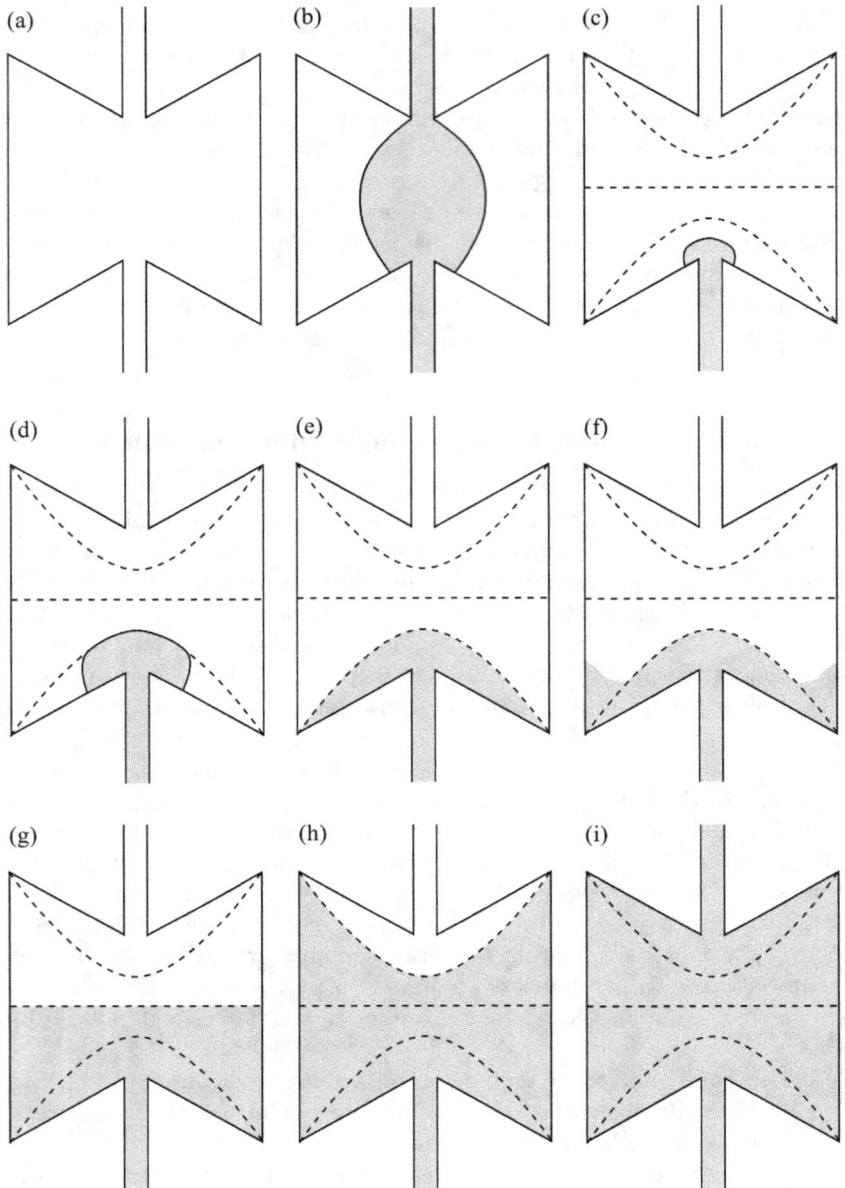

FIGURE 11.3.1. Filling a difficult to fill chamber without (b) and with (c-i) phase guides. Top view of a chamber.

these, the pinning mechanism is different. Some of these approaches are reviewed by Kalinin et al. (2009).

To fill the chamber with phase guides, we lay down the phase guides such that the phase guide line goes into the center of the corners (Figure 11.3.1c). We place the phase guide anywhere where we like the liquid to stop, and thus create subchambers in our chamber. As the liquid enters such a chamber from the inlet, the meniscus expands in a circular arc until it hits a phase guide (Figure 11.3.1d). At this point, there is less resistance to continue along the phase guide than to break the phase guide (pinning barrier). This is the first important point in designing the pinning barriers: the breaking pressure of the pinning barrier must be higher than pressure drop necessary to fill the entirety of the subchamber formed by the pinning barrier. The pressure drop necessary to fill the subchamber can be estimated, for example, from pressure drop in flow between plates, or for complicated geometries by direct numerical simulation. Estimating the breaking pressure of a pinning barrier is difficult, mostly due to uncertainty and sometimes complexity, (e.g., sharp corners) of the geometry. For a properly designed phase guide, the liquid will continue traveling along the phase guide, such that the normal to the meniscus is roughly parallel with the tangent of the phase guide. The liquid should continue to do this until the chamber is filled. The gas (air) is free to move across the phase guide and so is not trapped in the subchamber. (However, phase guided chambers can be designed to specifically trap gas in specific locations, for example for the purposes of creating well defined liquid-gas interfaces, for example, for well controlled interface reactions).

Once the liquid fills the subchamber (Figure 11.3.1e), we need to move it controllably over the phase guide. To do this, the pinning barrier's breaking pressure should be lower in one location relative to the breaking pressure along the majority of the barrier. An easy way to create this is by having a sharp corner. In Figure 11.3.1f the sharp corner serves as such a weakness point. Another way to create weakness points is by artificially creating corners in the pinning barrier such as by placing a V notch in the barrier. In Figure 11.3.1f we wanted the pinning barrier to break in the corners because we wanted the liquid travel from the corners towards the center of the chip, so that air is not trapped in the corners. If the weakness point was placed towards the center, the liquid would move from the center outward, trapping air in the corners. The placement of the weakness point is the second most important point in designing phase guided chambers. To move the liquid over the phase guide, the pressure typically needs to be slightly increased. If the chamber is filled at a constant flow rate (e.g., with a well damped syringe pump, or with an arbitrary pressure generator in a constant flow rate feedback loop mode), we would see a pressure increase as the liquid moves across the phase guide. This pressure increase due to phase guides can be used to monitor the position of the liquid meniscus in the chamber. We show the side of view of a liquid crossing one type of pinning barrier, a square bump, in Figure 11.3.2.

Once the liquid jumps the first phase guide (Figure 11.3.1f) it displaces the gas in the second subchamber, moves to the next phase guide, and stops there (Figure 11.3.1g). As the pressure is slightly increased, we can get the liquid to cross this phase guide, and fill the next subchamber. We do this until the entire chamber is filled. In a similar manner we can empty chambers as well, where now the meniscus travels over the pinning barrier in the reverse direction. Notice, that there is nothing that precludes fingering, as discussed in sections above, to form between phase guides. If we desire to avoid such fingers, we should place phase guides close enough to each other such that the distance between them is no more than then the largest finger to be tolerated. Notice that even if the fingering exists in a subchamber, as the meniscus reaches the next pinning barrier, the gas is expelled over the pinning barrier, while the liquid is held back and so the interface

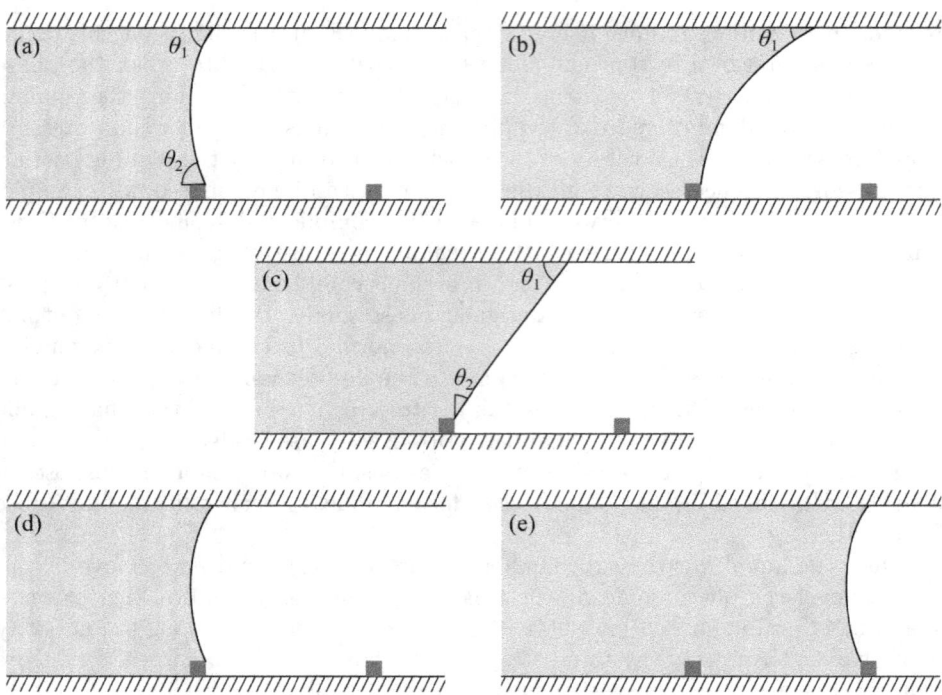

FIGURE 11.3.2. Side view of a phase guide in operation. Initially the liquid is pinned at the first pinning barrier (phase guide) (a). As the pressure increases, the portion of the meniscus not held by the pinning barrier (i.e., here that on the ceiling of the channel) moves past the location of the pinning barrier (b). As the pressure increases further, the meniscus momentarily straightens out (c) before it breaks. As it breaks, portion of the meniscus jumps to the next pinning barrier (e) while simultaneously a portion of the meniscus retreats to the original pinning barrier (d). As the space between the two pinning barriers is filled, the meniscus pinned at the first pinning barrier moves to the next pinning barrier.

is reset to being smooth at each of the pinning barriers. Thus, closely spaced phase guides retard the growth of fingers when filling or emptying chambers. In a similar manner, phase guides can be used to remove bubbles (often introduced in tubing connecting the microfluidic device to the macro world) and channel the gas in the bubbles to a dedicated part of the chip (or even out of the chip), thus removing the bubbles that may interfere with the rest of on chip processing. Another application of phase guides is to create specifically shaped liquid-gas interfaces for lensing of light in optofluidic devices.

11.4. Capillary burst valves

In the previous section we saw how to constrain liquid to certain parts of a chamber. It is of course also useful to constrain liquids between chambers, and for that two types of passive valves exists: hydrophobic valves, and hydrophilic valves.

As these rely on overcoming capillary pressure, they are often termed capillary burst valves. Active valves, such as mechanical obstruction type valves (membrane valves (e.g., most commonly implemented in PDMS, "Quake valves"), ball valves), as well as chemical or phase change valves, such as wax valves, sugar valves, and environmentally sensitive polymer valves can be used as well. The disadvantage of the active valves is that these often need moving parts, sometimes of exotic materials, to be placed inside the microfluidic chip. This is typically difficult and expensive from a manufacturing point of view. Additionally active valves need additional control infrastructure (pressure control channels, magnetic fields, heaters, lasers) to control their opening and closing. The advantage of active valves over passive valves is that they typically provide a better seal and separation of the valved off liquid from the rest of the device, and allow better control over the behavior of the valved liquids. In this section we will discuss passive capillary burst valves.

Hydrophobic capillary break valves are meant to stop an aqueous meniscus at a certain position in the channel by relying on differences in the contact angle between the channel material another material. They are formed by depositing a small strip of hydrophobic material, typically on one of the walls (e.g., ceiling) of the channel. To estimate the burst pressure of such a valve, let's consider the most ideal version of this valve: a circular capillary with radius R, made from a material with contact angle to our fluid θ_0, with a stripe of hydrophobic material, with contact angle θ_h painted in a complete ring on the inside of the capillary. Before the meniscus reached the hydrophobic ring, the capillary pressure across the meniscus was

$$(11.4.1) \qquad\qquad P_0 = 2\cos\theta_0/R$$

This is the pressure in the liquid just behind the meniscus, and is the pressure driving the liquid in the capillary. This driving pressure is balanced by the viscous dissipation in the liquid. For example, if the inlet of the liquid is at the same gas pressure as the gas ahead of the meniscus, combining (11.4.1) with the Hagen-Poiseuille equation, we obtain that the flow rate of the liquid is approximately

$$(11.4.2) \qquad\qquad Q = \frac{\pi R^4}{8\mu x} P_0 = \frac{\pi R^3}{8\mu x} 2\cos\theta_h$$

where x is the distance from the entrance to the meniscus. (Here we have approximated that the fluid flow is quasi steady). Notice that as x increases the flow rate drops. However, once the meniscus reaches the hydrophobic ring, the capillary pressure drops to

$$(11.4.3) \qquad\qquad P_h = 2\cos\theta_h/R$$

Since the driving pressure has decreased, or has gone negative (if $\theta_h > 90°$) the flow rate drops or even reverses until it is again in the hydrophobic section. For the latter case, the meniscus stops at the entrance to the ring. To get the fluid moving forward again, we need to add a pressure equivalent to P_h at the entrance. Thus P_h is an estimate of the burst pressure of our idealized hydrophobic valve.

We might notice from (11.4.3) that another way to drop the driving pressure in a channel is to increase the radius in our capillary. This is the basis of the function of the so called "hydrophilic valve" (or more correctly a valve made from the same material as the rest of the channel, but relies on geometry rather than contact angle for its function). While creating hydrophobic valves is easier in traditional capillaries (such as by lithographically patterning a material inside), creating hydrophilic valves is quite easy with chip based microfluidic devices. Here we will analyze the burst pressure of a hydrophilic valve following the derivations of Thio et al. (2013) and Chen et al. (2008) considering rectangular cross-section valves with geometric openings (as shown in Figure (11.4.1)) with the opening angle $\beta \leqslant 90°$. Similar analysis can be extended to such valves where $90° < \beta < 180°$, but we don't perform this analysis here.

We begin by recalling from Section 7.2 that the capillary pressure across a liquid gas interface scales as the change in total surface (interfacial) energy per change in the volume of the liquid, or

$$(11.4.4) \qquad P_c = -\frac{dG}{dV}$$

where G is the total interfacial energy of the solid-liquid-gas system and V is the volume of the liquid. The total interfacial energy is given by

$$(11.4.5) \qquad G = A_{sl}\gamma_{sl} + A_{sg}\gamma_{sg} + A_{lg}\gamma_{lg}$$

where A is the area, and the subscripts refer to solid-liquid interface, solid-gas interface, and liquid gas interface. Note we typically write γ_{lg} as just γ. Using Young's equation we can write,

$$(11.4.6) \qquad G = (A_{sl} + A_{sg})\gamma_{sg} + (A_{lg} - A_{sl}\cos\theta)\gamma$$

Note that since the area of the solid does not change, $A_{sl} + A_{sg}$ does not change as we pump more liquid into the channel, and so it is a constant. Thus, we can write,

$$(11.4.7) \qquad G = G_0 + (A_{lg} - A_{sl}\cos\theta)\gamma$$

$$(11.4.8) \qquad G_0 = (A_{sl} + A_{sg})\gamma_{sg}$$

Taking the derivative with respect to V of (11.4.6) and substituting this into (11.4.4), we obtain

$$(11.4.9) \qquad P_c = -\gamma\left(\frac{dA_{lg}}{dV} - \frac{dA_{sl}}{dV}\cos\theta\right)$$

Thus, to determine the capillary pressure, and so the pressure necessary to drive the fluid, all we have to do now is to determine how the areas of liquid gas interface and solid liquid interface change with increasing volume of liquid in the channel. We separate the liquid entering the capillary valve (channel widening) into 3 stages as shown in Figure 11.4.1. To determine these areas we use the geometrical constructions according to Thio et al. (2013) shown in Figure 11.4.1. Determining these areas is fairly laborious and the details are presented in Thio et al. (2013). Here will just present the intermediate and final results. We note that while Thio et al. derivation for the areas is only valid for channel opening angle $\beta \leqslant 90°$, using similar geometric construction relations can be found for $90° < \beta < 180°$ as well and likewise substituted into (11.4.9) to obtain the burst pressure of such a valve.

In the first stage (Figure 11.4.1a), the liquid interface is concave and the meniscus is approaching the diverging section of the channel. Using the 3D geometry of the channel and so the areas of the liquid-gas and liquid-solid interfaces, Thio et al. found that the capillary pressure at this point is

$$(11.4.10) \qquad P_{c,1} = \frac{2\gamma}{w}\left(\frac{w}{h} + 1\right)\cos\theta$$

where h is the height of the channel (into the page).

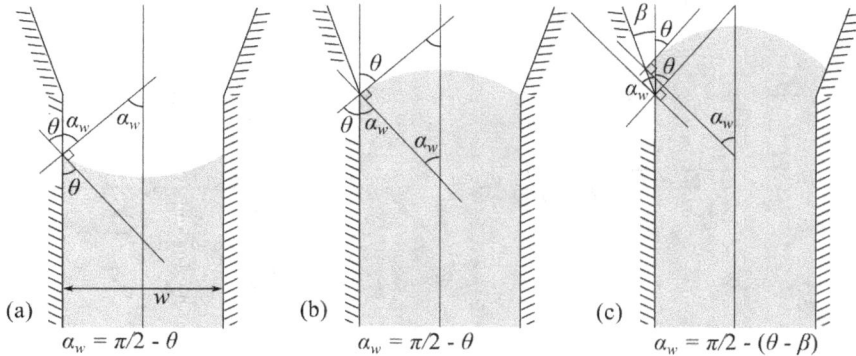

FIGURE 11.4.1. Schematic and geometric construction of a fluid in an expanding section capillary burst valve.

In the second stage (Figure 11.4.1b), the meniscus has reached the diverging section. Since the radius of curvature of the channel at this point increases, the capillary pressure decreases. Thus if we maintain the same driving pressure in the liquid, the interface inverts and becomes convex, as shown (Figure 11.4.1b). Once again, the 3D geometry of the channel and so the areas of the liquid-gas and liquid-solid interfaces, Thio et al, found that the capillary pressure at this point is

$$(11.4.11) \qquad P_{c,1} = \frac{2\gamma}{w}\left(\frac{w}{h} - 1\right)\cos\theta$$

In the third stage, the meniscus continues to travel up the divergence. This motion characterizes the bursting of this valve. Thio et al., found this to be

$$(11.4.12) \qquad P_{burst} = \frac{2\gamma}{w}\left(-\frac{w}{h}\cos\theta + \frac{\cos\theta - \alpha_w\sin\beta/\sin\alpha_w}{-\left(\cos\beta + \frac{\sin\beta}{\sin\alpha_w}\left(\frac{\alpha_w}{\sin\alpha_w} - \cos\alpha_w\right)\right)}\right)$$

This relation has been experimentally investigated by Thio et al. (2013) and found to correctly predict the overall scaling of the problem. However, assumptions in its derivation such as disregard for changes in the meniscus curvature in the plane orthogonal to the plane of the page; that the meniscus is always perfectly circular; that the meniscus is stage 3 moves over only an infinitesimal amount into the divergence; not considering cases with $\beta > 90°$ all leave room for improvement.

11.5. Motion of a lone long bubble in a tube

The flow of lone bubbles or bubble trains is widely encountered in microfluidic devices. Here we will explore key figures of merit of such flows: (a) the velocity of the bubble relative to the liquid flow, (b) the pressure drop to drive such flows, and (c) the thickness of the liquid film between the gas body of the bubble and the channel wall. We will begin exploring the problem of bubbly flow by considering the seminal approach of Bretherton (1961). While Bretherton's results do not perfectly agree with experiments over the entire range Ca numbers of interest, it agrees fairly well over some of the range and produces simple to use results. Furthermore, more modern and sophisticated approaches usually build upon Bretherton's approach or compare their results to it, so it is instructive to understand it. Furthermore,

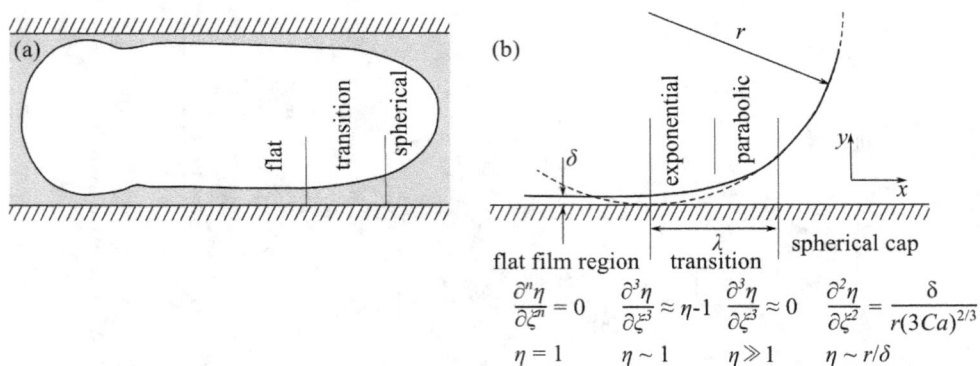

FIGURE 11.5.1. (a) Geometry of a long bubble (length > diameter) in a circular capillary. The bubble is moving from left to right (or in the frame of the bubble, the liquid around the bubble is moving from right to left). (b) Close up of the front of the bubble, which is subdivided into four regions: flat, exponential, parabolic and spherical.

Bretherton's approach is similar to the approach of Landau and Lebich (1942) in the problem of dragging of a liquid by a moving plate – another very important problem in fluid mechanics. After going through Bretherton's approach, we will briefly review more modern approaches, and present modern empirical fits for the key figures of merit, including the flow in the slugs between the bubbles.

For Bretherton's approach, we consider a lone long bubble in a small circular capillary (Figure 11.5.1). The bubble is several capillary diameters long, and thus considered a "long bubble". The Bond number is small, and so gravity has negligible influence on the shape of the bubble. The Weber number is also small, and so inertia plays a negligible role in the motion of the fluid or the bubble. Finally, the fluid in the bubble has negligible viscosity relative to the fluid around the bubble (e.g., an air bubble in water). To solve the flow around the bubble, we split the front of the bubble into four regions (Figure 11.5.1), solve for the shape of the liquid profile in each of the regions and then tweak the borders of each region just enough so that the solutions overlap and form a smooth profile. For convenience, we consider the problem in the reference frame moving with the bubble. Furthermore, we assume that the velocity component in the x-direction is the only one that is significant, that the flow is axisymmetric, and that the liquid is incompressible. This implies that,

$$(11.5.1) \qquad\qquad u\,(y = 0) = -U$$

where U is the velocity of the bubble. Furthermore,

$$(11.5.2) \qquad\qquad \frac{\partial u}{\partial y}\,(y = h) = 0$$

With these assumptions we simplify the Navier-Stokes momentum equations to

$$(11.5.3) \qquad\qquad \frac{\partial p}{\partial x} = \mu \frac{\partial^2 u}{\partial r^2}$$

And the continuity equation to

(11.5.4)
$$\frac{\partial h}{\partial t} + \frac{\partial q}{\partial x} = 0$$

where, h is the thickness of the flat film region, and

(11.5.5)
$$q = \int_0^h u \, dy$$

Basically, the change in the film thickness is only due to the differences between an inflow and the outflow in a particular region. Integrating the Navier-Stokes momentum equation, we obtain,

(11.5.6)
$$u = -U + \frac{\partial p}{\partial x}\left(\frac{y^2}{2\mu} - \frac{yh}{\mu}\right)$$

We assume that the pressure gradient in the x-direction is due to the Laplace pressure across the interface. Since this is the flat film region, we assume the slope of the film interface is small, and so via the small slope approximation, we can write that the pressure gradient is,

(11.5.7)
$$\frac{\partial p}{\partial x} = \frac{\partial}{\partial x}\left(y\frac{\partial h^2}{\partial y^2}\right)$$

Substituting this into (11.5.5) and integrating we obtain,

(11.5.8)
$$q = Uh - \frac{\gamma}{3\mu}h^3\frac{\partial^3 h}{\partial x^3}$$

Substituting this into the continuity equation,

(11.5.9)
$$\frac{\partial h}{\partial t} + U\frac{\partial h}{\partial x} - \frac{\gamma}{3\mu}\frac{\partial}{\partial x}\left(h^3\frac{\partial^3 h}{\partial x^3}\right) = 0$$

Since in our reference frame the bubble is at rest and its shape is not changing, $\partial h/\partial t$ is equal to zero, so

(11.5.10)
$$U\frac{\partial h}{\partial x} - \frac{\gamma}{3\mu}\frac{\partial}{\partial x}\left(h^3\frac{\partial^3 h}{\partial x^3}\right) = 0$$

Now let's consider the scaling of the problem. This will help us with generating the matching rules to piece together the solutions from the different regions. The horizontal axis $k\,(z)$ scales as the length of the transition region. The film thickness h scales as the characteristic film thickness δ. Thus from (11.5.10) we have the following scaling,

(11.5.11)
$$\frac{\delta}{\lambda} \sim \frac{\gamma}{3\mu U}\frac{1}{\lambda}\left(\delta^3\frac{\delta}{\lambda^3}\right)$$

Or

(11.5.12)
$$(3Ca)^{1/3} \sim \delta/\lambda$$

We can obtain another scaling from considering that the curvature scales as

(11.5.13)
$$\kappa \sim \frac{\partial^2 h}{\partial x^2} \sim \frac{\delta}{h^2}$$

But we also know that the curvature in the spherical cap region is order $\kappa = 1/r$, where r is the radius of the channel. Thus,

(11.5.14) $$\delta/\lambda^2 \sim 1/r$$

Combining this with (11.5.13), we see

(11.5.15) $$\delta \sim r(3Ca)^{2/3}$$

(11.5.16) $$\lambda \sim r(3Ca)^{1/3}$$

The scaling in (11.5.15) already gives us a feeling for the behavior of the liquid film thickness as a function of capillary radius and the capillary number of the flow. Let's now come back to (11.5.10). Since U is a constant, and the surface tension and liquid viscosity is a constant, we can rewrite (11.5.10) as

(11.5.17) $$\frac{\partial}{\partial x}\left(Uh - \frac{\gamma}{3\mu}h^3\frac{\partial^3 h}{\partial x^3}\right) = 0$$

This allows us to integrate this equation, and so obtain,

(11.5.18) $$Uh - \frac{\gamma}{3\mu}h^3\frac{\partial^3 h}{\partial x^3} = c$$

Where c is a constant to be determined. In the flat region, $h \approx \delta$ and $\partial^3 h/\partial x^3 = 0$ (because the interface is flat). Substituting this into (11.5.18) we find $c \approx \delta$. Now, we nondimensionalize (11.5.18) according to the scaling we have obtained. We let

(11.5.19) $$\eta = h/\delta$$

(11.5.20) $$\xi = \frac{(3Ca)^{1/3}}{\delta}x$$

and so scaling (11.5.18) obtain,

(11.5.21) $$\frac{\partial^3 \eta}{\partial \xi^3} = \frac{\eta - 1}{\eta^3}$$

Unfortunately, an analytical solution to this is not known, and so we have to simplify this equation and consider it in two regions. In the first region, we consider η to be of order unity, so

(11.5.22) $$\frac{\partial^3 \eta}{\partial \xi^3} \approx \eta - 1$$

This fortunately has a solution,
(11.5.23)

$$\eta = 1 + C_1\exp\left(\xi\right) + C_2\exp\left(-\xi/2\right)\cos\left(\frac{\sqrt{3}x}{2}\right) + C_3\exp\left(-\xi/2\right)\sin\left(\frac{\sqrt{3}x}{2}\right)$$

We see that because of the negative exponential the third and fourth (oscillatory) terms die out quickly. Thus,

(11.5.24) $$\eta = 1 + C_1\exp\left(\xi\right)$$

Since the solution in this region is exponential, we term this region the exponential region. Next, we consider the region where $\eta \ll 1$, so

(11.5.25) $$\frac{\partial^3 \eta}{\partial \xi^3} \approx 0$$

This has a solution,

$$(11.5.26) \qquad \eta = \frac{C_4}{2}\xi^2 + C_5\xi + C_6$$

Since the solution in this region is parabolic, we term this region the parabolic region. Now we match the two solutions. We begin by noting,

$$(11.5.27) \qquad \begin{aligned} \eta(0) &= 1 + C_1 \\ \eta'(0) &= C_1 \\ \eta''(0) &= C_1 \end{aligned}$$

We can use this these initial conditions to numerically solve (11.5.21) and then fit the constants in the two equations as to get the closest match to the numerical solution. From, this we obtain, in the exponential region,

$$(11.5.28) \qquad \eta = 1 + 0.001\exp(\xi)$$

and in the parabolic region,

$$(11.5.29) \qquad \eta = \frac{0.643}{2}\xi^2 + 2.79$$

Now we have to match the parabolic region to the spherical region. In the spherical region, we have an approximation that

$$(11.5.30) \qquad \frac{1}{r} = \frac{\partial^2 h}{\partial x^2}$$

Putting this into the non-dimensional form,

$$(11.5.31) \qquad \frac{\partial^2 \eta}{\partial \xi^2} = \frac{\delta}{r(3Ca)^{2/3}}$$

Substituting (11.5.29) into (11.5.31) we obtain that

$$(11.5.32) \qquad 0.643 = \frac{\delta}{r(3Ca)^{2/3}}$$

or

$$(11.5.33) \qquad \delta = 0.643r(3Ca)^{2/3}$$

This is the thickness of the liquid film surrounding the bubble. We are now ready to calculate the curvature of the spherical section. Putting the equation for the parabola back into dimensional variables,

$$(11.5.34) \qquad h = \left(\frac{x^2}{2r} + 1.79(3Ca)^{2/3}r\right)$$

We expect the spherical section to overlap with the parabolic at a position $x \approx 0$, where

$$(11.5.35) \qquad h \approx 1.79(3Ca)^{2/3}r$$

At this point, the distance from the center of the tube to the liquid film is $r - h$,

$$(11.5.36) \qquad r - h = r\left(1 - 1.79(3Ca)^{2/3}\right)$$

At this point the sphere has a tangent parallel to the wall. Based on this, the mean curvature of the spherical region is approximately

$$(11.5.37) \qquad \kappa = \frac{2}{r - h} = \frac{2}{r\left(1 - 1.79(3Ca)^{2/3}\right)}$$

We can simplify this further noting that we are concerned with situations where capillary number is very small,

$$(11.5.38) \qquad \kappa = \frac{2}{r} \frac{\left(1 + 1.79(3Ca)^{2/3}\right)}{\left(1 - \left(1.79(3Ca)^{2/3}\right)^2\right)} \approx \frac{2}{r}\left(1 + 1.79(3Ca)^{2/3}\right)$$

Thus the spherical section is composed of a sphere with radius of $r\left(1 + 1.79(3Ca)^{2/3}\right)$ and radius of curvature of $\frac{2}{r}\left(1 + 1.79(3Ca)^{2/3}\right)$.

Now we can similarly calculate what happens at the back of the bubble. The back of the bubble, similarly like the front, can be approximated by an exponential and a parabolic region. Calculating the numerical solution to (11.5.21) in this region, we find that the solutions here have oscillations. Non-intuitively, these oscillations are indeed experimentally observed. Going through a similar matching procedure as above, we find that the parabolic solution is

$$(11.5.39) \qquad \eta = \frac{0.643}{2}\xi^2 - 0.8$$

and the spherical section at the back of the bubble has a mean curvature of

$$(11.5.40) \qquad \kappa = \frac{2}{r}\left(1 - 0.46(3Ca)^{2/3}\right)$$

From the mean curvatures of the front and the back spherical caps of the bubble we can calculate the pressure drop across each using the Young-Laplace equation,

$$(11.5.41) \qquad \Delta p_{front} = \gamma\frac{2}{r}\left(1 + 1.79(3Ca)^{2/3}\right)$$

$$(11.5.42) \qquad \Delta p_{back} = \gamma\frac{2}{r}\left(1 - 0.46(3Ca)^{2/3}\right)$$

Putting these together, the total pressure drop across the bubble is

$$(11.5.43) \qquad \Delta p = \Delta p_{front} - \Delta p_{back} = 4.52(3Ca)^{2/3}\frac{\gamma}{r}$$

This is the pressure drop required to keep the bubble moving at the given capillary number.

Next we are interested in the velocity of the bubble relative to the average velocity of the liquid in the tube. It has been long known (e.g., Fairbrother and Stubbs (1935)) that the velocity of the bubble exceeds the velocity of the liquid. Let's define

$$(11.5.44) \qquad m = \frac{U - U_l}{U}$$

where U_l is the average velocity of the liquid far away from the bubble. In the frame of reference of the bubble, the flow rate behind the bubble is $(U_l - U)\,A$ where A is the cross-sectional area of the capillary. The flow rate in the liquid film surrounding the bubble in this frame of reference is $-UA_{film}$, where A_{film} is the cross-sectional area of the film. These of course must be equal by continuity. Thus,

$$(11.5.45) \qquad m = \frac{U - U_l}{U} = \frac{A_{film}}{A} = \frac{\pi r^2 - \pi(r - \delta)^2}{\pi r^2} = 1 - \left(1 - \frac{\delta}{r}\right)^2$$

Substituting in the relation for the thickness of the liquid film, (11.5.33), and neglecting the higher order terms in $(3Ca)^{2/3}$, we obtain

$$(11.5.46) \qquad m = \frac{U - U_l}{U} = 1.29(3Ca)^{2/3}$$

We see that while the difference between the bubble and the fluid velocity is typically small, the higher the capillary number the higher this difference. We will compare these relations for relative velocity, pressure drop, and film thickness in the next section.

11.6. Motion of long bubbles in tubes and bubble trains

Now having looked at Bretherton's classical result for a single long bubble, and obtained key figures of merit: film thickness, pressure drop and relative velocity, let's now take a look at experimental work and other theoretical work in this area, including other regimes. Here we will review key figures of merit for both lone bubbles and for bubble trains. The flow with bubble trains is also referred to as Taylor flow, after G.I. Taylor (Taylor (1961)). For this, we follow the review of Angeli and Gavriilidis (2008).

Let's begin with liquid film thickness around the bubble. This film thickness has important implications for mass transfer between liquid slugs that the bubble separates, and between the contents of the bubble and the wall of the capillary. It is also of great importance in electrokinetic experiments with trains of bubbles. The thickness of this film has been measured by conductimetric technique (Fairbrother and Stubbs (1935); Taylor (1961); Chen (1986); and Marchessault and Mason (1960)), by light absorption (Irandoust and Andersson (1989)), by direct video recording (Aussillous and Quere (2000)), and via volumetry by Bretherton himself (Bretherton (1961)). In the conductimetric technique, an electrolyte solution of known conductivity (such as potassium chloride solution) is used as a liquid phase. Typically the presence of potassium chloride has minimal impact on the liquid's surface tension and viscosity. As the bubble is traveling in the capillary, DC current is passed between silver-silver chloride reversible electrodes and voltage is measured. (Similar AC resistance measurements are also possible and a preferred when using non-reversible electrodes). From the resulting resistance measurement, the geometry of the bubble, and specifically the thickness of the film is calculated. For more details see Marchessault and Mason (1960). In the light absorption method, a capillary with bubbles is placed between a light source and a photodetector. The thickness is calculated from the light absorption in the liquid slug compared to that in the bubble, and using Beer-Lambert law to back calculate the film thickness. For more details see Irandoust and Andersson (1989). The techniques of direct video recording and volumetry are based on direct optical measurements of the length of the bubble and knowledge of the total volume of liquid in a given segment of the capillary, and from this back calculating the liquid thickness. We tabulate the empirical results of these authors in Table 1. We also plot these in Figure 11.6.1 and compare with Bretherton's result.

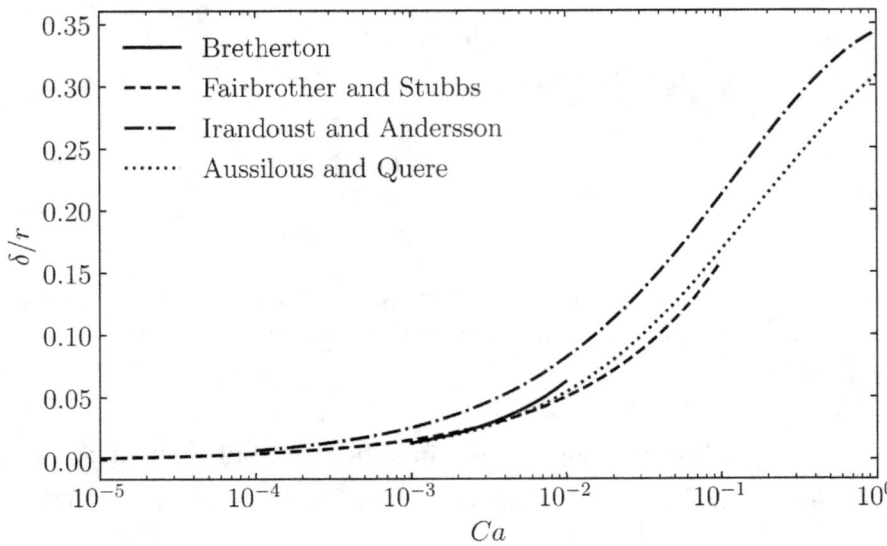

FIGURE 11.6.1. Liquid film thickness scaled by capillary radius as a function of capillary number for ranges of capillary number in which the relationship is believed to be valid.

TABLE 1. Empirical relations for film thickness

δ/r	Range of Ca	Ref.
$1.34Ca^{2/3}$	$10^{-3} \leqslant Ca \leqslant 10^{-2}$	[1]
$0.5Ca^{1/2}$	$5 \times 10^{-5} \leqslant Ca \leqslant 3 \times 10^{-1}$	[2]
$0.36\left(1 - \exp\left(-3.08Ca^{0.54}\right)\right)$	$9.5 \times 10^{-4} \leqslant Ca \leqslant 1.9$	[3]
$\dfrac{1.34Ca^{2/3}}{1+3.35Ca^{2/3}}$	$10^{-3} \leqslant Ca \leqslant 1.4$	[4]
$\left(0.89 - 0.05\Big/U_g^{1/2}\right)Ca^{2/3}$	$7 \times 10^{-6} \leqslant Ca \leqslant 2 \times 10^{-4}$	[5]

[1] volumetry, Bretherton (1961)
[2] conductimetry, Fairbrother & Stubbs (1935)
[3] light absorption, Irandoust & Andersson (1989)
[4] video recording, Aussilous & Quere (2000)
[5] conductimetry, Marchessault & Mason (1960)

U_g is in cm/s

Breatherton's original results considered bubbles at low Re and low Ca. However, as Re increases Edvinsson and Irandoust (1996) found by numerical calculations that the film thickness increases and the difference between the bubble and liquid velocity also increases. With increasing Ca the bubble shape at the back changes also from convex (as in Bretherton's result) to a concave shape! Aussillous and Quere (2000) experimentally found that for higher Re and $Ca > \mu_l/\gamma\rho_l r$, film thickness has a dependence on Weber number,

$$(11.6.1) \qquad \frac{\delta}{r} = \frac{Ca^{2/3}}{1 + Ca^{2/3} - We}$$

TABLE 2. Experimental correlations for relative bubble velocities.

m	Range of Ca	Ref.
$1.29(3Ca)^{2/3}$	$10^{-3} \leqslant Ca \leqslant 10^{-2}$	[1]
	$7.5 \times 10^{-5} \leqslant Ca \leqslant 0.014$	[2]
$Ca^{1/2}$		
$(\mu_l/\gamma)\left(-0.10 + 1.78\sqrt{U}\right)$	$7 \times 10^{-6} \leqslant Ca \leqslant 2 \times 10^{-4}$	[3]
0.6	$Ca > 1$	[4]
U/U_l		
1.7	$10^{-3} \leqslant Ca \leqslant 10^{-2}$	[5]
$\left(1 - 0.61 Ca_m^{0.33}\right)^{-1}$	$2 \times 10^{-4} \leqslant Ca_m \leqslant 0.39$	[6]

[1] Bretherton (1961)
[2] Fairbrother & Stubbs (1935)
[3] Marchessault & Mason (1960)
[4] Giavedoni & Saita (1997, 1999)
[5] bubble train, Laborie et al. (1999)
[6] bubble train, circular and square cross-section, Liu et al. (2005)

$Ca_m = \mu_l U_l / \gamma$

TABLE 3. Correlations for pressure drop in a bubble train flow.

f, Fanning friction factor	Range of Ca	Ref.
$f = \frac{16}{Re}\left(1 + \frac{d}{L_s}\frac{0.465}{Ca^{1/3}}\right)$	$Ca \ll 1,\ Re \ll 1$	[1]
$f = \frac{16}{Re}\left(1 + \alpha\frac{d}{L_s}\left(\frac{Re}{Ca}\right)^{1/3}\right)$	$2 \times 10^{-3} \leqslant Ca \leqslant 4 \times 10^{-2}$	[2]

[1] Bretherton (1961)
[2] Experimentally, $\alpha = 0.17$; numerically, $\alpha = 0.07$; For square capillaries 16 is replaced by 14.2. $Re = O(100)$ Kreutzer (2003)

$\Delta P/L = 2fU_l^2\rho_l\varepsilon_l/d$; ε_l : liquid phase volume fraction; d : channel diameter; L : channel length; L_s : slug length

$$(11.6.2) \qquad We = \rho_l U^2 (r - \delta)/\gamma$$

The next figure of merit of interest to us is the relative bubble velocities. We define the ratio of bubble and fluid velocities m as we have done in (11.5.46) and summarize the various correlations for m in Table 2. Particularly interesting is the result of Taylor (1961) and that of Giavedoni and Saita (1999), who obtained that for $Ca > 1$ m is a constant equal to 0.6. The next figure of merit of interest to us is the pressure drop in a bubble train flow. We summarize the models for this pressure drop due to Bretherton (1961) and that due to Kreutzer (2003) in Table 3, written in the form of the Fanning friction factor. For both models, the friction factor and so the pressure drop increases with decreasing capillary number. The pressure drop also increases as the spacing between bubbles, (i.e. L_s). As expected the friction factor collapses to the friction factor for ordinary pipe flow when $L_s \to \infty$

Another aspect of bubble flow of interest to us is the flow pattern inside the liquid slugs separating the bubbles. We show a qualitative picture of the flow pattern in Figure 11.6.2. Particularly interesting is the fact that as capillary number

(a) $Ca \gg 1$, $m > 0.5$

(c) $Ca \ll 1$, $m > 0.5$

(b) $0.6 < Ca < 0.69$, $m < 0.5$

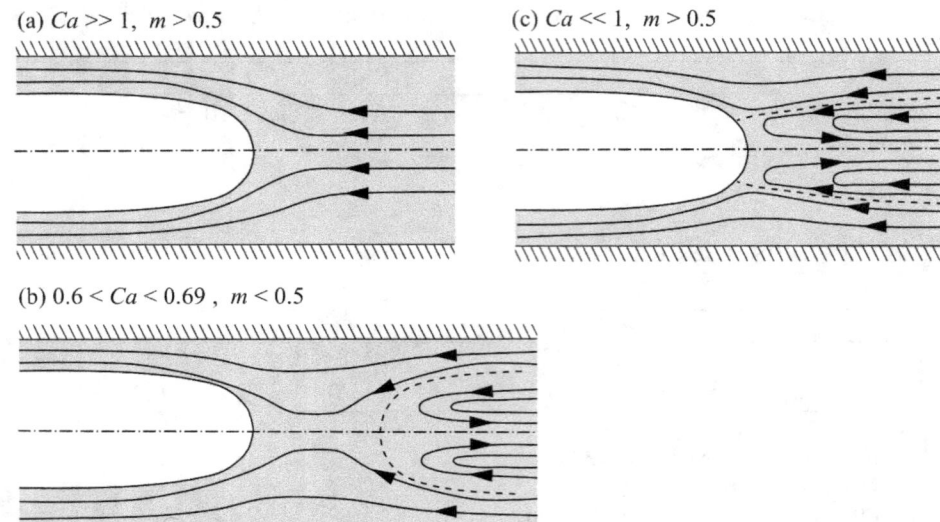

FIGURE 11.6.2. Flow pattern in front a bubble as predicted by Taylor (1961) and observed numerically by Giavedoni and Saita (1997). These patterns have also been observed experimentally. At high capillary numbers the liquid bypasses the bubble and a single stagnation point occurs on the nose of the bubble (a). As the capillary number is decreased, vortices and a zone of recirculation begins to appear in front of the bubble (b). When the capillary number is lowered further, the vortices and the recirculation zone moves closer to the bubble nose. For details see Giavedoni and Saita (1997).

is decreased a recirculation zone develops in the slug. Thus, as capillary number is decreased mixing in the slug is increased. Since such recirculation zones are not present in ordinary pipe flow, introducing slugs and ensuring low capillary number flow can be a way to introduce mixing into the flow, for example for controlling (or accelerating) reactions, including surface reactions on the walls of the channel. The change in flow patterns as a function of capillary number has also an impact on mass transfer between the individual bubbles. At high capillary number, due to the liquid flow in the slug, mass (e.g. chemical species) transport is predominantly from the surface of the front bubbles to that of the rear bubbles (in the direction from the "locomotive" to the "caboose"). However, at low capillary numbers the mass transfer is now bidirectional - chemical species are also transported from the rear bubbles to the front bubbles due to the recirculation vortices.

Another aspect of bubble flow of interest to us is the shape of the bubble in non-circular, i.e., rectangular channels, as rectangular cross-section channels are very often encountered in microfluidic chips. Taylor flow in square channels has been studied theoretically and experimentally by Kolb and Cerro (1991, 1993a,b). These authors found that at $Ca < 0.1$ the bubble is not axisymmetric but flattens out against the tube walls, leaving strings of liquid connecting the slugs in the channel corners, and very flat films of liquid connecting the fluid on the channel sides.

Kreutzer et al. (2005) and Kreutzer et al. (2005) found an empirical correlation for the ratio of bubble to channel diagonal

(11.6.3) $$\frac{d_{bubble}}{d_{channel}} = 0.7 + 0.5 \exp\left(-2.25 Ca^{0.445}\right)$$

At larger capillary numbers, with the transition around capillary number 0.4 to 1, the bubble becomes cylindrical and axisymmetric. The bubble end caps are typically spherical and convex. At high capillary numbers the back end cap changes it shape from convex to concave. The flow patterns in liquid slugs separating bubbles in a bubble train are similar to that for circular cross-sections. For example, Kolb and Cerro (1993a) found that the transition from recirculating flow (Figure 11.6.2c) to bypass flow (Figure 11.6.2a) occurs around $Ca = 0.6$.

CHAPTER 12

Surface tension gradient driven flows (Marangoni Effect)

Up until now we have considered systems where the surface tension (or interfacial tension) across the fluid boundary is constant. In this section, we will look into flows that are driven by a gradient in surface tension - so called Marangoni flows. Such gradient in surface tension can arise due to a gradient in temperature (due to surface tension dependence on temperature), due to a gradient in surfactant concentration, and due to electric fields. In this section, we will focus on the temperature gradient driven flows - thermocapillary Marangoni convection and surfactant gradient driven flows - solutocapillary flows. Since temperature gradients and (to a much lesser extent) surfactant concentration gradients are often encountered in microfluidic devices we have to be vigilant (and/or take advantage) of Marangoni flows in these devices.

Above we have considered two fluid systems where the force normal to the interface is given by the Young-Laplace (capillary) pressure as

$$\text{(12.0.1)} \qquad \mathbf{f}_n = \gamma \left(\frac{1}{R_1} + \frac{1}{R_2} \right) \mathbf{n}$$

However, surface tension does not have to be constant. If the surface tension has a gradient, it directly leads to a surface stress, which leads to a force tangential to the interface,

$$\text{(12.0.2)} \qquad \mathbf{f}_t = \nabla \gamma$$

Note that since surface tension is only defined at the interface, here ∇ denotes the surface gradient. The surface stress that results is often termed the Marangoni stress, after Carlo Marangoni who studied it in the late 1800s. (The effect however was studied before this, for example by James Thomson, and observed much earlier, as for example "wine tears" since antiquity.) Near the interface, this stress pulls liquid towards the region with larger surface tension. In absence of other motion, conservation of mass then dictates that the motion in the bulk of the liquid is then in the direction away from the region with larger surface tension - a circulation loop occurs in the liquid.

12.1. Mathematical formulation - thermocapillary flows

For our mathematical formulation and analysis of Marangoni effects, we follow that of Nepomnyashchy et al. (2001). For the thermocapillary Marangoni flows we begin with a slightly modified Navier-Stokes equations. Specifically we assume that our fluid is incompressible, and the slight density variations that do arise only contribute to the buoyancy flows. In other words we have used the incompressible assumption with the Boussinesq approximation. Furthermore we assume that the viscosity of each fluid is homogeneous. Thus with these approximations the continuity equation becomes

(12.1.1) $$\nabla \cdot \mathbf{v}_m = 0$$

and the momentum equation becomes

(12.1.2) $$\frac{\partial \mathbf{v}_m}{\partial t} + (\mathbf{v}_m \cdot \nabla) \mathbf{v}_m = -\frac{1}{\rho_m} \nabla p_m + \nu_m \nabla^2 \mathbf{v}_m + g \beta_m T_m \mathbf{e}$$

Here the subscript m signifies the phase (either first or second fluid, $m = 1, 2$). ν_m is the kinematic viscosity; z is the vertical coordinate where gravity points in the negative z direction. $\mathbf{e}$ is a unit vector directed upward (positive z direction). β is the thermal expansion coefficient. Here p_m is the hydrostatic pressure, where the total pressure is $p_m - \rho_m g z$. The last term on the right-hand side is the body force term accounting for buoyancy driven flow. Finally, we also have energy conservation equation, written in terms of temperature,

(12.1.3) $$\frac{\partial T_m}{\partial t} + (\mathbf{v}_m \cdot \nabla) T_m = \kappa_m \nabla^2 T_m$$

where here κ_m is the thermal diffusivity. Now we need to specify boundary conditions on the interface between the two fluids. We assume that the interface can be described by a surface represented by

(12.1.4) $$z = h(x, y, t)$$

We account for both the Laplace surface pressure and the thermocapillary (Marangoni) stresses on the interface via the condition,

(12.1.5)
$$\left((p_1 - \rho_1 g h) - (p_2 - \rho_2 g h) - \gamma \left(\frac{1}{R_1} + \frac{1}{R_2} \right) \right) \mathbf{n}_i - \frac{\partial \gamma}{\partial x_i} = (\tau'_{1,ik} - \tau'_{2,ik}) \mathbf{n}_k$$

where

(12.1.6) $$\tau'_{m,ik} = \mu_m \left(\frac{\partial v_{m,i}}{\partial x_k} + \frac{\partial v_{m,k}}{\partial x_i} \right)$$

is the viscous stress tensor for the mth fluid. Here $\mathbf{n}_i$ is the normal vector directed into the first fluid, and $\partial \gamma / \partial x_i$ is the surface gradient of the surface tension. Note the first term on the left-hand side of (12.1.5) represents the Laplace pressure, while the second term represents the Margangoni stress. We can also write the balance conditions for the normal and tangential stresses separately by introducing orthogonal tangential vectors $\tau^{(1)}$ and $\tau^{(2)}$ (tangential to the interface between the two fluids),

(12.1.7) $$\left((p_1 - \rho_1 g h) - (p_2 - \rho_2 g h) - \gamma \left(\frac{1}{R_1} + \frac{1}{R_2} \right) \right) = (\tau'_{1,ik} - \tau'_{2,ik}) \mathbf{n}_i \mathbf{n}_k$$

(12.1.8) $$(\tau'_{1,ik} - \tau'_{2,ik}) \tau_i^{(l)} \mathbf{n}_k + \frac{\partial \gamma}{\partial x_i} \tau_i^{(l)} = 0, \; l = 1, 2$$

Equation (12.1.8) defines the Marangoni stresses. Furthermore, at the interface we also have fluid continuity, which implies that the velocities at the interface in both fluids are equal,

(12.1.9) $$\mathbf{v}_1 = \mathbf{v}_2$$

and that the interface surface moves according to the kinematic condition,

$$(12.1.10) \qquad \frac{\partial h}{\partial t} + v_{1,x}\frac{\partial h}{\partial x} + v_{1,y}\frac{\partial h}{\partial y} = v_{1,z}$$

where we defined this condition based on the velocity of the first fluid. We did this arbitrarily, as we could have just as well write this kinematic condition in terms of the velocities of the second fluid.

Finally we come to the thermal conditions at the interface. We consider the interface to be quite thin and so the temperature is continuous across the interface,

$$(12.1.11) \qquad T_1 = T_2$$

and the normal heat fluxes are equal, by conservation of energy,

$$(12.1.12) \qquad \left(\lambda_1\frac{\partial T_1}{\partial x_i} - \lambda_2\frac{\partial T_2}{\partial x_i}\right)\mathbf{n}_i = 0$$

where λ_m is the thermal conductivity of each fluid. In special circumstances, it is also necessary to take into account the energy spent into deformation of the interface, as well as dissipation due to additional 'surface viscosity". For more on these mechanisms see Nepomnyashchy et al. (2001) but we shall neglect these mechanisms here.

The above equations are derived for two fluids contacting each other at the interface (e.g. oil-water interface). However, we often find ourselves in situations where one of the fluids is a gas, and the viscosity of the gas is significantly less than that of the neighboring liquid. Recall that the dynamic viscosity of water is about 500 times that of air! Hence we expect that the motion of the gas has negligible impact on the motion of the liquid. We call such liquid-gas interface a "free surface". We also largely ignore the temperature field in the gas and employ a convection coefficient to describe the heat transfer to the gas phase. With these assumptions, the above equations for the liquid phase can be written as

$$(12.1.13) \qquad \nabla \cdot \mathbf{v} = 0$$

$$(12.1.14) \qquad \frac{\partial \mathbf{v}}{\partial t} + (\mathbf{v} \cdot \nabla)\,\mathbf{v} = -\frac{1}{\rho}\nabla p + \nu\nabla^2\mathbf{v} + g\beta T\mathbf{e}$$

$$(12.1.15) \qquad \frac{\partial T}{\partial t} + (\mathbf{v} \cdot \nabla)\,T = \kappa\nabla^2 T$$

For the boundary conditions, we neglect the stresses caused by the gas on the liquid, and therefore obtain for the liquid phase

$$(12.1.16) \qquad (p - \rho gh) - \gamma\left(\frac{1}{R_1} + \frac{1}{R_2}\right) = \tau'_{ik}\mathbf{n}_i\mathbf{n}_k$$

$$(12.1.17) \qquad (\tau'_{ik})\,\tau_i^{(l)}\mathbf{n}_k + \frac{\partial\gamma}{\partial x_i}\tau_i^{(l)} = 0,\ l = 1, 2$$

The interface boundary condition stating that there must be the same fluid velocity at the boundary in both phases is no longer applicable as we don't track the gas phase velocity. The kinematic boundary condition (12.1.10) however remains the same,

$$(12.1.18) \qquad \frac{\partial h}{\partial t} + v_x\frac{\partial h}{\partial x} + v_y\frac{\partial h}{\partial y} = v_z$$

For heat transfer at the boundary, we substitute the conductive boundary conditions from before with a convective boundary condition,

$$(12.1.19) \qquad \lambda \frac{\partial T}{\partial x_i} \mathbf{n}_i = K \left(T - T_g \right)$$

where K is the convective heat transfer coefficient (typically empirically determined) and T_g is the local temperature in the gas (assumed, often as the ambient temperature or again empirically determined).

Now we are ready to nondimensionalize these equations as it would help us solve specific Marangoni flow cases. For this we let a be a characteristic length scale in the problem and θ be the characteristic temperature difference. From this, we scale time with a^2/κ, the heat diffusion time; velocity with κ/a, the heat diffusion velocity; and pressure with $\mu\kappa/a^2$. Thus, we obtain the following dimensionless equations (and we immediately drop the asterisks in the dimensionless equations for convenience)

$$(12.1.20) \qquad \nabla \cdot \mathbf{v} = 0$$

$$(12.1.21) \qquad \frac{1}{Pr} \left(\frac{\partial \mathbf{v}}{\partial t} + (\mathbf{v} \cdot \nabla) \, \mathbf{v} \right) = -\nabla p + \nu \nabla^2 \mathbf{v} + (Ra) \, T\mathbf{e}$$

$$(12.1.22) \qquad \frac{\partial T}{\partial t} + (\mathbf{v} \cdot \nabla) \, T = \nabla^2 T$$

Here Pr is the Prandlt number,

$$(12.1.23) \qquad Pr = \nu/\kappa$$

a ratio of viscous diffusion rate to thermal diffusion rate, or diffusion of momentum to the diffusion of heat. Ra is the Rayleigh number,

$$(12.1.24) \qquad Ra = \frac{\beta g a^3 \theta}{\nu \kappa}$$

When Rayleigh number is below a critical value for a particular fluid, heat transfer is dominated by conduction, whereas if the Rayleigh number is above a critical value, heat transfer is dominated by convection. Rayleigh number is the product of Grashof number (the ratio of buoyancy to viscous forces) and Prandlt number. Before we nondimensionalize the boundary conditions, let's assume, for simplicity a linear dependence of surface tension on temperature:

$$(12.1.25) \qquad \gamma = \gamma_0 - \alpha T$$

With this we obtain the dimensionless boundary conditions as

$$(12.1.26) \qquad p - hGa_* - Ca_*^{-1} \left(1 - \delta_a T \right) \left(\frac{1}{R_1} + \frac{1}{R_2} \right) = \tau'_{ik} \mathbf{n}_i \mathbf{n}_k$$

$$(12.1.27) \qquad \left(\tau'_{ik} \right) \tau_i^{(l)} \mathbf{n}_k - Ma \frac{\partial T}{\partial x_i} \tau_i^{(l)} = 0, \; l = 1, 2$$

$$(12.1.28) \qquad \frac{\partial h}{\partial t} + v_x \frac{\partial h}{\partial x} + v_y \frac{\partial h}{\partial y} = v_z$$

$$(12.1.29) \qquad \frac{\partial T}{\partial x_i} \mathbf{n}_i = -Bi \left(T - \bar{T}_g \right)$$

where Ga_* is the modified Galileo number,

$$(12.1.30) \qquad Ga_* = \frac{ga^3}{\nu\kappa}$$

Note, traditionally, the Galileo number is defined as

$$(12.1.31) \qquad Ga = \frac{ga^3}{\nu^2}$$

and is the ratio of gravitational forces to viscous forces. $Ca_*{}^{-1}$ is the modified capillary number

$$(12.1.32) \qquad Ca_*{}^{-1} = \frac{\mu\left(\kappa/a\right)}{\gamma_0}$$

where κ/a is the characteristic velocity scale of the problem. Ma is the Marangoni number,

$$(12.1.33) \qquad Ma = \frac{\alpha a^2 \theta}{\mu\kappa}$$

Bi is the Biot number defined as usual as

$$(12.1.34) \qquad Bi = \frac{aK}{\lambda}$$

Furthermore,

$$(12.1.35) \qquad \delta_a = \alpha\theta/\gamma_0 = Ma/Ca_*$$

and

$$(12.1.36) \qquad \bar{T}_g = T_g/\theta$$

Note that the above equations are only valid when the changes in density of the liquid are small compared to the characteristic density of the fluid (i.e., $\beta\theta \ll 1$), since we have made the Boussinesq approximation at the beginning of the derivation. Since $\beta\theta = Ra/Ga_*$, if both the Rayleigh number and Galileo number are of the same order, i.e., $Ra/Ga_* = O(1)$ we cannot use this result. Thus when we are interested in problems where buoyancy driven flow is important, e.g., when $Ra = O(1)$, we should verify that $Ga_* \gg 1$. For example, for water at room temperature, for 100 μm characteristic length Ga_* is 100. However, since both Galileo and Rayleigh number have the same dependence on length, kinematic viscosity and thermal diffusivity, we just have to ensure that $\beta\theta \ll 1$. The volumetric thermal expansion coefficient for water around room temperature is $207 \times 10^{-6}\,\mathrm{K}^{-1}$. Thus the order 10 K temperature difference that we may see in a microfluidic device (e.g., a PCR chip) would yield a $\beta\theta$ of order 0.002. Hence these equations should be valid for most common microfluidic situations.

When Galileo number is somewhat large, and capillary number small (as is very usual in microflows), we can simplify the boundary condition (12.1.26) by neglecting terms of order unity and keeping large terms, to

$$(12.1.37) \qquad h - Bo^{-1}\left(1 - \delta_a T\right)\left(\frac{1}{R_1} + \frac{1}{R_2}\right) = const.$$

where Bo is the Bond number,

$$(12.1.38) \qquad Bo = \frac{\rho g a^2}{\gamma_0} = Ga_* Ca$$

Another useful scenario is when both $\beta\theta$ is small and the characteristic length is small such that the Rayleigh number is small as well. In this case, buoyancy driven convection is not important and thermocapillary (Marangoni) convection is dominant. Note that already for 10 μm characteristic length, Galileo number is 0.1 and Rayleigh number 0.0002. For microfluidic devices with characteristic length scales starting at 100 μm ($Ra = 0.2$) and lower buoyancy driven convection is typically not important. We can see that for microfluidic devices with characteristic length scales 10 μm or less buoyancy driven convection is even less important.

12.2. Mathematical formulation - solutocapillary flows

In the previous section, we obtained a formulation for the general case of temperature gradient driven Marangoni flows. In this section, we will obtain a similar set of equations and boundary conditions now considering a gradient in surfactant driving a gradient in surface tension. We will consider two cases: one where we ignore the details of surface absorption-desorption, and one where we consider it.

We begin with a simpler case: a system where the interfacial absorption-desorption kinetics are so fast (compared to the fluid motion time scales in the system) that the interfacial concentration of surfactant Γ is immediately determined by the instantaneous volume (ordinary) concentration of the surfactant C near the interface. Notice that this is similar to the heat transfer situation we considered above - the temperature at the interface was determined by that in the nearby bulk. Thus, in this case, transport of surfactant and its influence on surface tension is mathematically the same as the transfer of heat (temperature) and its influence on buoyancy and surface tension at the interface. Thus, we can write our transport equations similarly,

$$(12.2.1) \qquad \nabla \cdot \mathbf{v}_m = 0$$

$$(12.2.2) \qquad \frac{\partial \mathbf{v}_m}{\partial t} + (\mathbf{v}_m \cdot \nabla)\,\mathbf{v}_m = -\frac{1}{\rho_m}\nabla p_m + \nu_m \nabla^2 \mathbf{v}_m + g\beta_{c,m} C_m \mathbf{e}$$

where

$$(12.2.3) \qquad \beta_c = -\frac{1}{\rho}\left(\frac{\partial \rho}{\partial C}\right)$$

Instead of the energy conservation equation, we have the mass (chemical species) conservation equation, which is mathematically similar,

$$(12.2.4) \qquad \frac{\partial C_m}{\partial t} + (\mathbf{v}_m \cdot \nabla)\,C_m = D_m \nabla^2 C_m$$

where D_m is the diffusion coefficient of the surfactant in the particular fluid. The boundary conditions are once again similar to the heat transfer ones, with the exception that α is replaced with

$$(12.2.5) \qquad \alpha_c = -\frac{\partial \gamma}{\partial C}$$

and that the concentration of surfactants on the interface of the two fluids is not equal but is instead proportional, that is,

$$(12.2.6) \qquad C_2 = kC_1$$

Note that $\partial\gamma/\partial C$ can be either positive or negative, while most commonly $\partial\gamma/\partial T$ is negative (although positive $\partial\gamma/\partial T$ does occur, and is knows as anomalous thermocapillarity). For liquid-gas interfaces, typically $\partial\gamma/\partial C$ is negative.

Now let's consider the case where the kinetics of adsorption and desorption at the interface are important. First we need to take into account the surface advection and diffusion of the surfactant at the interface. For this we write the surface advection and diffusion equation,

$$(12.2.7) \qquad \frac{\partial \Gamma}{\partial t} + \nabla_s \cdot (\Gamma \mathbf{v}_s - D_s \nabla_s \Gamma) = j$$

The subscripts s emphasize that these are surface quantities. E.g., D_s is the surface diffusivity of the surfactant, not the bulk diffusivity of the surfactant. j is the flux of the substance from the bulk to the interface. Next, we consider the force balance on the interface, boundary condition (12.1.5). For this case, we also consider the mass of the boundary, that is we consider that a chunk of interface of unit area has a mass $m\Gamma$ where m is the mass of the surfactant molecule. Thus, our equivalent boundary condition in this case is

$$(12.2.8) \quad m\frac{d}{dt}(\Gamma v_{1,i}) = -\left((p_1 - \rho_1 gh) - (p_2 - \rho_2 gh) - \gamma\left(\frac{1}{R_1} + \frac{1}{R_2}\right) \right)\mathbf{n}_i$$
$$+ \frac{\partial \gamma}{\partial x_i} + (\tau'_{1,ik} - \tau'_{2,ik})\mathbf{n}_k$$

where here d/dt denotes the Lagrangian (material) derivative. However, often the mass of the interface is relatively small or we can assume that the motion is quasi steady, and so we can neglect this term (left-hand side of (12.2.8)) and decouple (12.2.8) from (12.2.7).

Let's now consider the source term, j in (12.2.7). If the surfactant can be considered insoluble in one of the fluids, then there should not be any of it there and so there should not be any surfactant coming or going into that fluid. Thus, for that fluid, $j = 0$. If the surfactant is insoluble in both fluids, $j = 0$ in (12.2.7). If the surfactant is soluble in one or both fluids, then the interface flux is determined by the diffusion and advection of the surfactant to the surface. For simplicity sake, to avoid coupling the transport of surfactant to (12.2.2) we can assume that the transport takes place mainly by diffusion, and write

$$(12.2.9) \qquad j = -D\mathbf{n} \cdot \nabla C$$

The flux into the interface is also determined by the adsorption-desorption kinetics of the surfactant at the interface. Thus

$$(12.2.10) \qquad j = k_a C - k_d \Gamma$$

where C is the concentration of surfactant near the interface. The processes of transport and adsorption-desorption are serial to one another, and so we take j in (12.2.7) to be that of the slowest (rate determining) step. When the adsorption-desorption kinetics are fast compared to the time scales under consideration, we can assume that this process is always in equilibrium, that is $j = 0$. From (12.2.10) this implies that

$$(12.2.11) \qquad \Gamma = (k_a/k_d)\, C$$

Relation (12.2.11) is known as the ideal Gibbs adsorption isotherm, and is valid for very dilute solutions where interactions between molecules and effect of "sites" that can occupied by molecules on the interface are not felt. When surface concentration is high, the surfactant adsorption follows Langmuir adsorption isotherm, and when the surfactant concentration is very high it may even follow Hill adsorption isotherm (or other adsorption models). Here we will just consider Langmuir adsorption. In Langmuir adsorption the adsorption process follows

$$(12.2.12) \qquad j = k_a \left(1 - \Gamma/\Gamma_\infty\right) C - k_d \Gamma$$

where Γ_∞ represents the total number of available sites at the interface for the surfactant. Notice that the forward, adsorption kinetics are proportional to both the concentration of surfactant in the bulk and the fraction of sites for the surfactant currently available. Since we have a lot of surfactant, we have to now account for the space, (i.e. "sites") that they take up. Notice that when the surfactant amount is low, so even at equilibrium we expect $\Gamma \ll \Gamma_\infty$ and so in this case (12.2.12) collapses to the simpler case (12.2.10). When the adsorption-desorption kinetics are fast compared to the time scales under consideration, we can assume that this process is always in equilibrium, that is $j = 0$. From (12.2.12) this implies that

$$(12.2.13) \qquad \Gamma = \Gamma_\infty \frac{C}{k_d/k_a + C}$$

To predict the effect of surfactant adsorption on surface tension in this case, we can use the Szyszkowski equation

$$(12.2.14) \qquad \gamma - \gamma_0 = RT\Gamma_\infty \ln\left(1 + \frac{C}{k_d/k_a}\right)$$

See Meissner and Michaels (1949) for more details on this equation as well as a table of empirically derived k_d/k_a for common carboxylic acids, alcohols and esters in water. From (12.2.14) and (12.2.13) we can obtain that

$$(12.2.15) \qquad \Gamma = \left(-\frac{1}{RT}\frac{d\gamma}{dC}\right) C$$

and use this as a basis for finding concentration driven Marangoni convection.

12.3. Simple Marangoni flows

In this section we will use the mathematical formulation we just obtained and examine several simple situations involving surface tension driven flows.

12.3.1. Plane Marangoni flow. Consider a system shown in Figure 12.3.1. This system consists of a layer of fluid above another in a container with walls being very far away from the center. The system is being heated such that there is a linear temperature gradient setup in the system along the x direction, such that at the interface the temperature can be written as

$$(12.3.1) \qquad T_0 = Ax$$

where A is just a proportionality constant. We are concerned about the flow in a region in the center, which is far away from the walls and so we assume that the domain is infinite in both directions (in both x and y). Both layers are order a deep and we shall use this as the characteristic length scale of the problem. We use the Aa as the characteristic temperature scale in the problem and use this for nondimensionalization of temperature. For the rest of the quantities we nondimensionalize them as we did in Section 12.1, as we will be using the equations (12.1.20) -

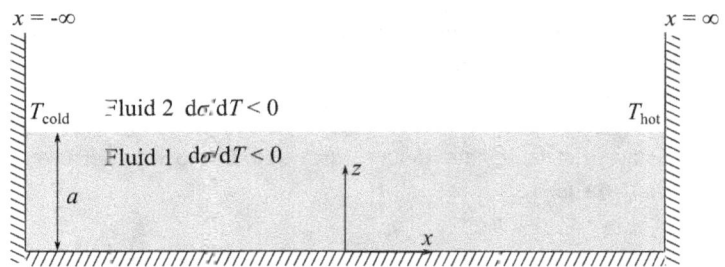

FIGURE 12.3.1. Schematic of a container containing two fluids in which we examine plane Marangoni flow. We are interested in the surface tension driven flow near $x = 0$, $y = 0$. The walls of the container are far way in both directions and are located at $x = \pm\infty$ and $y = \pm\infty$. The fact that the container has walls implies that the flow rate in each fluid across the entire depth of that fluid must sum up to zero - the fluid has nowhere to go. This implies that the flow must turn, and so at the walls there is finite velocity in the z direction. However, we assume near the center the flow in the z direction is negligible.

(12.1.22) from this section to solve this problem. The quantities we write below are nondimensionalized and we have dropped the asterisk for convenience. We assume that the boundary between the two fluids is flat (see Figure 12.3.1) and therefore we neglect any deformation of this boundary. Thus we can neglect the boundary conditions accounting for boundary curvature and deformation, (12.1.26) and (12.1.28). For boundary condition, (12.1.27), which we rewrite here for convenience,

$$(12.3.2) \qquad (\tau'_{ik}) \, \tau_i^{(l)} \mathbf{n}_k - Ma \frac{\partial T}{\partial x_i} \tau_i^{(l)} = 0, \ l = 1, 2$$

we assume that the temperature of the interface is fixed and is given by

$$(12.3.3) \qquad\qquad T = x$$

thus, this boundary condition becoming

$$(12.3.4) \qquad (\tau'_{ik}) \, \tau_i^{(l)} \mathbf{n}_k - Ma\tau_i^{(l)} = 0, \ l = 1, 2$$

This is the condition that drives the Marangoni flow.

The bottom of the chamber ($z = 0$) is rigid and we impose a no-slip and no-penetration boundary condition there,

$$(12.3.5) \qquad\qquad \mathbf{v} = 0$$

We also need to prescribe a boundary condition for temperature here. We assume that the lateral walls are heated uniformly along their length and so we can say that the same temperature gradient that occurs at the interface is also applicable at the container bottom,

$$(12.3.6) \qquad\qquad T = x$$

Furthermore, we can say that the container bottom in insulating, and thus that there is no heat flux across it,

$$(12.3.7) \qquad\qquad \frac{\partial T}{\partial z} = 0$$

The fact that the container has walls implies that the flow rate in each fluid across the entire depth of that fluid must sum up to zero - the fluid has nowhere to go. This implies that the flow must turn, and at the walls there is finite velocity in the z direction. However, we assume near the center the flow is only in the x direction. Furthermore, we assume that the thermal gradient has been there for a long time and that the situation is stationary, thus

$$(12.3.8) \qquad \mathbf{v} = (u(z), 0, 0)$$

There is also some convective heat transfer, which is inhomogeneous with respect to the vertical coordinate, z. This convective heat transfer generates a vertical temperature profile which we account for as

$$(12.3.9) \qquad T(x, z) = x + \tau(z)$$

Note that we assume that the temperature is independent of the y coordinate. The pressure is also independent of the y coordinate,

$$(12.3.10) \qquad p = p(x, z)$$

Substituting these simplifications, (12.3.8) - (12.3.10) into the continuity, momentum, and energy equations (12.1.20) - (12.1.22), we obtain

$$(12.3.11) \qquad 0 = -\frac{\partial p}{\partial z} + (x + \tau(z))\, Ra$$

$$(12.3.12) \qquad 0 = -\frac{\partial p}{\partial x} + \frac{d^2 u}{dz^2}$$

$$(12.3.13) \qquad u = \frac{d^2 \tau}{dz^2}$$

Substituting the simplifications into the boundary conditions, we obtain

$$(12.3.14) \qquad \frac{du}{dz} + Ma = 0 \text{ at } z = 1$$

$$(12.3.15) \qquad u = 0 \text{ at } z = 0$$

Recall that $z = 1$ is the location of the interface between the two fluids, while $z = 0$ is the bottom of the chamber. Finally, since the chamber has lateral walls and the fluid has nowhere to go laterally, we use the condition of zero horizontal flux of the fluid (applied in the lower fluid),

$$(12.3.16) \qquad \int_0^1 u(z)\, dz = 0$$

Since we have fixed the temperature on the fluid interface, we can assume

$$(12.3.17) \qquad \tau = 0 \text{ at } z = 1$$

and because of insulating boundary condition on the bottom of the chamber, (12.3.7), we obtain

$$(12.3.18) \qquad \frac{d\tau}{dz} = 0 \text{ at } z = 0$$

Note, that if instead of the insulating boundary condition we chose to keep the temperature of the bottom constant, then it that case, we would have had $\tau = 0$ at the bottom.

Now we can solve (12.3.11), and obtain

$$(12.3.19) \qquad p = xzRa + \int_0^z \tau(\zeta)d\zeta + p_0(x)$$

where p_0 is an unknown function, but is only a function of x. Substituting this result into (12.3.12), we obtain

$$(12.3.20) \qquad \frac{d^2u}{dz^2} = zRa + \frac{dp_0}{dx}$$

Observe that since u depends only on z, (12.3.20) implies that dp_0/dx should be a constant. Notice if dp_0/dx had dependence on x, then by (12.3.20) u would have a dependence on x, and we have previously stated that that is not the case. Thus dp_0/dx should be a constant

$$(12.3.21) \qquad dp_0/dx = C$$

Please do not be confused with the notation, in this problem C means constant, not the concentration of surfactant as there is no surfactant (or its gradient) in this problem. Thus,

$$(12.3.22) \qquad p_0(x) = Cx + C_1$$

and

$$(12.3.23) \qquad p = xzRa + \int_0^z \tau(\zeta)d\zeta + Cx + C_1$$

Next we integrate (12.3.20) using the boundary conditions (12.3.14) and (12.3.15) and obtain

$$(12.3.24) \qquad u = \left(\frac{z^3}{6} - \frac{z}{2}\right)Ra - zMa + C\left(\frac{z^2}{2} - z\right)$$

Finally to eliminate the constant C we employ the condition that there is zero horizontal flux, (12.3.16), which gives us

$$(12.3.25) \qquad C = -\left(\frac{5}{8}Ra + \frac{3}{2}Ma\right)$$

which we substitute into (12.3.24) to obtain

$$(12.3.26) \qquad u = \left(\frac{z^3}{6} - \frac{5z^2}{16} + \frac{z}{8}\right)Ra + \left(-\frac{3z^2}{4} - \frac{z}{2}\right)Ma$$

This is the (dimensionless) velocity profile for our surface flow near the center of the chamber. The first term on the right-hand side represents buoyancy driven flow, while second term represents surface tension gradient driven flow. Their relative strengths are determined by the magnitudes of the Rayleigh and the Marangoni numbers. We plot the resulting velocity profile in Figure 12.3.2.

Now we can also find the temperature profile in such a flow. To do this we integrate (12.3.13) subject to boundary condition at the interface (12.3.17), and either the insulating bottom boundary condition (12.3.18) or a constant temperature

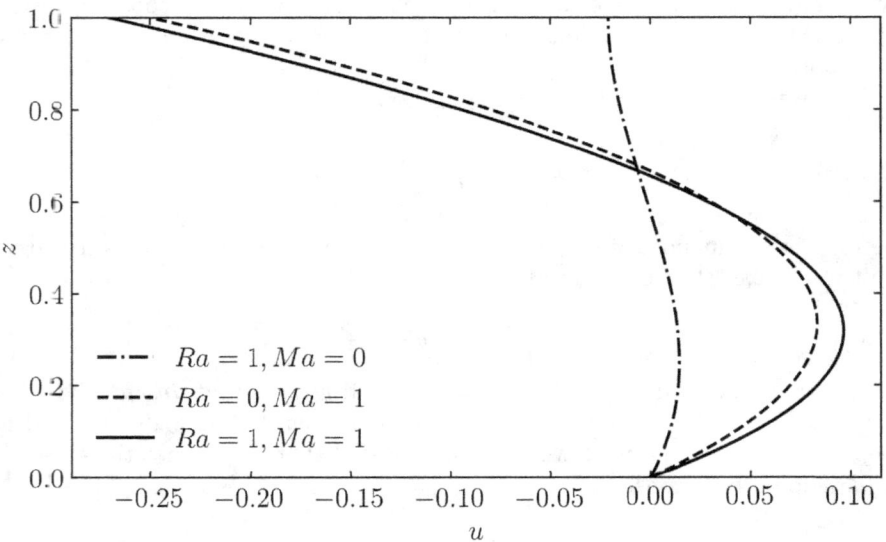

FIGURE 12.3.2. Dimensionless velocity profile in the bottom liquid for the plane Marangoni flow. Near the interface the flow is from the hot side to the cold side, and so is negative (in the negative x-direction). Near the bottom of the chamber is the return flow, which flows towards the positive x-direction. Note that when Rayleigh number and Marangoni number are both unity, the contribution of the surface tension gradient driven flow dominates the contribution from the buoyancy flow.

bottom condition, where $\tau = 0$ at the bottom. Before we proceed with the integration, let's observe that because we have a zero horizontal flux of fluid, (12.3.13) substituted into (12.3.16) implies

$$(12.3.27) \qquad \left. \frac{d\tau}{dz} \right|_{z=1} = \left. \frac{d\tau}{dz} \right|_{z=0}$$

Thus we obtain for the insulating bottom boundary condition that

$$(12.3.28) \quad \tau = \left(\frac{1}{120}z^5 - \frac{5}{192}z^4 + \frac{1}{48}z^3 - \frac{1}{320} \right) Ra - \left(\frac{1}{16}z^4 - \frac{1}{12}z^3 + \frac{1}{48} \right) Ma$$

and in the case of constant temperature bottom boundary condition that

$$(12.3.29) \quad \tau = \left(\frac{1}{120}z^5 - \frac{5}{192}z^4 + \frac{1}{48}z^3 - \frac{1}{320}z \right) Ra - \left(\frac{1}{16}z^4 - \frac{1}{12}z^3 + \frac{1}{48}z \right) Ma$$

Having found a solution above for the chamber that has lateral walls in the x direction, and so zero horizontal flux, it turns out it is also easy to find a solution where there are no walls in the x direction and so we do not have a zero horizontal flux of fluid. In this case, instead of (12.3.16) we have

$$(12.3.30) \qquad\qquad p = 0 \text{ at } z = 1$$

To solve (12.3.23) now we substitute into it (12.3.30), and obtain that

(12.3.31) $$C = -Ra$$

Thus from (12.3.24) we obtain

(12.3.32) $$u = \left(\frac{z^3}{6} - \frac{z^2}{2} + \frac{z}{2}\right)Ra - zMa$$

We can find the temperature distribution in this case as well. For the insulating bottom boundary condition, we find

(12.3.33) $$\tau = \left(\frac{1}{120}z^5 - \frac{1}{24}z^4 + \frac{1}{12}z^3 - \frac{1}{120}\right)Ra + \left(-\frac{1}{6}z^3 + \frac{1}{6}\right)Ma$$

and for the constant temperature bottom boundary condition,

(12.3.34) $$\tau = \left(\frac{1}{120}z^5 - \frac{1}{24}z^4 + \frac{1}{12}z^3 - \frac{1}{120}z\right)Ra + \left(-\frac{1}{6}z^3 + \frac{1}{6}z\right)Ma$$

Additionally, we can similarly find that for the temperature gradient boundary condition at the bottom of the chamber, where the temperature gradient is $T = x$, the temperature field is

(12.3.35) $$\tau = \left(\frac{1}{120}z^5 - \frac{1}{24}z^4 + \frac{1}{12}z^3 - \frac{1}{8}z\right)Ra + \left(-\frac{1}{6}z^3 + \frac{1}{2}z\right)Ma$$

12.3.2. Marangoni flow in a corner. Having studied the Marangoni flow near the center of the container, let's consider the Marangoni flow in a corner. We move the coordinate system so that the rigid lateral wall that forms the corner is now at $x = 0$. The fluid-fluid interface is located now at $z = h(x)$. However, as before we assume that this interface is flat and non-deformable. Now we also assume that the angle between this surface and the rigid lateral wall is equal to the contact angle θ_c. (This means we are considering the situation very near, i.e., less than a capillary length away from the rigid wall.) Furthermore, we assume that the Reynolds number of the flow is low, so that we can use the Stokes approximation for the momentum equation. Lastly, we assume the flow is steady. Thus, we write the Stokes approximation to the momentum equation as,

(12.3.36) $$\nabla^2\mathbf{v} - \nabla p = 0$$

We also assume that the flow is two dimensional (i.e. there is no variation in the y direction). Taking advantage of this, we can eliminate the pressure by taking the curl of the momentum equation (12.3.36) and by introducing a stream function, ψ, such that

(12.3.37) $$v_x = \partial\psi/\partial z$$
$$v_z = -\partial\psi/\partial x$$

This gives us the biharmonic equation in ψ,

(12.3.38) $$\nabla^4\psi = 0$$

Because the interface makes an arbitrary angle with the wall, it is more convenient to work in the polar coordinate system. Recall that in the vicinity of the corner point we assume that the surface is flat and that it makes an angle θ_c (contact angle) with the rigid lateral wall. In this coordinate system the boundary conditions are:

(12.3.39)
$$\psi = 0 \text{ at } \theta = 0$$
$$\frac{\partial \psi}{\partial \theta} = 0 \text{ at } \theta = 0$$

(12.3.40)
$$\psi = 0 \text{ at } \theta = \theta_c$$
$$\frac{1}{r^2}\frac{\partial^2 \psi}{\partial \theta^2} + Ma\frac{\partial T}{\partial r} = 0$$

The solution to this problem is given by Shevtsova et al. (1996). They found that

(12.3.41)
$$\psi = r^2(A\cos 2\theta + B\sin 2\theta + C\theta + D)$$

where

(12.3.42)
$$A = \frac{1}{4}Ma\frac{2\theta_c - \sin 2\theta_c}{\sin 2\theta_c - 2\theta_c \cos 2\theta_c}\left(\frac{\partial T}{\partial r}\right)_c$$

(12.3.43)
$$B = \frac{1}{4}Ma\frac{1 - \cos 2\theta_c}{\sin 2\theta_c - 2\theta_c \cos 2\theta_c}\left(\frac{\partial T}{\partial r}\right)_c$$

(12.3.44)
$$C = -2B$$

(12.3.45)
$$D - A$$

and where $(\partial T/\partial r)_c$ is the derivative of temperature calculated near the corner point and assumed to be constant.

12.3.3. Marangoni advection of spherical drops and bubbles. Another fundamentally important Marangoni flow problem is the motion of a drop or bubble due to the motion of its surface, specifically against gravity. The motion of the drop's surface here is driven by surface tension which arises due to temperature gradients (although surfactant gradient is also possible) and the resulting surface motion gives the drop an additional lift force. This lift is not to be confused with the Magnus lift of a spinning object. While both lift forces are superficially similar - the lift is generated by the motion of the object's surface, how the lift is generated is fundamentally different. In studying this problem, we follow the approach of Young et al. (1959). We begin with the steady Navier-Stokes momentum equation, simplified using the Stokes (creeping) flow assumption,

(12.3.46)
$$\mu_m \nabla^2 \mathbf{v}_m = \nabla(p_m + \rho g_m z)$$

and a continuity equation, assuming incompressible flow,

(12.3.47)
$$\nabla \cdot \mathbf{v}_m = 0$$

Here the subscript $m = 1$ refers to the outside of the bubble, and $m = 2$ to the inside of the bubble. As before, the temperature distribution inside and outside of the bubble is governed by heat convection-conduction equation,

(12.3.48)
$$\frac{\partial T_m}{\partial t} + (\mathbf{v}_m \cdot \nabla)T_m = \lambda_m \nabla^2 T_m$$

We again assume that the situation is steady and drop the first term on the left-hand side. Furthermore, we assume that the flow is slow enough, such that conduction dominates convection and so (12.3.48) simplifies to Laplace's equation,

$$(12.3.49) \qquad \nabla^2 T_m = 0$$

We assume that our flow is axisymmetric and the bubble is spherical. Furthermore, these equations must satisfy the basic boundary conditions, firstly, that the solutions are regular at the origin and that far away from the spherical bubble,

$$(12.3.50) \qquad \mathbf{v}_1 \to (0, 0, v_0) \text{ as } |\mathbf{r}| \to \infty$$

$$(12.3.51) \qquad p_1 \to -\rho_1 g z \text{ as } |\mathbf{r}| \to \infty$$

$$(12.3.52) \qquad T_1 \to T_0 + C z \text{ as } |\mathbf{r}| \to \infty$$

and that there is no-penetration condition at the bubble surface,

$$(12.3.53) \qquad u_{r,1} = u_{r,2} = 0 \text{ at } r = R$$

Here T_0 and C are constants describing the vertical temperature gradient in the system. The constant C represents the global vertical temperature gradient, $C = -dT/dz|_\infty$. We use the solution to the continuity and momentum equation from Rybczynski (1911) and Hadamard (1911b) (see Section 1.3.2). The radial and transverse components of the velocity, $u_{r,m}$ and $u_{\theta,m}$, as well as the pressure distribution in each of the phases that satisfy the above boundary conditions are

$$(12.3.54) \qquad u_{r,1} = \left((A_1/\mu_1) \left(r^{-1} - R^2 r^{-3} \right) + v_0 \left(1 - R^3 r^{-3} \right) \right) \cos\theta$$

$$(12.3.55) \qquad u_{\theta,1} = -\left((A_1/\mu_1) \left(r^{-1} - R^2 r^{-3} \right) + v_0 \left(1 + R^3 r^{-3}/2 \right) \right) \sin\theta$$

$$(12.3.56) \qquad u_{r,2} = (A_2/10\mu_2) \left(r^2 - R^2 \right) \cos\theta$$

$$(12.3.57) \qquad u_{\theta,2} = -(A_2/10\mu_2) \left(2r^2 - R^2 \right) \sin\theta$$

$$(12.3.58) \qquad p_1 = \left(A_1/r^2 \right) \cos\theta - \rho_1 g r \cos\theta$$

$$(12.3.59) \qquad p_2 = A_2 r \cos\theta - \rho_2 g r \cos\theta + B_2$$

where R is the bubble radius, and A_1, A_2, B_2 are constants to be found from evaluating further boundary conditions; v_0 is the vertical rising velocity of the bubble, also to be found later. The solutions to Laplace's equation in spherical coordinates are,

$$(12.3.60) \qquad T_1 = T_0 + C \left(r + k_1/r^2 \right) \cos\theta$$

$$(12.3.61) \qquad T_2 = T_0 + C k_2 r \cos\theta$$

where k_1, k_2 are constants to be determined from further the boundary conditions. These boundary conditions include (a), the continuity of transverse velocity at the bubble's surface,

$$(12.3.62) \qquad u_{\theta,1} = u_{\theta,2} \text{ at } r = R$$

(b) continuity of shear stress at the bubble's surface, (including the effects of surface tension gradient),

$$(12.3.63) \qquad \mu_2 \left(\frac{\partial u_{\theta,2}}{\partial r} - \frac{u_{\theta,2}}{r} \right) - \mu_1 \left(\frac{\partial u_{\theta,1}}{\partial r} - \frac{u_{\theta,1}}{r} \right) - \frac{1}{r} \frac{\partial \gamma}{\partial \theta} = 0 \text{ at } r = R$$

(c) continuity of normal stress at the bubble's surface,

$$(12.3.64) \qquad p_1 - p_2 + 2\mu_1 \frac{\partial u_{r,1}}{\partial r} - 2\mu_2 \frac{\partial u_{r,2}}{\partial r} - \frac{2\gamma}{R} = 0 \text{ at } r = R$$

(d) continuity of temperature at the bubble's surfaces,

$$(12.3.65) \qquad T_1 = T_1 \text{ at } r = R$$

and finally (e) the continuity of heat flux at the bubble's surface,

$$(12.3.66) \qquad \lambda_1 \frac{\partial T_1}{\partial r} - \lambda_2 \frac{\partial T}{\partial r} = 0 \text{ at } r = R$$

Substituting (12.3.54) through (12.3.61) into these boundary conditions, we eliminate the constants A_1, A_2, B_2 and k_1, k_2 find the vertical rising velocity of the bubble
(12.3.67)

$$v_0 = \frac{2\mu_1}{3} \left(\frac{1}{3\mu_1 + 2\mu_2} \right) \left((\mu_1 + \mu_2)(\rho_2 - \rho_1) gR^2 - \frac{3\mu_1 R}{2 + \lambda_2/\lambda_1} \frac{d\gamma}{dT} \frac{dT}{dz} \Big|_\infty \right)$$

where $d\gamma/dT$ is the temperature coefficient of surface tension. Note that the bubble has a constant velocity when the lift due to the Marangoni flow at the bubble's interface is balanced by the gravitational force and the drag force,

$$(12.3.68) \qquad F_\gamma + F_g + F_D = 0$$

The gravitational force on the bubble is

$$(12.3.69) \qquad F_g = \frac{4}{3} \pi R^3 g (\rho_2 - \rho_1)$$

and the drag force on the bubble from Rybczynski (1911) and Hadamard (1911b) is

$$(12.3.70) \qquad F_D = 4\pi \frac{1 + 3(\mu_2/\mu_1)/2}{1 + \mu_2/\mu_1} \mu_1 r v_0$$

Combining (12.3.67) through (12.3.70), we find that the Marangoni force on the bubble is then

$$(12.3.71) \qquad F_\gamma = -\frac{4\pi R^2}{(1 + \mu_2/\mu_1)(1 + \lambda_2/\lambda_1)} \frac{d\gamma}{dT} \frac{dT}{dz} \Big|_\infty$$

Typically, surface tension varies with temperature in a linear fashion according to

$$(12.3.72) \qquad \gamma = \gamma_0 + (d\gamma/dT)_0 (T - T_0)$$

where T_0 is some reference temperature. For this case, Marangoni flows at the drop or bubble surface force these particles to move from the cold region of the surrounding fluid to the hot region. However, there are cases where the surface tension varies quadratically with temperature,

$$(12.3.73) \qquad \gamma = \gamma_0 + k(T - T_0)^2$$

For example, Cini et al. (1972) found that even for water,

$$(12.3.74) \qquad \gamma = 75.668 - 0.1396T - 0.2885 \times 10^{-3} T^2$$

where for this formula T is in degrees Celsius, and γ is in mN/m. The quadratic case is particularly interesting as we can obtain bubble or drop focusing at a certain position in a fluid where $T = T_0$. Notice that at $T = T_0$, $(d\gamma/dT)_0 = 0$. Thus these particles are driven from both the cold side and the hot side to this location, at which they experience zero driving force and therefore stop. Generating such surface tension vs. temperature relationship allows us to thus concentrate droplets or bubbles.

12.4. Tears of Wine

Let now us turn to the most famous example of Marangoni flow: tears of wine. This phenomena consists of wine rising up the walls of a wine glass due to capillary action. The ethanol evaporates out of the wine film (especially in the region distant from the rest of the wine) creating a gradient in ethanol concentration and so in the surface tension of the liquid. This gradient causes Marangoni flow that bring up more liquid up the walls of the wine glass. This forms a ring of liquid at the edge of the film, which breaks up into droplets via the Rayleigh-Plateau instability. These droplets then fall down into the bulk wine under their own weight. The latter flow resembles tears, hence the name "tears of wine". The phenomena has been known since antiquity (Proverbs 23:31, Coogan et al. (2010)), and interestingly, the brief description of the phenomena in the Proverbs may be interpreted to suggest that it was even known that this phenomena is more vigorous when the alcohol content of the wine is higher. For wine (which has a typical maximum alcohol content of 15% v/v) this is indeed true.

In this section we will examine the Marangoni flow part of this phenomena. In our analysis, we will follow the hydrodynamic model of Venerus and Simavilla (2015). To begin, consider a solution film (e.g. ethanol-water mixture, wine) on a solid impermeable inclined plane (e.g., wall of a wine glass, or for easier modeling, martini glass) as shown in Figure 12.4.1. The film has a thickness $\delta(z)$ and the plane is inclined so that the angle with the gravity vector is β (as shown). We assume that the solution has a constant density ρ and viscosity μ. We consider that the inclined plane is infinite in the y direction (into the page), making our problem 2D. (We could have similarly, and more realistically modeled the problem in cylindrical coordinates, where the y direction would be the θ direction, and realistically assumed that the problem is axisymmetric, obtaining a 2D problem. However, to keep the analysis simpler we stick with the Cartesian coordinate system). Lastly, we assume that the solution is incompressible and that the problem is quasi-steady. Thus, the mass continuity equation for the liquid simplifies to

$$(12.4.1) \qquad \frac{\partial v_x}{\partial x} + \frac{\partial v_z}{\partial z} = 0$$

We note that the liquid film is observed to be thin, $\delta_0 \sim 30$ µm, travels quite a bit up the glass wall ($h \sim 10$ mm) and travels slowly. Thus we can assume that the typical Reynolds number is small and so we can use the Stokes approximation for the Navier-Stokes momentum equation. Furthermore, since δ_0/h is small, we can use the lubrication approximation, and so simplify the momentum equation to

$$(12.4.2) \qquad \mu \frac{\partial^2 v_z}{\partial x^2} = \frac{\partial p}{\partial z} + \rho g \cos \beta$$

At the surface of the inclined plane, we have the no-slip condition,

$$(12.4.3) \qquad v_x(0, z) = 0$$

and the no-penetration condition,

$$(12.4.4) \qquad v_z(0, z) = 0$$

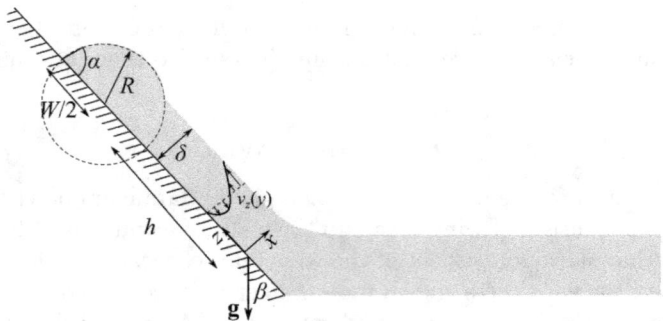

FIGURE 12.4.1. Schematic of an idealized tears of wine phenomena during the phase where an ethanol water mixture is climbing up a solid, impermeable inclined plane. The thickness of the liquid film is greatly exaggerated for clarity. Typically the film thickness, δ, is order 30 μm while the distance the film travels, h, is order 10 mm

At the liquid-gas interface, $x = \delta$, we have a momentum balance, as in this model we take into account the evaporation of the solution,

$$(12.4.5) \qquad \rho \left(v_x \left(\delta, z \right) - \frac{\partial \delta}{\partial z} v_z \left(\delta, z \right) \right) \approx Q_{evap}$$

where Q_{evap} is the mass flux into the gas phase due to the evaporation (for which we will introduce a model later). Additionally, we approximate the pressure jump across the interface due to capillary pressure using a simplified Laplace's equation,

$$(12.4.6) \qquad p \left(z \right) \approx \gamma \frac{\partial^2 \delta}{\partial z^2}$$

where we simplify Laplace's equation via the low slope approximation. Lastly, we account for the Marangoni stresses at the interface by considering the shear stress at the interface,

$$(12.4.7) \qquad \mu \frac{\partial v_z}{\partial x} \left(\delta, z \right) \approx \frac{\partial \gamma}{\partial z}$$

With these relations we can already obtain the expression for an equilibrium rise of a film on a practically vertical inclined plane. Since the film is at equilibrium we assume that flow is absent, the inclined plane is practically vertical ($\beta \ll 1$), we can combine (12.4.2) with (12.4.6) and (12.4.7) and obtain

$$(12.4.8) \qquad h_{eq} = \lambda_c \sqrt{2 \left(1 - \sin \alpha \right)}$$

where α is the contact angle defined in Figure 12.4.1, and $\lambda_c = \sqrt{\gamma/\rho g}$ is the capillary length. For ethanol water mixtures characteristic of wine, surface tension is around 55 mN/m, giving a capillary length of 2.4 mm. Ethanol-water mixtures typically wet clean glass well, and we can take the contact angle to be about 15°. This gives an equilibrium capillary rise of 3 mm, a correct order of magnitude.

Let's now obtain an expression for the characteristic fluid velocity in the thin film. In the film away from the ridge at the edge of the firm, the change in capillary pressure is much less than the change in gravitational pressure,

$$(12.4.9) \qquad \frac{\partial p}{\partial z} = \frac{\partial}{\partial z}\left(\gamma \frac{\partial^2 \delta}{\partial z^2}\right) \ll \rho g \cos \beta$$

Applying this condition to (12.4.2) and then integrating the result using the boundary conditions (12.4.4) and (12.4.7), we obtain

$$(12.4.10) \qquad v_z = \frac{\rho g \cos \beta \delta^2}{2\mu}\left(\left(\frac{x}{\delta}\right)^2 - 2\frac{x}{\delta}\right) - \frac{\delta}{\mu}\frac{\partial \gamma}{\partial z}\frac{x}{\delta}$$

We see that the first term, controlled by gravity trying to pull the liquid down, competes with the second term, the Marangoni stress trying to pull the liquid up. From (12.4.10) we see that for Marangoni forces to dominate over the gravitational forces,

$$(12.4.11) \qquad \frac{\partial \gamma}{\partial z} > \frac{\rho g \delta \cos \beta}{2}$$

For a vertical plate and typical film thickness, $\rho g \delta \cos \beta / 2$ is order 100 mPa. Typical temperature coefficient of surface tension of water, $\partial \gamma / \partial T$ is 0.14 mN/(m $\cdot$ K) (Cini et al. (1972)). Hence if this flow was driven purely by a temperature gradient effect (which it is not), it would need a temperature gradient of order 700 K/m or 0.7 K/mm to overcome the force of gravity. Such gradient is fairly simple to set up in typical microfluidic situations. Realistically, for this problem surface tension is both a function of temperature, and ethanol composition (mass fraction), w. Thus we can write the surface tension gradient as,

$$(12.4.12) \qquad \frac{\partial \gamma}{\partial z} = \frac{\partial \gamma}{\partial T}\frac{\partial T}{\partial z}(\delta, z) + \frac{\partial \gamma}{\partial w}\frac{\partial w}{\partial z}(\delta, z)$$

Our task now is to find the temperature and concentration gradients at the liquid-gas interface. We assume that the mass fraction of ethanol is governed by a simple steady advection diffusion equation,

$$(12.4.13) \qquad v_x \frac{\partial w}{\partial x} + v_z \frac{\partial w}{\partial z} = D\frac{\partial^2 w}{\partial x^2}$$

where we have also assumed that the diffusion in the z-direction is negligible compared to the advection in the z-direction (a usual assumption for thin film type geometries). We also assume that the temperature profile is governed by the simple steady thermal advection diffusion equation (i.e., the convection-conduction equation),

$$(12.4.14) \qquad \rho c_p \left(v_x \frac{\partial T}{\partial x} + v_z \frac{\partial T}{\partial z}\right) = \lambda \frac{\partial^2 T}{\partial x^2}$$

where λ is the thermal conductivity of the liquid, and c_p is the specific heat of the liquid (both assumed to be constant). Note that once again, we assumed that the conduction in the z-direction is negligible compared to the convection in the z-direction. Furthermore, in using these simple advection-diffusion and convection-conduction equations we have assumed that the Soret effect (thermomigration or motion in a temperature gradient due to differences in kinetic energy of bombarding particles of the surrounding fluid) and Dufour effect (reciprocal of the Soret effect, energy flux due to a mass concentration gradient) are negligible. Notice that (12.4.13) and (12.4.14) are coupled to the velocity distribution (12.4.10) through

(12.4.12). As for the boundary conditions for these equations, we state that at the entrance to the film both the temperature and concentrations are uniform:

$$(12.4.15) \qquad\qquad T(x,0) = T_0$$

$$(12.4.16) \qquad\qquad w(x,0) = w_0$$

Additionally, at the solid-liquid interface, we assume that the solid is a perfect insulator, and does not interact with the solution,

$$(12.4.17) \qquad\qquad \frac{\partial T}{\partial x}(0,z) = 0$$

$$(12.4.18) \qquad\qquad \frac{\partial w}{\partial x}(0,z) = 0$$

We chose the thermal boundary condition (perfect insulator) somewhat arbitrarily, more to match the form of the mass transport boundary condition, however the assumption of a perfect insulator at the solid-liquid interface is a reasonable one.

Finally, we need a thermal and mass transport boundary condition at the liquid-gas interface. At this interface we assume that the dominant heat transfer mechanism is due to evaporation, and so write,

$$(12.4.19) \qquad\qquad -\lambda \frac{\partial T}{\partial x}(\delta,x) \approx \Delta h_{vap} Q_{evap}$$

where Δh_{vap} is the enthalpy of evaporation, i.e., the enthalpy difference between the ethanol in vapor and liquid states. Similarly, we assume that the dominant mass transfer mechanism at this interface is evaporation, and so analogously write,

$$(12.4.20) \qquad\qquad -\rho D \frac{\partial w}{\partial x}(\delta,z) \approx (1 - w(\delta,z)) Q_{evap}$$

Now our task is to obtain a relation for the ethanol evaporation mass flux, Q_{evap}. We can reasonably assume that this mass flux is proportional to a mass transfer coefficient, k_g and the difference between the ethanol gas phase concentration at the liquid gas interface and the ethanol concentration in the bulk gas. We assume that the ethanol concentration in the bulk gas is negligible. We can use Raoult's law to relate the ethanol gas phase concentration near the liquid to that in the liquid. Thus we write the evaporation mass flux as,

$$(12.4.21) \qquad\qquad Q_{evap} = k_g \tilde{a} \frac{p_{vap}}{p} w(\delta,z)$$

where $\tilde{a}$ is the concentration dependent activity coefficient of ethanol (calculated according to standard methods such as that proposed by Fredenslund et al. (1975)) which is non-linearly dependent on mass fraction of ethanol; this activity coefficient decreases as ethanol fraction increases. p_{vap}/p is the ratio of pure ethanol vapor pressure to the total gas pressure above the liquid-gas interface. The mass transfer coefficient k_g depends on the gas phase flow, and estimates of it are tabulated for particular situations (e.g., natural convection flow).

Since we are mostly interested in the net transport of liquid in the liquid film and do not so much worry about the velocity distribution in the film, we can take an average across the thickness of the film of the transport equations to make the solution easier. We here define this average as,

$$(12.4.22) \qquad \langle (f) \rangle = \frac{1}{\delta} \int_0^\delta (f) \, dx$$

Combining equation (12.4.10) with (12.4.12) and performing the averaging, (12.4.22), we obtain

$$(12.4.23) \qquad \langle v_z \rangle = -\frac{\rho g \delta^2 \cos \beta}{3\mu} + \frac{\delta}{2\mu} \left(\frac{\partial \gamma}{\partial T} \frac{\partial \langle T \rangle}{\partial z} + \frac{\partial \gamma}{\partial w} \frac{\partial \langle w \rangle}{\partial z} \right)$$

To arrive at this result, we have also assumed that the temperature and ethanol concentration are approximately uniform in a given cross-section of the film, i.e., $T(\delta, z) \approx \langle T \rangle$ and $w(\delta, z) \approx \langle w \rangle$. Applying the film cross-section averaging to the continuity equation, (12.4.1) taken together with the boundary conditions (12.4.3) and (12.4.5), and the relation (12.4.21), we obtain

$$(12.4.24) \qquad \rho \frac{\partial (\delta \langle v_z \rangle)}{\partial z} = k_g \tilde{a} \frac{p_{vap}}{p} \langle w \rangle$$

Additionally, averaging (12.4.5) and the mass transport equation (12.4.13) using the mass transfer boundary conditions, (12.4.18) and (12.4.20), and employing (12.4.24) we obtain

$$(12.4.25) \qquad \rho \delta \langle v_z \rangle \frac{\partial \langle w \rangle}{\partial z} = -k_g \tilde{a} \frac{p_{vap}}{p} \langle w \rangle (1 - \langle w \rangle)$$

where we have also used the approximation, $\langle v_z w \rangle \approx \langle v_z \rangle \langle w \rangle$. Similarly averaging (12.4.5), and now the heat transfer equation (12.4.14), and so using the heat transfer boundary conditions, (12.4.17) and (12.4.19), we obtain,

$$(12.4.26) \qquad \rho c_p \delta \langle v_z \rangle \frac{\partial \langle T \rangle}{\partial z} = -k_g \tilde{a} \frac{p_{vap}}{p} \langle w \rangle \Delta h_{vap}$$

where we have again used the approximation, $\langle v_z T \rangle \approx \langle v_z \rangle \langle T \rangle$. Now we have a system of coupled ODEs rather than PDEs and we are ready to solve these. First, we combine (12.4.25) with (12.4.26) to obtain,

$$(12.4.27) \qquad \frac{\partial \langle w \rangle}{\partial z} = (1 - \langle w \rangle) \frac{c_p}{\Delta h_{vap}} \frac{\partial \langle T \rangle}{\partial z}$$

This allows us to compare the contributions of the temperature gradient and the concentration gradient to the overall flow. Using thermography Venerus and Simavilla (2015) found that in a typical wine glass with a typical wine (13% ethanol v/v), a temperature gradient of order 100 K/m occurs. Using (12.4.27) and representative specific heat of ethanol-water solutions and heat of vaporization of ethanol, we see that the mass fraction gradient is order 0.5 m^{-1}. Based on this estimated concentration gradient and length of a typical film, we can estimate that the change in ethanol concentration along the film is less than one percent. Using representative temperature and concentration coefficients of surface tension for water-ethanol mixtures, Venerus and Simavilla observe that both the temperature and the concentration gradients contribute similarly to the flow, and so neither can be neglected in favor of the other. However, since we know that the change in ethanol concentration along the film is small, we can seek a solution to this system of equations for a small change in w. To do this, we multiply (12.4.23) by $\langle v_z \rangle$ and combine the result with (12.4.25) and (12.4.26). Then in this result we set δ to δ_0, $\langle w \rangle$ to $\langle w \rangle_0$, and $\langle v_z \rangle$ to $\langle v_z \rangle_0$ (to the variable's value at the film entrance), and obtain a quadratic equation for the film averaged velocity at the film entrance, $\langle v_z \rangle_0$. Taking the positive solution, we obtain,

$$(12.4.28) \qquad \frac{\langle v_z \rangle_0}{V} = -\frac{1}{2} + \sqrt{\frac{1}{4} - C\tilde{a}\frac{p_{vap}}{p}w_0 \left(\frac{\Delta h_{vap}}{c_p}\frac{\partial \gamma}{\partial T} + (1 - w_0)\frac{\partial \gamma}{\partial w} \right)}$$

where

$$(12.4.29) \qquad V = \frac{\rho g \delta_0^2 \cos \beta}{3\mu}$$

is a characteristic velocity scale of the flow (for wine, typically 1 mm/s), and

$$(12.4.30) \qquad C = \frac{k_g}{2\rho\mu V^2}$$

Venerus and Simavilla found that quantity $\langle v_z \rangle_0 / V$ increases to maximum value of about 0.27 at $\langle w \rangle_0$ of about 0.1, and then steadily decreases. Interestingly, $\langle v_z \rangle_0 / V$ increases approximately proportional to $\langle w \rangle_0$ up to $\langle w \rangle_0$ of 0.05 ($\langle v_z \rangle_0 / V$ of 0.2). At low ethanol mass fractions the contribution from concentration gradient dominates, while at higher ethanol mass fractions, it is the contribution from temperature gradient that dominates the flow.

CHAPTER 13

Droplet generation

Droplets and so droplet generation - generation of monodisperse emulsions is important to many technologies from lab-on-a-chip devices (such as digital droplet PCR) to various e-ink displays. Production of monodisperse emulsions also allows us to produce highly monodisperse solid particles. These droplets can also serve as precision microreactors, isolating particular reactions from surrounding ones. Here we will discuss various, specifically microfluidic, ways of generating reproducible, monodisperse emulsions. Traditional, polydisperse emulsions such as the ordinary vinaigrette can be produced by mixing the two phases and an emulsifier by a sufficiently strong shear flow. However, here we will be more interested in generating monodisperse emulsions. We will begin by discussing the breakup and break off of droplets from a liquid jet, a phenomena known as Plateau–Rayleigh instability. We will then discuss the simplest droplet generators - membrane based droplet generators. Thirdly, we will discuss the common microfluidic droplet generators such as the co-flowing streams generator and the T-junction generator. Lastly, we will review surfactants used in droplet generation.

13.1. Droplets from a liquid jet: Plateau–Rayleigh instability

We commonly observe that a jet of water coming from a circular orifice (e.g., a faucet) does not retain its cylindrical form but will break up into droplets as shown in Figure 13.1.1a. As we will see shortly, oscillations occur on the surface of the jet, these oscillations grow in amplitude, and finally cause droplets to pinch off the jet. This phenomena can be explained by considering the capillary pressure variations in the jet. Let's consider a liquid cylinder such as that shown in Figure 13.1.1b. The surface of the cylinder is perturbed with a small periodic disturbance (greatly exaggerated in Figure 13.1.1b) of wavelength λ, so that its radius is

$$(13.1.1) \qquad r = a + b \cos\left(\frac{2\pi x}{\lambda}\right)$$

where $b \ll a$. The liquid cylinder will be stable if the pressure at the region of greatest radial extension, i.e. point A, is greater than that at the point of greatest radial contraction, point B. If this was the case, the liquid will then flow from a region of high pressure to low pressure, which will tend to restore the perfect cylindrical shape of the cylinder. If the case was the opposite, liquid would again flow the high pressure side to the low pressure side, but would just cause the radial extension to grow and the radial contraction to shrink and so break up into droplets.

To determine when each case occurs, let's consider that the pressure is solely determined by the Young-Laplace equation and that gravitational forces or other forces (e.g., electrical forces) are negligible. Thus, at any point on the surface the principal radii of curvature, R_1 and R_2 determine the excess pressure in the liquid

(a)

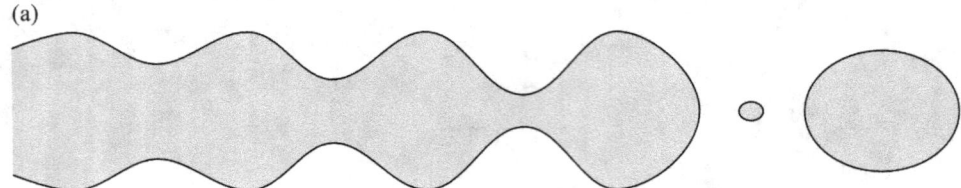

(b)

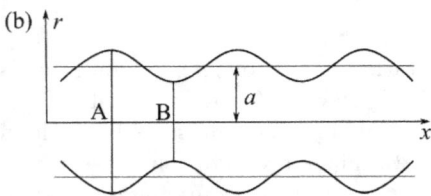

FIGURE 13.1.1. Break up of a cylindrical jet (a), and sinusoidal oscillations (greatly exaggerated for clarity) on a surface of a cylinder (b). The breakup of a liquid jet begins with a small perturbation in the cylinder's shape. This perturbation can be caused by, for example, an imperfection in the geometry of the orifice, vibration of the orifice, or vibrations in the air (e.g., ambient sounds). This causes the capillary pressure to squeeze down on the part on the narrowing of the jet and balloon out the widening of the jet. Under conditions described in the main text, this perturbation can grow in time and since the liquid in the jet is moving downstream, appear to grow as the liquid moves downstream. Eventually the capillary pressure in the neck causes the neck to pinch off generating a droplet or generating a large droplet and a smaller satellite droplet, as shown.

relative to the surrounding atmosphere and so determine the local flow of liquid inside the jet. At the point of maximum radial extension, A, the first radius of curvature is just

$$(13.1.2) \qquad\qquad R_1 = a + b$$

and at the point of maximum radial contraction, B, the first radius of curvature is just

$$(13.1.3) \qquad\qquad R_1 = a - b$$

The second radius of curvature (that which lies in the r-x plane) for both points is just given by

$$(13.1.4) \qquad\qquad R_2 = \frac{\left(1 + \left(\frac{dr}{dx}\right)^2\right)^{3/2}}{\frac{d^2r}{dx^2}}$$

(For details see Section 7.2.1). When deformations are small, $dr/dx \ll 1$, and we can simplify (13.1.4) to

$$(13.1.5) \qquad\qquad R_2 = \left(\frac{d^2r}{dx^2}\right)^{-1}$$

Substituting in the shape of the perturbed liquid jet, (13.1.1), at the point of maximum radial extension, A

$$R_2 = \frac{4\pi^2 b}{\lambda^2}$$

and at the point of greatest radial contraction, B,

$$R_2 = -\frac{4\pi^2 b}{\lambda^2}$$

Substituting these radii of curvature into the capillary equation, we obtain

(13.1.6) $$\Delta p_A = \gamma \left(\frac{1}{a+b} + \frac{4\pi^2 b}{\lambda^2} \right)$$

and

(13.1.7) $$\Delta p_B = \gamma \left(\frac{1}{a-b} - \frac{4\pi^2 b}{\lambda^2} \right)$$

The difference between these pressures is what drives the fluid locally, and so

(13.1.8) $$\Delta p_A - \Delta p_B = \gamma \left(\frac{1}{a+b} - \frac{1}{a-b} + \frac{8\pi^2 b}{\lambda^2} \right) = \gamma \left(\frac{8\pi^2 b}{\lambda^2} - \frac{2b}{a^2 + b^2} \right)$$

Since $b \ll a$,

(13.1.9) $$\Delta p_A - \Delta p_E = \gamma \left(\frac{8\pi^2 b}{\lambda^2} - \frac{2b}{a^2} \right) = \frac{2b\gamma}{a^2} \left(\left(\frac{2\pi a}{\lambda} \right)^2 - 1 \right)$$

As we explained above, when $\Delta p_A - \Delta p_B > 0$, the perturbation dies down and the cylinder returns to its perfect cylindrical shape. When $\Delta p_A - \Delta p_B < 0$ the perturbation grows and eventually the cylinder breaks into droplets. Thus from (13.1.9) we can see that the jet will be unstable when

(13.1.10) $$\lambda > 2\pi a$$

This condition was originally derived by Lord Rayleigh. Rayleigh (1892) also showed that the maximum instability occurs when $\lambda = 9.02a$. Furthermore, Kroesser and Middleman (1969) showed that the distance of the jet breakup from the orifice is given by

(13.1.11) $$\frac{L_b}{D} = 11 \left(We^{1/2} + \frac{3We}{Re} \right) = 11 We^{1/2} (1 + Oh)$$

where the Weber number, the Reynolds number, and the Ohnesorge number are defined based on the diameter of the jet at the position of the break up. (This differs from the orifice diameter as the jet narrows as it flows downstream). The time for break up is then

(13.1.12) $$t_b = 11 D^{3/2} \left(\frac{\rho}{\gamma} \right)^{1/2} (1 + Oh)$$

For low viscosity liquids, Grant and Middleman (1966) showed that the distance of the jet breakup from the orifice is given by

(13.1.13) $$\frac{L_b}{D} = 13.5 We^{1/2}$$

and the time for the break up to occur is then given by

$$(13.1.14) \qquad\qquad t_b = 13.5 D^{3/2} \left(\frac{\rho}{\gamma} \right)^{1/2}$$

where here D can be taken as the diameter of the office, since if viscosity is neglected, the tapering can also be neglected.

While this instability can be commonly observed by observing an ordinary faucet, this phenomena is also responsible for the correct functioning of common fuses, and is important in continuous inkjet (CIJ) printing, and flow cytometry. This phenomena is also leveraged in droplet generation when forcing a liquid through a mesh or a screen - a surface with many orifices. An example of this was the production of shot for muskets in shot towers, where molten lead was poured through a screen, turned to lead droplets and the lead droplets cooled while flying through a tower into a water basin below.

In common fuses the too high a current turns the wire in the fuse into a liquid column. This liquid column in turn becomes unstable and breaks up into metal drops, disconnecting the flow of current. For the purposes of crude, order of magnitude estimation, let's consider the original wire in the fuse was 100 µm diameter copper wire that heated up to 1400 K, and so had a density of 7962 kg/m^3 and viscosity of 3.74 mPa s (Assael et al. (2010)), and a surface tension of 1300 mN/m (Harrison et al. (1977)). The Ohnesorge number would be about 0.004. Equation (13.1.12) would predict that such a fuse would blow in about 1 ms after it got up to this temperature. Of course to predict the blowing of fuse more accurately, we have to consider the heating up of the wire, including various heat loss mechanisms, in a much more accurate fashion.

The size of the drops is dependent on the frequency of the perturbation, as we can intuitively infer from Figure 13.1.1. If we drive the orifice at a particular frequency, only drops of a particular size will be produced. (This can be easily demonstrated by touching an orifice with a particular vibrating tuning fork). If we are interested in producing different sized droplets simultaneously, we can drive the orifice with multiple frequencies. If we tilt the jet at an angle to the gravity vector, different sized drops will have different velocities and masses and so form different trajectories and so our differently sized drops can be separated from one another.

13.2. Membrane based droplet generators

A relatively simple way to generate a large number of droplets with a relatively monodisperse population quickly is via membrane emulsification. We show three types of most common membrane emulsification methods in Figure 13.2.1. Membrane emulsification can produce both water-in-oil and oil-in-water emulsions. The simplest membrane emulsification method is the direct membrane emulsification (DME). In DME we simplify push the phase to be dispersed (e.g., water in the water-in-oil droplet) through a membrane into the continuous phase (Figure 13.2.1a). The droplets are formed at the membrane-continuous phase interface and are stabilized by surfactants in the continuous phase (more on this later). Often (but not strictly necessary) a gentle shear flow is applied tangentially to the membrane to shear off the droplets. (We will discuss a model for this based on a microfluidic version of a "single pore" DME later as well). The shear flow should be gentle, so that it has high enough shear to detach the droplets but not to excessive

(a) Direct membrane
emulsification

(b) Premix membrane
emulsification without
phase inversion

(c) Premix membrane
emulsification with
phase inversion

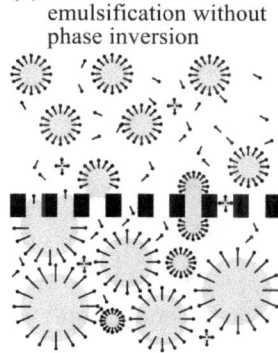
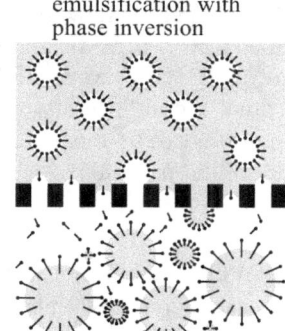

FIGURE 13 2.1. Schematic of most common membrane emulsification methods. In (a) and (b) gray color represents the dispersed phase, while in (c) it represents the continuous phase. Surfactant molecules (greatly exaggerated in size) are represented with a line with a dot. The line represents the hydrophobic tail whereas the dot represents the hydrophilic head.

to break up the droplets further. However, for small enough dispersed phase flux through the membrane, the drops will spontaneously detach even in the absence of a shear flow. In DME droplet size is largely governed by pore geometry, so to achieve a more monodisperse droplet population we must use a membrane with a highly monodisperse pore size. Many such membranes (e.g., anodic porous alumina, macroporous silicon) are commercially available as of this writing. Another consideration when choosing membranes for DME is that the pores should be far enough apart so that the droplets formed in neighboring pores do not interact sterically and coalesce. DME droplet generators are limited by the maximum dispersed phase flux through the membrane, typically $0.01 - 0 1 \, \mathrm{m^3/(m^2 h)}$. (For a typical syringe filter size, 12.5 mm diameter, this is 0.02 - 0.2 ml/min). Above this flux, the membrane transitions from a size stable droplet generation to a continuous outflow. At higher fluxes the membrane is able to produce droplets, but the population is highly polydisperse. DME have been demonstrated to produce emulsions of up to 30% (v/v) of dispersed phase and in size ranges from 200 nm to 100 μm.

Beyond simple direct membrane emulsification is the premix membrane emulsification (PME) (Figure 13.2.1b and c). Here we start by generating a starting polydisperse emulsion (such as by subjecting the two phases to a strong shear flow) and then pass this emulsion through a membrane. If on the other side of the membrane is the same continuous phase as in the coarse emulsion, a fine, significantly more monodisperse emulsion results. If on the other side of the membrane is the dispersed phase of the coarse emulsion, a phase inversion results and we obtain a fine, monodisperse emulsion where the now what used to be the continuous phase, now becomes the dispersed phase (Figure 13.2.1c). The advantage of PME is that we can use it to prepare a fine, concentrated emulsion from a low concentration coarse emulsion at high rates. In PME membrane fluxes above $1 \, \mathrm{m^3/(m^2 h)}$ are typically used, and 90% (v/v) of dispersed phase emulsions have been produced. (For a typical syringe filter size, 12.5 mm diameter, this translates to > 2 ml/min).

Furthermore, we can achieve smaller droplet sizes with PME than with DME. Additionally, shear flow upstream of the membrane is even less important (and largely unnecessary) in PME and driving pressure and surfactant properties are less important in PME. However, DME typically achieves higher monodispersity than PME. Droplet monodispersity can be improved in PME by successive passes through the membrane. For more information on membrane emulsification see Vladisavljevic and Williams (2005).

13.3. Common microfluidic droplet generators

While membrane based droplet generators are particularly simple to create and can generate relatively monodisperse populations of droplets sometimes we would like to have even more control over the droplets and droplet populations. For example, we might want to change the droplet composition, droplet size, or droplet spacing in a droplet train on the fly; or control the position of particular droplets in the droplet train. For this we can use three commonly used droplet generators: (a) the cross flow generator, (b) coflowing generator, and (c) the flow focusing generator. (See Figure 13.3.1 and Figure 13.3.2; Figure 13.3.3; Figure 13.3.4 respectively).

13.3.1. Cross flow (T-junction) droplet generator. We begin with the cross flow droplet generators as these are the simplest to fabricate and can even be created with off-the-shelf microfluidic chips having a simple T-junction. All we have to do is to flow the continuous phase through the straight (main) channel, and flow the phase to be dispersed through the orthogonal channel. There are two types of T-junction generators: the first with the main channel and the orthogonal channel having the same size - the confined generator; and the second with the orthogonal channel much smaller than the main channel - the unconfined generator. In confined droplet generators the depth of the channels is also typically comparable to the width of the channels. In the confined generator the dispersed phase enters the channel, reaches the distant wall, thus obstructing the continuous phase flow, apart from the thin region of continuous phase near the wall. The dispersed phase is confined by the distant wall, giving the droplet generator its name. Due to this obstruction, the pressure upstream of the droplet in the continuous phase increases and so additional force is now exerted on the continuous-dispersed phase interface, thus shearing off the droplet. Typically the droplets generated are longer than the width of their channel, and so are referred to as plugs. For this type of generator it has been shown that the capillary number based on the continuous phase can be as a high as 0.1 and droplets can be generated over for orders of magnitude in the flow rate ratio $\varphi = Q_d/Q_c$. ($Ca = \mu_c U_c/\gamma = \mu_c Q_c/\gamma w_c h$, where Q is the flow rate and the subscripts c and d refer to the continuous and dispersed phases respectively).

Note in these sections describing capillary forces between two liquids, γ is the interfacial tension between the two. Commonly one of the fluids is water or an aqueous solution. Demond and Lindner (1993) provide an extensive table of estimated surface tension between organic liquids and water.

T-junction generator of the confined type has been studied extensively by Garstecki et al. (2005) and our description of its function will follow the analysis of these authors. In this generator two immiscible fluids (often with a surfactant) form an interface at the junction of the inlet (orthogonal channel) and the main channel. The stream of the discontinuous phase first penetrates into the main channel and a droplet begins to grow (Figure 13.3.1a). The pressure gradient and the flow of the continuous phase in the main channel distorts the droplet in the downstream direction (Figure 13.3.1b). The interface on the upstream side of the droplet moves downstream and a neck begins to form (Figure 13.3.1c). This neck progressively narrows as the continuous phase pushes on the droplet of the discontinuous phase extending into the channel. When this neck narrows enough such that its leading edge approaches the downstream edge of the inlet (Figure 13.3.1d), the neck breaks and the droplet breaks off. The disconnected liquid droplets rear regains proper curvature as dictated by the capillary pressure and the droplet moves downstream in the main channel. Meanwhile, the tip of the discontinuous phase stream retracts to the end of the inlet. Then this cycle repeats again for the next drop.

In this generator Garstecki et al. (2005) observed two types of regimes: constant droplet length regime, and the linear droplet length regime. The constant droplet length regime occurred when the flow rate of the dispersed phase was low compared to that of the continuous phase. As the name implies, in this regime the resulting droplet length, L was independent of the dispersed to continuous flow rate ratio, and was largely dependent on the geometry of the main channel. In this regime, $L \sim w$, where w is the width of the main channel. This is the smallest droplet size that this type of generator is able to produce. The generator transitioned to the linear droplet length regime when the dispersed phase flow rate was increased above the continuous phase flow rate. In this regime, the resulting droplet length increased linearly as the flow rate ratio was increased. Importantly, both of these regimes occur only when the capillary number based on the continuous phase $Ca_c < 10^{-2}$. Larger capillary numbers we obtain either two co-flowing streams of "continuous" and "dispersed" phases or a single stream of continuous phase if the dispersed phase flow rate relative to the continuous phase flow rate is too low.

Let's now obtain a scaling for the final droplet length (and therefore size) to understand how to design and control this type of generator. To do this will consider the scaling for the three forces involved in droplet break off: capillary force, shear force, and force due to resistance. Let's begin with the capillary force. If we assume that the interface is quasi-static the capillary (Laplace) pressure across the dispersed-continuous phase boundary is

$$(13.3.1) \qquad \Delta p_L = \gamma \left(\frac{1}{r_a} + \frac{1}{r_r} \right)$$

where r_a is the radius of axial curvature (in the plane of the page, see Figure 13.3.1c), and r_a is the radius of radial curvature (orthogonal to the radius of axial curvature). During formation of the droplet the radial curvature is bounded by the height of the channel and so we can estimate that $r_r \approx h/2$ or less everywhere in the drop. Likewise we can estimate that the axial curvature at the tip of the drop is $r_{a,tip} \approx w/2$ and on the upstream side of the droplet $r_{a,up} \approx w$ (see Figure 13.3.1c). Thus, the capillary pressure on the downstream side of the droplet acts on the liquid inside the droplet with a pressure

$$(13.3.2) \qquad p_{L,dw} = -\gamma \left(\frac{2}{w} + \frac{2}{h} \right)$$

where, as a reminder, the negative sign indicates that the pressure points upstream. Meanwhile the pressure on the upstream side acts on the liquid inside the droplet

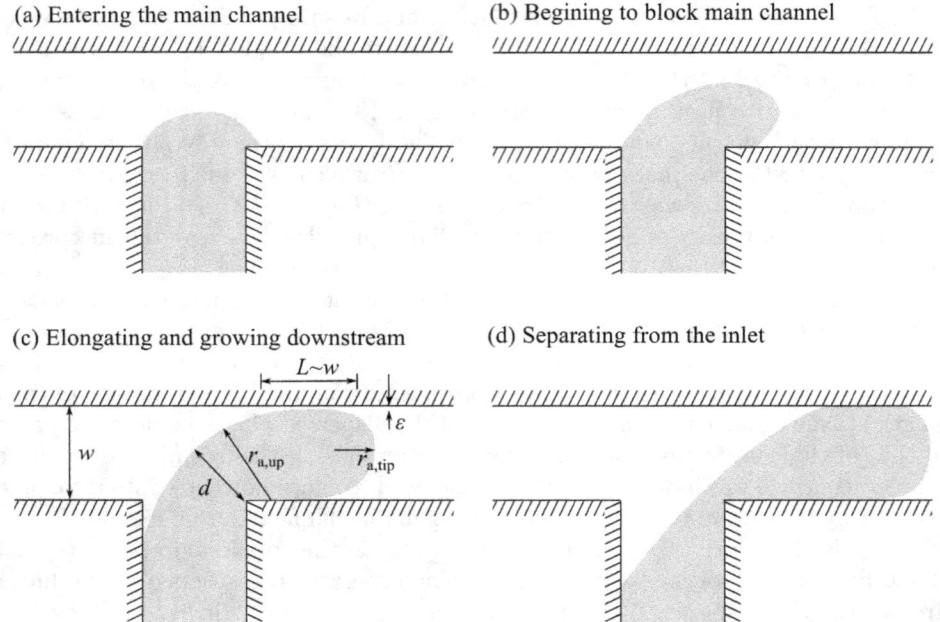

FIGURE 13.3.1. Schematic of droplet break off in T-junction generator of the confined type. The continuous phase (white) is flowing in the main (horizontal) channel from left to right, while the phase to be dispersed (gray) enters from the orthogonal channel from the bottom. Some of the features of the shape of the dispersed phase are exaggerated for clarity.

with a pressure of

$$(13.3.3) \qquad p_{L,dw} = \gamma\left(\frac{1}{w} + \frac{2}{h}\right)$$

Again, the positive sign indicates that the pressure points downstream. The sum of these pressures multiplied by the cross-sectional area of the channel approximately gives the capillary force on the droplet,

$$(13.3.4) \qquad F_\gamma \approx -\gamma h$$

The negative sign indicates that this force points upstream, which means that this force counteracts the force of the continuous phase trying to push the forming droplet downstream. This force resists the break off of the droplet. Next, let's consider the shear force on the droplet. We can approximate the shear stress exerted by the continuous phase on the tip of the droplet as

$$(13.3.5) \qquad \tau \approx \mu_c \frac{u_{gap}}{\varepsilon}$$

where $u_{gap} = Q_c/h\varepsilon$ is the mean velocity of the continuous phase through the gap between the distant wall of the main channel and the droplet. To estimate the shear force acting on the droplet in the gap, we multiply the shear stress by the area of the dispersed phase in the gap that sees this shear: $A_{gap} \approx hw$ (see Figure

13.3.1c). Thus the shear force that the droplet experiences is

$$(13.3.6) \qquad F_\tau \approx \mu_c Q_c \frac{w}{\varepsilon^2}$$

Since the force is positive, it points downstream, as expected. This force helps the break off of the drop. This relation however overestimates the shear force on the droplet. Firstly, we assumed that the dispersed phase in the drop is stationary, while in reality it moves downstream, hence the shear in the gap is actually caused by a net velocity of gap velocity minus the velocity of the droplet. Secondly, we assumed that the volumetric flow rate of the continuous phase around the droplet is equal to the externally imposed continuous phase flow rate. However, this is not the case - in reality, the continuous phase flow rate varies as a function of position in the cycle of the droplet break off.

Finally, let's estimate the force due to the resistance to flow caused by the dispersed phase occluding the channel. At the start of the droplet break off cycle, the gap between the dispersed phase and the distant wall of the main channel, ε is large and $\varepsilon \sim w$ (Figure 13.3.1a). Using Hagen-Poiseuille equation, we can estimate that the pressure drop over the top of the dispersed phase, length $\sim w$ is

$$(13.3.7) \qquad \Delta p \approx \mu_c Q_c \frac{w}{h^2 \varepsilon^2}$$

Near the end of the droplet break of cycle when the gap is small, $\varepsilon \ll w$, we can estimate (using lubrication theory) that the pressure drop is

$$(13.3.8) \qquad \Delta p \approx \mu_c Q_c \frac{w}{h \varepsilon^3}$$

Thus, for the end of the droplet break of cycle, the resistance pressure force is then

$$(13.3.9) \qquad F_R \sim \Delta p h w \sim \mu_c Q_c \frac{w^2}{\varepsilon^2}$$

Comparing the scaling for this force with the shear force, (13.3.6), we see that near the end of the droplet break off cycle when $\varepsilon \ll w$, the resistance force dominates the shear force. Thus the balance between the resistance force and the capillary force determines the droplet break off at this stage.

Now we are ready to derive a scaling for the size of the generated droplets. Recall, that as the tip of the discontinuous phase enters and blocks the main channel, that at this moment in the droplet break off cycle the length of the drop approximately equals the width of the main channel (see Figure 13.3.1c). At the same time the increased pressure upstream of the drop in the continuous phase squeezes the neck (which has a characteristic width d. The width of the neck decreases approximately at a rate equal the mean speed of the continuous fluid,

$$(13.3.10) \qquad u_{sq} \approx Q_c / h w$$

Thus the time to pinch off the neck is roughly, d/u_{sq}. Meanwhile, during this pinching off, the drop elongates at a rate of governed by the flow of the dispersed phase,

$$(13.3.11) \qquad u_{gr} \approx Q_d / h w$$

Putting this together, we find that the final length of the drop is

$$(13.3.12) \qquad L \approx w + \frac{d}{u_{sq}} u_{gr} = w + d \frac{Q_d}{Q_c}$$

Here the first term on the left right-hand side is the length gained during the phase where the continuous phase blocks the channel, and the second term is due to the growth of the droplet length during the pinch-off phase. We can nondimensionalize this equation,

$$(13.3.13) \qquad\qquad\qquad \frac{L}{w} \approx 1 + \alpha \frac{Q_d}{Q_c}$$

where $\alpha = d/w$. Since our estimation is only approximate, and that in reality, for example, some of the continuous phase bypasses the droplet and does not contribute to squeezing, it is reasonable to treat α as a fitting parameter of order one, as Garstecki et al. (2005) do. In fact for their oil-water combination and their set of viscosity ratio and their geometry, they found that $\alpha = 1$ fits their data well. Furthermore, notice that this scaling predicts well the behavior that at low dispersed phase flow rates the droplet size is governed solely by the geometry, while at higher dispersed phase flow rates, the droplet size increases linearly with dispersed phase flow rate. We also see that in the geometry controlled regime droplets should be very monodisperse as the geometry practically does not change even when generating a large number of droplets. This is in fact the reason why microfluidic droplet generators are able to generate highly monodisperse droplet populations, while droplet populations created with membrane processes are more polydisperse. The microfluidic droplet generators are always generating the droplet with the same geometry and if operated in geometry controlled regime the droplets should all be the same size. Meanwhile for membrane based droplet generators, there are inherent differences between the membrane pores hence even if operated in geometry controlled mode, some polydispersity will arise. Thus membranes with tight pore distributions are desired for monodisperse membrane droplet generators.

Having considered a T-junction generator of the confined type, let's now consider droplet generation in a T-junction generator of the unconfined type, as shown in Figure 13.3.2. The characteristic difference in the unconfined type vs. the confined type is that the dispersed phase inlet is much narrower than the main channel and so the droplet in its growth phase never reaches the distant wall of the main channel. To the droplet it is as if this distant wall is infinitely far away. Hence, this type of droplet generator is also a good model for the membrane droplet (emulsion) generators with cross flow.

In the unconfined T-junction droplet generator the dispersed phase first enters the main channel (Figure 13.3.2a) As the droplet expands into the main channel, the drag from the flowing continuous phase pushes it over to one side, while interfacial tension holds it attached to the dispersed phase channel (Figure 13.3.2b). As the droplet grows in size, the drag on it increases and necking emerges (Figure13.3.2c). Once the drag overcomes the interfacial tension, the droplet breaks off (Figure 13.3.2d). In our description of droplet generation in this generator, we follow that of Husny and Cooper-White (2006).

To obtain a scaling for droplet size, we need to find a scaling for the drag force and interfacial tension force on the droplet. To find the drag force, we assume a parabolic profile in the continuous channel and from a Hagen-Poiseuille type equation find that at the center of the drop diameter d (i.e. $d/2$ from the lower

(a) Entering the main channel

(b) Expanding into main channel

(c) Necking emerging

(d) Separating from the inlet

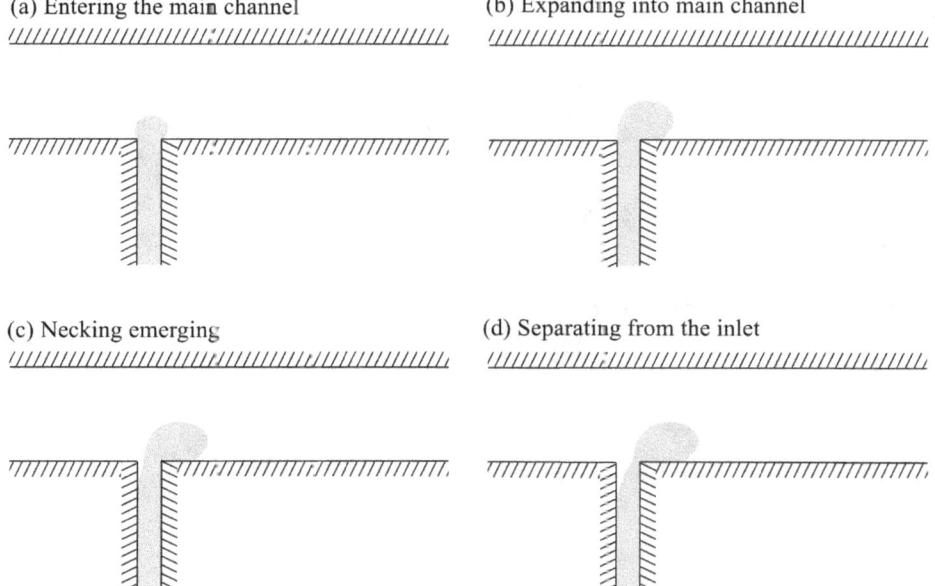

FIGURE 13.3.2. Schematic of droplet break off in T-junction generator of the unconfined type. The continuous phase (white) is flowing in the main (horizontal) channel from left to right, while the phase to be dispersed (gray) enters from the orthogonal channel from the bottom. Some of the features of the shape of the dispersed phase are exaggerated for clarity.

wall) the continuous phase velocity is

$$(13.3.14) \qquad u_{d/2} \approx 2u_c \left(1 - \left(\frac{d_c - d}{d_c} \right)^2 \right)$$

where d_c is the hydraulic diameter of the continuous phase channel. For most commonly encountered rectangular channel geometry $d_c = 2w_c h/(w_c + h)$, where w_c is the width of the continuous phase channel. We consider that the droplet generator is operating in a low Reynolds number regime and so assume that the droplet experiences Stokes' drag. If the droplet was a solid sphere the drag force on the droplet would be

$$(13.3.15) \qquad F_D \approx 3\pi\mu_c d \left(u_{d/2} - u \right)$$

where u is the drop velocity. $u_{d/2} - u$ accounts for the reduction of the drag due to the drop's motion relative to the continuous phase. This relation is only approximate as it does not account for the sphere's interaction with the wall. Furthermore, we have assumed that the drop is a solid. For a liquid drop with a viscosity different from the continuous phase, we use the corrected Stokes' drag due to Rybczynski (1911) and Hadamard (1911b) and so

$$(13.3.16) \qquad F_D \approx 3\pi\mu_c d \left(u_{d/2} - u \right) \frac{1 + 2\alpha/3}{1 + \alpha}$$

where $\alpha = \mu_c/\mu_d$. The interfacial tension force on the breaking off droplet is even more difficult to estimate, due to the complicated geometry of the droplet neck. As a scaling, we estimate the interfacial tension force to be of order

$$(13.3.17) \qquad\qquad F_\gamma \sim \pi\gamma w_d$$

where w_d is the width of the dispersed phase channel. Substituting (13.3.14) into (13.3.16) and equating with (13.3.17) we find a scaling for the resulting drop diameter to be,

$$(13.3.18) \qquad 3\pi\mu_c d\left(2u_c\left(1-\left(\frac{d_c-d}{d_c}\right)^2\right)-u\right)\beta - \pi\gamma w_d = 0$$

$$(13.3.19) \qquad\qquad \beta = \frac{1+2\alpha/3}{1+\alpha}$$

Next we nondimensionalize this equation, such that

$$d^* = d/d_c$$

$$(13.3.20) \qquad\qquad u_c^* = u_c\frac{3\mu_c}{\gamma}\beta\frac{d}{w_c}$$

$$u^* = u\frac{3\mu_c}{\gamma}\beta\frac{d}{w_c}$$

and so obtain

$$(13.3.21) \qquad\qquad d^{*2} - 2d^* + \frac{1+u^*}{2u_c^*} = 0$$

This has a solution of

$$(13.3.22) \qquad\qquad d^* = 1 \pm \sqrt{1 - \frac{1+u^*}{2u_c^*}}$$

We note that when both u^* and u_c^* are order unity, this model predicts that $d^* \approx 1$, or that the drop diameter is about the size of the hydraulic diameter of the continuous phase channel. This is not the typical operation of this droplet generator. Furthermore, we assume that we cannot generate a drop with a diameter larger than the hydraulic diameter of the continuous phase channel and so simplify (13.3.22) to

$$(13.3.23) \qquad\qquad d^* = 1 - \sqrt{1 - \frac{1+u^*}{2u_c^*}}$$

Furthermore, when u_c^* is much less than unity we cannot rely on this model for accurate predictions as this model predicts d^* to be complex. Similarly, when u^*/u_c^* is greater than unity, we cannot rely on this model for accurate predictions as this model predicts d^* to be complex again. However, when u^* is much less than unity,

$$(13.3.24) \qquad\qquad d^* \approx 1 - \sqrt{1 - 1/2u_c^*}$$

and so the droplet size becomes independent of dispersed phase flow rate. In more general cases the droplet size is controlled by both the geometry, viscosity, and flow rates of the continuous and dispersed phase channels.

13.3.2. Co-flowing droplet generator. Having discussed cross flowing droplet generators let's turn to the co-flowing droplet generator. Such generator typically consists of an internal capillary carrying the phase to be dispersed surrounded by the flowing continuous phase (Figure 13.3.3). Droplet generation in such generator can be broken down into four stages: initial growth, late stage growth, start of separation, and late stage separation. Our description of droplet generation in this generator follows that of Umbanhowar et al. (2000). In the growth part of the cycle, the dispersed phase tip has the shape of a section of a sphere. At the beginning of separation a neck develops between the channel feeding the dispersed phase and the drop (Figure 13.3.3c). As the separation continues, this neck stretches and narrows, until it eventually breaks. Capillary force then forces the droplet into a spherical shape, while the dispersed phase recoils back towards its channel. For stable drop growth at the tip of the dispersed phase channel, the dispersed phase flow rate must be low enough so that a jet does not develop. We can find a conservative estimate (a lower bound) for the maximum tolerable flow rate before jetting occurs, by equating the kinetic energy per unit length of the dispersed phase

$$(13.3.25) \qquad \rho_d \pi d_i^2 u_d^2 / 8$$

with the interfacial surface energy per unit length $\pi d_i \gamma$ where u_d is the area averaged velocity of the dispersed phase, $u_d = 4Q_d / \pi d_i^2$. Conservatively, jetting will not occur as long as the interfacial surface energy is larger than the kinetic energy of the drop, or

$$(13.3.26) \qquad Q_d < \sqrt{\frac{\pi^2 d_i^3 \gamma}{2\rho_d}}$$

However, practically the danger of jetting typically is not what limits the dispersed phase flow rate, but rather the requirement that the flow rate be slow enough such that the spacing between the drops is on the order of at least the drop diameter so that the drops don't tend to coalesce.

In this droplet generator drops grow spherically from the dispersed phase capillary tip until the drag force on the drop exceeds the capillary force holding the drop to the capillary. When the drag force exceeds the capillary force the drop detaches, and interfacial tension rapidly forces the drop into a spherical shape as well as returns the meniscus to the dispersed phase capillary. To obtain a scaling for drop size as function of key parameters, we estimate that the surface tension force scales as

$$(13.3.27) \qquad F_\gamma \sim \pi \gamma d_i$$

Here we assume low Reynolds number operation and so the drag the droplet experiences can be estimated by a modified Stokes' drag, accounting for the "shielding" of the droplet by the dispersed phase channel,

$$(13.3.28) \qquad F_D \approx 3\pi \mu_c (d - d_o)(u_c - u)$$

where d is the drop diameter (see Figure 13.3.3), u_c is the area averaged continuous phase velocity, and u is the mean drop velocity, $u \sim Q_c / \pi d$. The $d - d_o$ term accounts for the "shielding" of the droplet by the dispersed phase channel. The $u_c - u$ accounts for the reduction of the drag due to the drop's motion relative to the continuous phase. Assuming $d_o \approx d_i$ and equating (13.3.27) and (13.3.28) we

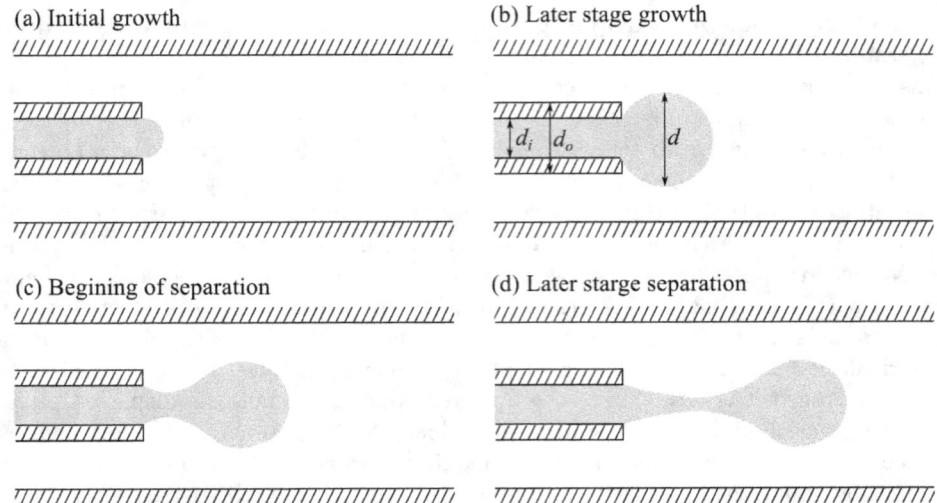

FIGURE 13.3.3. Schematic of droplet break off in a simple co-flowing stream generator. The continuous phase (white) is flowing in the top and bottom sheath channels left to right, while the phase to be dispersed (gray) flows from the center channel. Such generator can also be implemented by inserting a smaller capillary carrying the dispersed phase into a larger capillary carrying the continuous phase; or by placing the dispersed phase capillary in a spinning bath of the continuous phase such that the flow in the bath is roughly parallel to the dispersed phase capillary.

obtain

$$(13.3.29) \qquad 3\pi\mu_c \left(d - d_o \right) \left(u_c - Q_c/\pi d \right) - \pi\gamma d_i = 0$$

Next we nondimensionalize this equation, such that

$$(13.3.30) \qquad \begin{aligned} d^* &= d/d_i \\ u^* &= u_c \frac{3\mu_c}{\gamma} \\ Q^* &= Q_d \frac{3\mu_c}{\pi d_i^2 \gamma} \end{aligned}$$

and obtain

$$(13.3.31) \qquad u^* d^{*3} - \left(u^* + 1 \right) d^{*2} - Q^* d^* + Q^* = 0$$

a cubic equation for the droplet diameter. Notice u^* has the same scaling as the capillary number. While we can solve this analytically, the result is cumbersome. However, for small Q^*, the equation reduces to

$$(13.3.32) \qquad d^* = 1 + 1/u^* = 1 + \left(3Ca \right)^{-1}$$

Thus, for low dispersed phase flow rates droplet size only depends on continuous phase capillary number and the geometry of the device. This result agrees well with Umbanhowar et al. (2000) experimental results. Furthermore, Umbanhowar et al. found that they could achieve polydispersities of less than 3% with this type of droplet generator.

13.3.3. Focusing co-flowing droplet generator.

An improvement on the simple co-flowing droplet generator discussed above is the focusing co-flowing generator (see Figure 13.3.4). The improvement comes from placing an orifice in front of the dispersed phase channel so that the dispersed phase becomes focused in this orifice (Figure 13.3.4a) and the necking and separation become more reproducible. In describing droplet generation in this generator we follow the description of Garstecki et al. (2005). While Garstecki et al. developed this description for gas-in-liquid dispersions, they claim that their mechanism is also applicable for liquid-in-liquid dispersions as well. The description is valid for low capillary number (based on continuous phase). This droplet generator works by having the dispersed phase thread advance into the orifice and restrict the flow of continuous phase into the outlet channel (Figure 13.3.4a). This creates an increase in the pressure upstream and in the orifice, which in turn squeezes the dispersed phase thread into a neck shape. Once the neck becomes narrow enough it breaks, pushing the drop upstream of the orifice and with the rest of the dispersed phase returning upstream of the orifice. The volume of dispersed phase in the neck contributes negligibly to the final size of the droplet.

Looking more closely at the necking stage of the droplet generation we observe that it occurs in three substages. Firstly, as the portion of the dispersed phase upstream of the orifice enters the orifice, initially the width of the neck is set by the width of the orifice. We note that the geometry of the device must be such that the orifice must be close enough to the dispersed phase channel opening so that the tread does not become unstable prematurely by the Rayleigh-Plateau type instability. Next, the neck gradually thins, but in a rectangular channel the dispersed phase nominally stays in contact with the top and bottom walls of the structure (we say nominally, as there is a thin wetting film of the continuous phase separating it and the walls). Thirdly, a portion of the dispersed phase detaches from the top and bottom walls and forms an axisymmetric shape. This shape thins further and breaks according to Rayleigh-Plateau type instability.

We can estimate the final volume of the drop by considering the time scale to break off the droplet, and keeping in mind that this is all the time we have to fill the droplet at the flow rate of the dispersed phase. The time for break of is inversely proportional to the continuous phase flow rate as the higher the flow rate the quicker the droplet is sheared off. Hence, the volume of the droplet scales as

$$V_d \sim Q_d/Q_c$$

We can write this scaling also in terms of the pressures supplied to the two channels, assuming that the flow rate scales via the Hagen-Poiseuille relation, $Q \sim p/\mu$, as

$$V_d \sim \frac{p_d}{p_c} \frac{\mu_c}{\mu_d}$$

In this droplet generator, in this regime, the monodispersity of the droplets is mainly controlled by the steadiness of the flow rate or pressure generator supplying the system. Other droplet generation regimes are also possible in this geometry, including the tread formation, dripping and jetting regimes (Figure 13.3.4b-d). For more information on these regimes, see Anna and Mayer (2006).

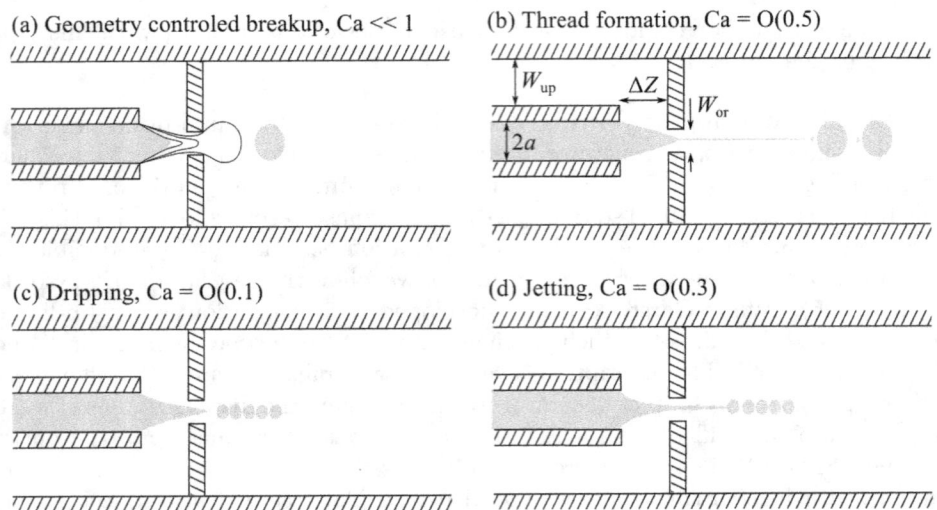

FIGURE 13.3.4. Schematics of different regimes of a focusing droplet generator. The continuous phase (white) is flowing in the top and bottom sheath channels left to right, while the phase to be dispersed (gray) flows from the center channel. For the geometry controlled break regime (a) we also outline the phase interface showing droplet growth and breaking off. In the jetting regime (d) a liquid column develops, becomes unstable via what appears to be Plateau–Rayleigh instability and breaks up into droplets. Some of the features of the shape of the dispersed phase are exaggerated for clarity.

13.4. Surfactants and their uses in droplet generation

So far, we have discussed various mechanisms for droplet generation. However, once the droplet has been generated we are often interested in having the droplets not coalesce quickly so that they can fulfill their role as microreactors, for example. Consider that a dispersion of a liquid in another liquid is a system that is out of thermodynamic equilibrium, as the area of the interface between the two phases is much larger than the minimum area possible between the two phases. Thus, coalescence of droplets is normally thermodynamically favorable process. We can decrease the chances of droplets coalescing by adding a surfactant. The surfactant locates itself at the interface between the continuous phase and the dispersed phase. By doing so the surfactant creates an energy barrier that needs to be overcome to lower the overall interfacial area and thus stabilizes the dispersion in a metastable state. Surfactants achieve this by three mechanisms: (a) droplets are prevented from coalescing by steric repulsion of surfactant molecules; (b) droplets are prevented from coalescing by electrical repulsion of charge groups between surfactant molecules; (c) presence of surfactants enables surfactant and so interfacial tension gradients, driving Marangoni flows that inhibit drainage of the continuous phase between the droplets as the droplets approach each other. When the dispersed phase is even slightly soluble in the continuous phase another emulsion aging mechanism occurs. Small droplets in the emulsion have higher capillary pressure than larger droplets and so tend to dissolve, while larger droplets are able to take up the dissolved molecules and grow. This effect is known as Ostwald ripening. We can reduce this effect by adding a surfactant layer between the dispersed and continuous phase, reducing the chance that the dispersed phase dissolves in the continuous phase. We summarize various popular emulsion systems and the surfactants used to stabilize them in Table 1. More information on popular emulsion systems and their surfactants, including information on compatibility with biological constructs such as cells and proteins can be found in Baret (2012). Water-in-fluorinated oil emulsions and associated fluorinated surfactants appear to be especially compatible with nucleic acids, proteins, and cells.

When adding surfactants we must pay attention how the surfactant will influence the wettability of the surfaces of the droplet generating device. For example, for membrane emulsification, for preparation of oil-in-water emulsion hydrophilic membranes should be used. This avoids the spreading of the disperse phase (oil) on the membrane surface, which interferes with emulsification. Similarly, for preparing water-in-oil emulsions hydrophobic membranes should be used, again to avoid the spreading of the dispersed phase (water) on the membrane surface. This strategy of course also applies to on-chip droplet generators. Hence, for example, for oil-in-water emulsions, it is important to choose surfactants that do not carry the charge opposite to that of the membrane (or chip) surface so that the membrane retains its hydrophilicity. Furthermore, it is also advantageous to add the surfactant to the continuous phase. If the surfactant is added to the dispersed phase, it increases the chances that the surfactant will coat the membrane (or chip) allowing the dispersed phase to wet the chip, which interferes with emulsification.

TABLE 1. Popular water-in-oil emulsion systems including surfactant used to stabilize the system.

Continuous phase	Viscosity (mPa s)	Surface tension (mN/m)	Density (g/ml)	Surfactant	Application, Notes	Ref
Water-in-silicon oil						
AR20 (Polyphenyl-methylsiloxane)	20		1.01	Triton X-100	PCR, popular	1
				ABIL EM90		2
DC200 (PDMS)	1, 10, 100, 1000	20-22	~1	None		3
Water-in-hydrocarbon						
Hexadecane	3.5	27.5	0.77	Span 80	Very popular	4
				Tween20/80		4, 5
				Synperonic		6
				SDS		4,6
Tetra / octa /dodecane	2.13 / ??? / 1.36	26.6 / 27.0 / 25.4	0.76 / 0.78 / 0.78	Span 80		7
				Phospholipids	Lipid bilayers	8
Mineral oil	Varies, 0.3-72,000	~29	~0.85	Span 80	Very popular	9
				Oleic acid		10
				C12E8		11
				n-Butanol		12
				ABIL EM90	PCR	13
				None	PCR	14
Isopar M	3.8	25	0.79	Span 80		15
Squalane	~12	28	0.8	Monolein	Monolayers	16
Sunflower oil	36	7.4	0.91	None		17
Water-in-fluorinated oil						
HFE/Novec, e.g., 7100	0.45	13.6	1.52	PFPE-COOH		18
				PFPE-COONH4		19
				PFPE-PEG	PCR, very popular	14
				PFPE-DMP	Cells in droplets	20
FC 40	4.1	16	1.86	PF-octanol		21
				PF-decanol	PCR	22
				PFPE-COOH		23
				PFPE-COONH4	Cells in droplets	24
				PFPE-PEG	Cells in droplets, very popular	24
				PFPE-DMP	Bacteria in droplets	25

TABLE 2. Abriviated references for Table 1

#	Reference
1	P. Kumaresan, C. J. Yang, S. A. Cronier, R. G. Blazej and R. A. Mathies, Anal. Chem., 2008, 80, 3522–3529.
2	B. M. Paegel and G. F. Joyce, Chem. Biol., 2010, 17, 717–724.
3	V. Trivedi, A. Doshi, G. K. Kurup, E. Ereifej, P. J. Vandevord and A. S. Basu, Lab Chip, 2010, 10, 2433–2442.
4	L. Shui, A. van den Berg and J. C. T. Eijkel, Lab Chip, 2009, 9, 795–801.
5	K. Wang, Y. Lu, J. Xu and G. Luo, Langmuir, 2009, 25, 2153–2158.
6	M. L. J. Steegmans, A. Warmerdam, K. G. F. H. Schroen and R. M. Boom, Langmuir, 2009, 25, 9751–9758.
7	H. Willaime, V. Barbier, L. Kloul, S. Maine and P. Tabeling, Phys. Rev. Lett., 2006, 96, 054501
8	S. Punnamaraju and A. J. Steckl, Langmuir, 2011, 27, 618–626.
9	P. S. Dittrich, M. Jahnz and P. Schwille, ChemBioChem, 2005, 6, 811–814.
10	T. Ward, M. Faivre and H. Stone, Langmuir, 2010, 26, 9233–9239.
11	S. L. Anna and H. C. Mayer, Phys. Fluids, 2006, 18, 121512.
12	J. D. Martin and S. D. Hudson, New J. Phys., 2009, 11, 115005.
13	Y. Schaerli, V. Stein, M. M. Spiering, S. J. Benkovic, C. Abell and F. Hollfelder, Nucleic Acids Res., 2010, 38, e201.
14	D. Pekin, Y. Skhiri, J.-C. Baret, D. L. Corre, L. Mazutis, C. B. Salem, F. Millot, A. E. Harrak, J. B. Hutchison, J. W. Larson, D. R. Link, P. Laurent-Puig, A. D. Griffiths and V. Taly, Lab Chip, 2011, 11, 2156–2166.
15	C. Priest, S. Herminghaus and R. Seemann, Appl. Phys. Lett., 2006, 89, 134101–3.
16	S. Thutupalli, S. Herminghaus and R. Seemann, Soft Matter, 2010.
17	M. Belloul, W. Engl, A. Colin, P. Panizza and A. Ajdari, Phys. Rev. Lett., 2009, 102, 194502.
18	A. R. Abate and D. A. Weitz, Lab Chip, 2011, 11, 1911–1915.
19	A. R. Abate and D. A. Weitz, Small, 2009, 5, 2030–2032.
20	P. Abbyad, P.-L. Theraux, J.-L. Martin, C. N. Baroud and A. Alexandrou, Lab Chip, 2010, 10, 2505–2512.
21	C. King, E. Walsh and R. Grimes, Microfluid. Nanofluid., 2007, 3, 463–472.
22	M. Chabert, K. D. Dorfman, P. de Cremoux, J. Roeraade and J.-L. Viovy, Anal. Chem., 2006, 78, 7722–7728.
23	M. Zagnoni, G. L. Lain and J. M. Cooper, Langmuir, 2010, 26, 14443–14449.
24	J. Clausell-Tormos, D. Lieber, J.-C. Baret, A. El-Harrak, O. J. Miller, L. Frenz, J. Blouwolff, K. J. Humphry, S. Koester, H. Duan, C. Holtze, D. A. Weitz, A. D. Griffiths and C. A. Merten, Chem. Biol., 2008, 15, 427–437.
25	P. Marcoux, M. Dupcy, R. Mathey, A. Novelli-Rousseau, V. Heran, S. Morales, F. Rivera, P. Joly, J. Moy and F. Mallard, Colloids Surf., A, 2011, 377, 54–62.

Part 3

Electrokinetics

We now come to the last fundamental topic of this book: the influence of electric fields on the motion of fluids and particles and molecules within the fluids - electrokinetics. Some readers, especially those that design devices without the intention of placing an electric field in their device may find surprising that electrokinetics is still important to the design of their device. As we will see in the next sections that even in these devices electric effects can play significant roles: through free charge at boundaries, and through gradients in concentration, conductivity and permittivity.

We will start by considering the two fundamental electrical forces that arise in a fluid and derive the so called fluid force equation. We will then look at free charge at interfaces and explore the electrical double layer. Thirdly, we will explore electrokinetic phenomena enabled by presence of free charge. Fourthly, we will explore electrokinetic phenomena enabled by dipoles and polarization. Lastly, we will consider electrokinetic phenomena coupled with capillarity.

CHAPTER 14

Electrical forces on fluids

The two electrical forces that can be applied on fluid are due to free charge, and due to polarization. The force due to free charge can be considered from the point of view of Coulomb's law - force on a free charge is proportional to the magnitude of the charge and the electric field placed on it. This free charge (ions in a liquid is of highest interest to us here) then collides with non-charged molecules in the fluid, imparting momentum to them and "dragging" them, and so the whole fluid along. Thus, the force density (i.e., force per unit volume) on the fluid due to free charge can be written as

(14.0.1) $$\mathbf{f}_f = \rho_f \mathbf{E}$$

where ρ_f is the local density of free charge in the fluid, and $\mathbf{E}$ is the electric field, as usual. The situation with the force due to polarization is more subtle. Hence we will first consider a very simple case of polarization forces in parallel plate electrodes and only then derive the general relation for the polarization force.

To derive the polarization force, let's consider polarization forces on an isotropic incompressible media arranged between parallel plate electrodes (Figure 14.0.1). Majority of situations of interest (e.g., bulk aqueous solutions) are indeed isotropic and incompressible. As we see in Figure 14.0.1 the interface between the two media can be arranged either to lie normal to the surfaces of the electrodes or tangentially to the surfaces of the electrodes. The first case resembles two ideal parallel plate capacitors in parallel, while the second resembles two parallel plate capacitors in series. Our goal will be to find the electrical forces on the media and we will employ conservation of energy arguments to do this.

We define electrical energy stored in a dielectric (or free space) as W and the total force on the dielectric in the z direction as F. Conservation of energy tells us that in an increment of time dt the electrical power entering our capacitor either increases the stored electrical energy or does mechanical work on the dielectric. Thus, we can write that, by conservation of energy

(14.0.2) $$V dq = dW + F d\xi$$

where V is the potential between the two plates, and q is the total charge on the electrode connected to the positive electrode. As usual $V dq$ represents the electrical energy and $F dz$ the mechanical work on the dielectric. It is easier to think that the electrode plates are constrained to have constant energy instead of constant charge (and this is easier to implement experimentally). Hence we can use the following transformation: $W' = qV - W$. W' is termed co-energy. Differentiating this gives $dW' = dqV + qdV - dW$. Combining this with (14.0.2) gives

(14.0.3) $$dW' = qdV + F d\xi$$

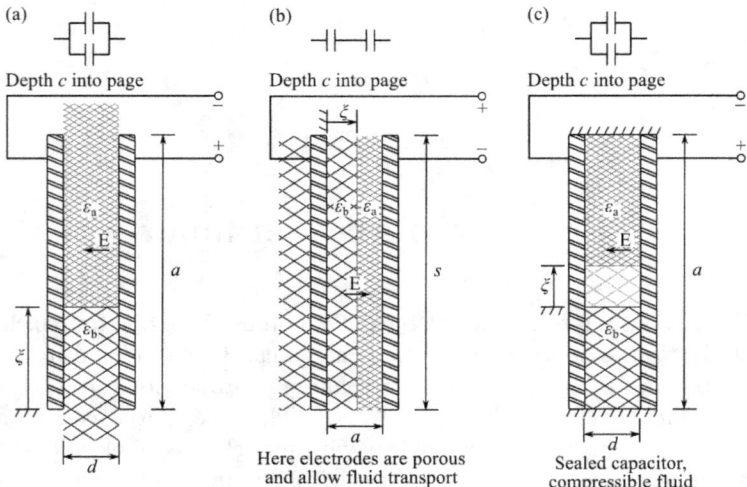

FIGURE 14.0.1. Schematics of simple parallel plate capacitor arrangements to explore forces on an inhomongeneous dielectric inside a capacitor. In (a) the capacitor is open in the vertical direction and the dielectric is able to vertically move in the capacitor; the dielectric is incompressible. The gradient (here jump) in permittivity is normal to the electric field. In (b) the capacitor is no longer open in the vertical direction, but has idealized porous electrodes (i.e. allow fluid dielectric through, but still appear like a flat sheet), and allow the dielectric to move in the horizontal direction. The permittivity gradient here is parallel to the field. In (c) the capacitor is sealed from all sides. However, the dielectrics are compressible. In all cases ξ is the distance traveled by the interface from the reference shown.

Note that W' is a function of both the voltage on the electrodes V and the position of the interface z. To find W' we first integrate over z, keeping in mind that initially there is no potential on the electrodes, and so there is no electrical field between the electrodes, and so moving the interface of the dielectrics in z contributes no energy, as there is no force on the dielectrics. Next, we integrate over V and that gives us

$$(14.0.4) \qquad\qquad W' = \int q dV$$

We have assumed that the capacitor is ideal, and so ignore the fringing fields (this is a pretty good approximation when the length of the plates is large compared to the gap and the interface is far from the ends). Then the charge on the capacitor in the capacitors in parallel case is

$$(14.0.5) \qquad\qquad q_p = V\frac{c}{d}\varepsilon_0\left[(a-\xi)\,\varepsilon_a + \xi\varepsilon_b\right]$$

and that in the serial case is

$$(14.0.6) \qquad\qquad q_s = \frac{csV\varepsilon_0\varepsilon_a}{a - \xi\left(1 - \varepsilon_a/\varepsilon_b\right)}$$

where c is the depth of the capacitor plates into the page, ε_0 is the permittivity of free space, ε_a and ε_b are the dielectric constants of the respective materials, and the other dimensions are as labeled in Figure 14.0.1a and b. When we fix the potential on the electrode plates, V then, W' is only a function of the position of the interface, z. Thus

$$(14.0.7) \qquad dW' = \frac{\partial W'}{\partial \xi} d\xi$$

and we can rewrite (14.0.3) for this particular situation (fixed potential, i.e. $dV = 0$) as

$$(14.0.8) \qquad \left(\frac{\partial W'}{\partial \xi} - f \right) d\xi = 0$$

Since this relation must hold for all values of z, then

$$(14.0.9) \qquad \frac{\partial W'}{\partial \xi} - f = 0$$

We now can calculate the force F using (14.0.9) by calculating W' via (14.0.4) using the either (14.0.5) or (14.0.6) for the appropriate case. For the parallel case we obtain

$$(14.0.10) \qquad F_p = \frac{cd}{2} \left(\frac{V}{d} \right)^2 \varepsilon_0 \left(\varepsilon_b - \varepsilon_a \right)$$

and for the serial case we obtain

$$(14.0.11) \qquad F_s = \frac{cs\varepsilon_0\varepsilon_a}{2} \frac{(1 - \varepsilon_a/\varepsilon_b)\, V^2}{((a - \xi) + \xi\varepsilon_a/\varepsilon_b)^2}$$

The forces for both of these cases are due to the inhomogeneity of the material between the electrodes. Notice that despite the fact that in one case the material interface was parallel with the field (capacitors in parallel case) and perpendicular to the field in the other (serial case) the force scales similarly - as the square of the potential between the plates. For both cases, notice that the force (in the direction of z) is positive when $\varepsilon_b > \varepsilon_a$. In other words, the capacitor draws in the more polarizable material and forces out a less polarizable material. We shall informally term this force "material inhomogeneity force."

Next, we consider another incarnation of the capacitor problem, which we sketch in Figure 14.0.1c. In this case, the parallel plate capacitor is filled with two fluids, but the fluids in the capacitor are now confined inside. In other words, the capacitor is fully closed, and no new fluid may enter. However, the fluids now are somewhat compressible and so movement of the interface in the positive direction shown is due to compression of fluid a and expansion of fluid b. We again assume that the fringing fields are negligible. This case resembles the capacitors in parallel case from the previous problem, and so the charge on the electrodes is again

$$(14.0.12) \qquad q_p = V \frac{c}{d} \varepsilon_0 \left[(a - \xi)\, \varepsilon_a + (b + \xi)\, \varepsilon_b \right]$$

Again, applying (14.0.4) we obtain that

$$(14.0.13) \qquad W' = \frac{1}{2} cd \left(\frac{V}{d} \right)^2 \varepsilon_0 \left[(a - \xi)\, \varepsilon_a + (b + \xi)\, \varepsilon_b \right]$$

The magnitude of permittivity of a material represents how well it can be polarized, and so can be expected to scale with the density of dipoles inside of it. If we increase the density of the material, we can expect to increase the density of dipoles as well. Hence changing the density of fluids in this example will change the permittivity of each fluid. Thus, in this example, permittivity is a function of interface position z. We take this into account and use (14.0.9) to find the force on the interface

$$(14.0.14) \qquad F = \frac{cd}{2}\left(\frac{V}{d}\right)^2 \varepsilon_0 \left[(\varepsilon_b - \varepsilon_a) + \left[(a - \xi)\frac{\partial \varepsilon_a}{\partial \xi} + (b + \xi)\frac{\partial \varepsilon_b}{\partial \xi}\right]\right]$$

Observe that (14.0.14) is identical to (14.0.10) with the exception of an additional term accounting for change in permittivity as a function of position. To simplify (14.0.14) we employ mass conservation and assume that the density in each phase is uniform. Note, that of course the density can change as the location of the interface changes, i.e. as z changes. Considering the mass of each fluid as the interface moves is constant, we can write

$$(14.0.15) \qquad cd\rho_a (a - \xi) = const$$

$$(14.0.16) \qquad cd\rho_b (b + \xi) = const$$

Differentiating (14.0.15) and (14.0.16) with respect to z

$$(14.0.17) \qquad \frac{\partial \rho_a}{\partial z}(a - \xi) - \rho_a = 0$$

$$(14.0.18) \qquad \frac{\partial \rho_b}{\partial z}(b + \xi) + \rho_b = 0$$

Let's now assume that permittivity is only a function of density in this system. Thus

$$(14.0.19) \qquad \begin{aligned} \frac{\partial \varepsilon_a}{\partial \xi} &= \frac{\partial \varepsilon_a}{\partial \rho_a}\frac{\partial \rho_a}{\partial \xi} \\ \frac{\partial \varepsilon_b}{\partial \xi} &= \frac{\partial \varepsilon_b}{\partial \rho_b}\frac{\partial \rho_b}{\partial \xi} \end{aligned}$$

Combining this with (14.0.17) and (14.0.18)we obtain

$$(14.0.20) \qquad \begin{aligned} \frac{\partial \varepsilon_a}{\partial \xi} &= \frac{\partial \varepsilon_a}{\partial \rho_a}\frac{\rho_a}{a - \xi} \\ \frac{\partial \varepsilon_b}{\partial \xi} &= -\frac{\partial \varepsilon_b}{\partial \rho_b}\frac{\rho_b}{b + \xi} \end{aligned}$$

We combine this result with (14.0.14) to obtain the force on the interface

$$(14.0.21) \qquad F = \frac{cd}{2}\left(\frac{V}{d}\right)^2 \varepsilon_0 \left[(\varepsilon_b - \varepsilon_a) + \left[\rho_a\frac{\partial \varepsilon_a}{\partial \rho_a} - \rho_b\frac{\partial \varepsilon_b}{\partial \rho_b}\right]\right]$$

By adding the ability of the material to be compressible and have a finite change of permittivity with density we observe another electrical force that can be applied to the fluid. This force is sometimes termed "electrostrictive force".

So far, we have limited ourselves to uniform electric fields. Next we shall consider the case where the electric field is not uniform, by generalizing the parallel plate capacitor problem we just solved. Consider a situation sketched in Figure 14.0.2, where we dip arbitrarily shaped ideally conducting electrodes into a generally

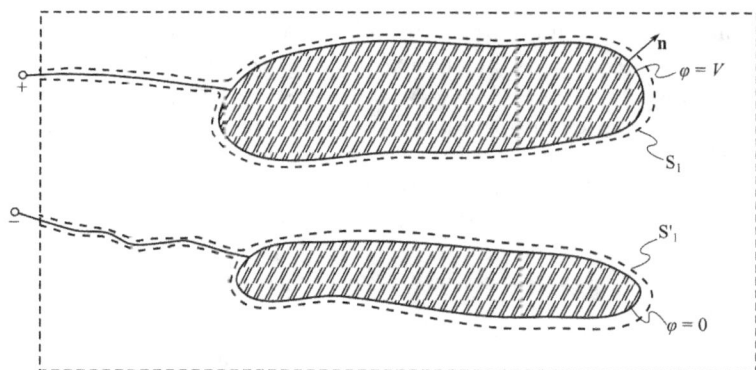

FIGURE 14.0.2. Schematic of two arbitrarily shaped electrodes surrounded by a medium with generally inhomogeneous dielectric constant. We draw a control volume rectangular box U around our region of interest. This volume includes all of the dielectric media, but excludes the electrodes.

inhomogeneous dielectric fluid. We assume that in this case the effects of magnetic induction are negligible and so the electric field is irrotational, and so we define a potential φ such that, as per usual definition, $\mathbf{E} = -\nabla\varphi$. Our arbitrarily shaped electrodes are connected by infinitesimally thin wires to the outside of the control volume. We use these wires to place a potential $\varphi = V$ on the upper electrode and $\varphi = 0$ on the lower electrode.

To find the force density in the dielectric media, we proceed as we have done before. That is we employ conservation of energy, and so we write the three dimensional analogue of (14.0.3) as

$$(14.0.22) \qquad \int_{S_1} \sigma_f \delta\varphi \, da = \int_U \delta w' dU - \int_U (\mathbf{f} \cdot \delta\xi) \, dU$$

where the term on the left corresponds to qdV, the first term on the right corresponds to dW' and the second term on the right corresponds to $Fd\xi$. We find the free charge on the positive electrode by integrating the charge density σ_f over the surface of the entire electrode. We then integrate the co-energy and the mechanical work in deforming dielectric locally $(\mathbf{f} \cdot \delta\xi)$ by a small $\delta\mathbf{z}$, where $\mathbf{f}$ is an electrical force density. To simplify this equation, we assume that our dielectric is linear, such that the displacement field $\mathbf{D} = \varepsilon_0\varepsilon\mathbf{E}$. From Gauss's law, the free charge on the surface of the electrode is $\sigma_f = \varepsilon_0\varepsilon\mathbf{E}\cdot\mathbf{n}$ where $\mathbf{n}$ is a unit normal as shown in Figure 14.0.2. The volume U over which we integrate encloses all of the region occupied by the dielectric media (and so the electric field) but excludes that occupied by the electrodes, as shown in Figure 14.0.2.

Note that on the lower electrode the potential is set to zero and the rectangular boundary of our domain is assumed to be in the field free region. Hence we can now apply our latest set of assumptions and simplify equation (14.0.22) using the divergence theorem.

$$(14.0.23) \qquad \int_U \nabla \cdot [\varepsilon_0\varepsilon\nabla\varphi\delta\varphi] dU = \int_U \delta w' dU - \int_U (\mathbf{f} \cdot \delta\xi) \, dU$$

(Note volume U has a unit normal opposite that drawn in Figure 14.0.2, hence when substituting in $\mathbf{E} = -\nabla\varphi$, the negatives cancel out). As before, to perform the integration, we first consider that we put the system together mechanically first (i.e. insert the electrodes into the fluid) and only then apply a voltage to the top electrode. Hence, just after the first step, the voltage is zero on both electrodes and so there is no electrical force on the medium. Thus, the second term on the right-hand side is zero. Next we expand the left-hand side term as

$$(14.0.24) \qquad \int_U \nabla \cdot [\varepsilon_0 \varepsilon \nabla\varphi \delta\varphi] dU = \int_U [\delta\varphi (\nabla \cdot \varepsilon_0 \varepsilon \nabla\varphi) + \varepsilon_0 \varepsilon \nabla\varphi \cdot \nabla (\delta\varphi)] dU$$

Note that $\nabla \cdot \varepsilon_0 \varepsilon \nabla\varphi = -\rho_f$, the free charge density. But, we know that free charge density is zero in the control volume U, and so

$$(14.0.25) \qquad \int_U [\varepsilon_0 \varepsilon \nabla\varphi \cdot \nabla (\delta\varphi)] dU = \int_U \delta w' dU$$

Next we exchange the order of differentiation, and recalling that ε is fixed as long as $\mathbf{z}$ is fixed, we obtain that

$$(14.0.26) \qquad \int_U \delta \left[\frac{\varepsilon_0 \varepsilon \nabla\varphi \cdot \nabla\varphi}{2} \right] dU = \int_U \delta w' dU$$

This is true when

$$(14.0.27) \qquad w' = \frac{1}{2} \varepsilon_0 \varepsilon \nabla\varphi \cdot \nabla\varphi$$

Next, as in the parallel plate capacitor example, we find the force density $\mathbf{F}$ by considering a process where a potential is held constant on the top electrode and our dielectric material undergoes a small change in displacement $\delta\mathbf{z}$. For this case, in (14.0.22) $\delta\varphi$ is zero as the top electrode potential is fixed. Therefore for this case (14.0.22) simplifies to material inhomogeneity force

$$(14.0.28) \qquad \int_U [\delta w' - (\mathbf{f} \cdot \delta\xi)] \, dU = 0$$

To obtain the force density $\mathbf{F}$, we place $\delta w'$ in the form of $(\ldots) \cdot \delta\mathbf{z}$. We start by taking the derivative and expanding the expression for coenergy (14.0.27) as

$$(14.0.29) \qquad \delta w' = \frac{1}{2} \varepsilon_0 [\varepsilon \delta (\nabla\varphi \cdot \nabla\varphi) + (\nabla\varphi \cdot \nabla\varphi) \delta\varepsilon]$$

In our case, the electrode potential is fixed and so $\delta (\nabla\varphi \cdot \nabla\varphi) = 0$. In our case, the small changes in coenergy have to come from small changes in permittivity. Thus we need to find how a change in material displacement $\delta\mathbf{z}$ relates to $\delta\varepsilon$. As we have seen in the simple capacitor problems, this can take place if the material between the electrodes is inhomogeneous in permittivity (our first two capacitor examples) or due to the compressibility of the material (our third example). To find this, we consider that a chunk of fluid was initially at position $\mathbf{r} - \delta\xi$ and moves to a position $\mathbf{r}$ after the displacement $\delta\xi$. Hence if after displacement $\varepsilon = \varepsilon(\mathbf{r})$, then before displacement

$$(14.0.30) \qquad \varepsilon = \varepsilon(\mathbf{r} - \delta\xi) \approx \varepsilon(\mathbf{r}) - \delta\xi \cdot \nabla\varepsilon$$

by Taylor expansion. Thus

(14.0.31) $\delta\varepsilon = -\delta\xi \cdot \nabla\varepsilon$

We can substitute this result into (14.0.29) and then substitute that result into (14.0.28) and recall that we defined $\mathbf{E} = -\nabla\varphi$ and obtain

(14.0.32) $\int_U \left[\frac{1}{2}\varepsilon_0\mathbf{E}\cdot\mathbf{E}\nabla\varepsilon + \mathbf{f}\right]\cdot\delta\xi dU = 0$

Since $\delta\mathbf{z}$ is arbitrary, to make the integral zero

(14.0.33) $\frac{1}{2}\varepsilon_0\mathbf{E}\cdot\mathbf{E}\nabla\varepsilon + \mathbf{f} = 0$

Let's compare this force to that obtained in our first two capacitor problems. Firstly, in both cases, the force scales as the square of the voltage difference between the electrodes, as expected. In fact, now we can clearly see that this force scales as the square of the electric field. Secondly, in both cases, as expected, the force is in the direction of a negative gradient of permittivity. That is the force acts to move a higher permittivity more polarizable material into the field. This is our familiar "material inhomogeneity force".

Now let's find the electrostrictive force. In a compressible fluid, as we saw before, displacement of a fluid lead to changes in density, which in turn, lead to change in permittivity (e.g., due to change in the density of the dipoles). Therefore, when a cube with dimensions $\Delta x_1, \Delta x_2, \Delta x_3$ is displaced by $\delta\mathbf{z}$ its volume changes to

(14.0.34) $\Delta x_1\Delta x_2\Delta x_3\left(1 + \nabla\cdot\delta\xi\right)$

However, we require that mass be conserved, and so

(14.0.35) $\rho\Delta x_1\Delta x_2\Delta x_3 = (\rho + \delta\rho)\left(1 + \nabla\cdot\delta\xi\right)\Delta x_1\Delta x_2\Delta x_3$

Simplifying this and taking the first order terms,

(14.0.36) $\delta\rho = -\rho\nabla\cdot\delta\xi$

Since we have assumed $\varepsilon = \varepsilon(\rho)$ then

(14.0.37) $\delta\varepsilon = -\rho\frac{\partial\varepsilon}{\partial\rho}\nabla\cdot\delta\xi$

We now combine this with (14.0.31) and substitute this into (14.0.29) and then into (14.0.28) and obtain

(14.0.38) $\int_U \left[\frac{1}{2}\varepsilon_0 E^2\nabla\varepsilon\cdot\delta\xi + \frac{1}{2}\varepsilon_0 E^2\rho\frac{\partial\varepsilon}{\partial\rho}\nabla\cdot\delta\xi + \mathbf{f}\cdot\delta\xi\right]dU = 0$

The second term of the integrand is still not in the form that we require but we can use the divergence theorem and the distributive identity $\psi\nabla\cdot A = \nabla\cdot A\psi - A\cdot\nabla\psi$ to rewrite the term as

(14.0.39)

$\int_U \left[\frac{1}{2}\varepsilon_0 E^2\rho\frac{\partial\varepsilon}{\partial\rho}\nabla\cdot\delta\xi\right]dU = \int_S \left[\frac{1}{2}\varepsilon_0 E^2\rho\frac{\partial\varepsilon}{\partial\rho}\delta\xi\cdot\mathbf{n}\right]da - \int_U \left[\nabla\left(\frac{1}{2}\varepsilon_0 E^2\rho\frac{\partial\varepsilon}{\partial\rho}\right)\cdot\delta\xi\right]dU$

With the surface integral we are quite in luck. Recall that the surface S, the skin of volume U is such that the outside (the rectangular boundary) portion lies in a field free region (far enough away from the electrodes); meanwhile the inside portion lies on the surface of the rigid electrodes. Thus $\delta\xi \cdot \mathbf{n}$ is zero in both places, and the surface integral is zero. Substituting this into (14.0.38), we obtain

$$(14.0.40) \qquad \int_U \left[\frac{1}{2}\varepsilon_0 E^2 \nabla\varepsilon - \nabla\frac{1}{2}\left(\varepsilon_0 E^2 \rho \frac{\partial\varepsilon}{\partial\rho} \right) + \mathbf{f} \right] \cdot \delta\xi dU = 0$$

Again, to make the integral zero, the term in the square brackets must be zero. Hence

$$(14.0.41) \qquad \mathbf{f} = -\frac{\varepsilon_0 E^2}{2}\nabla\varepsilon + \nabla\left(\frac{\varepsilon_0 E^2}{2}\rho\frac{\partial\varepsilon}{\partial\rho} \right)$$

Let's compare this force to the force we obtained in our third capacitor problem. In both cases we properly recover the "material inhomogeneity force" term. In addition, we have our electrostrictive force term, which once again scales as the voltage between the electrodes squared and is in the direction of the gradient in $\rho\partial\varepsilon/\partial\rho$. This term arises because we allowed the fluid to be compressible and so our fluid has to be compressible for us to expect this term to arise. Furthermore, let's expand this relation to obtain

$$(14.0.42) \qquad \mathbf{f} = \frac{\varepsilon_0}{2}\left[-E^2\nabla\varepsilon + \rho\frac{\partial\varepsilon}{\partial\rho}\nabla\left(E^2 \right) + E^2\frac{\partial\varepsilon}{\partial\rho}\nabla\rho + \rho E^2\nabla\left(\frac{\partial\varepsilon}{\partial\rho} \right) \right]$$

This way, we clearly see that in a compressible media, electrical polarization force may arise due to four different inhomogeneities. These forces are sometimes called ponderomotive forces. It may arise due to a gradient in (a) dielectric constant, (b) electric field squared, (c) density, (e) change of dielectric constant with density. These gradients may arise naturally, e.g., due to several different media being present in the electric field, such as in our capacitor problems. Secondly, these may also be imposed, e.g., by changing the temperature of some area of the media in the field, say with a laser. Thirdly, these may be imposed by the electric field passing through the media via uneven heating of the media, and thus causing gradients in these properties. Similarly, if large enough pressures are allowed to build up in the system, the electrical forces may exert pressures on the fluid, and in turn cause changes in both density and dielectric constant. Depending on the material, gradients in temperature and pressure may also produce additional forces. Combining this force equation with the force due to free charge,

$$(14.0.43) \qquad \mathbf{f} = \rho_f\mathbf{E} + \frac{\varepsilon_0}{2}\left[-E^2\nabla\varepsilon + \rho\frac{\partial\varepsilon}{\partial\rho}\nabla\left(E^2 \right) + E^2\frac{\partial\varepsilon}{\partial\rho}\nabla\rho + \rho E^2\nabla\left(\frac{\partial\varepsilon}{\partial\rho} \right) \right]$$

This is known as the Korteweg-Helmholtz equation, and represents the electrical forces in an ordinary, compressible linear media. Just as a reminder, $\mathbf{f}$ is a force density, i.e. force per unit volume.

14.1. Maxwell stress tensor and surface forces

We notice that if we apply polarization force equation (14.0.42) directly to our capacitor problems, we will see that at the boundaries of materials, where there is an abrupt change in properties, we will find that the force there is singular, as the gradients in the properties are infinite. This poses quite a significant problem, especially since the problem of a field through immiscible media, and so a field with an equivalent media with abrupt changes in properties are frequently encountered,

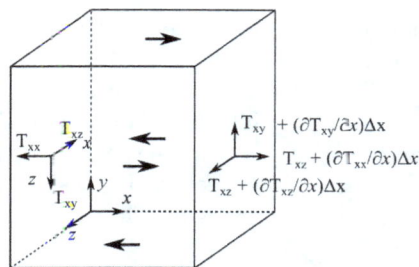

FIGURE 14.1.1. Unit cube, Δx by Δy by Δz, illustrating stresses in a material. Each face of the cube has three stresses: one normal to the face and two tangential. For each stress, we write the first subscript to designate the normal direction of the face, and the second subscript to designate the direction of the stress. For clarity, we only drew the stresses on the x direction face of the cube. With bold arrows we designate additional stresses that are relevant to the force in the x direction on the cube.

e.g., air-liquid interface in digital microfluidics. Hence to handle this, we make a stress-tensor representation of the force density. Here we briefly review what a stress tensor is and then derive that for the electric forces.

Recall that we can define an infinitesimal cube, with three stresses on each of its surfaces, corresponding to each of the axis. We show such a cube in Figure 14.1.1. On each face, there is a normal stress, pointing in the direction normal to that particular face of the cube and two shear stresses pointing in the other two dimensions. For each stress, we write the first subscript to designate the normal direction of the face, and the second subscript to designate the direction of the stress.

To find the force, for example, in the x direction we simplify sum the stresses in that direction, multiplying by the surface area of the face:

$$(14.1.1) \quad f_x \Delta x \Delta y \Delta z = \left(T_{xx} + \frac{\partial T_{xx}}{\partial x} - T_{xx} \right) \Delta y \Delta z$$

$$+ \left(T_{yx} + \frac{\partial T_{yx}}{\partial x} - T_{yx} \right) \Delta x \Delta z + \left(T_{zx} + \frac{\partial T_{zx}}{\partial x} - T_{zx} \right) \Delta x \Delta y$$

where f_x as before is the force density (hence the multiplication by the unit volume to obtain the force). We can simplify this to

$$(14.1.2) \qquad\qquad f_x = \frac{\partial T_{xx}}{\partial x} + \frac{\partial T_{yx}}{\partial y} + \frac{\partial T_{zx}}{\partial z}$$

Similarly, we can write,

$$f_y = \frac{\partial T_{xy}}{\partial x} + \frac{\partial T_{yy}}{\partial y} + \frac{\partial T_{zy}}{\partial z}$$

$$(14.1.3)$$

$$f_y = \frac{\partial T_{xz}}{\partial x} + \frac{\partial T_{yz}}{\partial y} + \frac{\partial T_{zz}}{\partial z}$$

Thus

$$(14.1.4) \qquad\qquad \mathbf{f} = \nabla \cdot T$$

the force density is equal to the divergence of the stress tensor. For ordinary media (e.g. Newtonian fluids) with which we are concerned here, the stress tensor is symmetric, i.e., $T_{ij} = T_{ji}$. This is a consequence of conservation of angular momentum, i.e., the fact that at equilibrium summation of moments with respect to an arbitrary point is zero. We can now write the force density in tensor notation, where i designates the direction of the vector. Just as a reminder, repeated index designates summation (even if the index is in the denominator),

$$(14.1.5) \qquad\qquad f_i = \frac{T_{ij}}{\partial x_j}$$

Now we are ready to write the force density due to polarization, i.e., equation (14.0.41) as a tensor equation. We first expand E^2 as $\mathbf{E} \cdot \mathbf{E}$ and obtain

$$(14.1.6) \qquad \mathbf{f} = -\frac{\varepsilon_0 \left(\mathbf{E} \cdot \mathbf{E}\right)}{2} \nabla \varepsilon + \nabla \left(\frac{\varepsilon_0 \left(\mathbf{E} \cdot \mathbf{E}\right)}{2} \rho \frac{\partial \varepsilon}{\partial \rho} \right)$$

We then write this in tensor notation. Just as a reminder, $\mathbf{E} \cdot \mathbf{E} = E_j E_j = E_x E_x + E_y E_y + E_z E_z$. The repeated index can be assigned arbitrarily, i.e., $E_j E_j = E_k E_k$. Thus

$$(14.1.7) \qquad f_i = -\frac{\varepsilon_0}{2} E_j E_j \frac{\partial \varepsilon}{\partial x_i} + \frac{\partial}{\partial x_i} \left(\frac{\varepsilon_0}{2} E_k E_k \rho \frac{\partial \varepsilon}{\partial \rho} \right)$$

Since we are only considering the polarization force, the divergence of the displacement field $\mathbf{D} = \varepsilon \varepsilon_0 \mathbf{E}$ is zero. Furthermore, since we before assumed that magnetic fields are not varying (and really are weak, and negligible), the curl of electric field is zero. We can write the curl of electric field being zero as

$$(14.1.8) \qquad\qquad \frac{\partial E_j}{\partial x_i} - \frac{\partial E_i}{\partial x_j} = 0$$

Multiplying this zero by the displacement field and expanding

$$(14.1.9) \qquad 0 = -D_j \left(\frac{\partial E_j}{\partial x_i} - \frac{\partial E_i}{\partial x_j} \right) = -\varepsilon \varepsilon_0 E_j \frac{\partial E_j}{\partial x_i} + D_j \frac{\partial E_i}{\partial x_j}$$

$$-\varepsilon \varepsilon_0 E_j \frac{\partial E_j}{\partial x_i} + D_j \frac{\partial E_i}{\partial x_j} = -\varepsilon \varepsilon_0 E_j \frac{\partial E_j}{\partial x_i} + \left(\frac{\partial}{\partial x_j} \left(E_i D_j \right) - E_i \frac{\partial D_j}{\partial x_j} \right)$$

$$(14.1.10) \qquad = -\varepsilon \varepsilon_0 \left(\frac{1}{2} \frac{\partial}{\partial x_i} \left(E_j E_j \right) \right) + \left(\frac{\partial}{\partial x_j} \left(E_i D_j \right) - E_i \frac{\partial D_j}{\partial x_j} \right)$$

$$= -\varepsilon \varepsilon_0 \left(\frac{1}{2} \frac{\partial}{\partial x_i} \left(E_j E_j \right) \right) + \left(\frac{\partial}{\partial x_j} \left(E_i D_j \right) \right)$$

Note $\partial D_j / \partial D_j = \nabla \cdot \mathbf{D} = 0$. Hence, we can add this to (14.1.7) so that
(14.1.11)

$$f_i = -\frac{\varepsilon_0}{2} E_j E_j \frac{\partial \varepsilon}{\partial x_i} - \varepsilon \varepsilon_0 \left(\frac{1}{2} \frac{\partial}{\partial x_i} \left(E_j E_j \right) \right) + \left(\frac{\partial}{\partial x_j} \left(E_i D_j \right) \right) + \frac{\partial}{\partial x_i} \left(\frac{\varepsilon_0}{2} E_k E_k \rho \frac{\partial \varepsilon}{\partial \rho} \right)$$

We now recall the properties of the Kronecker delta:

$$(14.1.12) \qquad\qquad \delta_{ij} = \left\{ \begin{array}{l} 0 \ \ i \neq j \\ 1 \ \ \ \ i = j \end{array} \right.$$

$$(14.1.13) \qquad\qquad \delta_{ij} = \delta_{ji}$$

$$(14.1.14) \qquad\qquad a_i = \delta_{ij} a_j$$

(14.1.15)
$$\frac{\partial x_i}{\partial x_j} = \delta_{ij}$$

Expanding the first term using the product rule and using the definition of displacement field

(14.1.16)
$$f_i = -\frac{1}{2}\left(\frac{\partial}{\partial x_i}\left(\varepsilon_0\varepsilon E_j E_j\right) - \varepsilon_0\varepsilon\frac{\partial}{\partial x_i}\left(E_j E_j\right)\right)$$
$$-\varepsilon\varepsilon_0\left(\frac{1}{2}\frac{\partial}{\partial x_i}\left(E_j E_j\right)\right) + \left(\frac{\partial}{\partial x_j}\left(\varepsilon\varepsilon_0 E_i E_j\right)\right) + \frac{\partial}{\partial x_i}\left(\frac{\varepsilon_0}{2}E_k E_k \rho\frac{\partial\varepsilon}{\partial\rho}\right)$$

and canceling out the terms

(14.1.17) $\quad f_i = -\frac{1}{2}\left(\frac{\partial}{\partial x_i}\left(\varepsilon_0\varepsilon E_j E_j\right)\right) + \left(\frac{\partial}{\partial x_j}\left(\varepsilon\varepsilon_0 E_i E_j\right)\right) + \frac{\partial}{\partial x_i}\left(\frac{\varepsilon_0}{2}E_k E_k \rho\frac{\partial\varepsilon}{\partial\rho}\right)$

and applying (14.1.15) to the first and third terms term we obtain:
(14.1.18)
$$f_i = -\frac{1}{2}\left(\frac{\partial}{\partial x_j}\left(\delta_{ij}\varepsilon_0\varepsilon E_j E_j\right)\right) + \left(\frac{\partial}{\partial x_j}\left(\varepsilon\varepsilon_0 E_i E_j\right)\right) + \frac{\partial}{\partial x_j}\left(\delta_{ij}\frac{\varepsilon_0}{2}E_k E_k \rho\frac{\partial\varepsilon}{\partial\rho}\right)$$

We compare this to the force density written in terms of tensor components (14.1.5), note that $E_j E_j = E_k E_k$ and obtain that

(14.1.19)
$$T_{ij} = -\delta_{ij}\frac{\varepsilon_0}{2}\varepsilon E_k E_k + \varepsilon\varepsilon_0 E_i E_j + \delta_{ij}\frac{\varepsilon_0}{2}E_k E_k \rho\frac{\partial\varepsilon}{\partial\rho}$$

which we clean up to obtain

(14.1.20)
$$T_{ij} = \varepsilon\varepsilon_0 E_i E_j - \frac{1}{2}\delta_{ij}\varepsilon\varepsilon_0 E_k E_k + \frac{1}{2}\delta_{ij}\varepsilon_0 E_k E_k \rho\frac{\partial\varepsilon}{\partial\rho}$$

This is the Maxwell stress tensor for a compressible fluid without free charge. Notice that the third term came from the force due to compressibility of the fluid, whereas the first and second terms are from the "material inhomogeneity force."

Now we will obtain the Maxwell stress tensor for when free charge is present, e.g., from equation (14.0.43). We begin by writing (14.0.43) in a more compact form, similar to (14.1.6):

(14.1.21)
$$\mathbf{f} = \rho_f\mathbf{E} - \frac{\varepsilon_0\left(\mathbf{E}\cdot\mathbf{E}\right)}{2}\nabla\varepsilon + \nabla\left(\frac{\varepsilon_0\left(\mathbf{E}\cdot\mathbf{E}\right)}{2}\rho\frac{\partial\varepsilon}{\partial\rho}\right)$$

We then apply the fact that $\rho_f = \nabla\cdot\mathbf{D}$

(14.1.22)
$$\mathbf{f} = \mathbf{E}\left(\nabla\cdot\mathbf{D}\right) - \frac{\varepsilon_0\left(\mathbf{E}\cdot\mathbf{E}\right)}{2}\nabla\varepsilon + \nabla\left(\frac{\varepsilon_0\left(\mathbf{E}\cdot\mathbf{E}\right)}{2}\rho\frac{\partial\varepsilon}{\partial\rho}\right)$$

Expanding this in indicial notation

(14.1.23)
$$f_i = E_i\frac{\partial D_j}{\partial x_j} - \frac{\varepsilon_0}{2}E_j E_j\frac{\partial\varepsilon}{\partial x_i} + \frac{\partial}{\partial x_i}\left(\frac{\varepsilon_0}{2}E_k E_k \rho\frac{\partial\varepsilon}{\partial\rho}\right)$$

We recall that we are assuming linear media, so $\mathbf{D} = \varepsilon\varepsilon_0\mathbf{E}$ and so

(14.1.24)
$$f_i = \varepsilon_0 E_i\frac{\partial}{\partial x_j}\left(\varepsilon E_j\right) - \frac{\varepsilon_0}{2}E_j E_j\frac{\partial\varepsilon}{\partial x_i} + \frac{\partial}{\partial x_i}\left(\frac{\varepsilon_0}{2}E_k E_k \rho\frac{\partial\varepsilon}{\partial\rho}\right)$$

And using the product rule to expand the first and second terms

$$(14.1.25) \quad f_i = \varepsilon_0 \left(\frac{\partial}{\partial x_j}(\varepsilon E_i E_j) - \varepsilon E_j \frac{\partial E_i}{\partial x_j} \right) - \frac{\varepsilon_0}{2} \left(\frac{\partial}{\partial x_i}(\varepsilon E_j E_j) - \varepsilon \frac{\partial}{\partial x_i}(E_j E_j) \right)$$
$$+ \frac{\partial}{\partial x_i} \left(\frac{\varepsilon_0}{2} E_k E_k \rho \frac{\partial \varepsilon}{\partial \rho} \right)$$

and apply the product rule again to the fourth term

$$(14.1.26) \quad f_i = \varepsilon_0 \left(\frac{\partial}{\partial x_j}(\varepsilon E_i E_j) - \varepsilon E_j \frac{\partial E_i}{\partial x_j} \right) - \left(\frac{\varepsilon_0}{2} \frac{\partial}{\partial x_i}(\varepsilon E_j E_j) - \varepsilon_0 \varepsilon E_j \frac{\partial E_j}{\partial x_i} \right)$$
$$+ \frac{\partial}{\partial x_i} \left(\frac{\varepsilon_0}{2} E_k E_k \rho \frac{\partial \varepsilon}{\partial \rho} \right)$$

We then apply equation (14.1.8) to the second term

$$(14.1.27) \quad f_i = \varepsilon_0 \left(\frac{\partial}{\partial x_j}(\varepsilon E_i E_j) - \varepsilon E_j \frac{\partial E_j}{\partial x_i} \right) - \left(\frac{\varepsilon_0}{2} \frac{\partial}{\partial x_i}(\varepsilon E_j E_j) - \varepsilon_0 \varepsilon E_j \frac{\partial E_j}{\partial x_i} \right)$$
$$+ \frac{\partial}{\partial x_i} \left(\frac{\varepsilon_0}{2} E_k E_k \rho \frac{\partial \varepsilon}{\partial \rho} \right)$$

and so eliminate the second and fourth terms

$$(14.1.28) \quad f_i = \varepsilon_0 \left(\frac{\partial}{\partial x_j}(\varepsilon E_i E_j) \right) - \left(\frac{\varepsilon_0}{2} \frac{\partial}{\partial x_i}(\varepsilon E_j E_j) \right) + \frac{\partial}{\partial x_i} \left(\frac{\varepsilon_0}{2} E_k E_k \rho \frac{\partial \varepsilon}{\partial \rho} \right)$$

We then apply (14.1.15) to the second and third terms and obtain

$$(14.1.29) \quad f_i = \varepsilon_0 \left(\frac{\partial}{\partial x_j}(\varepsilon E_i E_j) \right) - \left(\frac{\varepsilon_0}{2} \delta_{ij} \frac{\partial}{\partial x_j}(\varepsilon E_j E_j) \right) + \delta_{ij} \frac{\partial}{\partial x_j} \left(\frac{\varepsilon_0}{2} E_k E_k \rho \frac{\partial \varepsilon}{\partial \rho} \right)$$

Comparing this result to (14.1.5), we obtain that

$$(14.1.30) \qquad T_{ij} = \varepsilon \varepsilon_0 E_i E_j - \frac{1}{2} \delta_{ij} \varepsilon \varepsilon_0 E_k E_k + \frac{1}{2} \delta_{ij} \varepsilon_0 E_k E_k \rho \frac{\partial \varepsilon}{\partial \rho}$$

This is the Maxwell stress tensor for a compressible fluid with free charge. The third term again is due to compressibility of the fluid, while the first two terms are due to both the free charge density and inhomogeneity in the permittivity of the material. Comparing this to (14.1.20) we see that the stress tensor is the same whether there is free charge in the material or there isn't! Thus considering forces due to presence of an interface between two materials (i.e. a jump in permittivity) it is enough to consider the permittivity gradient term, and not worry about the free charge.

14.2. Maxwell stresses at an interface

We are now ready to apply the concept of the stress tensor the problem where two immiscible media (e.g., fluids) are touching each other at an interface and so the gradient of permittivity is zero in the bulk of the fluids and infinite at the interface. Thus we consider an interface shown in Figure 14.2.1. Recall that the force per unit area in ith direction at an interface is

$$(14.2.1) \qquad T_i = n_j (T_{ij,a} - T_{ij,b})$$

where the subscript behind the comma indicates the appropriate phase that meet at the interface, and n_j is the component of the normal vector in the jth direction.

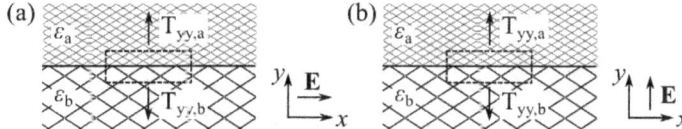

FIGURE 14.2.1. Stresses on an interface between two fluids with different permittivities, when the electric field is directed parallel to the interface (a) and perpendicular to the interface (b).

Here we define the normal vector to point from phase b to phase a. Let's use this method to derive the forces on fluids in a parallel plate capacitor where the electric field is parallel and orthogonal to the interface and assuming that the fluid is incompressible (see Figure 14.0.1a,b). Here the force is only due to the gradient of permittivity.

Let's consider first the case where the electric field is parallel to the interface (Figure 14.2.1a). Using (14.1.20) after dropping the force due to compressible fluid term,

(14.2.2)
$$T_{yy,a} = -\frac{1}{2}\varepsilon_0\varepsilon_a E^2$$
$$T_{yy,b} = -\frac{1}{2}\varepsilon_0\varepsilon_b E^2$$

Note, that the electric field is the same in both phases (imagine capacitor plates to the left and right of the material in Figure 14.2.1). Therefore

(14.2.3)
$$T_y = -\frac{1}{2}\varepsilon_0\left(\varepsilon_a - \varepsilon_b\right)E^2$$

Comparing this to what we derived for the parallel case capacitor, equation (14.0.10) we see that it is identical to the result we have obtained. Next, we write the force per unit area for the case where the electric field is perpendicular to the interface:

(14.2.4)
$$T_{yy,a} = \frac{1}{2}\varepsilon_0\varepsilon_a E_a^2$$
$$T_{yy,b} = \frac{1}{2}\varepsilon_0\varepsilon_b E_b^2$$

Here the electric fields are different in each of the phases. To find the relation between electric fields, we note that the displacement field must be continuous across the interface by Gauss's law, as we assume there is no free charge at the interface. Since there is no free charge, by Gauss's law, the divergence of the displacement field is zero in the control volume at the interface drawn in Figure 14.2.1. Hence, $E_b = (\varepsilon_a/\varepsilon_b) E_a$ and so

(14.2.5) $$T_y = \frac{1}{2}\varepsilon_0\left(\varepsilon_a E_a^2 - \varepsilon_b E_b^2\right) = \frac{1}{2}\varepsilon_0\varepsilon_a E_a^2\left(1 - (\varepsilon_a/\varepsilon_b)\right)$$

Again, comparing this to what we derived for the parallel case capacitor, equation (14.0.11), after rearrangement of (14.0.11), can be shown to be identical to our current result. Hence we see that the Maxwell stress tensor is very useful for calculating forces in problems with discontinuities in material properties.

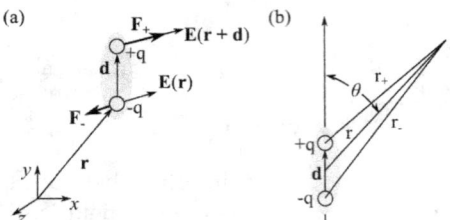

FIGURE 14.3.1. (a) Schematic of a dipole in an inhomogeneous electric field and the associated forces. (b) Schematic of a dipole and the associated construction geometry to find the field caused by the dipole at an arbitrary point.

14.3. Forces on point inhomogeneities in a material

Here we consider another example of forces on discontinuities in material properties, but where the object causing the discontinuities can be assumed to be infinitely small. An example of such situation is a dilute suspension of particles (<1% v/v) where the size of the particle is much smaller than the size of the spacing between electrodes (at least 10x). For this analysis we treat the particle as an ideal, infinitesimal dipole. If the particle is charged, we also linearly superimpose a charge onto the particle, and its associated effects.

Consider the dipole in Figure 14.3.1a. If the electric field is non-uniform (as shown) then the two charges $+q$ and $-q$ will experience different magnitudes and directions of the electric fields and so a dipole will experience a net force, the sum of the forces on the individual charges in the dipole. This force is

$$(14.3.1) \qquad \mathbf{F} = q\mathbf{E}\left(\mathbf{r}+\mathbf{d}\right) - q\mathbf{E}\left(\mathbf{r}\right)$$

where $\mathbf{r}$ is the position vector of location of $-q$. Assuming $\mathbf{d}$ is very small compared to $\mathbf{r}$ and expanding the electric field as a Taylor expansion,

$$(14.3.2) \qquad \mathbf{E}\left(\mathbf{r}+\mathbf{d}\right) = \mathbf{E}\left(\mathbf{r}\right) + \mathbf{d}\cdot\nabla\mathbf{E}\left(\mathbf{r}\right) + O\left(\mathbf{d}^2\right)$$

Combining this with (14.3.1), we obtain that

$$(14.3.3) \qquad \mathbf{F} = q\mathbf{d}\cdot\nabla\mathbf{E}\left(\mathbf{r}\right)$$

where terms of the second order in $\mathbf{d}$ and higher have been neglected. We then define the dipole moment $\mathbf{p} \equiv q\mathbf{d}$ (as it is usually defined) and obtain

$$(14.3.4) \qquad \mathbf{F} = \mathbf{p}\cdot\nabla\mathbf{E}\left(\mathbf{r}\right)$$

Note, that there is no force on the dipole unless the field is non-uniform. Similarly, we can find the torque on an infinitesimal dipole, again from Figure 14.3.1a:

$$(14.3.5) \qquad \mathbf{T} = \frac{\mathbf{d}}{2}\times q\mathbf{E}\left(\mathbf{r}+\mathbf{d}\right) + \frac{-\mathbf{d}}{2}\times -q\mathbf{E}\left(\mathbf{r}\right)$$

and again using Taylor expansion for the electric field (14.3.2),

$$(14.3.6) \qquad \mathbf{T} = q\mathbf{d}\times\mathbf{E}(\mathbf{r}) + q\frac{\mathbf{d}}{2}\times(\mathbf{d}\cdot\nabla\mathbf{E}(\mathbf{r})) + O\left(\mathbf{d}^3\right)$$

Since again we assume $\mathbf{d}$ is small, we neglect terms of $\mathbf{d}$ higher than the first order and obtain,

$$(14.3.7) \qquad \mathbf{T} = q\mathbf{d}\times\mathbf{E}\left(\mathbf{r}\right) = \mathbf{p}\times\mathbf{E}\left(\mathbf{r}\right)$$

Note that an infinitesimal dipole can experience a torque even in a uniform electric field. Note also the nature of the approximation if the field is not uniform: this approximation is valid when the product $\mathbf{d} \cdot \nabla \mathbf{E} \ll \mathbf{E}$, i.e., the field is changing slowly on the scale of the particle. All that is left now is to find what is the particular dipole moment for a particular particle.

14.3.1. Effective dipole on a particle. Dipoles on particles can arise due to a permanent separation of charges (electret particles) or be induced by the applied field. Electret materials are fairly common in nature. Examples include naturally occurring silicon dioxide (e.g., quartz) or materials synthesized from fluoropolymers, polypropylene, polyethyleneterephthalate, and other polymers. Here, however we are interested in finding dipoles induced by the imposed electric fields. Consider a particle in a uniform electric field, where the field polarizes the particle. The particle's effective dipole moment is therefore aligned with the imposed field. This field is shaped by the particle's presence near the particle. However, by symmetry (imposed field vs. arising in the particle) the field appears the same as if it arose from an equivalent point dipole immersed in the same dipole and positioned at the location of the center of the particle. We can derive the field due to a point dipole in a uniform potential with the aid of Figure 14.3.1b. Consider two point charges as shown in Figure 14.3.1b immersed in a linear dielectric with permittivity $\varepsilon_0 \varepsilon_f$ (where the subscript f designates "surrounding fluid") The charges are located near the origin, a distance $\pm d/2$ from it for the positive and negative charge respectively. The charge distribution is axisymmetric and so the electric potential φ is only a function of the radial and azimuthal coordinates, r and θ. We take the limit where $d \to 0$ and $q \to \infty$, making the product $p = qd$ finite. However, we also take into account the finite spacing between the two (finite) charges, and by superposition and Coulomb's law to obtain a net potential

$$(14.3.8) \qquad \varphi(r, \theta) = \frac{q}{4\pi\varepsilon_0\varepsilon_f r_+} - \frac{q}{4\pi\varepsilon_0\varepsilon_f r_-}$$

Next, from geometry we can relate the distance r_+ and r_- to r, θ and d by

$$(14.3.9) \qquad \begin{aligned} \frac{r}{r_+} &= \left(1 + \left(\frac{d}{2r}\right)^2 - \frac{d}{r}\cos\theta\right)^{-1/2} \\ \frac{r}{r_-} &= \left(1 + \left(\frac{d}{2r}\right)^2 + \frac{d}{r}\cos\theta\right)^{-1/2} \end{aligned}$$

We can expand this using the following Maclaurin series for the right-hand side

$$(14.3.10) \qquad (1+x)^{-1/2} = 1 - \frac{x}{2} + \frac{3x^2}{8} - \frac{5x^3}{13} + \ldots$$

to obtain

$$\frac{r}{r_+} = 1 - \frac{1}{2}\left[\left(\frac{d}{2r}\right)^2 - \frac{d}{r}\cos\theta\right] + \frac{3}{8}\left[\left(\frac{d}{2r}\right)^2 - \frac{d}{r}\cos\theta\right]^2$$

$$- \frac{5}{16}\left[\left(\frac{d}{2r}\right)^2 - \frac{d}{r}\cos\theta\right]^3 + ...$$

(14.3.11)

$$\frac{r}{r_-} = 1 - \frac{1}{2}\left[\left(\frac{d}{2r}\right)^2 + \frac{d}{r}\cos\theta\right] + \frac{3}{8}\left[\left(\frac{d}{2r}\right)^2 + \frac{d}{r}\cos\theta\right]^2$$

$$- \frac{5}{16}\left[\left(\frac{d}{2r}\right)^2 + \frac{d}{r}\cos\theta\right]^3 + ...$$

This in turn can be simplified to
(14.3.12)

$$\frac{r}{r_+} = P_0\left(\cos\theta\right) + \left(\frac{d}{2r}\right)P_1\left(\cos\theta\right) + \left(\frac{d}{2r}\right)^2 P_2\left(\cos\theta\right) + \left(\frac{d}{2r}\right)^3 P_3\left(\cos\theta\right) + ...$$

$$\frac{r}{r_-} = P_0\left(\cos\theta\right) - \left(\frac{d}{2r}\right)P_1\left(\cos\theta\right) + \left(\frac{d}{2r}\right)^2 P_2\left(\cos\theta\right) - \left(\frac{d}{2r}\right)^3 P_3\left(\cos\theta\right) + ...$$

where P_n are the Legendre polynomials with $\cos\theta$ as the independent variable, where

(14.3.13)

$$P_0\left(x\right) = 1$$
$$P_1\left(x\right) = x$$
$$P_2\left(x\right) = \tfrac{1}{2}\left(3x^2 - 1\right)$$
$$P_3\left(x\right) = \tfrac{1}{2}\left(5x^3 - 3x\right)$$

Combining this result with (14.3.8), we obtain

(14.3.14) $$\varphi\left(r, \theta\right) = \frac{q}{4\pi\varepsilon_0\varepsilon_f}\frac{d}{r^2}P_1\left(\cos\theta\right) + \frac{q}{4\pi\varepsilon_0\varepsilon_f}\frac{d^3}{4r^4}P_3\left(\cos\theta\right) + ...$$

Here the first term is due to the dipole, and the second term is an octupolar correction ($n = 3$), i.e., the dipole expanded as an octupole, an object with 4 charges. We can write the nth term of this series as

(14.3.15) $$\varphi_n = \frac{p^{(n)}}{4\pi\varepsilon_0\varepsilon_f r^{n+1}}P_n\left(\cos\theta\right)$$

where we write the general expression for the multipolar moment of order n as

(14.3.16) $$p^{(n)} = n!q_n d_n^n$$

where q_n and d_n represent the unit charge and the separation of the point charges constituting the nth linear multipolar distribution respectively.

We only take the first term in this expansion as we let $d \to 0$ and so the assume that the higher terms in the expansion are negligible. Thus we write (using the definitions of Legendre polynomials, (14.3.13)) that the electrostatic potential due to a point dipole is

(14.3.17) $$\varphi\left(r, \theta\right) = \frac{qd\cos\theta}{4\pi\varepsilon_0\varepsilon_f r^2} = \frac{p\cos\theta}{4\pi\varepsilon_0\varepsilon_f r^2}$$

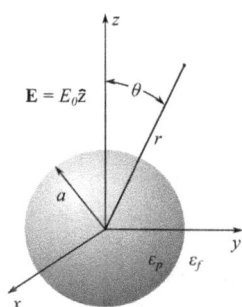

FIGURE 14.3.2. Dielectric sphere radius a and relative permittivity ε_p in a dielectric medium with relative permittivity ε_f with an imposed uniform electric field, E_0 along the z-axis.

To find the effective dipole for a finite particle, we solve the appropriate boundary value problem and compare the result to (14.3.17).

For example, consider an insulating dielectric sphere relative permittivity ε_p and radius a in dielectric medium with relative permittivity ε_f with an imposed uniform electric field, E_0 along the z-axis (Figure 14.3.2). We assume that there are no free charge anywhere in the sphere or the dielectric fluid. Thus, the electrostatic potential satisfies Laplace's equation,

$$(14.3.18) \qquad \nabla^2 \varphi = 0$$

Laplace's equation for similar geometry has a well known solution (for example see Stratton (1941)) in the form of

$$(14.3.19) \qquad \varphi_{out}(r, \theta) = -E_0 r \cos\theta + A \frac{\cos\theta}{r^2}$$

$$(14.3.20) \qquad \varphi_{in}(r, \theta) = -B r \cos\theta$$

for the potential outside the sphere (subscript out, $r > a$) and inside the sphere (subscript in, $r < a$); A and B are the unknown coefficients to be determined from the boundary conditions. The first term in (14.3.19) comes from the imposed electric field, while the second term in the induced dipole term is due to the presence of the particle. At the surface of the particle, $r = a$, the electrostatic potential must be continuous, thus

$$(14.3.21) \qquad \varphi_{out}(r = a, \theta) = \varphi_{in}(r = a, \theta)$$

Secondly, the normal component of the displacement flux must be continuous across the interface, due to Gauss's law:

$$(14.3.22) \qquad -\varepsilon_0 \varepsilon_p \left.\frac{\partial \varphi_{in}}{\partial r}\right|_{r=a,\theta} = -\varepsilon_0 \varepsilon_f \left.\frac{\partial \varphi_{out}}{\partial r}\right|_{r=a,\theta}$$

Combining these boundary conditions with (14.3.19) and (14.3.20), we obtain that

$$(14.3.23) \qquad A = \frac{\varepsilon_p - \varepsilon_f}{\varepsilon_p + 2\varepsilon_f} a^3 E_0$$

$$(14.3.24) \qquad B = \frac{3\varepsilon_f}{\varepsilon_p + 2\varepsilon_f} E_0$$

Comparing the second term of (14.3.19), that is due to the induced dipole to the potential from a dipole (14.3.17), we see that

$$(14.3.25) \qquad p = 4\pi\varepsilon_0\varepsilon_f \frac{\varepsilon_p - \varepsilon_f}{\varepsilon_p + 2\varepsilon_f} a^3 E_0$$

We define a function

$$(14.3.26) \qquad K = \frac{\varepsilon_p - \varepsilon_f}{\varepsilon_p + 2\varepsilon_f}$$

This function is typically referred to as the Clausius-Mossotti function. This function provides a measure of the strength of polarization of a particle as a function of the permittivities of the particle and the surrounding media. Notice that the larger the difference in permittivities between the particle and the media, the greater the polarization of the particle, and so the force on it. Secondly, notice that K is a bounded function, that is when $\varepsilon_f \ll \varepsilon_p$ then $K = 1$. When $\varepsilon_f \gg \varepsilon_p$ then $K = -\frac{1}{2}$. Thirdly, notice that when $\varepsilon_f \ll \varepsilon_p$ the dipole in the same direction as the imposed electric field whereas when $\varepsilon_f \gg \varepsilon_p$ the two are antiparallel. Furthermore, notice that the dipole strength is proportional to the permittivity of the fluid. Thus, a force on a point dipole dielectric spherical particle in a non-uniform electric field, which can be assumed approximately uniform locally near the particle is

$$(14.3.27) \qquad \mathbf{F} = \mathbf{p} \cdot \nabla\mathbf{E} = \left(4\pi\varepsilon_0\varepsilon_f \frac{\varepsilon_p - \varepsilon_f}{\varepsilon_p + 2\varepsilon_f} a^3 \mathbf{E} \right) \cdot \nabla\mathbf{E}$$

or after applying the product rule $\nabla\left(\mathbf{E}^2\right) = \nabla\left(\mathbf{E} \cdot \mathbf{E}\right) = 2\mathbf{E} \cdot \nabla\mathbf{E}$

$$(14.3.28) \qquad \mathbf{F} = 2\pi\varepsilon_0\varepsilon_f a^3 \frac{\varepsilon_p - \varepsilon_f}{\varepsilon_p + 2\varepsilon_f} \nabla\left(\mathbf{E}^2\right) = 2\pi\varepsilon_0\varepsilon_f a^3 K \nabla\left(\mathbf{E}^2\right)$$

Comparing this for the force on objects of finite size, equation (14.0.43), we see again a dependence on the gradient of the square of the electric field, but now without the need for the objects to be compressible. This equation is the fundamental equation of the phenomena of dielectrophoresis. While here it was derived for a very special case and it is actually a first order approximation to a force, we will derive a similar equation for more general cases of forces on point like particles, such as a conducting particles, lossy dielectrics, and shelled particles in later sections. For these cases, in general the scaling will remain as is in (14.3.28) but the Clausius-Mossotti function will differ.

CHAPTER 15

Free charge at interfaces: the electrical double layer

So far, we have considered electrical forces on objects in a fluid without considering the details at what occurs that interface between the object and the surroundings. While the results that we obtained are correct in predicting the high level phenomena that occurs - the motion of objects under consideration, we have neglected some of the "lower level" phenomena that produces many interesting and useful effects. Much of this phenomena occurs due to presence of free charge at these interfaces.

15.1. Thermodynamic origin of interfacial charge

To understand why free charge occurs at the interfaces between objects of different material we need to take a look at the interface from a thermodynamic point of view. Let's consider the electrochemical Gibbs free energy at the interface. Recall that "regular", chemical Gibbs free energy is the energy needed to create a system and make space for it, minus the energy that we got from the environment that the system is in due to heat transfer between the environment and the system. Gibbs free energy is also the greatest amount of mechanical work, which can be obtained from the system, without increasing the system's total volume or allowing heat to pass to or from external bodies. Processes with a negative change in Gibbs free energy are thermodynamically favorable. Electrochemical free energy, $\overline{G}$ differs from the chemical free energy, G, by inclusion of the effects from the large-scale electrical environment (e.g., the energy of the electric field imposed on the system by electrodes of the power supply, not the local electric field of an ion).

Suppose we bring into contact two different phases, α and β (e.g. two different materials) to form an interface. At the interface, they are separated by area A. We show the side view of the interface in Figure 15.1.1. The regions between lines AA' and BB' is the interfacial zone, in which we are interested in. To the left of AA' is pure phase α and to the right of BB' is pure phase β. We consider that in the zone between the dividing plane (dashed line) and AA' is a region with properties almost like that of pure α, but with a perturbation; similarly for the region to the right of the dashed line for β. We are exactly concerned with these perturbations, and we call these excesses if the variable of interest is higher there (in the interfacial region) than in the "parent" phase, and deficiencies, if the variable is lower there than in the parent phase. Let's also consider an (imaginary) reference system, in which the phases have not yet been brought to touch or influence one another. That is phase α extends all the way to the dividing line, where it meets phase β, but each has no influence on the other (as they are not "touching"). The differences between the quantities of various species in the interface region (in the real system) and that

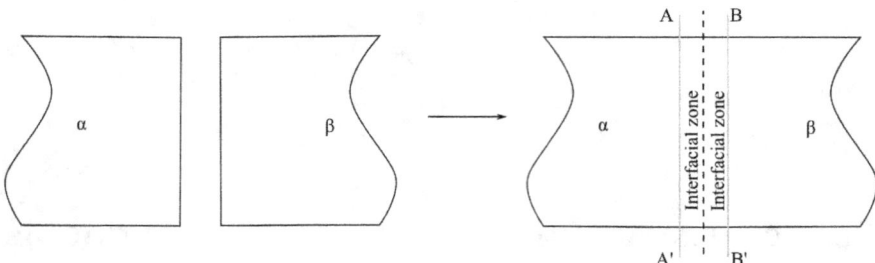

FIGURE 15.1.1. Contacting of two phases of α and β and the formation of an interfacial zone upon contact.

in the parent regions are called surface excess quantities. For example, the surface excess in number of moles of say chloride ions would be

$$(15.1.1) \qquad n_i^\sigma = n_i^S - n_i^R$$

where n_i^σ is the excess quantity and n_i^S and n_i^R are the numbers of the moles of species i (e.g. chloride) in the interfacial region of the actual system S, and of a reference, i.e. parent system R.

For the reference system the electrochemical free energy depends on temperature, pressure, and molar quantities of all components: $\overline{G}^R = \overline{G}^R\left(T, p, n_i^R\right)$. The surface area has no effect on the relative surface because in the reference system the phases do not interact. On the other hand, the free energy of the real system does depend on the area. (For example, real systems have a tendency to minimize interfacial energy, and so area). Thus, $\overline{G}^S = \overline{G}^S\left(T, p, A, n_i^R\right)$. From this, the total differentials are therefore

$$(15.1.2) \qquad d\overline{G}^R = \left(\frac{\partial \overline{G}^R}{\partial T}\right)dT + \left(\frac{\partial \overline{G}^R}{\partial P}\right)dP + \sum_i \left(\frac{\partial \overline{G}^R}{\partial n_i^R}\right)dn_i^R$$

$$(15.1.3) \qquad d\overline{G}^S = \left(\frac{\partial \overline{G}^S}{\partial T}\right)dT + \left(\frac{\partial \overline{G}^s}{\partial P}\right)dP + \left(\frac{\partial \overline{G}^s}{\partial A}\right)dA + \sum_i \left(\frac{\partial \overline{G}^S}{\partial n_i^S}\right)dn_i^S$$

We will consider here only experiments at constant temperature and pressure and so we drop the first two terms of the right-hand side of both equations. The quantity called electrochemical potential, $\overline{\mu}_i$ is typically defined as the partial derivative $\partial \overline{G}/\partial n_i$. That is

$$(15.1.4) \qquad \overline{\mu}^\alpha{}_i \equiv \left(\frac{\partial \overline{G}}{\partial n_i}\right)_{T,P,n_{j\neq i}} \equiv \mu^\alpha{}_i + z_i F \varphi^\alpha = \left(\frac{\partial G}{\partial n_i}\right)_{T,P,n_{j\neq i}} + z_i F \varphi^\alpha$$

where z_i is the charge of the charge carrier (e.g., ionic species), F is the Faraday's constant, and φ is the local electrostatic potential. The superscript α indicates the phase (e.g., particle or fluid, or solid or liquid). From this definition, the electrochemical potential has the following properties:

(1) For an uncharged species: $\overline{\mu}_i^\alpha = \mu_i^\alpha$
(2) For any substance: $\mu_i^\alpha = \mu_i^{0\alpha} + RT \ln\left(a_i^\alpha\right)$ where $\mu_i^{0\alpha}$ is the standard chemical potential (which is tabulated for many substances), R is the gas constant, T is absolute temperature, and a_i^α is the activity (the activity

coefficient times the concentration) of species i in phase α. (Activity coefficients are factors used to account for deviations from ideal behavior of molecules at appreciable concentrations).

(3) For a pure phase at unit activity (e.g., solid metals, solid salts): $\mu_i^\alpha = \mu_i^{0\alpha}$

(4) For electrons in a metal ($z = -1$): $\overline{\mu}_e^\alpha = \mu_e^\alpha - F\varphi^\alpha$. Activity coefficient can be disregarded because the electron concentration never changes appreciably.

(5) For equilibrium of species i between phases α and β the electrochemical potentials are equal, by definition of equilibrium: $\overline{\mu}_i^\alpha = \overline{\mu}_i^\beta$. At equilibrium, free energy in both phases is the same, or in other words $\Delta\overline{G}$ between the two phases is zero. This is a very important point and will be used shortly!

Continuing with our analysis of two phases in contact, we assume that enough time passes that equilibrium has been achieved in the interfacial zone. Therefore, the electrochemical potential is constant for any given phase throughout the entire system (pure "parent" phases and the interfacial zone). (Note, that before the two materials touched, the pure "parent" phases existed by themselves. Their existence constitutes our reference system, superscript R. Once we brought the two parent phases together, resulting in an interface between the two, we created our real system, superscript S.) Since the electrochemical potential is the same in the interfacial zone as in the pure zones, then

$$(15.1.5) \qquad \overline{\mu}_i = \frac{\partial \overline{G}^R}{\partial n_i^R} = \frac{\partial \overline{G}^S}{\partial n_i^S}$$

Surface tension, γ, the measure of energy needed to produce a unit area of a new surface, is typically defined as $(\partial G/\partial A)_{T,p,n_i}$ (for details see Section 7.1). Recall that to produce a new surface requires that atoms or molecules previously in the bulk be brought to the new surface where they may have fewer binding interactions with their neighbors in the original phase, and potentially new interactions with neighbors of the different phase. Thus the surface tension depends on the chemical identity of both phases α and β.

We now define differential excess free energy as

$$(15.1.6) \qquad d\overline{G}^\sigma = d\overline{G}^S - d\overline{G}^R = \gamma dA + \sum_i \overline{\mu}_i d\left(n_i^S - n_i^R\right)$$

and from (15.1.1) we have

$$(15.1.7) \qquad d\overline{G}^\sigma = \gamma dA + \sum_i \overline{\mu}_i dn_i^\sigma$$

This shows that under our assumption of constant temperature and pressure, the interfacial free energy can be described by the extensive variables of surface area and molar amounts of species. This allows us to invoke Euler's theorem, as $\overline{G}$ is a linear and homogenous function of A and n_i. (If A or n_i double, so does $\overline{G}$). Hence,

$$(15.1.8) \qquad \overline{G}^\sigma = \frac{\partial \overline{G}^\sigma}{\partial A} A + \sum_i \left(\frac{\partial \overline{G}^\sigma}{\partial n_i^\sigma}\right) n_i^\sigma$$

or

$$(15.1.9) \qquad \overline{G}^\sigma = \gamma A + \sum_i \overline{\mu}_i n_i^\sigma$$

We then take the total differential of this equation

(15.1.10)
$$d\overline{G}^{\sigma} = \gamma dA + A d\gamma + \sum_i \overline{\mu}_i dn_i^{\sigma} + \sum_i n_i^{\sigma} d\overline{\mu}_i$$

and compare it to that in (15.1.7). Since these must be the same, this implies

(15.1.11)
$$A d\gamma + \sum_i n_i^{\sigma} d\overline{\mu}_i = 0$$

If we define excess surface concentration as $\Gamma_i = n_i^{\sigma}/A$, we can rewrite (15.1.11) as

(15.1.12)
$$-d\gamma = \sum_i \Gamma_i d\overline{\mu}_i$$

which is the Gibbs adsorption isotherm.

We can now also consider the case that is the most important to us, where the area of contact between the two phases cannot change. This is the case of a liquid in contact with a solid, for example at a wall of a channel. In this case, (15.1.7) simplifies to

(15.1.13)
$$d\overline{G}^{\sigma} = \sum_i \overline{\mu}_i dn_i^{\sigma}$$

We can integrate this across the interface

(15.1.14)
$$\int_{\overline{G}(\alpha)}^{\overline{G}(\beta)} d\overline{G}^{\sigma} = \int_{\alpha}^{\beta} \left(\sum_i \overline{\mu}_i dn_i^{\sigma} \right) dx$$

and assume that the interface is at equilibrium and so the interfacial energy is zero,

(15.1.15)
$$0 = \int_{\alpha}^{\beta} \left(\sum_i \overline{\mu}_i dn_i^{\sigma} \right) dx$$

and so using the definition of electrochemical potential, obtain

(15.1.16)
$$0 = \int_{\alpha}^{\beta} \left(\sum_i \left(\mu_i + z_i F \varphi \right) dn_i^{\sigma} \right) dx$$

This result implies that differences in chemical potentials of materials meeting at an interface should be balanced by an electrical potential. The electrical potential may bring new charged species into the interface, partially decreasing this electrical potential. In this book, we will be concerned with what occurs only on one side of the interface - what happens in the fluid phase, and we don't concern ourselves with what happens in the solid phase. The differences in chemical potentials at the interface result in free charge accumulation in the fluid phase.

15.2. Chemical origin of interfacial charge

In the previous section we found the thermodynamic reason why free charges can accumulate at the fluid side of a liquid-solid interface. Here, we consider where these charges come from - their chemical origins. At an interface between a liquid and a solid free charge may arise due to the following: (a) ionization of surface, (b) specific ion adsorption, (c) adsorption of amphoteric substances, (d) isomorphic

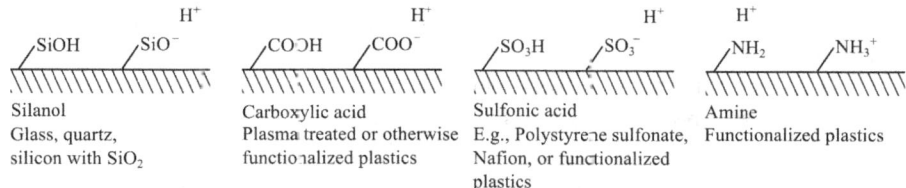

FIGURE 15.2.1. Schematic of acquisition of surface charge by ionization of common surfaces.

substitution, (e) differential desorption of ions from surfaces of sparingly soluble
salts, and (f) appearance of charged crystal surfaces.

Ionization of surfaces is the most common: an example is ionization of surface
silanol groups on glass, and the ionization of carboxylic acids, sulfonic acids, and
amines on plastics (see Figure 15.2.1). These surface charges are very much dependent on the pH of the solution, with certain surfaces altering their charge from
negative to positive based on pH. The point at which the surface charge is neutral
is termed point of zero charge.

The second common source of free charge is specific ion adsorption. Examples
include surfactants adsorbing on hydrophobic surfaces (e.g., sodium deodecyl sulfate to generate net negative charge), cationic surfactants adsorbing to negatively
charged surfaces to give net positive charge (e.g. polylysine). Surfactants are often specifically added to particle suspensions increase surface charge and prevent
particle aggregation.

Next source of free charge is the adsorption of amphoteric substances such as
proteins. The most common example of this is the (typically purposeful) coating
of microfluidic chambers with bovine serum albumin (BSA), a cheap and widely
available protein. Chambers are typically coated with BSA so that other proteins
(that are for example expected to participate in a purposeful reaction) do not stick
to the walls and become inhibited or denatured; these include proteins even on
surfaces of cells. Proteins typically contain a very large number of both amine and
carboxylic groups from side groups of the amino acids in their sequences and thus
their net charge is a strong function of pH and amino acid composition. Proteins
may also have hydrophobic domains allowing them to stick to hydrophobic surfaces
creating a charged hydrophilic surface.

Other surface charge mechanisms are rarer, but are encountered in specific applications. An example of isomorphic substitution is Al^{3+} replacing Si^{4+} in the
surface of clay, leading to a net negative surface charge. An example of differential desorption of ions from sparingly soluble salts is when a silver iodide crystal
is placed in water. Ions dissolve from the salt crystals until the product of ionic
concentrations of each is equal to the solubility product. If equal amounts of silver
and iodide ions dissolve, the surface would be uncharged. However, typically silver
ions dissolve preferentially, creating a negatively charged surface. However, if we
add silver ions, say by adding silver nitrate to the solution, since product of ionic
concentrations of silver and iodide is equal to the solubility product, we push iodide ions to dissolve preferentially, and so even create a positively charged surface!
Since the solubility product of silver iodide is well characterized, you can obtain
particles of very precise and tunable surface charge using this method. Lastly, an

example of free charge due to charged crystal surfaces occur when a crystal is broken and surfaces with different properties are exposed. An example of this occurs when kaolinite clay platelet is broken and the exposed edges containing aluminum hydroxide take up protons to give a positively charged surface.

15.3. Spatial distribution of charge at the interface

From the previous two sections we now understand that at the interface between a solid and a fluid (typically liquid) there is free charge and an electric field directed towards the bulk of the fluid. We can imagine that these charges attract towards themselves charges of opposite charge, termed counterions. Thus there are two layers, a layer of original ions strongly attached to the wall of the solid, and a layer of ions of opposite charge - counterions. Thus this structure is termed a double layer.

The simplest model for this is to treat the situation as a parallel plate capacitor, with original ions "sitting" on one plate, and the counterions "sitting" on another plate. This model was originally due to Helmholtz and Perin. However, we quickly realize that the counterions are not likely to be lined up into a tight plane, but rather diffuse about while being attracted to the surface of the solid by the electric field. This model accounting for a counterbalance between diffusion and the electric field of the charge not fixed to the surface is originally due to Gouy (1910) and Chapman (1913) (independently). This is a fairly sophisticated and useful model and is largely enough to predict most of the phenomena discussed here. However, this model fails when the surface charge is high, and so was upgraded by Stern to include both a layer of fixed counterion charge right next to the solid (Stern layer), as well as a diffuse layer of counterions. The next model is due to Grahame, who proposed that both charged and uncharged species can penetrate the Stern layer, including solvent molecules. While the Stern model treats ions as mathematical points, Grahame's ions have finite radius. This model separates the double layer into three regions: (1) ions that are specifically adsorbed to the electrode; (2) solvated ions; and (3) diffuse layer of ions. A so called "inner Helmholtz plane" runs through the centers of the specifically adsorbed ions; and "outer Helmholtz plane" runs through the centers of solvated ions at a distance of their closest approach to the electrode; beyond this plane lies the diffuse layer of ions. The upgrade to this model is the Bockris, Devanthan, Muller (BDM) model. This model assumes that there is a layer of solvent molecules directly in contact with the electrode and that these solvent molecules are oriented (recall the dipole nature of a water molecule). Thus the permittivity of the media in the double layer is not constant. In this model, the inner Helmholtz plane passes through the centers of these solvent molecules. Additionally, partially adsorbed solvated ions are also in this layer. Fully solvated counterions are located beyond the inner Helmholtz plane. Through the centers of these passes the outer Helmholtz plane. Beyond the outer Helmholtz plane, like in the previous model, lies the diffuse layer of ions. These models are schematically pictured in Figure 15.3.1.

The next level models, specifically that of Trassatti and Buzzanca, that of Conway, and that of Marcus account for charge transfer between ions in solution and the solid. These models are useful for describing electrochemical capacitors and various supercapacitors. However, the extra complexity of charge transfer is typically not necessary for describing phenomena discussed in this book and so will

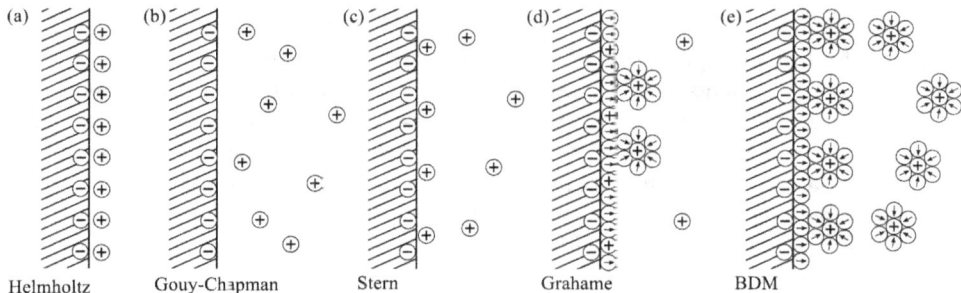

FIGURE 15.3.1. Qualitative schematic of different models of the double layer. Helmholtz and Gouy-Chapman treat their ions as point charges; Stern treats the counterions absorbed onto the wall as having finite volume but the rest (those in the diffuse layer) as point charges. Grahame and Bockris, Devanthan, Muller (BDM) begin to treat their ions as having finite volume and begin to account for presence of solvent molecules having a dipole (circle with arrow) and solvation shells around ions.

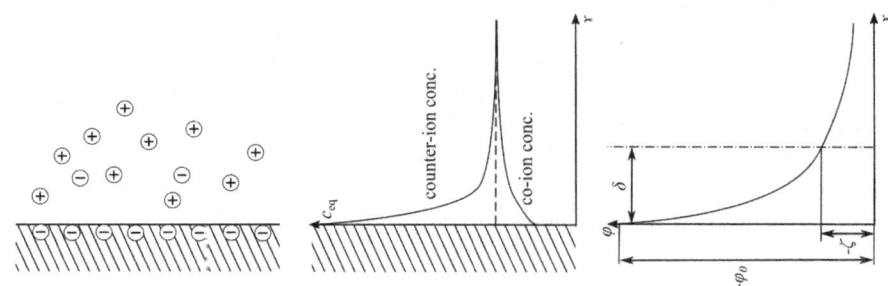

FIGURE 15.3.2. Schematic representation of the Gouy-Chapman picture of the double layer, a qualitative profile of the counter and co-ion concentrations, and the potential profile in the double layer. φ_0 is the potential at the wall, whereas ζ is the potential at the so-called "slip plane" - the plane where fluid motion begins. There appears to be a layer of fluid with thickness δ that does not move in electrophoresis and other experiments used to determine the corresponding potentials.

not be discussed further. For more details on all of these models see Bockris et al. (2000).

15.3.1. Gouy-Chapman model of the double layer. Let's begin with the Gouy-Chapman model (Figure 15.3.2). Imagine a geometrically smooth plane of the solid, to which there is attached a uniform layer of charge. This layer has no thickness. Swimming in the solution next to it are point charges of the opposite charge along with a few charges of the same charge. These again have no volume. These point charges shield the field from the charged plane of the solid and so the electric field drops off as the distance from the solid increases. Thus, at the surface of the solid there is a certain constant potential and this potential decays to zero at the outer edge of the double layer. If this was not the case there would be a constant flux of ions in or out of the double layer and thus an equilibrium state would never be reached. This also implies that the counterions fully shield the surface charge and so there is overall net neutrality in the double layer - the sum of all negative and positive charges is zero. The fact that the electric field drops off as the distance from the surface of the solid increases, also implies that counterions are less attracted to those regions and so the concentration of counterions decreases away from the surface. On the other hand, ions with the same charge as the surface, so called, co-ions are less repulsed from the surface as the distance from the surface increases. Thus their concentration should increase as the distance from the surface increases. Thus at the edges of the double layer, concentration of excess charge will approach to zero. We can write the balance of diffusive forces and electromigration forces on ions as

(15.3.1)
$$-D^+ \frac{dc^+}{dx} - \mu^+ c^+ \frac{d\varphi}{dx} = 0$$
$$-D^- \frac{dc^-}{dx} + \mu^- c^- \frac{d\varphi}{dx} = 0$$

where D is the diffusion coefficient, c is the equilibrium concentration, x is the distance from the surface of the solid, μ is the electrophoretic mobility of the ion (the ratio between the velocity of an ion and the electric field applied to it), and φ is the equilibrium electrostatic potential. The superscripts designate if the ion is a cation (positive) or an anion (negative). The first term of the equation represents the diffusive flux, where the second term represents the electromigration flux. We can simplify these equations somewhat by using the Einstein–Smoluchowski relation, an approximation relating diffusivity to electrophoretic mobility

(15.3.2)
$$\mu = \frac{zeD}{k_B T} = \frac{zFD}{RT}$$

where k_B is the Boltzmann constant, T is the absolute temperature, z is the ion valance, and F is the Faraday's constant (see Section 16.7.6 for details). Applying this to (15.3.1)

(15.3.3)
$$-D^+ \left(\frac{dc^+}{dx} + \frac{F}{RT} z^+ c^+ \frac{d\varphi}{dx} \right) = 0$$
$$-D^- \left(\frac{dc^-}{dx} - \frac{F}{RT} z^- c^- \frac{d\varphi}{dx} \right) = 0$$

This simplification allows us to calculate the distribution of the ions in the double layer without knowing the diffusivity or electrophoretic mobility of those ions! Additionally, we write the boundary conditions far from the solid that the concentration of both co- and counterions approaches that of the bulk, and the potential decays to zero:

(15.3.4)
$$c^+ \big|_{x \to \infty} = c_\infty^+$$
$$c^- \big|_{x \to \infty} = c_\infty^-$$

(15.3.5)
$$\varphi \big|_{x \to \infty} = 0$$

Integrating (15.3.3) and applying the boundary conditions (15.3.4) and (15.3.5) gives us

(15.3.6)
$$c^+ = c_\infty^+ \exp\left(-z^+ e\varphi / k_B T\right)$$
$$c^- = c_\infty^- \exp\left(+z^- e\varphi / k_B T\right)$$

which is the Boltzmann distribution. We can define the charge density as

(15.3.7)
$$\rho = N_A \sum_i z_i e c_i$$

where e is electron charge. Then we can use the Poisson's equation in order to eliminate the potential term in (15.3.6):

(15.3.8)
$$\nabla \cdot (\varepsilon_0 \varepsilon \nabla \varphi) = -\rho$$

or in one dimension

(15.3.9)
$$\frac{d}{dx}\left(\varepsilon_0 \varepsilon(x) \frac{d\varphi}{dx}\right) = -\rho(x)$$

Combining (15.3.9) with (15.3.7) and (15.3.6) we obtain

(15.3.10)
$$\frac{d}{dx}\left(\varepsilon_0 \varepsilon(x) \frac{d\varphi}{dx}\right) = e N_A \sum_i z_i c_{i\infty} \exp\left(\frac{-z_i e\varphi}{k_B T}\right)$$

So far we have allowed that the dielectric constant be a function of position relative to the solid. However, now to easily integrate (15.3.10) we can assume that the dielectric constant is constant and equal to that in the bulk solvent, the same assumption Gouy and Chapman made. Performing this integration we obtain that

(15.3.11)
$$\frac{d\varphi}{dx} = \pm \sqrt{\frac{2RT}{\varepsilon_0 \varepsilon} \sum_i c_{i\infty} \left(\exp\left(\frac{-z_i e\varphi}{k_B T}\right) - 1\right)}$$

where we have used the fact that the $d\varphi/dx \to 0$ as $x \to \infty$. To perform the next integration and obtain the potential directly as a function of the distance away from the surface, we either need to integrate (15.3.11) numerically, or add additional assumptions. For example, Gouy and Chapman in their model assumed that

(15.3.12)
$$\left|\frac{z_i e\varphi}{k_B T}\right| \ll 1$$

in other words, that $|z_i \varphi| \ll 25\,\text{mV}$ at room temperature. This is neither a great assumption, but neither it is terrible, as most naturally charged surfaces have a potential of 10 - 100 mV and with an electrode surface one can apply a potential up to several volts in an appropriate solvent. Thus, using the Taylor expansion for the exponential function

(15.3.13)
$$\exp(x) = 1 + x + O(x^2)$$

and only keeping the first term, we obtain a much simpler ordinary differential equation

(15.3.14)
$$\frac{d\varphi}{dx} = \pm \sqrt{\frac{2RT}{\varepsilon_0 \varepsilon} \sum_i c_{i\infty} \left(\frac{-z_i e\varphi}{k_B T}\right)}$$

TABLE 1. Debye length for a 1:1 electrolyte (e.g., KCl, or NaCl) as at 298 K.

Concentration	Debye length, κ^{-1}
1 M	0.307 nm
150 mM (approx. saline solution)	0.792 nm
100 mM	0.971 nm
10 mM	3.07 nm
1 mM	9.71 nm
100 µM	30.7 nm
10 µM (approx. pH 5, ISO Grade 3 DI)	97.1 nm
1 µM	307 nm
100 nM (18 MΩ cm DI)	971 nm

This approximation is often termed the Debye approximation. Integrating this, we obtain

(15.3.15)
$$\varphi(x) = \varphi_0 \exp(-\kappa x)$$

where

(15.3.16)
$$\kappa = \sqrt{e^2 N_A \frac{\sum_i c_{i\infty} z_i^2}{\varepsilon_0 \varepsilon k_B T}}$$

Here φ_0 is the potential at the solid surface. κ^{-1} is termed the Debye length and is a measure of the thickness of the double layer. For solutions of monovalent electrolytes at room temperature, the Debye length is approximately

(15.3.17)
$$\kappa^{-1} = 0.3 \ \text{nm}/\sqrt{c}$$

where c is the concentration of electrolyte in moles/liter. For example, for 10 mM NaCl, c is 0.01, and the Debye length is 3 nm. Note for comparison that the average carbon-carbon single bond is about 0.15 nm. Very pure deionized water has a total concentration of ions of roughly of 1×10^{-7} M, giving a Debye length of almost 1 µm. 1 mM NaCl produces a Debye length of almost 10 nm. Double layers are fairly thin! Because double layers are typically so thin, even double layers on nanoparticles, (e.g., 100 nm diameter) can be described with this model since on the scale of double layer, the particle looks flat - the curvature of the particle is much larger than the double layer!

We can milk the Gouy-Chapman assumption some more and additionally assume that the electrolyte is a symmetric binary electrolyte, i.e., an electrolyte for which $|z^+| = |z^-| = z$. This combined approximation is sometimes referred to as the Debye-Huckel approximation. For this case, the free charge density (using equations (15.3.6) and (15.3.7)) is

(15.3.18)
$$\rho = N_A e\left(z_+ c_+ + z_- c_-\right) = z e N_A c_\infty \left(\exp\left(-z^+ e\varphi/k_B T\right) - \exp\left(+z^- e\varphi/k_B T\right)\right)$$

or

(15.3.19)
$$\rho = -2 z e N_A c_\infty \sinh\left(z e \varphi / k_B T\right)$$

Combining this with (15.3.9) and explicitly assuming that the permittivity has no spatial dependence, we obtain that

$$(15.3.20) \qquad \frac{d^2\varphi}{dx^2} = \frac{2zeN_Ac_\infty}{\varepsilon_0\varepsilon}\sinh\left(ze\varphi/k_BT\right)$$

Equation (15.3.20) is often termed the Poisson-Boltzmann equation as it combines Poisson's equation for electric potential with the Boltzmann distribution for free charge at a surface. Now let's apply the assumption that the potential is small, (15.3.12), so that we can assume $\sinh\left(z\varphi/k_BT\right) \approx z\varphi/k_BT$ and so simplify the above equation to

$$(15.3.21) \qquad \frac{d^2\varphi}{dx^2} = \frac{2z^2e^2N_Ac_\infty}{\varepsilon_0\varepsilon k_BT}\varphi = \kappa^2\varphi$$

where κ^{-1} is again the Debye length, defined as before, but for this particular case

$$(15.3.22) \qquad \kappa = \sqrt{\frac{2z^2e^2N_Ac_\infty}{\varepsilon_0\varepsilon k_BT}}$$

Equation (15.3.21) is quite simple and therefore makes the Debye-Huckel model useful in many applications that we will discuss further, such as electroosmosis. Here we solve this for two important boundary conditions. Firstly, for a fluid above an infinite plane, with a Debye-Huckel double layer on the plane and secondly, for a particular case of fluid confined between two flat plates, with a Debye-Huckel double layer on each of the infinite flat plates. For the first case the boundary conditions are:

$$(15.3.23) \qquad \varphi = \varphi_0 \text{ at } y = 0$$

$$(15.3.24) \qquad \frac{d\varphi}{dy} = 0 \text{ at } y \to \infty$$

For the second case. we move the coordinate system (walls are now at $y = \pm h$), and the boundary conditions are:

$$(15.3.25) \qquad \varphi = \varphi_0 \text{ at } y = \pm h$$

$$(15.3.26) \qquad \frac{d\varphi}{dy} = 0 \text{ at } y = 0$$

The first condition just sets the potential at the walls of the channel to equal to a known, constant potential. As we will see later, φ_0 is actually a potential not at the wall, but at a position where the fluid begins to move or slip, (at the so called slip plane), and is termed ζ (zeta) potential. The second condition states that the potential does not vary with distance far from the double layer (at infinity in the first case, and right at the point of symmetry in the second). For the first case, the solution is

$$(15.3.27) \qquad \varphi = \varphi_0 \exp\left(-\kappa y\right)$$

and for the second case, the solution is

$$(15.3.28) \qquad \varphi = \varphi_0 \frac{\cosh\left(\kappa y\right)}{\sinh\left(\kappa y\right)}$$

For these two cases, the free charge density (see equation (15.3.8) and then (15.3.21)) is

(15.3.29) $$\rho_f = -\varepsilon_0 \varepsilon \kappa^2 \varphi$$

and so for the first case the free charge density is

(15.3.30) $$\rho_f = -\varepsilon_0 \varepsilon \kappa^2 \varphi_0 \exp\left(-\kappa y\right)$$

and the second case

(15.3.31) $$\rho_f = -\varepsilon_0 \varepsilon \kappa^2 \varphi_0 \frac{\cosh\left(\kappa y\right)}{\sinh\left(\kappa y\right)}$$

Another interesting and less restrictive case to consider is the case of just symmetric binary electrolytes, i.e., electrolytes for which $|z^+| = |z^-| = z$. For this case, (15.3.11) simplifies to

(15.3.32)
$$\frac{d\varphi}{dx} = \pm \sqrt{\frac{2RT}{\varepsilon_0 \varepsilon} c_\infty \left(\exp\left(\frac{ze\varphi}{k_B T}\right) + \exp\left(\frac{-ze\varphi}{k_B T}\right) - 2 \right)}$$

$$\frac{d\varphi}{dx} = \sqrt{\frac{8 k_B N_A T c_\infty}{\varepsilon_0 \varepsilon}} \sinh\left(\frac{ze\varphi}{k_B T}\right)$$

which can then be integrated to obtain

(15.3.33) $$\Phi = 2 \ln \left[\frac{1 + \exp\left(-\kappa x\right) \tanh\left(\Phi_0/4\right)}{1 - \exp\left(-\kappa x\right) \tanh\left(\Phi_0/4\right)} \right]$$

where

(15.3.34)
$$\Phi = \frac{ze\varphi}{k_B T}$$
$$\Phi_0 = \frac{ze\varphi_0}{k_B T}$$

This solution is valid for all potential value surfaces, but is generally applicable for a symmetric electrolyte and is significantly more complicated than the result (15.3.15). Substituting this result into (15.3.6) and (15.3.7) and integrating over the entire double layer (to infinity) we can obtain the total charge per unit area inside the double layer. As we stated earlier, we expect net neutrality to exist in the system and so this charge should equal to the charge density (charge per area) of the solid wall. This is given by

(15.3.35) $$\sigma = \sqrt{8\varepsilon_0 \varepsilon k_B T N_A c_\infty} \sinh\left(\Phi_0/2\right)$$

and the concentrations of the positive and negative ions are

(15.3.36)
$$\frac{c^+}{c_\infty} = \exp\left(-\Phi_0 \exp\left(-\kappa x\right)\right)$$
$$\frac{c^-}{c_\infty} = \exp\left(\Phi_0 \exp\left(-\kappa x\right)\right)$$

From this we can find using (15.3.7) that

$$\rho = eN_A \left(z_+ c_+ + z_- c_-\right) = zeN_A c_\infty \left(\exp\left(-\Phi_0 \exp\left(-\kappa x\right)\right) - \exp\left(\Phi_0 \exp\left(-\kappa x\right)\right)\right)$$

or

(15.3.37) $$\rho = -2zeN_A c_\infty \sinh\left(\Phi_0 \exp\left(-\kappa x\right)\right)$$

As we shall see later in the chapter on electrophoresis, this model predicts various free charge electrokinetic phenomena such as electrophoresis, electroosmosis and others fairly well. However, it has some limitations in predicting actual structure of the double layer, i.e. where the ions actually are. For example, for a 0.1 M solution of sodium chloride, for a 200 mV surface potential (easily achievable on an electrode, and potentially found on some solids), this model predicts a concentration of counterions near the wall of 300 M! This is not physically reasonable. Consider that the concentration of water molecules in water is only about 55 M! This wild prediction is associated with the fact that the Gouy-Chapman model treats ions as point charges, which of course have no volume and so can pack as densely as they like, hence a packing of 300 M is possible. Real ions of course have finite volume, and cannot pack at such high densities.

15.3.2. Stern correction to Gouy-Chapman model. In the previous section we noticed that we need to account for the finite volume of ions, at least where they are likely to pack the tightest - at the wall. Furthermore, we may hypothesize that once an ion, especially that charged oppositely of the wall charge approaches the wall so closely, it is not unreasonable that it may become adsorbed to the wall. However, the forces of adsorption are fairly weak, and only extend up to a distance of a monolayer of these ions This is exactly what Stern proposed. Thus he separated the fluid side of the double layer into an inner layer, consisting of adsorbed ions, and an outer layer of diffusely situated ions. In the outer layer, where ions are not densely packed, the ions are again, just like in Gouy-Chapman model are treated as point particles. Hence, if you draw an imaginary wall after the layer of the adsorbed ions and call the potential of that wall φ_d that is lower than the wall potential φ_0 you can directly use all of the results obtained in the previous section, just replacing φ_d in place of φ_0. This new potential is of course lower than the wall potential because of the adsorbed ions screen the wall charge. Here the subscript d stands for the distance from the wall where our new imaginary wall is located - the size of the adsorbed ions. We show a schematic of the Stern model of the double layer in Figure 15.3.3.

Having these two, currently unknown to us potentials is actually not enough. Turns out that another, third potential is needed - the ζ (zeta) potential. Turns out that the motion of the fluid relative to a solid wall (or equally the motion of a particle relative to the fluid) does not start at exactly the solid wall, or at our imaginary wall separating the diffuse layer for the adsorbed layer. It starts a little further into the diffuse layer, at the so-called "slip plane", and the potential where it starts is termed the ζ potential. It is the potential that can be actually found in electrophoresis, electroosmosis, and related experiments.

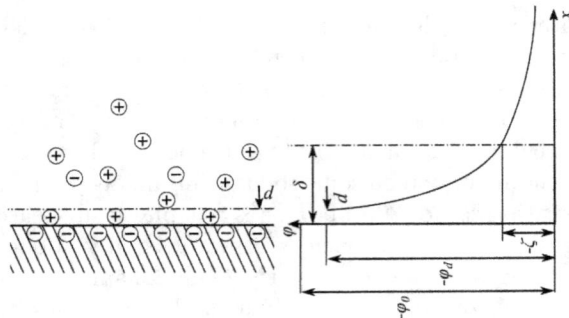

FIGURE 15.3.3. Schematic representation of the Stern picture of the double layer and the potential profile in the double layer. The Stern double layer consists of the ions of the solid, the adsorbed layer of counterions, and a diffuse layer of counterions.

CHAPTER 16

Free charge electrokinetic phenomena

16.1. DC electroosmotic flow

Let's begin with the most striking consequence of electrical double layers (free charge at a channel wall) - DC electroosmotic flow. Firstly, to clear some misnomer in the naming. "Osmotic" typically refers to a flow of liquid solvent driven by an imposed solute concentration gradient, where the solute gradient is often maintained by an appropriate membrane. Osmosis has many applications, including in water desalination and in biology. In biology, osmosis drives water from a hypotonic solution into a cell across the cell's membrane. The cell's membrane ensures that there is higher concentration of solute inside the cell, before the water permeates in and the solute concentration either equilibrates or the cell lyses. In microfluidics there are also osmotic pumps (not to be confused with electroosmotic pumps). In these pumps there is a compartment filled with solute separated from the outside of the pump by a semi-permeable membrane, and internally separated from the fluid to be pumped by a flexible but impermeable membrane (Figure 16.1.1). When this pump is immersed in solution (e.g. implanted into a mouse or human) water rushes into the solute chamber, expanding impermeable membrane, and pushing the fluid to be pumped out. Since the rate of the water permeating the semi-permeable membrane is controlled by the size of the membrane pores and the surface area of the membrane, the flow rate of fluid to be pumped can be set. Such pumps offer a number of advantages: it has no moving parts, it is quiet (highly desired feature for implantables), usually highly reliable, and most importantly highly scalable. Such pumps can be built on a cm scale as well as on a 100 µm scale. Their main disadvantage is they are not actively controlled and the flow rate depends on the media the pump is immersed in. While we show an implantable version of a pump in Figure 16.1.1, we can imagine that a similar structure can be used to drive fluid on a lab-on-a-chip system and elsewhere.

Electroosmotic pumps, (and electroosmotic phenomena) in general have nothing to do with osmotic pumps! They do not rely on a solute gradient and a semi-permeable membrane to move fluid, but rather on the electric double layers on channels walls and an externally imposed electric field along the axial direction of the channel. Typically electroosmosis refers to the flow of liquid in a porous media under the influence of the electric field parallel to the direction of the flow. One of the simplest case of electroosmosis is the movement of liquid between two infinite parallel plates (Figure 16.1.2) and so we examine this case first.

Let's consider the case of electroosmotic flow as shown in Figure 16.1.2, and specifically the ions in the double layer near the wall. The situation of course is described by the full Navier-Stokes equations, but for convenience we immediately assume that the flow is incompressible and is a low Reynolds number flow. Thus, we

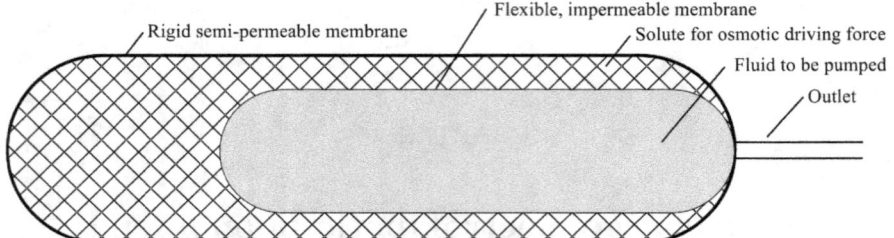

FIGURE 16.1.1. Schematic of a cross-section of an osmotic pump (not to be confused with electroosmotic pump). The pump consists of an outer container with rigid semi-permeable membrane (typically a rigid plastic with holes or a porous plastic sandwiched with a true semi-permeable membrane) and an inner container with a flexible impermeable membrane wall and an outlet connection. The outer container contains a solute for creating an osmotic driving force (sometimes encased in a gel matrix, sometimes dissolved). The inner container contains the fluid to be pumped.

begin with the simplified Stokes momentum equation that takes these assumptions into account

$$(16.1.1) \qquad \mu \nabla^2 u = \nabla p + \rho_f \nabla \varphi$$

or in Cartesian coordinates

$$(16.1.2) \qquad \mu \left(\frac{\partial^2 u_x}{\partial x^2} + \frac{\partial^2 u_x}{\partial y^2} + \frac{\partial^2 u_x}{\partial z^2} \right) = \frac{\partial p}{\partial x} + \rho_f \frac{\partial \varphi}{\partial x}$$

$$(16.1.3) \qquad \mu \left(\frac{\partial^2 u_y}{\partial x^2} + \frac{\partial^2 u_y}{\partial y^2} + \frac{\partial^2 u_y}{\partial z^2} \right) = \frac{\partial p}{\partial y} + \rho_f \frac{\partial \varphi}{\partial y}$$

$$(16.1.4) \qquad \mu \left(\frac{\partial^2 u_z}{\partial x^2} + \frac{\partial^2 u_z}{\partial y^2} + \frac{\partial^2 u_z}{\partial z^2} \right) = \frac{\partial p}{\partial z} + \rho_f \frac{\partial \varphi}{\partial z}$$

to which we add body force term for the electrical body force on free charge. Just like the gravitational body force (which we omit here) the electrical body force scales as the product of the density of free charge (instead of mass density) and the electric field (instead of gravitational field). To simplify these equations further, we first assume that the velocity and electric field in the z direction is zero, and thus eliminate (16.1.4) all together. Second, we note that we don't expect a significant fluid velocity in the y-direction and so set $u_y = 0$ to obtain that

$$(16.1.5) \qquad 0 = \frac{\partial p}{\partial y} + \rho_f \frac{\partial \varphi}{\partial y}$$

Thirdly, we don't expect u_x to vary at all in the z direction, and also to vary negligibly in the x direction, and so obtain

$$(16.1.6) \qquad \mu \frac{\partial^2 u_x}{\partial y^2} = \frac{\partial p}{\partial x} + \rho_f \frac{\partial \varphi}{\partial x}$$

This is result is what we must solve to obtain the velocity profile due to both the electroosmotic flow (last term on the right-hand side) and pressure driven flow (first

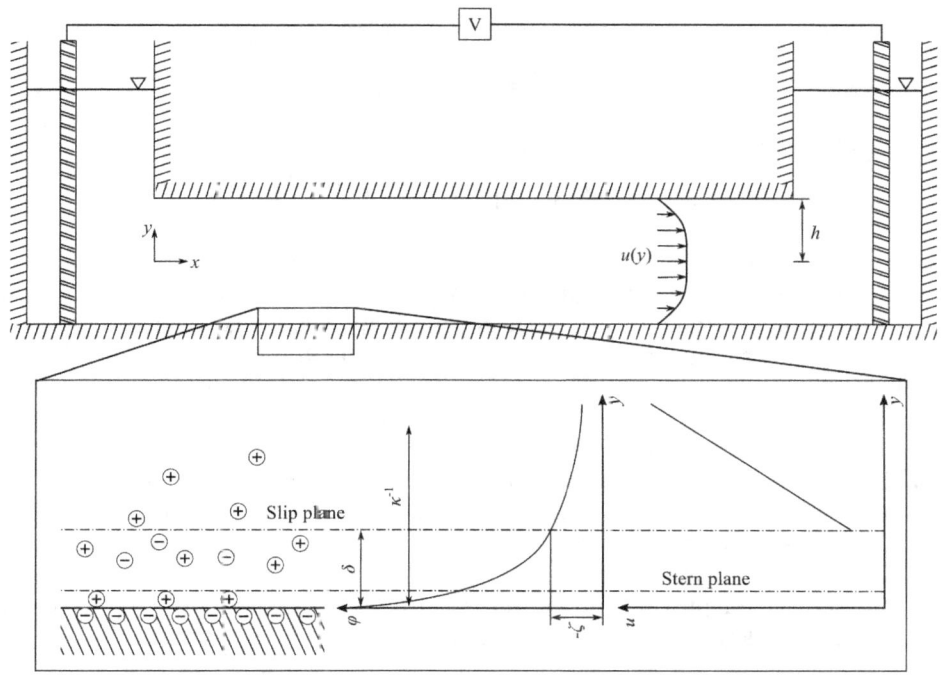

FIGURE 16.1.2. Cross-section view of a parallel plate microfluidic device in which electroosmotic and pressure driven flow can
be setup. The device consists of a microchannel formed between
two parallel plates connected to two reservoirs. Each reservoir has
an electrode, across which a potential can be setup to drive the
electroosmotic flow. The inset shows a zoomed in view of the situation near the wall the charge distribution in the double layer,
the potential distribution in the double layer, and the velocity distribution in the double layer very near the wall.

term on the right-hand side). For convenience we set

$$(16.1.7) \qquad\qquad p_x = -\frac{dp}{dx}$$

$$(16.1.8) \qquad\qquad E_x = -\frac{\partial \varphi}{\partial x}$$

which are typically imposed by the user of the device and so are known. Furthermore, we use the relation for free charge we obtained earlier for a similar situation
(equation (15.3.31)) and so obtain

$$(16.1.9) \qquad \mu \frac{\partial^2 u_x}{\partial y^2} = -p_x + \varepsilon_0 \varepsilon \kappa^2 \zeta \frac{\cosh(\kappa y)}{\sinh(\kappa y)} E_x$$

where ζ is the potential at the slip plane, the effective wall in our case. To obtain
the velocity profile, we then integrate this equation subject to the no-slip boundary
conditions at the wall:

$$(16.1.10) \qquad\qquad u_x = 0 \text{ at } y = \pm h$$

and a symmetry conditions at the center of the channel

(16.1.11) $$\frac{du_x}{dy} = 0 \text{ at } y = 0$$

Integrating we obtain

(16.1.12) $$u_x(y) = \frac{h^2 p_x}{2\mu}\left[1 - \left(\frac{y}{h}\right)^2\right] - \frac{\varepsilon_0 \varepsilon E_x \zeta}{\mu}\left[1 - \frac{\cosh(\kappa y)}{\cosh(\kappa h)}\right]$$

The reader should recognize that the first term is the Hagen-Poiseuille velocity profile for pressure driven flow in between two parallel plates. The second term on the other hand is due to the electroosmotic flow. Notice the dependence on the transverse coordinate y for both cases. For the pressure driven term the dependence is the familiar parabolic. For the electroosmotic component however, the profile can be fairly flat. To give a flavor, we plot the salient part of the relation for the electroosmotic flow, representing the shape of the profile in Figure 16.1.3 as a function of the ratio between channel size and double layer thickness. Note that this analysis is only valid when the channel size is large compared to the double layer thickness, or $\kappa h > 1$. Notice, that as the double layer thickness becomes small compared to channel size, the velocity profile moves from being almost parabolic (like in the case of pressure driven flow) to flat! This is unlike pressure driven flow, where the flow profile is always parabolic. Because pressure driven flow and electroosmotic flow are superimposed on one another, their combined velocity profile can be quite complex, even such where velocity is positive in the center of the channel and negative near the walls. Also note that for most common situations, channel height is order 10 to 100 μm, while the double layer thickness is typically order 1 to 10 nm (0.1 M to 0.001 M solutions respectively); hence κh is large and so the electroosmotic profile can be approximated as uniform in the y-direction.

Next, we calculate the volumetric flow rate due to both pressure driven and electroosmotic flows. In a parallel plate microchannel the volumetric flow rate per unit width is given by

(16.1.13) $$Q = 2\int_0^h u_x(y)dy$$

substituting the velocity profile from (16.1.12) and integrating we obtain

(16.1.14) $$Q = \frac{2h^3 p_x}{3\mu} - \frac{2h\varepsilon_0 \varepsilon E_x \zeta}{\mu}\left[1 - \frac{\tanh(\kappa h)}{\kappa h}\right]$$

Here as before, the first term is due to the pressure driven flow and the second due to electroosmotic flow. Note that as κh becomes large, $\tanh(\kappa h)$ approaches unity and so the flow rate can be written as

(16.1.15) $$Q \approx \frac{2h^3 p_x}{3\mu} - \frac{2h\varepsilon_0 \varepsilon E_x \zeta}{\mu}$$

Firstly, notice that the larger the κh more efficient the electroosmotic flow. Secondly, notice that electroosmotic flow rate depends on h to the first power, while the pressure driven flow depends on h to the third power. This means that for the same driving force (same p_x and E_x), as we half the channel dimension h the pressure driven flow drops by eight, while the electroosmotic flow drops only by half. Thus at small dimensions for the same driving force pressure driven flow becomes

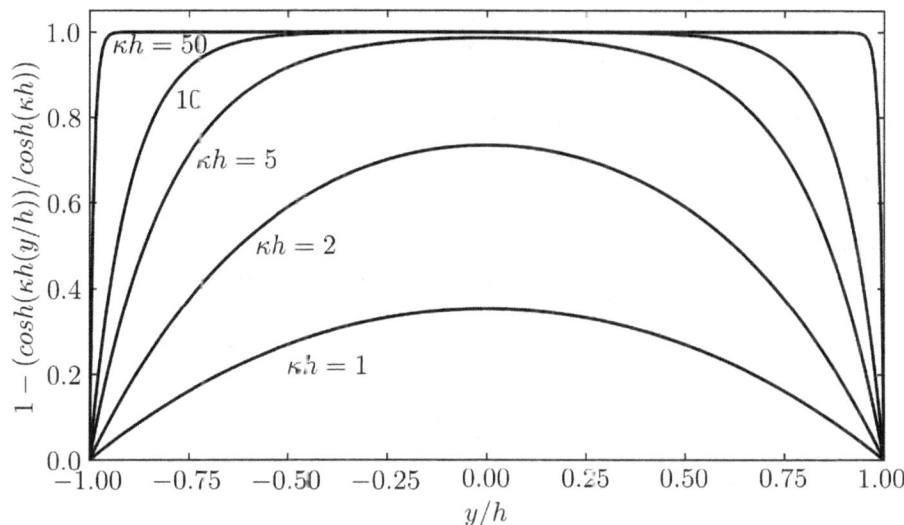

FIGURE 16.1.3. Shape of electroosmotic flow profile as a function of ratio of channel size to double layer thickness. When the channel is large compared to the double layer thickness (κh is large) the profile is fairly flat and the velocity can be assumed constant in the y-direction to a good approximation.

less important and electroosmotic flow becomes more important. This is why we typically see electroosmotic flow implemented in small channels and pressure driven flow implemented in large pipes.

16.1.1. Canceling out electroosmotic flow with pressure driven flow.
Before moving onto other electrokinetics phenomena, lets discuss one interesting application of combined pressure driven and electroosmotic flow - an apparatus for canceling out electroosmotic flow in measurements of electrokinetic phenomena, namely electrophoresis. Say for example you are interesting in measuring how fast a particle moves in an electric field due to electrophoresis (to be described in detail in Section 16.7). If you placed a particle directly in a regular channel, applied an electric field, and began to observe the particle (say with a microscope) you would see the particle move both due to the electric field and due to the electroosmotic flow. So you would need to first measure the electroosmotic flow to find the particle's motion due to electric field alone - that can be inconvenient. What you can do instead is have pressure driven flow counteract electroosmotic flow. To do this, you simply close your electrophoresis channel from both sides! This way, there is no net flow in the channel and an axial pressure gradient is setup to counteract the electroosmotic flow. In equation form this is

$$(16.1.16) \qquad Q = \frac{2h^3 p_x}{3\mu} - \frac{2h\varepsilon_0 \varepsilon E_x \zeta}{\mu}\left[1 - \frac{\tanh{(\kappa h)}}{\kappa h}\right] = 0$$

or

(16.1.17)
$$p_x = \frac{3\varepsilon_0 \varepsilon E_x \zeta}{h^2} \left[1 - \frac{\tanh(\kappa h)}{\kappa h} \right]$$

Substituting this result into the equation for the velocity profile, (16.1.12) gives

(16.1.18)
$$u_{xQ0}(y) \frac{\mu}{\varepsilon_0 \varepsilon E_x \zeta} = \frac{3}{2} \left[1 - \frac{\tanh(\kappa h)}{\kappa h} \right] \left[1 - \left(\frac{y}{h} \right)^2 \right] - \left[1 - \frac{\cosh(\kappa y)}{\cosh(\kappa h)} \right]$$

This equation has two zeros, i.e. points where pressure driven flow cancels out electroosmotic flow. These are the points at which you should be observing your electrophoresing particles. For general κh this equation can be solved numerically. For the most common case where κh is large (>100), the equation has zeros at

(16.1.19)
$$\frac{y}{h} = \pm \frac{1}{\sqrt{3}} = \pm 0.577$$

In general the position of these zero points varies comparatively little, varying from roughly 0.45 at $\kappa h \approx 1$ (and tending towards zero) to 0.58 at κh greater than 100. Thus, by observing the velocities of particles only at one of these planes, for example using an objective with a low depth of field or a confocal microscope, one can accurately measure the particles electrophoretic velocity without worrying about electoosmotic flow.

16.1.2. Electric current in electroosmotic flow and streaming current.
Another important aspect to understand is how much electric current will flow (and so must be supplied) for an electroosmotic flow to occur and what this depends on. Recall that electric current is defined as the flow of charge carriers, i.e. the amount of charge flowing per time. For electrolyte solutions which we are interested in, the charge carriers are ions. So to determine the electric current we ought to tally up how many of these ions are flowing and at what velocity. Ions can have a velocity mainly for three reasons: (a) they are moved along with the flow, (b) they are diffusing due to a concentration gradient (moving randomly, but appear to be moving into the region of low concentration in the net, because less of them are moving randomly from the low concentration as they are less of them there to begin with), and (c) moving due to a Coulomb force due to an electric field imposed on them. However, it is more convenient to work with current per area, or current density then straight current. Thus, all this can be placed in equation form, in order, and written as

(16.1.20)
$$\mathbf{i} = e\mathbf{u} \sum_k z_k n_k - e \sum_k D_k z_k \nabla n_k - e \nabla \varphi \sum_k z_k \mu_{ek} n_k$$

where $\mathbf{i}$ is the current density vector, e is elementary charge, $\mathbf{u}$ is the net flow velocity vector, z_k is the charge of the ion (e.g., +1), n_k is the number density of the ion (concentration * Avagradro's number), D_k is the diffusion coefficient, φ is the electric potential, and μ_{ek} is the ion's electrophoretic mobility (the ratio between the velocity of an ion and the electric field applied to it). Next we will simplify (16.1.20) by eliminating the electrophoretic mobility, by using the Einstein–Smoluchowski relation, (15.3.2) which we rewrite here for convenience

(16.1.21)
$$\mu_e = \frac{zeD}{k_B T} = \frac{zFD}{RT}$$

and rewrite (16.1.20) as

$$(16.1.22) \qquad \mathbf{i} = e\mathbf{u} \sum_k z_k n_k - e \sum_k D_k z_k \nabla n_k - \frac{e^2 \nabla \varphi}{k_B T} \sum_k z_k^2 D_k n_k$$

Since in our case the flow is unidirectional, we only take the x term of the above equation,

$$(16.1.23) \qquad i_x = e u_x \sum_k z_k n_k - e \sum_k D_k z_k \frac{\partial n_k}{\partial x} - \frac{e^2 \nabla \varphi}{k_B T} \frac{\partial \varphi}{\partial x} \sum_k z_k^2 D_k n_k$$

Most commonly there is no concentration gradient in the x direction (unless in a special case one is introduced), and so the second term is zero, and thus

$$(16.1.24) \qquad i_x = e u_x \sum_k z_k n_k - \frac{e^2}{k_B T} \frac{\partial \varphi}{\partial x} \sum_k z_k^2 D_k n_k$$

The total current flow is just the integral of this current per area over the entire area of the channel, and so the current per unit depth (into the page, similarly to our definition of flow rate (16.1.13)) is

$$(16.1.25) \qquad I = 2 \int_0^h i_x dy$$

and so

$$(16.1.26) \qquad I = 2e \int_0^h u_x \sum_k z_k n_k dy + \frac{2e^2 E_x}{k_B T} \int_0^h \sum_k z_k^2 D_k n_k dy$$

where u_x is given by (16.1.12)

$$(16.1.27) \qquad u_x(y) = \frac{h^2 p_x}{2\mu} \left[1 - \left(\frac{y}{h}\right)^2 \right] - \frac{\varepsilon_0 \varepsilon E_x \zeta}{\mu} \left[1 - \frac{\cosh(\kappa y)}{\cosh(\kappa h)} \right]$$

The first term in (16.1.26) is due to the convective net transport of ions, i.e. transport of ions in regions where the electroneutrality term, $\sum_k z_k n_k$ (which is a function of y) is non-zero. But, you may ask, if the electroneutrality term is non-zero, surely this will setup up an electric field to bring ions from a far to make it zero; and if we started from a net electroneutrality solution in the channel, how can this be?! Well, a better question to ask, is "Where can this be?". At the wall we know there are fixed charges (which before introduction of the solution, generally should be neutral, but say if they were acidic groups, gave up a proton to the solution) and so they can balance the non-electroneutrality condition in the solution phase. In fact, this presence of non-mobile charges fixed to the wall, is what creates this non-electroneutrality in the solution. Recall that this is the concept of the electrical double layer. So, say if mechanical pressure drives fluid along the channel, this drags charge along the wall, setting up a net current! This is typically referred to as streaming current, I_s. Note that since in our analysis we assume motion begins at the shear plane, we tacitly assume that there is no movement of ions attached to the wall and that there is no motion of ions in the Stern layer. You can measure the streaming current to determine the flow rate of a known electrolyte through a channel of know material (and so ζ potential) and dimension. You can also use it, in combination with a flow measurement (e.g., a differential pressure flow meter, or a calorimetric flow meter) to measure zeta potential. One may think to use this to generate electrical power, and while it is highly scalable to very small dimensions, the overall efficiency of this method is pretty low. In the case when electric field is applied, the ions near the shear layer move, and this is also called a streaming current (second term of (16.1.27)). However, this case is a lot less useful, as in addition to the streaming current, there is another current, second term of (16.1.26).

The second term in (16.1.26) is due electrical conduction within the channel, and is referred to as conduction current I_c. To get an appreciation for the conduction current, let's consider a special case of a symmetric electrolyte $(z : z)$ where the species have approximately the same diffusion coefficients. An example of such electrolyte is a potassium chloride solution. Then the conduction current part of (16.1.26) becomes

$$(16.1.28) \qquad I_c = \frac{2e^2 z^2 D E_x}{k_B T} \int_0^h (n_+ + n_-)\, dy$$

From Section 15.3 we know that charge as a function of the distance from the wall follows a Boltzmann distribution, (15.3.6) and so inserting that, we obtain

$$(16.1.29) \qquad I_c = \frac{4e^2 z^2 D n_\infty E_x}{k_B T} \int_0^h \cosh\left(ze\varphi/k_B T\right) dy$$

Now to find the conduction current all we need is to obtain the potential distribution as a function of the distance from the wall, φ. We can obtain this from the Poisson-Boltzmann equation, or to keep the equations simple, using the Debye-Huckel approximation for low potentials. To do this, we can Taylor expand the cosh term, keeping only the leading order terms, to be consistent with the Debye-Huckel approximation. Taylor expanding the cosh term

$$(16.1.30) \qquad I_c = \frac{4e^2 z^2 D n_\infty E_x}{k_B T} \int_0^h \left(1 + \frac{1}{2}\left(\frac{ze\varphi}{k_B T}\right)^2\right) dy$$

Substituting in the Debye-Huckel solution

$$(16.1.31) \qquad I_c = \frac{4e^2 z^2 D n_\infty E_x}{k_B T} \int_0^h \left(1 + \frac{1}{2}\left(\frac{ze\zeta}{k_B T}\right)^2 \frac{\cosh^2(\kappa y)}{\cosh^2(\kappa h)}\right) dy$$

We now integrate this and obtain

$$(16.1.32) \qquad I_c = 2\sigma^\infty h E_x \left[1 + \frac{1}{4}\left(\frac{ze\zeta}{k_B T}\right)^2 \left(\frac{\tanh(\kappa h)}{\kappa h} + \frac{1}{\cosh^2(\kappa h)}\right)\right]$$

$$(16.1.33) \qquad f_{nn} = \left[1 + \frac{1}{4}\left(\frac{ze\zeta}{k_B T}\right)^2 \left(\frac{\tanh(\kappa h)}{\kappa h} + \frac{1}{\cosh^2(\kappa h)}\right)\right]$$

where

$$(16.1.34) \qquad \sigma^\infty = \frac{2e^2 z^2 D n_\infty}{k_B T}$$

is the conductivity of the bulk (using the definition of solution conductivity and Einstein–Smoluchowski relation). Notice that equation (16.1.32) is just Ohm's law, with just a fudge factor in the square brackets. This fudge factor accounts for the

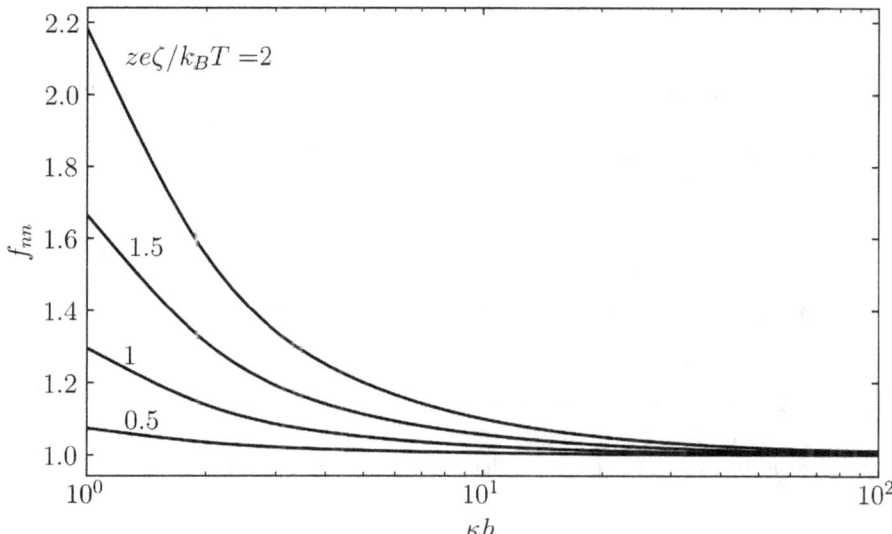

FIGURE 16.1.4. Non-neutrality factor contributing to the additional conductivity of the solution due to the bound charge at the wall as a function of scaled zeta potential and scaled double layer thickness. For thin double layers (large κh, i e. high solution ionic strength) and low zeta potentials, this factor becomes negligible and the conductivity approaches that predicted by Ohm's law.

non-electroneutrality of the electrolyte solution due to the presence of the charged wall. Because the solution is weakly, but still somewhat non-neutral, it can carry a bit more current than as a solution (or a conductor) that is perfectly electroneutral. That is why the fudge factor is slightly more than one. We plot the magnitude of this fudge factor, f_{nn} (nn for non-neutral) in Figure 16.1.4. In deriving this result, we have assumed that the scaled zeta potential, $ze\zeta/k_B T$ was smaller than unity, but the result is still somewhat accurate for scaled zeta potentials not significantly larger than unity. You can see that for scaled zeta potential less than unity, you can set the value of this fudge factor to be about unity already for κh of 5 or greater. For scaled zeta potentials of about 2, we can set this fudge factor to unity for $\kappa h > 50$. Note that many plastics including cyclic olefin copolymer (COC), a commonly used plastic in microfluidics, zeta potential ranges from 0 to -50 mV (Tandon et al. (2008)) while for glass it ranges from -40 to -80 mV (Gu and Li (2000)). At room temperature the scaled zeta potential is then about 0 to -2 for COC, and -1.6 to 3.2 for glass. As we discussed before for typical solutions and channel dimensions in microfluidics, κh is larger than 50 and so this fudge factor can be neglected. We only have to pay attention to it when channels become small (around 1 µm or smaller), when the solutions are very dilute (around 10 µM or smaller), and/or when zeta potential is unusually high (100 mV or larger).

One can quickly observe that conductivity and Debye length are related. For a simple case of 1:1 electrolyte where the ions have nearly same mobility (or diffusivity) we here derive a simple relation using the equation for Debye length (15.3.22)

TABLE 1. Debye length and conductivity for a 1:1 electrolyte KCl ($D = 1.96 \times 10^{-9}\,\mathrm{m^2/s}$) at 298 K, $\kappa h \to \infty$.

Concentration	Debye length, κ^{-1}	Conductivity
1 M	0.307 nm	15 S/m
150 mM (approx. saline solution)	0.792 nm	2.21 S/m
100 mM	0.971 nm	1.5 S/m
10 mM	3.07 nm	150 mS/m
1 mM	9.71 nm	15 mS/m
100 μM	30.7 nm	1.5 mS/m
10 μM (approx. pH 5, ISO Grade 3 DI)	97.1 nm	150 μS/m
1 μM	307 nm	15 μS/m
100 nM (18 MΩ cm DI)	971 nm	1.5 μS/m *

Diffusion coefficient from Harned and Nuttall (1947); dilute solution assumed, even for 1 M condition: conductivity overpredicted. * This is nonsensical as here there is 100 nM of protons contributing to the conductivity.

and our simple (16.1.34) equation for conductivity

$$(16.1.35) \qquad \sigma^{\infty} = \kappa^2 \varepsilon \varepsilon_0 D$$

or

$$(16.1.36) \qquad \kappa^{-1} = \sqrt{\frac{\varepsilon \varepsilon_0 D}{\sigma^{\infty}}}$$

Now, having dealt with electrical conduction current in detail, let's turn our attention to the convection transport (streaming) current, rewriting it here again for convenience:

$$(16.1.37) \qquad I_s = 2e \int_0^h u_x \sum_k z_k n_k \, dy$$

Recall from equation (15.3.7) (and that $c = n/N_A$) that charge density

$$(16.1.38) \qquad \rho = \sum_i z_i e n_i$$

and so

$$(16.1.39) \qquad I_s = 2 \int_0^h u_x \rho \, dy$$

combining this with Poisson's equation (15.3.8)

$$(16.1.40) \qquad I_s = -2\varepsilon\varepsilon_0 \int_0^h u_x \frac{d^2\varphi}{dy^2} \, dy$$

then expanding this integral via integration by parts

$$(16.1.41) \qquad I_s = -2\varepsilon\varepsilon_0 \left(\left[u_x \frac{d\varphi}{dy} \right]_{y=0}^{y=h} - \int_0^h \frac{d\varphi}{dy} \, du_x \right)$$

The first term is zero since

(16.1.42)
$$\frac{d\varphi}{dy} = 0 \text{ at } y = 0$$
$$u_x = 0 \text{ at } y = h$$

and so

(16.1.43)
$$I_s = 2\varepsilon\varepsilon_0 \int_0^h \frac{d\varphi}{dy} du_x = 2\varepsilon\varepsilon_0 \int_0^h \frac{du_x}{dy} \frac{d\varphi}{dy} dy$$

Taking the first derivative of velocity with respect to y from the integral of (16.1.9)

(16.1.44)
$$I_s = \frac{2\varepsilon\varepsilon_0}{\mu} \int_0^h \left(-p_x y + \varepsilon\varepsilon_0 E_x \frac{d\varphi}{dy} \right) \frac{d\varphi}{dy} dy$$

(16.1.45)
$$I_s = \frac{2\varepsilon\varepsilon_0}{\mu} \int_0^h \left(-p_x y \frac{d\varphi}{dy} + \varepsilon\varepsilon_0 E_x \left(\frac{d\varphi}{dy} \right)^2 \right) dy$$

where we evaluate each term in the integral to be

(16.1.46)
$$\int_0^h y \frac{d\varphi}{dy} dy = h\zeta \left(1 - \frac{\tanh(\kappa h)}{\kappa h} \right)$$

(16.1.47)
$$\int_0^h \left(\frac{d\varphi}{dy} \right)^2 dy = \frac{h\zeta^2}{2} \left(\tanh(\kappa h) - \frac{\kappa h}{\cosh^2(\kappa h)} \right)$$

and so
(16.1.48)
$$I_s = -\frac{2\varepsilon\varepsilon_0 h\zeta}{\mu} p_x \left(1 - \frac{\tanh(\kappa h)}{\kappa h} \right) + \frac{(\varepsilon\varepsilon_0)^2 \kappa\zeta^2}{\mu} E_x \left(\tanh(\kappa h) - \frac{\kappa h}{\cosh^2(\kappa h)} \right)$$

Thus, we now have expressions for both the conduction current and the streaming current, which added together give the total current

(16.1.49)
$$I = I_c + I_s$$

16.1.3. Non-equilibrium thermodynamics perspective.
Having obtained the relation for total current and total flow rate in the previous sections, we want to observe some similarities between them, and so we shall cast these in a convention of non-equilibrium thermodynamics. Firstly, to make equations look cleaner, let's define variables

(16.1.50)
$$\alpha_1 = 1 - \frac{\tanh(\kappa h)}{\kappa h}$$

(16.1.51)
$$\alpha_2 = \frac{\tanh(\kappa h)}{\kappa h} - \frac{1}{\cosh^2(\kappa h)}$$

Then we can write the total current as

(16.1.52)
$$I = \left[\frac{(\varepsilon\varepsilon_0)^2 \kappa^2 h\zeta^2}{\mu} \alpha_2 + 2\sigma^\infty h f_{nn} \right] E_x + \left[-\frac{2\varepsilon\varepsilon_0 h\zeta}{\mu} \alpha_1 \right] p_x$$

We can similarly write the total flow rate as

$$(16.1.53) \qquad Q = \left[-\frac{2\varepsilon\varepsilon_0 h\zeta\alpha_1}{\mu} \right] E_x + \left[\frac{2h^3}{3\mu} \right] p_x$$

Which we can write even cleaner as

$$(16.1.54) \qquad \begin{aligned} I &= L_{11}E_x + L_{12}p_x \\ Q &= L_{21}E_x + L_{22}p_x \end{aligned}$$

or in matrix form

$$(16.1.55) \qquad \begin{bmatrix} L_{11} & L_{12} \\ L_{21} & L_{22} \end{bmatrix} \begin{bmatrix} E_x \\ p_x \end{bmatrix} = \mathbf{L} \begin{bmatrix} E_x \\ p_x \end{bmatrix} = \begin{bmatrix} I \\ Q \end{bmatrix}$$

where

$$(16.1.56) \qquad L_{11} = 2\sigma^\infty h \left[\frac{(\varepsilon\varepsilon_0)^2 \kappa^2 \zeta^2}{2\sigma^\infty \mu} \alpha_2 + f_{nn} \right]$$

$$(16.1.57) \qquad L_{12} = \left[-\frac{2\varepsilon\varepsilon_0 h\zeta}{\mu} \alpha_1 \right]$$

$$(16.1.58) \qquad L_{21} = \left[-\frac{2\varepsilon\varepsilon_0 h\zeta}{\mu} \alpha_1 \right]$$

$$(16.1.59) \qquad L_{22} = \left[\frac{2h^3}{3\mu} \right]$$

Thus we can clearly see a similarity between pressure and electric field and current and fluid flow - both pressure and electric field can be the driving force for current and flow. Notice that they are coupled through the coupling matrix $\mathbf{L}$ and this coupling is caused by the presence of the double layer. It is important to understand that if the double layer was absent and so the presence of free charge near the wall was absent, flow would be independent of electric field and vice versa. However, alas, in real situations the double layer is present, even if it is thin, and so electrokinetic phenomena can creep in to situations when one does not expect it.

16.2. Streaming potential

When liquid is pumped through a channel using pressure driven flow, the mobile ions in the double layer will move with the fluid, which of course causes streaming current. However, this implies that there will be a depletion of such charges at the entrance of the channel and an excess of such charges at the exist. Thus a potential is set up across the channel - the streaming potential. This potential in turn causes a conduction current to flow in the direction opposite of the streaming current, such that the sum of the two currents, i.e. the total current is zero. Thus setting this condition for equation (16.1.55) we find that

$$(16.2.1) \qquad \frac{E_x}{p_x} = -\frac{L_{12}}{L_{11}} = \frac{\varepsilon\varepsilon_0\zeta \left(1 - \tanh\left(\kappa h\right)/\kappa h \right)}{\mu\sigma^\infty \left((\varepsilon\varepsilon_0)^2 \kappa^2 \zeta^2 \alpha_2 \big/ 2\sigma^\infty \mu + f_{nn} \right)}$$

We can Taylor expand this in terms of the scaled zeta potential, $z e\zeta / k_B T$ and obtain that to the first order

$$(16.2.2) \qquad \frac{E_x}{p_x} \approx \frac{\varepsilon\varepsilon_0\zeta}{\mu\sigma^\infty} \left(1 - \frac{\tanh\left(\kappa h\right)}{\kappa h} \right) + O\left(\left(z e\zeta / k_B T\right)^3 \right)$$

or

(16.2.3)
$$-\frac{\Delta V}{L} = E_x = \frac{\varepsilon \varepsilon_0 \zeta}{\mu \sigma^\infty} \left(1 - \frac{\tanh\left(\kappa h\right)}{\kappa h} \right) \left(-\frac{\Delta p}{L} \right)$$

where ΔV is the streaming potential, Δp is the pressure drop, and L is the length of the channel. Interestingly the streaming potential is largely independent of the transverse dimension (i.e. h) of the channel when the double layers are thin ($\kappa h \to \infty$). To get a sense for streaming potential, let's examine the flow of kerosene jet fuel as if it is to be pumped into an aircraft. While kerosene is highly non-conductive (σ^∞ typically is 50 pS/m) and so has larger Debye lengths compared to aqueous solutions, the hose used to fuel aircraft is also typically large (several centimeters in diameter). Furthermore, while equation (16.2.3) was derived for a sandwich of infinitely flat planes geometry, we will assume it is valid within an order of magnitude for a circular tube cross-section. Thus, let's assume we are pumping kerosene at a flow rate of 50,000 gal/hr (typical jet airliner fuel capacity is around 50,000 gal) through a 5 m long, 5 cm diameter insulating hose, with a zeta potential of 70 mV. Kerosene has the following properties: dynamic viscosity, 1.64 mPa s; density, 800 kg/m^3; electrical conductivity 50 pS/m; dielectric constant, 1.8. Using Darcy-Weisbach equation, we get a pressure drop of 3.5 atm. Then using (16.2.3) we get a streaming potential of 5 MV! To put this in perspective, breakdown field for air is about 3 MV/m, so if the airplane was parked about a meter away from the fuel truck, with no electrical connections between them or grounding ("perfect storm" scenario), then surely there will be a spark between the fuel truck and the airplane wing! If the fuel-air mixture of the fuel vapors was right, this would cause a rapid unplanned disassembly of both, fuel truck and airplane - a highly undesirable circumstance. While a scenario described here is only illustrative, sparking and explosions from pumping hydrocarbons have caused airplane explosions during WWII, plagued the petroleum industry in the 1950s, and are a significant concern when pumping petroleum products today. To alleviate such problems, better grounding is implemented, and improved conductivity hoses are used. Furthermore conductivity improvers are added to the hydrocarbon liquid to decrease streaming potential. For more discussion on this, see Touchard (2001) and Gieras (2013). Also, note that streaming potential can occur in two phase flows such as flows of foams and wet steam, where the streaming potential can be much larger than that for a single phase liquid flow. For more discussion on this topic, see Sprunt et al. (1994).

Lastly notice that this electrokinetic effect arose without any external electrical power sources, and without desire to have electrokinetic effects. Electrokinetic effects can be important and need to be accounted for even when electrokinetic effects are not desired or planed for!

16.3. Electroviscous effect

Another surprising electrokinetic effect that occurs when such are neither desired nor planned for is the electroviscous effect. In short, it is the increase in apparent viscosity of the fluid due to the presence of electrical double layers. Just like the streaming potential, it occurs without an external potential applied to the flow channel, but rather when a pressure is a applied to the channel. It is in fact a result of streaming potential. When a potential is set up across the channel it induces both the conduction current, and also it inhibits the streaming current, and thus inhibits the flow that is driven by the pressure in the channel. This inhibition to the streaming current and so the fluid flow is most evident when the double layers are fairly thick, i.e. κh is order unity. If we substitute the electric field due to a streaming potential (equation (16.2.1)) into the general equation for flow rate, (16.1.53)), we obtain that the flow rate in the case of streaming potential,

$$(16.3.1) \qquad Q = \frac{2h^3}{3\mu} p_x \left[1 - \frac{3(\varepsilon\varepsilon_0)^2 \kappa^2 \zeta^2}{\mu \sigma^\infty (\kappa h)^2} \frac{\alpha_1^2}{\left((\varepsilon\varepsilon_0)^2 \kappa^2 \zeta^2 \alpha_2 \Big/ 2\sigma^\infty \mu + f_{nn} \right)} \right]$$

The first term in the brackets (unity) represents the flow rate the due to pressure driven flow in a situation where electrokinetic effects do not exist. The second term in the brackets by quick inspection is always positive. Thus the flow rate is always smaller in a situation when the double layer exists, then in the idealized case where it does not! This reduced flow rate at the same pressure appears to the experimenter as an additional resistance, a phantom viscosity, and so has been termed electroviscous effect. We can estimate how much the magnitude of this effect, approximately, by again Taylor expanding the term in the bracket in terms of the scaled zeta potential, $ze\zeta/k_B T$ and taking the leading order term. We obtain

$$(16.3.2) \qquad Q \approx \frac{2h^3}{3\mu} p_x \left[1 - \frac{3(\varepsilon\varepsilon_0)^2 \kappa^2 \zeta^2}{\mu \sigma^\infty (\kappa h)^2} \alpha_1^2 \right]$$

To a naive experimenter, the flow rate as a function of pressure should be

$$(16.3.3) \qquad Q = \frac{2h^3}{3\mu_a} p_x$$

where μ_a is the apparent viscosity. Thus, to find the ratio of apparent viscosity to actual viscosity we set the flow rates in (16.3.2) and (16.3.3) equal, and find

$$(16.3.4) \qquad \frac{\mu_a}{\mu} = \left[1 - \frac{3(\varepsilon\varepsilon_0)^2 \kappa^2 \zeta^2}{\mu \sigma^\infty (\kappa h)^2} \alpha_1^2 \right]^{-1}$$

and after using a Taylor expansion for $(1-x)^{-1}$ and keeping the first term, we obtain

$$(16.3.5) \qquad \frac{\mu_a}{\mu} \approx \left[1 + \frac{3(\varepsilon\varepsilon_0)^2 \kappa^2 \zeta^2}{\mu \sigma^\infty (\kappa h)^2} \alpha_1^2 \right] = \left[1 + \frac{\beta \alpha_1^2}{(\kappa h)^2} \right]$$

$$(16.3.6) \qquad \beta = \frac{3(\varepsilon\varepsilon_0)^2 \kappa^2 \zeta^2}{\mu \sigma^\infty}$$

We plot μ_a/μ as a function of κh and β in Figure 16.3.1.

Notice that electroviscous flow is significant when κh is order unity and β is large. When pumping low conductivity liquids (e.g. kerosene) through small channels (κh order unity) can result electroviscous effect. One should expect to encounter this effect petroleum extraction (hydrocarbon flow through porous media) and chromatography (organic solvent flow in porous media). One should also be careful measuring viscosity of low conductivity fluids (e.g., 1-10 pS/m) using narrow bore (0.2 mm) viscometers, since for these fluids Debye length can often be of order 10-100 µm or more, and so κh of order unity.

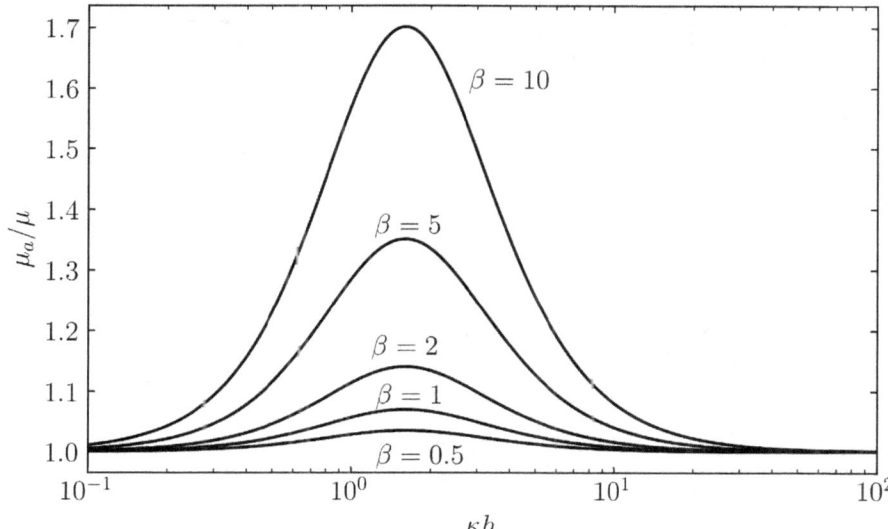

FIGURE 16 3.1. Apparent viscosity scaled by the viscosity of the solvent as a function of scaled channel size κh and the electroviscous parameter β. The apparent viscosity peaks around κh equal to unity, where the retardation of the streaming current by the built up streaming potential is the highest. From (16.3.5) we see that the scaled apparent viscosity is linear with β.

16.4. Results for electroosmotic phenomena in circular channels

Here we present just the resulting pertinent equations for electroosmotic flow now in circular channels. These are similar to the ones we derived for the slit channel, but just with the derivation occurring in a cylindrical coordinate system instead of a Cartesian one. For details of the derivations see Rice and Whitehead (1965) (channels with low surface potential) and Olivares et al. (1980) and Levine et al. (1975) (high surface potentials).

For a symmetric $(z:z)$ electrolyte and when the surface potential is small, i.e., $ze\varphi/k_B T < 1$ or the surface potential is less than 25 mV at room temperature, we can use the Debye-Hückel approximation and obtain a linearized Poisson-Boltzmann equation. Integrating that like we have done in Section 15.3.1 we obtain that the free charge density distribution is

$$(16.4.1) \qquad \rho_f = -\varepsilon\varepsilon_0 \kappa^2 \zeta \frac{I_0\left(\kappa r\right)}{I_0\left(\kappa a\right)}$$

where ζ is the zeta potential of the wall, a is the channel radius, and I_0 is the zeroth-order modified Bessel function of the first kind. To remind the reader the properties of the modified Bessel function of the first kind we plot it in Figure 16.4.1.

As previously, we substitute this result into the simplified Navier-Stokes equations (with the Stokes flow assumption and free charge body force term) and assume

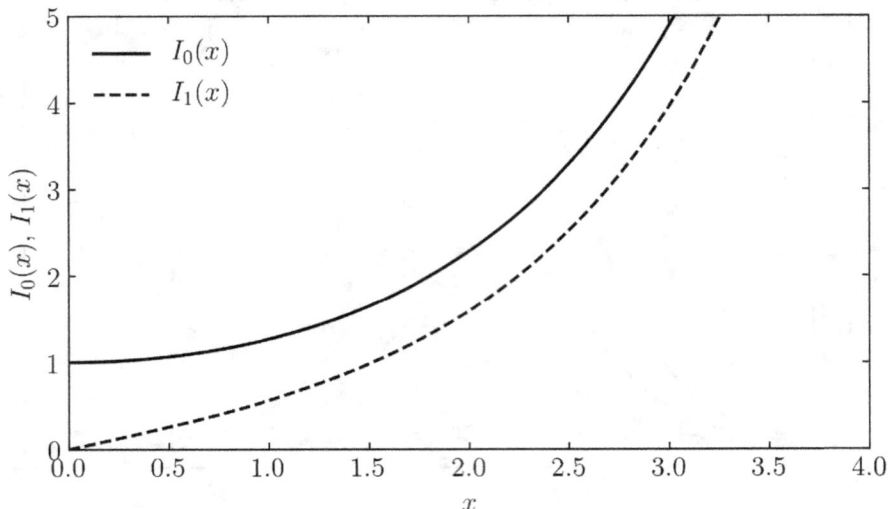

FIGURE 16.4.1. Modified Bessel function of the first kind of zeroth order I_0 and first order I_1. Both functions quickly approach infinity for large x.

a uniform axially directed field inside the capillary. We integrate the resulting equation and find that the total velocity due to pressure driven and electroosmotic flow is

$$(16.4.2) \qquad u_x(r) = \frac{ap_x}{4\mu}\left(1 - \left(\frac{r}{a}\right)^2\right) - \frac{\varepsilon\varepsilon_0\zeta E_x}{\mu}\left(1 - \frac{I_0(\kappa r)}{I_0(\kappa a)}\right)$$

This is similar to equation (16.1.12) we found for the slit channel. We now explore a few special cases. When there is no pressure gradient and the double layers are thin ($\kappa a \gg 1$) $I_0(\kappa r)/I_0(\kappa a) \approx 0$ and so

$$(16.4.3) \qquad u_{x,\kappa a\gg 1}(r) \approx -\frac{\varepsilon\varepsilon_0\zeta}{\mu}E_x$$

This result is known as the Helmholtz-Smoluchowski equation. Notice that the velocity is independent of the radius - the flow moves as a plug, as if the fluid slips at the wall. This result is quite useful as (an often highly accurate) first approximation of electroosmotic flow velocity as quite often in microchannels with realistic electrolytes double layers are quite thin (see Table 1 for typical Debye lengths vs. concentration). In the case of thick double layers as can occur in nanochannels ($\kappa a \ll 1$) and zero pressure gradient, we can expand the Bessel function and obtain

$$(16.4.4) \qquad u_{x,\kappa a\ll 1}(r) \approx -\frac{\varepsilon\varepsilon_0\zeta}{\mu}E_x\frac{(\kappa a)^2}{4}\left(1 - \left(\frac{r}{a}\right)^2\right)$$

Note that as κa approaches zero so does the electroosmotic flow in this case. This model predicts that as the double layers significantly overlap electroosmotic flow goes to zero. For moderately overlapping double layers the model predicts that the velocity profile is parabolic. This is due to the fact that when the double layers are

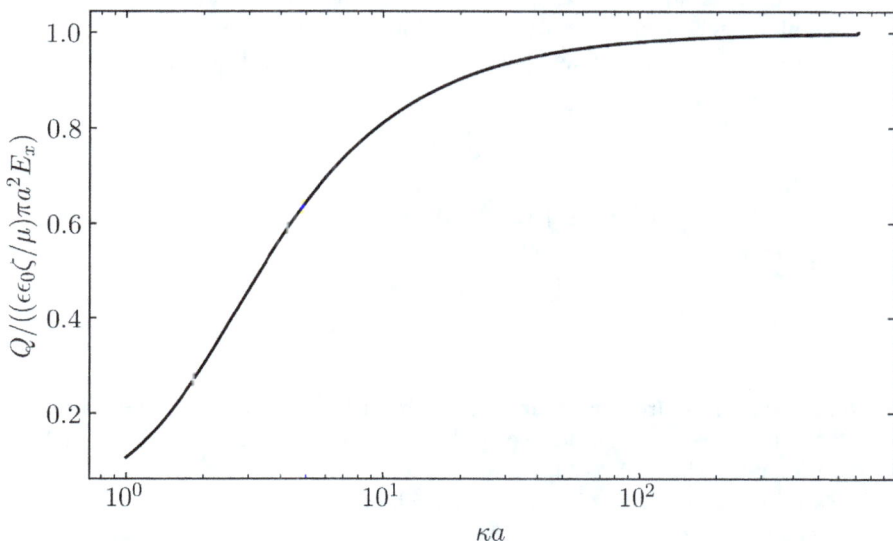

FIGURE 16.4.2. Scaled electroosmotic flow rate as a function of the relative double layer thickness κa for circular capillaries. This scaled electroosmotic flow approaches unity for the thin double layers (most common case), and so high ionic strength solutions are most efficient at producing electroosmotic flow.

moderately overlapping free charge is roughly uniform in the radial direction, and thus the body force is independent of the radial direction. This is reminiscent of the pressure term in the Navier-Stokes equation and thus produces the characteristic parabolic profile.

We can integrate equation (16.4.2) and obtain that the total flow rate is

$$(16.4.5) \qquad Q = \frac{\pi a^4}{8\mu} p_x - \frac{\varepsilon \varepsilon_0 \zeta}{\mu} \pi a^2 E_x \left(1 - \frac{2 I_1(\kappa a)}{\kappa a I_0(\kappa a)} \right)$$

For the case of zero pressure gradient, we plot the scaled flow rate as a function of relative double layer thickness, κa, in Figure 16.4.2. We can interpret this plot as the efficiency of electroosmotic flow as a function of the electrolyte used. We can see that the thinner the double layer, the greater the electroosmotic flow per channel area.

Next let's consider what kind of pressures can the electroosmotic flow generate. The maximum pressure can be achieved when there is zero flow rate and so from (16.4.5)

$$(16.4.6) \qquad p_{x,\text{max}} = \frac{8 \varepsilon \varepsilon_0 \zeta}{a^2} \left(1 - \frac{2 I_1(\kappa a)}{\kappa a I_0(\kappa a)} \right) E_x$$

Notice that the maximum pressure is independent of viscosity. When double layers are thin, i.e. $\kappa a \gg 1$, this result simplifies to

$$(16.4.7) \qquad p_{x,\text{max}} = \frac{8 \varepsilon \varepsilon_0 \zeta}{a^2} E_x$$

TABLE 2. Radius in a circular capillary scaled by the radius of the capillary where the electroosmotic flow is canceled by the built up pressure driven flow when the capillary is sealed at both ends (zero net flow rate).

κa	r_0/a
1	0.579
10	0.635
100	0.7
500	0.706
∞	0.707 ($\sqrt{2}/2$)

Since maximum pressure scales inversely with the square of capillary radius, to obtain large pressures we should use both small radius capillaries and a high ionic strength buffer (so that $\kappa a \gg 1$).

We can find the positions in the circular channel where the pressure driven flow cancels out electroosmotic flow, just as we have done before for the slit channel in Section 16.1.1. We substitute (16.4.6) into (16.4.2) and find the velocity profile in the closed channel,

$$(16.4.8) \qquad u_x = \frac{\varepsilon\varepsilon_0\zeta}{\mu}E_x \left(2\left(1-\left(\frac{r}{a}\right)^2\right)\left(1-\frac{2I_1(\kappa a)}{\kappa a I_0(\kappa a)}\right) - \left(1-\frac{I_0(\kappa r)}{I_0(\kappa a)}\right)\right)$$

When double layers are thin, i.e. $\kappa a \gg 1$, this result simplifies to

$$(16.4.9) \qquad u_x = \frac{\varepsilon\varepsilon_0\zeta}{\mu}E_x\left(1-2\left(\frac{r}{a}\right)^2\right)$$

Solving (16.4.8) for zero velocity, we find the radius at which electroosmotic flow is cancelled out by pressure driven flow. We list these in Table 2.

We can also obtain an equation for current in electroosmotic flow just as we have in Section 16.1.2. Here we just write the result for a symmetric ($z : z$) electrolyte and low surface potentials ($ze\varphi/k_B T \ll 1$):

$$(16.4.10)$$
$$I = \pi a^2 \sigma^\infty f_{nn}\left(1-\left(\frac{\varepsilon\varepsilon_0\zeta}{\mu}\right)^2\frac{\mu\kappa^2}{\sigma^\infty f_{nn}}\left(1-\frac{2}{\kappa a}\frac{I_1(\kappa a)}{I_0(\kappa a)}-\left(\frac{I_1(\kappa a)}{I_0(\kappa a)}\right)^2\right)\right)E_x$$
$$-\frac{\varepsilon\varepsilon_0\zeta}{\mu}\pi a^2\left(1-\frac{2}{\kappa a}\frac{I_1(\kappa a)}{I_0(\kappa a)}\right)p_x$$

$$(16.4.11) \qquad f_{nn} = 1 + \left(\frac{ze\zeta}{k_B T}\right)^2\frac{1}{I_0^2(\kappa a)}\int_0^1 I_0^2(\kappa a R)R\,dR$$

where R is r/a. Using the relations for flow rate and current we can obtain the relation for streaming potential, just as we did in Section 16.2. Streaming potential is

$$(16.4.12) \qquad \frac{\Delta V}{L} = E_x = \frac{\varepsilon\varepsilon_0\zeta}{\mu}\frac{1}{\sigma^\infty}\left(1-\frac{2}{\kappa a}\frac{I_1(\kappa a)}{I_0(\kappa a)}\right)g(\kappa a, \beta, f_{nn})\,p_x$$

(16.4.13)
$$g\left(\kappa a, \beta, f_{nn}\right) = \cfrac{1}{f_{nn} - \beta\left(\left(1 - \frac{2}{\kappa a}\frac{I_1(\kappa a)}{I_0(\kappa a)} - \left(\frac{I_1(\kappa a)}{I_0(\kappa a)}\right)^2\right)\right)}$$

(16.4.14)
$$\beta = \left(\frac{\varepsilon\varepsilon_0\zeta}{\mu}\right)^2\frac{\mu\kappa^2}{\sigma^\infty}$$

This can be Taylor expanded in terms of $ze\zeta/k_BT$ and if only the leading order is kept,

(16.4.15)
$$-\frac{\Delta V}{L} = E_x \approx \frac{\varepsilon\varepsilon_0\zeta}{\mu}\frac{1}{\sigma^\infty}\left(1 - \frac{2}{\kappa a}\frac{I_1(\kappa a)}{I_0(\kappa a)}\right)p_x$$

which for the most common case of thin double layers, $\kappa a \gg 1$, simplifies further to

(16.4.16)
$$-\frac{\Delta V}{L} = E_x \approx \frac{\varepsilon\varepsilon_0\zeta}{\mu}\frac{1}{\sigma^\infty}p_x$$

We can similarly obtain a relation for the electroviscous effect as we have done in Section 16.2. The ratio of apparent viscosity to the viscosity of the fluid is then

(16.4.17)
$$\frac{\mu_a}{u} = \left(1 - \frac{8\beta}{(\kappa a)^2}\left(1 - \frac{2}{\kappa a}\frac{I_1(\kappa a)}{I_0(\kappa a)}\right)^2 g\left(\kappa a, \beta, f_{nn}\right)\right)^{-1}$$

which can then be simplified using the Taylor expansion for $1/(1-x) = 1 + x + O\left(x^2\right)$ to

(16.4.18)
$$\frac{\mu_a}{\mu} \approx 1 + \frac{8\beta}{(\kappa a)^2}\left(1 - \frac{2}{\kappa a}\frac{I_1(\kappa a)}{I_0(\kappa a)}\right)^2 g\left(\kappa a, \beta, f_{nn}\right)$$

and if we neglect contributions from $g\left(\kappa a, \beta, f_{nn}\right)$,

(16.4.19)
$$\frac{\mu_a}{\mu} \approx 1 + \frac{8\beta}{(\kappa a)^2}\left(1 - \frac{2}{\kappa a}\frac{I_1(\kappa a)}{I_0(\kappa a)}\right)^2$$

This function peaks around $\kappa a = 2.5$ and the apparent viscosity can be 3 times higher than the actual viscosity at this peak value for β of 10. As we have seen in the slit microchannel geometry, electroviscous effect is again important around κa of unity.

16.5. Results for electroosmotic flow in porous media

To a first (and often pretty good) approximation porous media (including packed beads) can be modeled as a bundle of tortuous circular capillaries. For this model, Yao and Santiago (2003) derived the set of relations for current and flow rate in electroosmosis in a porous media using the assumption that we have used in the past two sections: symmetric electrolyte and low surface potentials (Debye-Huckel approximation). For these relations, A is the cross-sectional area of the porous structure, τ is the tortuosity of the porous structure (typically 1.5) and ψ is the porosity (ratio of void to total volume). We list these here in the form of non-equilibrium thermodynamic expressions in Table 3, along with those for slit channel and circular capillary.

TABLE 3. Non-equilibrium thermodynamic expressions for electroosmotic phenomena in slit channel, circular capillary, and porous media geometries. $I = L_{11}E_x + L_{12}p_x$, $Q = L_{21}E_x + L_{22}p_x$. Notice that the off-diagonal elements are the same, i.e. the coupling matrix is symmetric.

Coefficients L	Equation
Slit channel, d here is depth of the real slit	
L_{11}	$2\sigma^\infty hd \left(\frac{(\varepsilon\varepsilon_0)^2 \kappa^2 \zeta^2}{2\sigma^\infty \mu} \alpha_2 + f_{nn} \right) f_{nn} = 1 + \frac{1}{4}\left(\frac{ze\zeta}{k_B T}\right)^2 \left(\frac{\tanh(\kappa h)}{\kappa h} + \frac{1}{\cosh^2(\kappa h)} \right)$
	$\alpha_2 = \frac{\tanh(\kappa h)}{\kappa h} - \frac{1}{\cosh^2(\kappa h)}$
L_{12}	$-\dfrac{2\varepsilon\varepsilon_0 hd\zeta}{\mu}\alpha_1$
	$\alpha_1 = 1 - \dfrac{\tanh(\kappa h)}{\kappa h}$
L_{21}	$-\dfrac{2\varepsilon\varepsilon_0 hd\zeta}{\mu}\alpha_1$
L_{22}	$\dfrac{2h^3 d}{3\mu}$
Circular capillary	
L_{11}	$\pi a^2 \sigma^\infty f_{nn} \left(1 - \left(\frac{\varepsilon\varepsilon_0\zeta}{\mu}\right)^2 \frac{\mu\kappa^2}{\sigma^\infty f_{nn}} \left(1 - \frac{2}{\kappa a}\frac{I_1(\kappa a)}{I_0(\kappa a)} - \left(\frac{I_1(\kappa a)}{I_0(\kappa a)}\right)^2 \right) \right) f_{nn} =$
	$1 + \left(\frac{ze\zeta}{k_B T}\right)^2 \frac{1}{I_0^2(\kappa a)} \int\limits_0^1 I_0^2(\kappa a R) R dR \quad R = r/a$
L_{12}	$-\frac{\varepsilon\varepsilon_0\zeta}{\mu}\pi a^2 \left(1 - \frac{2}{\kappa a}\frac{I_1(\kappa a)}{I_0(\kappa a)} \right)$
L_{21}	$-\frac{\varepsilon\varepsilon_0\zeta}{\mu}\pi a^2 \left(1 - \frac{2}{\kappa a}\frac{I_1(\kappa a)}{I_0(\kappa a)} \right)$
L_{22}	$\frac{\pi a^4}{8\mu}$
Porous media	
L_{11}	$-\frac{\psi A}{\sqrt{\tau}}\sigma_\infty \left(\begin{array}{c} \beta\left(\frac{I_1^2(\kappa a) - I_0(\kappa a) I_2(\kappa a)}{I_0^2(\kappa a)} \right) + 1 \\ -\zeta^* \frac{\Lambda_+ - \Lambda_-}{\Lambda} \frac{I_1(\kappa a)}{\kappa a I_0(\kappa a)} + \frac{\zeta^{*2}}{2}\left(1 - \frac{I_1^2(\kappa a)}{I_0^2(\kappa a)} \right) \end{array} \right)$
	$\zeta^* = \frac{ze\zeta}{kT}; \; \beta = \frac{(\varepsilon\varepsilon_0)^2 \xi^2 \kappa^2}{\mu\sigma_\infty}$
L_{12}	$-\frac{\varepsilon\varepsilon_0\zeta A}{\mu}\frac{\psi}{\sqrt{\tau}} f(\kappa a), \; f(\kappa a) = \left(1 - \frac{2}{\kappa a}\frac{I_1(\kappa a)}{I_0(\kappa a)} \right)$
L_{21}	$-\frac{\varepsilon\varepsilon_0\zeta A}{\mu}\frac{\psi}{\sqrt{\tau}} f(\kappa a), \; f(\kappa a) = \left(1 - \frac{2}{\kappa a}\frac{I_1(\kappa a)}{I_0(\kappa a)} \right)$
L_{22}	$\frac{Aa^2}{8\mu}\frac{\psi}{\sqrt{\tau}}$

16.6. AC electroosmotic flow

Having covered DC electroosmotic flow in considerable detail, let's turn to an-other electroosmotic phenomena - AC electroosmotic flow. While studying DC electroosmosis one might ask, what happens when an AC field instead of a DC field is applied to the capillary? The answer to this is that the double layer will move back and forth, and as a result of ion movement the solution in the channel will heat up, but nominally there will be no net flow. We say "nominally" because (a) on a time scale less than the period of the oscillation there certainly will be regular "DC" electroosmotic flow - hence the back and forth motion; and (b) if the double layer has some particular non-uniformity in the voltage drop across the double layer (and so non uniformity charge in the double layer parallel to the direction of the electric field) there actually might be flow in a particular direction. Nonuniformi-ties in adjacent double layers coupled with an electric field is in-fact what causes AC electroosmotic flow. The main advantage of AC electroosmotic flow over DC electroosmotic flow is that electrode reactions can be avoided as well as the associ-ated metal ion generation and gas generation. Avoiding gas generation is especially important as in microfluidic systems the generated gas can generate its own flows and interfere with desired processes. We will discuss here these nonuniformities, how to set these up, and how to create such flows, or how such are setup naturally as a byproduct of trying to do something else.

The easiest way to set up a non-uniformity in the double layer is by setting up completely different polarity double layers on two adjacent electrodes by apply-ing an electric field between them (see Figure 16.6.1). With an AC field, at any one time, one electrode is polarized positive, and another polarized negative. The polarization of the electrodes switches as the fields oscillates. Such electrode con-figuration is quite common in microfluidic devices - it is found in dielectrophoresis devices (to be discussed later), and in impedance and electrochemical sensing. Thus AC electroosmotic flow is encountered in situations when it is unexpected and often undesired. AC electroosmotic flow is also sometimes setup to pump fluid and to mix fluid near a surface, so that for example, a surface reaction is not diffusion limited.

When one applies a potential to adjacent electrodes such as ones shown in Fig-ure 16.6.1 and obtains oppositely polarized double layers, one also obtains electric field lines going from one electrode to the next as shown in Figure 16.6.1a. The electric field near the electrode (in the double layer) thus has both an upward com-ponent (normal to the electrode surface) and a side ward component (tangent to the electrode surface). This tangential component places a force on the charge in the double layer, and so moves the charge in the direction tangential to the electrode. This charge in turn drags with it the surrounding fluid, generating the flow.

Let's examine the moment where the left electrode is charged positive relative to the right electrode. (Figure 16.6.1a, left panel). As soon as the left electrode is charged this way, negative ions from the solution flock towards the left electrode to form a negative ion double layer above it. While these negative ions partially screen the positive charge of the electrode, they don't shield this charge completely, and so the positive charges in the electrode put out an electric field from the left electrode towards the right electrode. Meanwhile, of course, on the right electrode the negative charges attract positive ions to the right electrode and so a positively charged double layer is formed there. The electric field that is formed between

(a) Electric field lines and force on the double layer as the AC field oscilates the charging of the double layer

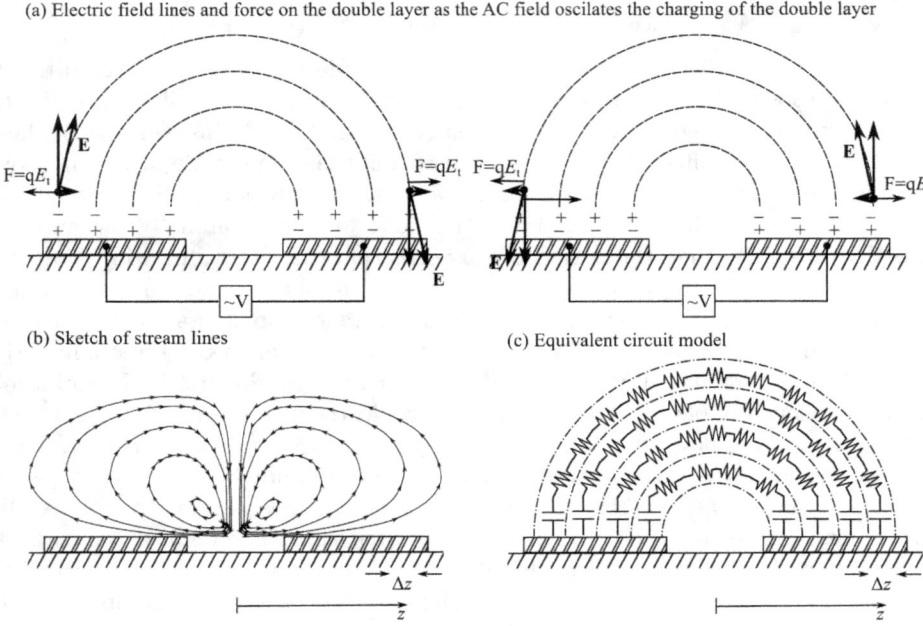

(b) Sketch of stream lines

(c) Equivalent circuit model

FIGURE 16.6.1. AC electroosmotic flow on parallel strip electrodes, where an AC field is applied between the two electrodes.

the left and right electrodes has a rightward tangential component. Above the left electrode, this places a leftward force on the negative ions in the left electrode's double layer, thus forcing the fluid leftward there. Above the right electrode, the rightward tangential electric field places a rightward force on the positive charge in the double layer there, and so these charges move rightward and drag the fluid rightward there. Thus the fluid gets dragged from the center between the electrodes, outward.

Notice what happens when the polarization of the electrodes gets flipped, when the field switches. (Figure 16.6.1a, right.) Now the left electrode is negatively charged and so attains a positive double layer, while the right electrode is positively charged and attains a negative double layer. Thus, now the field is from the right electrode towards the left electrode, and the tangential part of the electric field is in the leftward direction - the opposite from before. The leftward electric field forces the positive charges in the left electrode's double layer leftward, and the negative charges of right electrode's double layer rightwards, dragging their respective fluids in these directions. Once again, fluid gets dragged from the center between the electrodes, outward! Thus a two rolls form dragging the fluid from up above the center between the electrodes outwards, as shown on Figure 16.6.1b. In Figure 16.6.1b we show a sketch of the streamlines of such flow. This sketch is based on experimental visualization of these streamlines via microscopic particles by Green et al. (2000).

Let's now consider AC electroosmotic flow from a quantitative point of view. To obtain an approximate electric field above the electrodes, we model the situation as a number of discrete resistor-capacitor (RC) circuits forming arcs from one electrode to another, i.e. current tubes (Figure 16.6.1c). The capacitors in the RC circuit model the capacitance of the double layer, and the resistors in the circuit model the resistance of the bulk solution. The width of such thin current tube in our model Δz and the location of its center is z from the origin. While we show only four current tubes in Figure 16.6.1c, we can assume in fact that such current tubes are infinitesimal in width. Thus the tubes have a path length of πz. Thus if the electrodes have a depth d into the page a resistance of one of the current tubes at position z is

$$(16.6.1) \qquad R = \frac{\pi z}{\sigma d \Delta z}$$

where σ is the conductivity of the bulk solution. To obtain the capacitance, here we will use the Gouy-Chapman model of the double layer with a Debye-Huckel simplification (i.e. assumption of low surface potential and a symmetric binary electrolyte, see Section15.3.1). From that model we obtained that the charge density of the double layer was given by (15.3.30),

$$(16.6.2) \qquad \rho_f = -\varepsilon_0 \varepsilon \kappa^2 \varphi_0 \exp\left(-\kappa y\right)$$

This charge has of course to be balanced exactly by the bound charge on the surface. To find that charge, we can integrate the (16.6.2) from zero to infinity in the y, to find the amount of charge sits above a particular area of the wall surface. Performing this integration, we find that the wall charge is

$$(16.6.3) \qquad q_s = \varepsilon_0 \varepsilon \kappa \varphi_0$$

where, as a reminder, φ_0 is the surface potential of the wall. This looks like a charge on a simple parallel plate capacitor with a dielectric permittivity $\varepsilon_0 \varepsilon$, plate separation κ^{-1} and charged to a voltage of φ_0. Thus we may treat the double layer as a parallel plate capacitor element with a capacitance

$$(16.6.4) \qquad C = \frac{\varepsilon \varepsilon_0 d \Delta z}{\kappa^{-1}}$$

where the area of plates is the width times the depth of the current tube. Now having relations for the individual circuit elements and the layout of the circuit, we can find that the voltage drop across one such double layer is

$$(16.6.5) \qquad V_d = \frac{V_0}{1 + j\omega\pi\varepsilon\varepsilon_0\kappa z/\sigma}$$

where V_0 is the voltage applied between such electrodes. Note that the tangential electric field is

$$(16.6.6) \qquad E_t = -\frac{\partial V_d}{\partial z}$$

From equation (16.1.27) we know that the purely electrically driven flow (with no pressure gradient) in the limit of thin double layers is

$$(16.6.7) \qquad u_z = -\frac{\varepsilon_0 \varepsilon V_d}{\mu} E_t$$

where instead of zeta potential, we have used the potential drop across the double layer, V_d. We note since the field is oscillatory, and thus so is the velocity, we are interested in the time averaged velocity, averaged over aver a time scale larger than the time scale of the field oscillation

$$(16.6.8) \qquad \langle u_z \rangle = \frac{1}{2} \text{Re}\left[-\frac{\varepsilon_0 \varepsilon V_d}{\mu} E_t^*\right]$$

Substituting into this the relation for tangential electric field (16.6.6) and (16.6.5), and then performing the time averaging we find that

$$(16.6.9) \qquad \langle u_z \rangle = \frac{\varepsilon \varepsilon_0 V_0^2 \Omega^2}{8 \mu z (1 + \Omega^2)^2}$$

$$(16.6.10) \qquad \Omega = \frac{\omega}{\frac{\sigma}{\varepsilon \varepsilon_0} \frac{\kappa^{-1}}{z} \frac{2}{\pi}}$$

where Ω is the scaled frequency. Note that here the frequency of the field is scaled both by the relaxation frequency $\sigma/\varepsilon\varepsilon_0$ but also by the ratio of the Debye length to the characteristic length. We plot the scaled AC electroosmotic velocity

$$(16.6.11) \qquad U = \langle u_z \rangle \frac{32 \mu z}{\varepsilon \varepsilon_0 V_0^2} = \frac{4 \Omega^2}{(1 + \Omega^2)^2}$$

as a function of Ω in Figure 16.6.2. AC EO velocity is maximum when Ω is unity and decays rapidly when Ω is large or small. Note that already at $\Omega = 0.1$ or $\Omega = 10$ the scaled velocity is less than 5% of the maximum.

At low frequencies (small Ω) electrical charges (ions) have plenty of time to arrive from the bulk and screen the electric charge in the electrode thus the electric field from electrode to the other electrode is zero. Thus there is no tangential electric field to move the double layers, and so there is no observable flow. The time scale for these frequencies are above the charge relaxation time and the charge can "relax". In this relaxed state the charges are not under "tension" to move to a new place because there is a deviation from net neutrality there.

At high frequencies (large Ω) electrical charges (ions) do not have time to follow the field and so the electrode does not have a double layer above it. Thus while there is a strong field from electrode to electrode, there is practically no charges for that field to move, and so there is no flow that is generated. Maximum AC electroosmotic flow occurs at a happy medium - where the charges have just enough time to move to the electrode to try to form a double layer, but not enough time to completely shield the charge and so the field coming from one electrode to the other. We list Ω as a function of conductivity of the solution in, for a KCl solution assuming $z = 10$ and 100 µm in Table 4.

We have just derived a velocity relation for AC electroosmotic flow for a case where double layers are dramatically differently charged next to each other - the case of two oppositely charged planar parallel electrodes. However, we can obtain AC electroosmotic flow even if the double layers next to each other are even slightly differently charged. We just need a gradient in the voltage drop (and so charging) of the double layer to obtain flow. For example, for the case of thin double layers compared to electrode dimensions (the most common case) and sinusoidal field applied to the electrode, Ramos et al. (2003) derived that the velocity at the electrode surface

$$(16.6.12) \qquad \mathbf{u} = -\frac{\varepsilon \varepsilon_0}{4 \mu} \Lambda \nabla_s \left(|\Delta V|^2 \right)$$

$$(16.6.13) \qquad \Lambda = \frac{C_S}{C_S + C_D} = \frac{1}{1 + \frac{h/\kappa^{-1}}{\varepsilon_c/\varepsilon}}$$

Here, Λ is the ratio of capacitances of the Stern layer C_S to that of the total capacitance of the double layer, including the capacitance of the diffuse layer C_D.

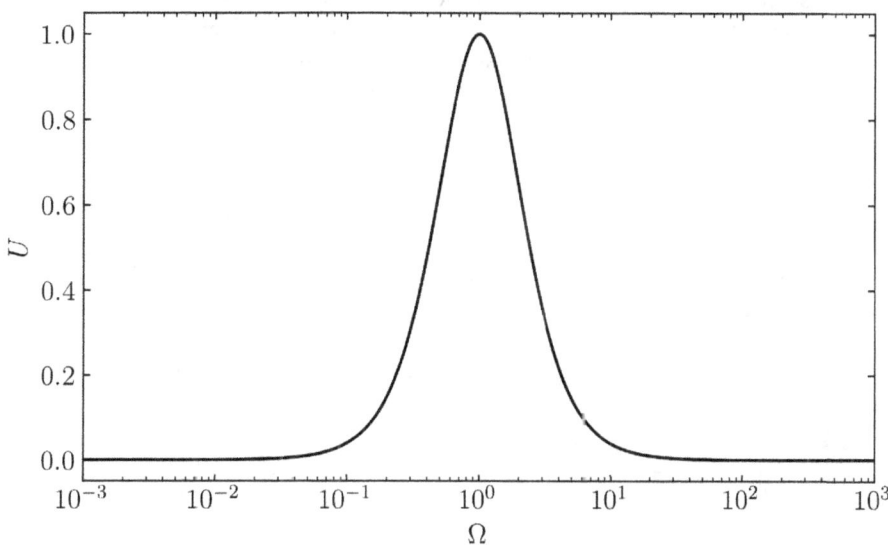

FIGURE 16.6.2. Scaled AC electroosmotic velocity as a function
of scaled frequency.

Here h is the thickness of the Stern layer, and ε_c. Here $\Lambda \leqslant 1$ and $\Lambda = 1$ when
the Stern layer is neglected. ∇_s is a surface (2D) gradient over the surface of the
electrode, and ΔV is the potential drop across the double layer. How can we set
up a gradient in the potential drop across the double layer? We can do this by
(a) having very thin electrodes so that the potential drop through the metal of the
electrode is significant; this effect can be increased by having also thin electrodes
with non-uniform metal thickness. We can also do this by (b) having an array
of electrodes with slightly different potential applied to them, and (c) fouling, or
purposefully coating the electrodes with material that alters double layer formation
on the surface.

16.6.1. AC Electroosmotic pumping. So far we have described AC Electroosmotic flow over a commonly occurring structure, two parallel electrodes lying
in a single plane. As we saw from the streamlines looking end-on onto the structure,
such structure would make a fine stirrer or mixer, but not a very good pump. In
this section we will consider slight modifications to that structure so that while the
structure is still easy to fabricate and implement in a microfluidic device, now it
will make a decent pump. We will consider two such modifications: (a) asymmetric
electrode array and (b) traveling wave electrode array. The advantage of the first
one is that with slight modification of geometry (see Figure 16.6.3) we can produce flow with a conventional single channel sinusoidal waveform generator. The
advantage of the second one is that with a standard uniform geometry (see Figure
16.6.4) we can produce flow that is easily reversible, but with a more specialized,
multichannel sinusoidal waveform generator. This flow has less recirculation than
the flow from the asymmetric electrode array. While the uniform geometry itself is
easier to fabricate, even at very small scales (e.g., with Lloyd's mirror technique)
for practical implementation it requires a multi layer chip (e.g., with vias) to route

TABLE 4. Debye length, conductivity, solution relaxation frequency (roughly corresponding to a minimum velocity for electrothermal flow) and frequency corresponding to maximum AC electroosmotic flows for 10 and 100 μm electrodes; calculated for a 1:1 electrolyte KCl ($D = 1.96 \times 10^{-9}\,\mathrm{m^2/s}$) at 298 K, $\kappa h \to \infty$

Concentration	Debye length, κ^{-1}	Conductivity	$\frac{\sigma}{\varepsilon\varepsilon_0}$	$\Omega = 1$ $z = 10\,\mu m$	$\Omega = 1$ $z = 100\,\mu m$
1 M	0.307 nm	15 S/m	21 GHz	1.0 MHz	100 kHz
150 mM (approx. saline solution)	0.792 nm	2.21 S/m	3.1 GHz	390 kHz	39 kHz
100 mM	0.971 nm	1.5 S/m	2.1 GHz	320 kHz	32 kHz
10 mM	3.07 nm	150 mS/m	210 MHz	100kHz	10 kHz
1 mM	9.71 nm	15 mS/m	21 MHz	32 kHz	3.2 kHz
100 μM	30.7 nm	1.5 mS/m	2.1 MHz	10 kHz	1.0 kHz
10 μM (approx. pH 5, ISO Grade 3 DI)	97.1 nm	150 μS/m	210 kHz	3.2 kHz	320 Hz
1 μM	307 nm	15 μS/m	21 kHz	1.0 kHz	100 Hz
100 nM (18 MΩ · cm DI)	971 nm	1.5 μS/m *	2.1 KHz	320 Hz	32 Hz

Diffusion coefficient from Harned and Nuttall (1947); dilute solution assumed, even for 1 M condition: conductivity overpredicted. * This is nonsensical as here there is 100 nM of protons contributing to the conductivity.

each electrode to the appropriate voltage. However, as the flow is in the direction of the traveling wave, and so can be easily reversed by just applying electric field to the electrodes in different order (unlike that for the asymmetric geometry).

We show the setup and representative streamlines for the asymmetric electrode array in Figure 16.6.3. The geometry consists of an array of parallel co-planar electrodes, with the large electrode order 6 times larger than the small electrode. For example, in the geometry described by Brown et al. (2000) the small electrode was 4.2 μm, the gap between the small and the large electrode 4.5 μm; the large electrode 25.7 μm, and the gap between the large electrode and the small electrode 15.6 μm. All the small electrodes are connected to one terminal of a sinusoidal signal generator, while the large electrodes are connected to another. When the field is applied to the structure, the flow over the electrodes in principle is similar to that described in Figure 16.6.1. However the flow over the larger electrode is so much higher than the flow over the small electrode that it dominates the overall flow field and the flow illustrated in Figure 16.6.3 results. The "remnants" of the original flow are preserved by the vortex over the small electrode. You can see that the flow is still away from the gap between the oppositely polarized electrodes, but the flow above the small electrode is smaller, and gets bent into a vortex. Thus, since the flow over the large electrode dominates, the net flow is in the direction from the small electrode to the large electrode. The presence of such vortices decreases the overall efficiency of such pumping. For more details on such AC electroosmotic pumps see Brown et al. (2000) and Ramos et al. (2003).

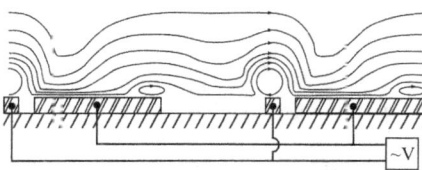

FIGURE 16.6.3. Streamlines above an asymmetric electrode array for AC electroosmotic pumping.

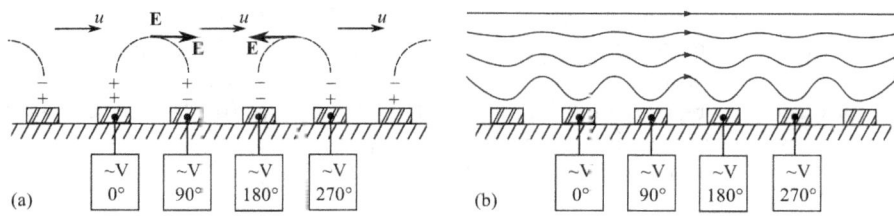

FIGURE 16.6.4. Electric field and force on the fluid (a) and the resulting representative streamlines (b) in traveling wave AC electroosmotic flow.

We show the setup and representative streamlines for the traveling wave electrode array in Figure 16.6.4. The geometry consists of an array of uniform electrodes where each forth electrode is actuated 90° apart. This results in a traveling wave. The pumping results due to a lag between the potential applied to the electrode and the charge of the double layer that is above it. In this traveling wave, if the electrodes are actuated at a frequency near the maximum AC electroosmotic frequency ($\Omega = 1$) the charging of the double layer lags the potential of the electrodes. For example, consider the double layer above the 0° electrode (Figure 121a). The electrode has just turned positive and the double layer has not had a chance to respond and so it is still charged positive. The next electrode (90° electrode) was negative a moment before and is still negative, and so it has a positive double layer above it. Thus, the product of charge and tangential electric field for both of these is to the right. The next electrode (180° electrode) has just turned negative, the double layer has not had a chance to respond and so it is still charged negative, while the next electrode after it, 270° electrode has been positive the moment before and still is positive, and so has a negative double layer above it. Once again the product of charge and tangential electric field for both of these is to the right. Note that this double layer lagging behavior is essential. Just like in standard AC electroosmotic flow, if the frequency is too low or too high, no effective traveling AC electroosmotic pumping will take place. The product of the double layer charge and the tangential electric field is always in same direction - the direction of the traveling wave, here left to right (Figure 16.6.4a). Thus, the flow is also in the direction of the traveling wave. Here lies the main advantage of this method of generating AC electroosmotic flow - flow direction can be easily reversed by just changing the timing of your power supply. For more details on such AC electroosmotic pumps see Cahill et al. (2004) and Ramos et al. (2005).

16.7. Electrophoresis

Having discussed in great detail electroosmosis we now turn to the next common electrokinetic phenomena: electrophoresis. Electrophoresis is the movement of particles in an electric field due to surface charge on the particle. Actually electrophoresis is just another manifestation of electroosmosis. Consider that both are just movement of a solid relative to a fluid in an electric field due to presence of charge at the interface of the solid and fluid. In fact, from the point of view of the particle, it is actually the fluid that it is moving around it! Therefore, in this section we will transfer our coordinates into the frame of reference of the particle (Galilean transformation) to take advantage of the techniques we have used for deriving relations for the electroosmotic phenomena. In this section we will first develop a qualitative understanding of electrophoresis, and then derive relations for electrophoresis of small particles, of large particles (e.g., cells), discuss the electrophoresis of intermediately sized particles, and then connect electrophoretic mobility to the familiar phenomena of diffusion.

16.7.1. Qualitative description of electrophoresis.
Let's consider a spherical particle radius a with a total charge Q_s suspended in a dielectric fluid infinitely far from any walls, such that there are no free charges (e.g., ions) in the fluid. From Coulomb's law we know if we apply an electric field $\mathbf{E}_\infty$ across the fluid the particle will experience a force

$$(16.7.1) \qquad \mathbf{F_E} = Q_s\mathbf{E}_\infty$$

However, as soon as the particle starts to move, it will begin to experience Stokes' drag, assuming the particle is fairly small and it moves slow enough, such that its Reynolds's number is substantially less than unity. The Stokes' drag force it will experience is

$$(16.7.2) \qquad \mathbf{F_H} = 6\pi\mu a\mathbf{U}$$

Eventually the particle will reach a steady state velocity where the hydrodynamic force will equal the electrophoretic force. This steady state velocity (also known as the terminal drift velocity) is then

$$(16.7.3) \qquad \mathbf{U} = \frac{Q_s\mathbf{E}_\infty}{6\pi\mu a}$$

In previous sections we have used a very useful concept of electrophoretic mobility, the ratio between the velocity of an ion and the electric field applied to it. Here we have a chance to expand on this. According to our definition, the electrophoretic mobility for the particle is then

$$(16.7.4) \qquad \mu_e = \frac{Q_s}{6\pi\mu a}$$

In this text we use μ as a symbol for both electrophoretic mobility, fluid viscosity, and electrochemical potential as this is the symbol commonly used in literature for these variables. Typically it is obvious from context which variable we are referring too. However, in situations where it is confusing we will distinguish one from the other using a subscript or change the variable all together (and alert the reader beforehand). Here, of course, μ_e is the electrophoretic mobility.

While the definition of electrophoretic mobility as "the ratio between the velocity of an ion and the electric field applied to it" is the most common one, it is certainly not the only one. Another common definition is "the ratio of the particles drift velocity to the applied force" for which our particles mobility would be

(16.7.5)
$$\mu_e' = \frac{1}{6\pi\mu a}$$

The reader should be aware of these differences, especially when comparing results from literature, and should be aware of the definition used in a particular work. In this text, we will stick to the first definition.

We began by considering immersing the particle in an ideal dielectric fluid. But real fluids, including even dielectric oils, and especially the most commonly encountered aqueous electrolytes, are certainly not ideal dielectrics. The real liquids that we would like to consider do contain free charge (ions), even though on the whole they are typically electroneutral. When we apply an electric field across such a homogeneous liquid, since it is electroneutral, and so the volumetric free charge density ρ_f is zero, there is no net force on the liquid, and so there will be no flow in the liquid. (We stress that here the liquid we consider is homogeneous, i.e., the application of electric field does not cause inhomogeneities in the liquid such as by heating the liquid and locally altering its permeability and conductivity, and so causing, for example, electrothermal flows).

On the other hand, if we introduce our spherical particle into the liquid a double layer will form at the interface between the particle and the liquid. Thus there now will be a spatially varying volumetric free charge density around the particle in the fluid. Thus when we apply an electric field across the fluid now there will be a superimposition of the two electric fields: (a) the electric field that we have imposed on the fluid (probably generally unidirectional), and (b) the electric field of the double layer of the particle, which will be generally spherically symmetric. Due to these two fields we expect (a) the particle to move in the direction as before (b) the fluid, having free charge density of opposite charge to the particle, to move in the opposite direction of the particle; (c) flow of ions due the presence of the two superimposed fields.

The presence of a double layer around the particle leads two three phenomena that generally slow down the travel of the particle and thus decrease its mobility. These are: (a) electrophoretic retardation, (b) charge relaxation, and (c) surface conductance. Electrophoretic retardation is when the particle moves in the electric field, the fluid, having free charge density of opposite charge to the particle, moves in the opposite direction of the particle. This places additional drag force on the particle and slows down the particle, hence retardation.

Charge relaxation describes the phenomena that the electric double layer around the particle is not symmetric. When an electric field is placed across the particle, the charges in the double layer move in response to the field, and so there is excess of charges in the double layer on one side of the particle and a deficiency of charges on the other. This charge screens the surface charge of the particle, and so also slows down the particle. When the external field is turned off, a certain time is necessary for this asymmetrical double layer to return to its spherically symmetric state - the relaxation time, hence the name charge relaxation. This effect is not important when the particle is small relative to the thickness of the double layer - the asymmetry then is negligible.

Surface conductance describes the phenomena that the externally applied electric field will drive the flow of ions in the mobile portion of the double layer of the particle. The current density around the particle will be higher than that in the bulk because typically there is a higher density of ions in the mobile part of the double layer than in the bulk and so higher conductivity. Thus, the energy of the electric field will be directed into driving this current (surface current), instead of moving the particle. When the zeta potential of the particle and the particle size relative to its double layer is small then the conductivity of the mobile part of the double layer is close to that of the bulk and so this effect diminishes.

16.7.2. Electrophoresis of small particles $(\kappa a \ll 1)$. Now let's turn to electrophoresis of small particles, including ions, where by small we mean that the particle's radius is small compared to its Debye length. To derive the equation for the velocity of the particle and its mobility we will employ the same equations as we have done for electroosmosis. Namely, we will Poison's equation to obtain electric field (15.3.8), Navier-Stokes equations (including both the momentum (16.1.1) and continuity equations) for incompressible flow in the Stokes regime to obtain the fluid flow, and current conservation (Nernst-Planck) (16.1.20) to obtain the flow of ions. We begin by making our usual simplifying assumptions: symmetric $(z : z)$ electrolyte and low surface potential (Debye-Huckel limit). Since we are interested in situations where the particle's radius is small compared to its Debye length, we are interested in the curvature of the particle and its double layer and so we use the spherical coordinate system. To begin, as before for electroosmotic flow, we start with the Poisson-Boltzmann equation, (15.3.20). Recall that when we derived the Boltzmann distribution, we equated diffusion flux with electromigration flux and neglected any flux associated with advective flow. Thus we tacitly assumed in our derivation for electroosmotic flow that the flow of electrolyte does not significantly affect the distribution of charges (ions) in the double layer. We will continue this assumption here. Poisson-Boltzmann distribution in spherical coordinates is

$$(16.7.6) \qquad \frac{1}{r^2} \frac{d}{dr} \left(r^2 \frac{d\varphi}{dr} \right) = \frac{2zeN_A c_\infty}{\varepsilon_0 \varepsilon} \sinh\left(ze\varphi / k_B T \right)$$

Here we have assumed that the charge in double layer is spherically symmetric and so only varies with the distance from the center of the sphere. Assuming that the surface potential is small, and so $\sinh\left(z\varphi / k_B T \right) \approx z\varphi / k_B T$, we simplify (16.7.6) to

$$(16.7.7) \qquad \frac{1}{r^2} \frac{d}{dr} \left(r^2 \frac{d\varphi}{dr} \right) = \frac{2zeN_A c_\infty}{\varepsilon_0 \varepsilon} \frac{ze}{k_B T} \varphi = \kappa^2 \varphi$$

where κ is the Debye length as usual. To solve this ODE we let $\xi = r\varphi$, which simplifies (16.7.7) to

$$(16.7.8) \qquad \frac{d^2 \xi}{dr^2} = \kappa^2 \xi$$

which has a general solution of

$$(16.7.9) \qquad \xi = A \exp\left(-\kappa r \right) + B \exp\left(\kappa r \right)$$

We apply the usual boundary conditions: constant potential at the particle surface and zero potential far away from the particle:

$$(16.7.10) \qquad \begin{aligned} \varphi &= \zeta \text{ at } r = a \\ \varphi &= 0 \text{ at } r \to \infty \end{aligned}$$

and so obtain that the potential distribution

(16.7.11)
$$\varphi = \xi \frac{a}{r} \exp\left(-\kappa\left(r-a\right)\right)$$

Now we can obtain the total charge on the particle, Q_s. By electroneutrality, the charge on the particle must equal to the sum of all the charges in the double layer in magnitude and be opposite in sign,

(16.7.12)
$$Q_s = -\int_a^\infty 4\pi r^2 \rho_f \, dr$$

We obtain an expression for the free charge density as usual from Poisson's equation, which in spherical coordinates is

(16.7.13)
$$\frac{1}{r^2}\frac{d}{dr}\left(r^2\frac{d\varphi}{dr}\right) = -\frac{\rho_f}{\varepsilon\varepsilon_0}$$

Combining these, we obtain

(16.7.14)
$$Q_s = 4\pi\varepsilon\varepsilon_0 \int_a^\infty r^2 \left(\frac{1}{r^2}\frac{d}{dr}\left(r^2\frac{d\varphi}{dr}\right)\right) dr = 4\pi\varepsilon\varepsilon_0 \left[r^2\frac{d\varphi}{dr}\right]_a^\infty$$

We recall that $\varphi = 0$ at $r \to \infty$, and so also

(16.7.15)
$$\frac{d\varphi}{dr} = 0 \text{ at } r \to \infty$$

Thus,

(16.7.16)
$$Q_s = -4\pi\varepsilon\varepsilon_0 a^2 \left.\frac{d\varphi}{dr}\right|_{r=a}$$

But we can obtain $d\varphi/dr$ from differentiating (16.7.11) and so

(16.7.17)
$$Q_s = 4\pi\varepsilon\varepsilon_0 a^2 \zeta \frac{(1+\kappa a)}{a}$$

Note that up till now we have not used the fact that the double layers are thin or thick, just that both the particle and its double layer is spherically symmetric. Thus (16.7.17) is valid for both small and large particles. Using (16.7.17) we can find the initial force on such a particle according to (16.7.1) to be

(16.7.18)
$$\mathbf{F_E} = 4\pi\varepsilon\varepsilon_0 a\zeta\left(1+\kappa a\right)\mathbf{E_\infty}$$

As soon as the particle begins to move, a drag force will apply to the particle. As we have noted above this drag force will be both due to just the particle translating through the fluid and due to the fact that the electrolyte around the particle will be moving in the opposite direction to the particle (electrophoretic retardation). To obtain the final velocity, we need a relation for velocity field due to an arbitrary force field in a Stokes regime. Luckily the Navier-Stokes equations under Stokes flow assumption are linear, and thus its solutions are linearly superimposable. This means that if we can decompose the arbitrary force field into a sum of point forces (which we can), we can add up the point force solutions to obtain the resulting velocity field. Here the point forces are due to infinitesimal charge dQ equal to the product of spatially varying charge density ρ_f and an infinitesimal volume dV. Thus the total force is $E_\infty dQ$. Employing the point force solution, we can obtain that the velocity component along the electric field direction of the entire charge cloud is

$$(16.7.19) \qquad U_E = \int_V \frac{E_\infty \rho_f}{8\pi\mu r} \left(1 + \cos^2\theta\right) dV$$

where V is the entire volume of the charge cloud. Employing the charge density from Poisson's equation as we have done above and evaluating the integral, we obtain

$$(16.7.20) \qquad U_E = -\frac{2}{3} \frac{\varepsilon\varepsilon_0\zeta}{\mu} \kappa a E_\infty$$

Since this electrophoretic retardation velocity slows down the overall particle, the total hydrodynamic drag on the particle is

$$(16.7.21) \qquad F_H = 6\pi\mu a \left(U - U_E\right)$$

which gives us that the velocity of the particle is

$$(16.7.22) \qquad U = \frac{F_H}{6\pi\mu a} + U_E$$

Now setting the hydrodynamic force equal to the electrical force on the particle and substituting in (16.7.18) and (16.7.20), we find that

$$(16.7.23) \qquad U = \frac{4\pi\varepsilon\varepsilon_0 a \left(1 + \kappa a\right) E_\infty}{6\pi\mu a} - \frac{2}{3}\frac{\varepsilon\varepsilon_0\zeta}{\mu}\kappa a E_\infty$$

or

$$(16.7.24) \qquad U = \frac{2}{3}\frac{\varepsilon\varepsilon_0\zeta}{\mu} E_\infty$$

This solution was originally derived by Huckel (1924) and is often termed the Huckel solution. We can also write the electrophoretic mobility of such particle as per our usual definition:

$$(16.7.25) \qquad \mu_e = \frac{U}{E_\infty} = \frac{2}{3}\frac{\varepsilon\varepsilon_0\zeta}{\mu}$$

Notice that the electrophoretic mobility is proportional to the zeta potential of the particle, but is independent of the particle size! Differently sized particles with the same zeta potential cannot be separated by electrophoresis! Indeed electrophoretic mobility of DNA is roughly constant after a certain size of DNA ($\sim$300 base pairs) and cannot be separated by free solution electrophoresis (electrophoresis of DNA in an electrolyte solution) (Stellwagen et al. (1997)). Around this DNA length, DNA is in a coiled ball configuration, with roughly constant zeta potential. As the DNA size increases further, the radius of the ball increases, but the zeta potential remains roughly the same. Therefore we cannot separate different sized DNA coils from each other. Hence, to separate DNA electrophoretically one either has to force the DNA to migrate through a gel or a polymer solution - i.e. introduce another drag force on the DNA that is highly size dependent. This is why gel electrophoresis is so popular for DNA separations.

16.7.3. Electrophoresis of large particles $(\kappa a \gg 1)$. Having derived electrophoretic velocity and mobility for small particles, we now derive this for large particles. For large particles the double layer is thin compared to the radius of the particle and so locally at the scale of the double layer it appears that the particle is flat. Thus, in the reference frame of the particle the problem becomes that of an electrolyte flowing past a planar surface with the electric field parallel to the surface. For this geometry the Navier-Stokes equations in the Stokes limit and when the pressure term is neglected simplify to

$$(16.7.26) \qquad \mu \frac{d^2 u_x}{dy^2} = -\rho_f E_\infty$$

meanwhile Poisson's equation is

$$(16.7.27) \qquad \varepsilon \varepsilon_0 \frac{d^2 \varphi}{dy^2} = -\rho_f$$

Combining these we obtain

$$(16.7.28) \qquad \mu \frac{d^2 u_x}{dy^2} = \varepsilon \varepsilon_0 E_\infty \frac{d^2 \varphi}{dy^2}$$

In this situation the hydrodynamic boundary conditions are that there is no-slip at the particle surface, and that there is a constant free stream velocity far away from the particle. Since we are operating in the particle's frame of reference, the free stream velocity is in the opposite direction to the particle's velocity (which is what we are trying to find). We can write this as

$$(16.7.29) \qquad u_x = 0 \text{ at } y = 0$$

$$(16.7.30) \qquad u_x = -U \text{ at } y \to \infty$$

and as a consequence

$$(16.7.31) \qquad \frac{du_x}{dy} = 0 \text{ at } y \to \infty$$

The electrical boundary conditions are that there is a constant ζ (zeta) potential at the particle surface and that the potential approaches zero far away from the particle. We can write this as

$$(16.7.32) \qquad \varphi = \zeta \text{ at } y = 0$$

$$(16.7.33) \qquad \varphi = 0 \text{ at } y \to \infty$$

and as a consequence

$$(16.7.34) \qquad \frac{d\varphi}{dy} = 0 \text{ at } y \to \infty$$

Integrating (16.7.28) from infinity to an arbitrary y position subject to the boundary conditions (16.7.29) to (16.7.34), we find that

$$(16.7.35) \qquad U = \frac{\varepsilon \varepsilon_0 \zeta}{\mu} E_\infty$$

and the electrophoretic mobility

$$(16.7.36) \qquad \mu_e = \frac{U}{E_\infty} = \frac{\varepsilon \varepsilon_0 \zeta}{\mu}$$

This is often referred to as the Helmholtz-Smoluchowski or the Smoluchowski solution. Once again, the particle velocity and electrophoretic mobility is independent of particle size!

16.7.4. Electrophoresis of intermediate particles. We have seen in the previous two sections that the difference between electrophoretic mobility of small particles (compared to their double layers) and large particles is just a factor of $2/3$. We expect that for particles of intermediate size the mobility lies between the values for large and small particles, and it indeed does. The solution for the general (intermediate) situation is quite more involved than the limits. One of the earliest examples of such a solution is due to Henry. In this solution it is assumed that the double layers are not disturbed by the flow, the electric field due to the surface charge and the external field are linearly superimposable and that the surface potential is low (Debye-Huckel assumption). We will not present the whole solution here but we will just state the result:

$$(16.7.37) \qquad U = \frac{2}{3}\frac{\varepsilon\varepsilon_0\zeta}{\mu}f(\kappa a)E_\infty$$

where

$$(16.7.38) \qquad f(\kappa a) = \frac{3}{2} - \frac{1}{2\left(1 + 0.072(\kappa a)^{1.13}\right)}$$

is a fit to the numerically evaluated analytical integral equation.

16.7.5. Electrophoretic velocities for constant charge particles. Relations for electrophoretic velocity and mobility derived so far are expressed as a function of constant surface (zeta) potential. Here we convert these into a form in terms of constant surface charge. Recall that for a particle having constant charge the force on it is

$$(16.7.39) \qquad \mathbf{F_E} = Q_s\mathbf{E}_\infty$$

For small particles ($\kappa a \ll 1$) recall that the drag force is given by (16.7.21) and the charge in terms of zeta potential is given by (16.7.17). Combining these we obtain that

$$(16.7.40) \qquad U = \frac{(1 + \kappa a)\,Q_s E_\infty}{6\pi\mu a}$$

Now we see that for particles of constant charge, their velocity is inversely proportional to their size. The higher the charge to size ratio, the larger the electrophoretic mobility. Now for the large particles we have an equation for velocity given by (16.7.35). Once again we employ (16.7.17),

$$(16.7.41) \qquad Q_s = 4\pi\varepsilon\varepsilon_0 a^2\zeta\frac{(1 + \kappa a)}{a}$$

and simplify it for $\kappa a \gg 1$

$$(16.7.42) \qquad Q_s = 4\pi\varepsilon\varepsilon_0\,(\kappa a)\,a\zeta$$

Combining this with (16.7.35) we obtain that

$$(16.7.43) \qquad U = \frac{Q_s E_\infty}{4\pi\mu a\,(\kappa a)}$$

Notice now that for a large particles with constant charge, the velocity is both inversely proportional to their size and the ratio of particle radius to double layer thickness. We can use (16.7.17) on Henry's solution for intermediate particles, (16.7.37) and obtain

$$(16.7.44) \qquad U = \frac{(1 + \kappa a)\, Q_s E_\infty}{6\pi\mu a} f\left(\kappa a\right)$$

16.7.6. Diffusion and electrophoretic mobility. In the first decade of the twentieth century Sutherland (1905), Einstein (1905) and independently Smoluchowski (1906) found an unexpected connection between a particle's diffusivity and the ratio of particle's steady state velocity to a force applied to it (our second definition of mobility). Strictly, this connection was supposed to hold for particles undergoing Brownian motion, i.e. particles larger than the solvent molecules However, its use has been extended (including in this book) to the relation between diffusion of small molecules and ions (sometimes smaller than the solvent molecules) and their electrophoretic mobility. It turns out that the relationship for small molecules and ions is not perfect, but holds pretty well. It is a good first approximation for finding either the diffusivity or electrophoretic mobility of a species as typically only one of them has been previously measured. It is also a good first approximation for simplifying the diffusion - electrophoretic equation (e.g., (15.3.1)) so that we have to deal with only one unknown (either diffusivity or mobility).

Solvent molecules randomly move in appropriate degrees of freedom - this motion is where they "store" their thermal energy as kinetic energy. The strength of such motion correlates with the macroscale property - the temperature of the solvent. Since the solvent molecules move randomly, they randomly collide with our particle of interest and move it about in random direction. Turns out how much they can move the particle, and how much another force can move the particle (i.e., the particle's mobility) is, not surprisingly, well connected. This connection is the Einstein-Smoluchowski equation.

To derive this equation we need to more quantitatively to understand diffusion and its more macroscale cousin Brownian motion, and we follow Probstein (2005) in the derivation. In fact it is much easier to understand Brownian motion and extrapolate these results to diffusion, which we will do. We begin this by postulating that at the same temperature all suspended particles regardless of their size have the same translational kinetic energy of $k_B T/2$ per degree of freedom. In general a particle has 6 degrees of freedom: translational movement of up-down, left-right, and front-back; as well as 3 rotational degrees of freedom. Particle doublets and more complicated particles may have even more degrees of freedom. Thus, the time averaged translational kinetic energy is

$$(16.7.45) \qquad \frac{1}{2}m\left\langle U^2\right\rangle = \frac{3}{2}k_B T$$

where the angle brackets represent time averaging, and m is the mass of the particle. However, if we observe particles in solution under a microscope we will find that their average velocity is much less than that predicted here by (16.7.45)! Why is this? The answer to this is collisions. The velocity obtained here is if the particle had no solvent - it was free move around without collisions. With the particle undergoing collisions the particle behaves as if it is going on a "random walk" or a "drunkard's walk". In other words it goes forward, something bumps it, and it changes direction. For a particle constrained to move in one dimension, one can model its walk by stipulating that the particle has a step size l and after each step a fair coin is flipped to determine if the particle will go forward or backwards. It can be shown that after n such steps the probability that the particle is now between x and $x + dx$ is a Gaussian distribution,

$$(16.7.46) \qquad P\left(n, x\right) dx = \frac{1}{\sqrt{2\pi n l^2}} \exp\left(\frac{-x^2}{2 n l^2}\right) dx$$

Note, that by the central limit theorem, the normalized sum of any independent random variables tends towards a Gaussian distribution. The number of steps is proportional to time, $n = Kt$. Furthermore, if we now have a lot of such particles, all starting at the origin, the concentration of these particles at a particular time and location would be proportional to the probability of finding one such particle at a particular location. The proportionality constant is of course the initial concentration at the origin, since the probability of finding a particle at the origin at zero time is unity. Thus we can write

$$(16.7.47) \qquad c(x, t) = \frac{c_0}{\sqrt{2\pi K t l^2}} \exp\left(\frac{-x^2}{2 K t l^2}\right)$$

Turns out the exponential is a solution to the one dimensional diffusion equation

$$(16.7.48) \qquad \frac{\partial c}{\partial t} = D \frac{\partial^2 c}{\partial x^2}$$

and if we substitute (16.7.47) into (16.7.48) we find that $K = 2D/l^2$. And so

$$(16.7.49) \qquad c(x, t) = \frac{c_0}{2\sqrt{\pi D t}} \exp\left(\frac{-x^2}{4Dt}\right)$$

From the solutions of the three dimensional diffusion equation we know that the concentration is given by

$$(16.7.50) \qquad c(r, t) = \frac{c_0}{8(\pi D t)^{3/2}} \exp\left(\frac{-r^2}{4Dt}\right)$$

where r is the distance from the origin. Going backwards, to find the probability of a particle at a distance between r and $r + dr$ at a time t we can multiply the solution for c/c_0, (16.7.50), by the volume of the spherical shell $4\pi r^2 dr$ and obtain

$$(16.7.51) \qquad P(r, t) dr = \frac{4\pi}{8(\pi D t)^{3/2}} \exp\left(\frac{-r^2}{4Dt}\right) r^2 dr$$

The mean displacement of the particle is of course zero because the Gaussian distribution is symmetric. Thus to get a feeling for the motion of the particle under our process we use root mean square displacement, which we write as $\sqrt{\langle r^2 \rangle}$. We evaluate root mean square displacement by integrating the square of the displacement multiplied by the probability of the displacement over all possible displacements:

$$(16.7.52) \qquad \langle r^2 \rangle = \int_0^\infty r^2 P\left(r, t\right) dr$$

This is a shortcut to evaluating the mean displacement over all possible times because over all possible times the particle would sample all possible displacements.

Substituting in our expression for the probability of finding a particle under diffusion in a 3D space in a particular location and time, (16.7.51), and performing the integration, we find

(16.7.53)
$$\langle r^2 \rangle = 6Dt$$

Next, we assume that the particle when it moves experiences a Stokes type drag

(16.7.54)
$$\mathbf{F} = -f\mathbf{U} = -f\frac{d\mathbf{r}}{dt}$$

where the force is proportional to the velocity by a mean translation friction coefficient f. The motion of the particle is then described by the Langevin equation

(16.7.55)
$$m\frac{d^2\mathbf{r}}{dt^2} = \mathbf{G}(t) - \mathbf{F}$$

derived from a force balance on the particle and Newton's second law. Here $\mathbf{G}(t)$ is the force associated with the collisions of the solvent molecules with the particle, occurring at a time scale of 10^{-13} s (for water); $\mathbf{F}$ is the Stokes' drag on the particle, which occurs on a much slower time scale. Substituting in (16.7.54)

(16.7.56)
$$m\frac{d^2\mathbf{r}}{dt^2} = \mathbf{G}(t) - f\frac{d\mathbf{r}}{dt}$$

and then multiplying by $\mathbf{r}$ and employing the product rule to simplify we obtain

(16.7.57)
$$\frac{m}{2}\frac{d^2\left(\mathbf{r}^2\right)}{dt^2} - m\left(\frac{d\mathbf{r}}{dt}\right)^2 = \mathbf{r}\cdot\mathbf{G}(t) - \frac{f}{2}\frac{d\left(\mathbf{r}^2\right)}{dt}$$

Time averaging this over a time scale than many collisions, we obtain

(16.7.58)
$$\frac{m}{2}\frac{d^2\langle\mathbf{r}^2\rangle}{dt^2} - m\left\langle\left(\frac{d\mathbf{r}}{dt}\right)^2\right\rangle = \langle\mathbf{r}\cdot\mathbf{G}(t)\rangle - \frac{f}{2}\frac{d\langle\mathbf{r}^2\rangle}{dt}$$

We notice that over many collisions the random fluctuation force averages to zero and so $\langle\mathbf{r}\cdot\mathbf{G}(t)\rangle = 0$. Furthermore, we notice that

(16.7.59)
$$m\left\langle\left(\frac{d\mathbf{r}}{dt}\right)^2\right\rangle = m\mathbf{U}^2$$

is just the kinetic energy in three dimensions and so with kinetic energy in one dimension being equal to $k_B T/2$

(16.7.60)
$$m\left\langle\left(\frac{d\mathbf{r}}{dt}\right)^2\right\rangle = \frac{3}{2}k_B T$$

Thus (16.7.58) simplifies down to

(16.7.61)
$$\frac{m}{2}\frac{d^2\langle\mathbf{r}^2\rangle}{dt^2} + \frac{f}{2}\frac{d\langle\mathbf{r}^2\rangle}{dt} = \frac{3}{2}k_B T$$

Integrating this equation once we obtain

(16.7.62)
$$\frac{d\langle\mathbf{r}^2\rangle}{dt} = \frac{6k_B T}{f}\left(1 - \exp\left(-\frac{f(t_0 - t)}{m}\right)\right)$$

where t_0 is the time from which the particle displacement is measured. The characteristic time to obtain the steady state mean square displacement of $6k_B T/f$ is the time constant m/f. This time is often referred to as the viscous relaxation time. For times longer then the viscous relaxation time

$$(16.7.63) \qquad \frac{d\langle \mathbf{r}^2 \rangle}{dt} = \frac{6k_BT}{f}$$

and upon integration

$$(16.7.64) \qquad \langle \mathbf{r}^2 \rangle = \frac{6k_BT}{f}t$$

Comparing this result that we obtained from the solution of the diffusion equation and the random (drunkard's) walk model, (16.7.53), we see that

$$(16.7.65) \qquad \langle \mathbf{r}^2 \rangle = \frac{6k_BT}{f}t = 6Dt$$

or that

$$(16.7.66) \qquad D = \frac{k_BT}{f}$$

In our case we assume that the particle undergoes a Stokes' drag force for a spherical particle, and so

$$(16.7.67) \qquad \mathbf{F_H} = 6\pi\mu a \mathbf{U}$$

or

$$(16.7.68) \qquad f = 6\pi\mu a$$

Combining (16.7.66) and (16.7.68) we obtain a relation between particle diffusivity and particle size

$$(16.7.69) \qquad D = \frac{k_BT}{6\pi\mu a}$$

This relation is commonly known as the Stokes-Einstein relation and is useful for estimating particle diffusivities. For particles of non-spherical geometry we could have also substituted in their drag equation, for example from Section 1.3, and obtained an appropriate Stokes-Einstein relation for that type of particle. We list diffusivities for various representative biological particles in Table 5. We also recall that for a particle undergoing the simplest electrophoresis (Section 16.7.1) we defined the electrophoretic as

$$(16.7.70) \qquad \mu_e = \frac{Q_s}{6\pi\mu a} = \frac{Q_s}{f}$$

Thus,

$$(16.7.71) \qquad \mu_e = \frac{Q_sD}{k_BT}$$

or written in another notation,

$$(16.7.72) \qquad \mu_e = \frac{zeD}{k_BT} = \frac{zFD}{RT}$$

where z is the ion valance, e is the electron (elementary) charge, R is the universal gas constant, and F is Faraday's constant. Thus how much an electric force can move a particle and how much random collisions (diffusion, Brownian motion) can move a particle are related, and so we have derived the widely used Einstein-Smoluchowski relation. We list diffusion coefficients and absolute values of

TABLE 5. Diffusion coefficients of proteins and viruses.

Particle	Size [kDa]	Radius of gyration [nm]	Diffusion coefficient $\times 10^{-9}$ [m^2/s]
Ribonuclease (bovine pancreas)	12.6	1.48	0.131
α-Lactalbumin (bovine milk)	13.3	1.45	0.106
Lysozyme (chicken egg white)	13.9	1.43	0.112
Hemoglobin (human)	63	2.48	0.069
Bovine serum albumin	65.4	2.98	0.062
γG-Immunoglobulin (IgG)	156	3.25	0.04
DNA dependent RNA polymerase	360	6.03	0.033
β-Casein (cow, aggregated)	1,200	13.5	0.014
Peruvate dehydrogenase (E. coli)	3,780	15.7	0.012
Satellite tobacco necrosis virus	1,700	6.8	0.02
Tobacco bushy stunt virus	10,700	12	0.015
Bacteriophage λ (full head)	56,000	25	0.006
Tobaco mosaic virus	31,000	92.4	0.005

Data from Tyn and Gusek (1990), which has an extensive table of diffusion coefficients for proteins and viruses.

electrophoretic mobilities for many commonly encountered solutes including small molecule acids, amino acids, and common inorganic ions in Table 6. As an exercise for the reader, the reader may want to look up electrophoretic mobilities of multivalent ions (such as in those found in PeakMaster, Jaros et al. (2004)), calculate the ratio of electrophoretic mobility at nth valance compared to that the first valance. Einstein-Smoluchowski relation predicts that this ratio should be equal to the valance z, as diffusivity of the molecule is only a function of its size as per Stokes-Einstein relation and to a first approximation this does not change much with valance. The reader should compare this calculated ratio to the valance. It should be pretty close for most molecules, but it will not be perfect.

16.8. Diffusiophoresis and liquid junction potential

Diffusiophoresis is the electrophoretic migration of particles induced by an electric field setup by a concentration gradient of solutes found in the fluid. The electric field setup by the concentration gradient of the solutes is termed the junction potential. It is easiest to understand the junction potential when such gradient is high - when the junction is sharp. We can achieve such relatively sharp junction for example in an H-filter type geometry. In Figure 16.8.1 we sketch there types of liquid junctions. In the first type both streams (phases) have the same solute but at different concentrations. In the second type, the concentrations of the solute is the same, but one of the ions is different between the two solutes. The third type is the general type: both the concentrations and solutes differ between the phases.

For simplicity, let's first examine the first type. At the junction there is a high concentration gradient of both the hydrogen ion and the chloride ion and hence both diffuse from right to left. From Table 6 we know that the diffusivity of hydrogen ion

TABLE 6. Diffusion coefficients and absolute values of electrophoretic mobilities for common solutes.

Solute	Diffusion coefficient $\times 10^{-9}$ [m^2/s]	Absolute electrophoretic mobility $\times 10^{-9}$ [m^2/V·s]
Formic acid	1.41	56.6
Acetic acid	1.21	42.4
Propionic acid	1.06	37.1
Butyric acid	0.87	33.8
Benzoic acid	1	33.6
Fluorescein	0.54	-
Methanol	1.5	-
Glycerol	0.94	-
Glucose	0.67	30
Sucrose	0.52	30
Glycine	1.1	37.4 / 39.5
Glutamine	0.76	28.8 / 28.1
Alanine	0.91	32.2 / 34
Leucine	0.73	26.4 / 27.6
Serine	0.88	33.6 / 32
Valine	0.83	28.4 / 29.6
Urea	1.38	-
H+ *, **	9.31	362.4
Li+	1.03	40.1
K+ **	1.96	76.2
Na+ **	1.33	51.9
NH4+	1.97	76.2
OH- *, **	5.26	205
Cl-	2.03	79.1
NO3- **	1.9	74.1
HCO3-	1.18	46.1
SO42-	1.06	82.9
H2PO4-	0.88	34.6

Diffusion coefficients from: Cunningham et al. (2011); * from Mortimer (2000); ** Haynes (2017). Electrophoretic mobility at infinite dilution from PeakMaster (Jaros et al. (2004) or ** Haynes (2017)). For amino acids: absolute mobility is given for anion form / cation form. Diffusivity of hydrogen ion is dependent on the ionic strength of the solution: it decreases from its pure water value (listed) to about 85% of that value in order 0.1 M solutions and 70-80% of that value in order 1 M solutions (Woolf (1960)). Diffusivities and electrophoretic mobilities are typically concentration dependent (typically decrease with concentration) are a listed here as extrapolated to infinite dilution. A large table of diffusion coefficients D and ionic conductivities λ (from which absolute electrophoretic mobility μ can be calculated: $\mu = \lambda/F$, where F is Faraday's constant) can be found in Haynes (2017).

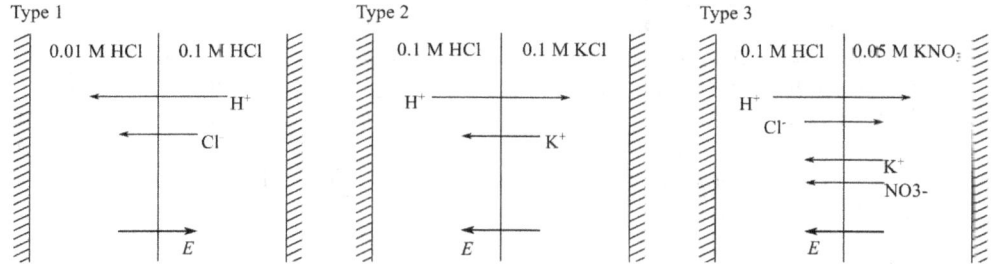

FIGURE 16.8.1. Three types of liquid junctions. Skinny arrows show the direction of net movement of the ions. Fat arrow indicates the direction of the electric field. The line down the center is just an imaginary partition between the two solutions. Such partition may occur in an H-filter microfluidic chip between the two laminar flowing streams containing different solutes.

(in pure water) is over 4 times that of chloride and so initially there would be a much higher flux of hydrogen ions into the left phase than chloride ions. This excess of positive charge setups up an electric field from left to right, retarding the migration of hydrogen ions (and actually forcing their migration back into the right section) and accelerating the migration of chloride ions as to maintain electroneutrality in both phases. The potential between the two phases is often referred to as a junction or diffusion potential.

To calculate the junction potential let's imagine that we place platinum/hydrogen electrodes into each of the phases as if to measure the junction potential. We use platinum/hydrogen electrodes in this example because they are fully reversible and their electrochemical potential is defined (see Section 15.1). As an aside. while in principle we can dip platinum/hydrogen electrodes into an H-filter type setup, that would be highly impractical to do in practice as platinum/hydrogen electrodes are difficult to work with. Let's also call the left phase α and the right phase β. The events in this system can now be divided into chemical transformations at the metal-solution interfaces and charge transport at the liquid junction. At the first electrode the chemical transformations at the metal-solution interface can be written as:

(16.8.1)
$$\frac{1}{2}H_2 \longrightarrow H^+(\alpha) + e(Pt)$$

and at the second (prime) electrode:

(16.8.2)
$$H^+(\beta) + e\left(Pt'\right) \rightleftharpoons \frac{1}{2}H_2$$

and the charge transport at the liquid junction (Figure 16.8.1, Type 1):
(16.8.3)
$$\frac{|\mu_+|}{|\mu_+| + |\mu_-|}H^+(\alpha) + \frac{|\mu_-|}{|\mu_+| + |\mu_-|}Cl^-(\beta) \rightleftharpoons \frac{|\mu_+|}{|\mu_+| + |\mu_-|}H^+(\beta) + \frac{|\mu_-|}{|\mu_+| + |\mu_-|}Cl^-(\alpha)$$

where μ_+ and μ_- is the electrophoretic mobility of hydrogen ion and chloride ion respectively. We didn't include H+ or Cl- in the subscript this time to condense the notation. This equation represents the movement of ions in the system at steady state. At steady state hydrogen ions move from the left (α) phase to the right (β)

phase, while the chloride ions move from the β into the α phase. The amount of their movement is proportional to how much current they carry (as there must be current continuity across the interface and in each phase by current conservation as no charge is generated). The amount of current each ion carries at a given location is proportional to the ratio of its absolute mobility to the sum of the absolute mobilities of all other ions at that location. This ratio is often referred to as the transference number:

$$(16.8.4) \qquad t_i = \frac{|\mu_i|}{\sum\limits_j |\mu_j|}$$

Based on this definition, we can see that the sum of transference numbers in a particular location is unity. To condense the notation and to decrease the confusion between the notation for electrophoretic mobility and for electrochemical potential we will work with transference numbers in this section. Thus, we rewrite (16.8.3)

$$(16.8.5) \qquad t_+ H^+(\alpha) + t_- Cl^-(\beta) \rightleftharpoons t_+ H^+(\beta) + t_- Cl^-(\alpha)$$

When we placed our platinum/hydrogen electrodes (i.e. standard hydrogen electrodes) into the two phases to measure the potential between them, we connected a voltmeter across these. Since the voltmeter has a very high impedance (even decent hand-held multimeters have impedance > 1 MΩ) no current flows in the circuit. At the null-current condition the two electrode reactions are in strict equilibrium and so the electrochemical free energy change for each of the reactions is zero. This is also of course true for the sum of (16.8.1) and (16.8.2):

$$(16.8.6) \qquad H^+(\beta) + e\left(Pt'\right) \rightleftharpoons H^+(\alpha) + e(Pt)$$

Again, since no net current flows through the system, the electrochemical free energy change for the entire process is zero. The sum of (16.8.1), (16.8.2), and (16.8.5) represents the overall system operation. Since electrochemical free energy change for the entire process is zero, and it is zero for the electrode processes, we must conclude that it is also zero for the charge transport represented in (16.8.5). Since the electrochemical free energy change for (16.8.5) is zero, following the properties of electrochemical potential (Section 15.1), we may write this as

$$(16.8.7) \qquad \bar{\mu}^\beta_{H+} + \bar{\mu}^{Pt'}_e = \bar{\mu}^\alpha_{H+} + \bar{\mu}^{Pt}_e$$

where overbar and the use of both subscript and superscript designates electrochemical potential, to differentiate it from electrophoretic mobility. (As a reminder, the sum of electrochemical potential on the left side of the equation minus the sum of electrochemical potential on the right side is the total change in electrochemical free energy, which in our case is zero). Using the properties of electrochemical potential (electrons in a metal), we can expand this as

$$(16.8.8) \qquad \bar{\mu}^\beta_{H+} + \left(\mu^{0Pt'}_e - F\varphi^{Pt'}\right) = \bar{\mu}^\alpha_{H+} + \left(\mu^{0Pt}_e - F\varphi^{Pt}\right)$$

where μ^{0Pt}_e and $\mu^{0Pt'}_e$ are the standard chemical potential of electrons in platinum in the platinum in the left and right phases respectively. The zero in the superscript denotes that it is standard chemical potential. Since they are the same platinum, they are equal, and so cancel out of (16.8.8). This leaves us with

$$(16.8.9) \qquad F\left(\varphi^{Pt'} - \varphi^{Pt}\right) = \bar{\mu}^\beta_{H+} - \bar{\mu}^\alpha_{H+}$$

Using the definition of electrochemical potential, (15.1.4), which we rewrite here for convenience:

(16.8.10) $$\bar{\mu}_i^\alpha \equiv \mu_i^\alpha + z_i F \varphi^\alpha$$

we expand (16.8.9) to

(16.8.11) $$F\left(\varphi^{Pt'} - \varphi^{Pt}\right) = \left(\mu_{H+}^\beta + F\varphi^\beta\right) - \left(\mu_{H+}^\alpha + F\varphi^\alpha\right)$$

and then using second property of electrochemical potential
(16.8.12)
$$F\left(\varphi^{Pt'} - \varphi^{Pt}\right) = \left(\mu_{H+}^{0\beta} + RT\ln\left(a_{H+}^\beta\right) + F\varphi^\beta\right) - \left(\mu_{H+}^{0\alpha} + RT\ln\left(a_{H+}^\alpha\right) + F\varphi^\alpha\right)$$

where a is the activity of the species. We define $V = \varphi^{Pt'} - \varphi^{Pt}$ as the total potential measured across the platinum electrodes with our voltmeter and rearrange (16.8.12) to

(16.8.13) $$V = \varphi^{Pt'} - \varphi^{Pt} = \frac{RT}{F}\ln\left(\frac{a_{H+}^\beta}{a_{H+}^\alpha}\right) - \left(\varphi^\beta - \varphi^\alpha\right)$$

noting that the standard chemical potential of hydrogen ion in both phases is of course the same and so cancels out. The first term on the right-hand side of (16.8.12) represents the Nernst relation for reversible chemical change while the second term is the potential difference between the β and α phases, i.e., the liquid junction potential. In general for an electrochemical chemically reversible system through which no current flows (such as in our case) the total potential is equal to the sum of the Nernst potential and the junction potential

(16.8.14) $$V = V_N + V_J$$

To obtain the junction potential we evaluate the electrochemical potentials for equation (16.8.5)

(16.8.15) $$t_+\bar{\mu}_{H+}^\alpha + t_-\bar{\mu}_{Cl-}^\beta = t_+\bar{\mu}_{H+}^\beta + t_-\bar{\mu}_{Cl-}^\alpha$$

rearranging this

(16.8.16) $$t_+\left(\bar{\mu}_{H+}^\alpha - \bar{\mu}_{H+}^\beta\right) + t_-\left(\bar{\mu}_{Cl-}^\beta - \bar{\mu}_{Cl-}^\alpha\right) = 0$$

and performing similar expansions using the properties of electrochemical potential to those we performed above, we obtain that
(16.8.17)
$$t_+\left(RT\ln\left(\frac{a_{H+}^\alpha}{a_{H+}^\beta}\right) + F\left(\varphi^\alpha - \varphi^\beta\right)\right) + t_-\left(RT\ln\left(\frac{a_{Cl-}^\beta}{a_{Cl-}^\alpha}\right) + F\left(\varphi^\alpha - \varphi^\beta\right)\right) = 0$$

Since it is difficult to measure activity coefficients of single ions, these are typically equated to a mean ionic activity coefficient. For our case of a strong acid this is fairly straight forward. We let $a_{H+}^\alpha = a_{Cl-}^\alpha = a_\alpha$ and $a_{H+}^\beta = a_{Cl-}^\beta = a_\beta$. (For much of back of the envelope analysis it is also quite useful to let activity coefficients to be unity and use concentrations instead of activities.) And so we rewrite (16.8.17) as

(16.8.18) $$RT\ln\left(\frac{a_\alpha}{a_\beta}\right)(t_+ - t_-) + F\left(\varphi^\alpha - \varphi^\beta\right)(t_+ + t_-) = 0$$

Using the fact that the sum of transference numbers at a particular location is always equal to unity,

$$(16.8.19) \qquad V_J = \varphi^\beta - \varphi^\alpha = (t_+ - t_-) \frac{RT}{F} \ln\left(\frac{a_\alpha}{a_\beta}\right)$$

We can also expand this in terms of electrophoretic mobilities using the definition of transference numbers

$$(16.8.20) \qquad V_J = \varphi^\beta - \varphi^\alpha = \frac{|\mu_+| - |\mu_-|}{|\mu_+| + |\mu_-|} \frac{RT}{F} \ln\left(\frac{a_\alpha}{a_\beta}\right)$$

This is the junction potential for type 1 junctions with 1:1 electrolytes. For the type 1 junction illustrated in Figure 16.8.1, for a quick and dirty calculation, we can use electrophoretic mobilities from Table 6 and the concentrations given in the figure as the activities (basically assuming very dilute solutions). We obtain that the junction potential is -38 mV, the Nernst potential, 59 mV, and so the overall potential as measured by our electrodes should be 21 mV. Notice that firstly the absolute value of the junction potential is actually greater than the absolute value of the overall potential. Secondly the absolute value of the junction potential is substantial, more than the thermal voltage RT/F of 25.7 mV at room temperature. This begins to hint that if particles were placed in a junction situation like this, they would electrophoretically migrate from one region to the other. This migration is termed diffusiophoresis. One situation where this is of concern is the design of cell chemotaxis assays, a fairly common assay in biology. These assays are used to evaluate the cell's chemotactic ability - the ability to travel towards or away a particular chemical (such as a signaling molecule). In these assays a gradient of a chemical of interest is set up and the cell of interest is observed to propel itself up or down this gradient. It is imperative to stress that the cell is supposed to propel itself in this situation, for example using flagella, filopodia, or a number of other cellular propulsion mechanisms. The cell is not supposed to electrophoretically migrate! However, as we have just seen on a simple example, when you have a chemical gradient you will develop a junction potential, and so setup diffusiophoresis. It is imperative when designing chemotaxis assays and equipment to minimize or account for junction potentials, otherwise we can obtain incorrect conclusions regarding the cell's true chemotactic ability.

Let's now obtain the equation for junction potential for the general case. In the derivation above we conveniently assumed that the transference numbers for all the cations and all the ions (and so their mobilities) were the same in our system. For our system were both phases contained the same cations and anions (albeit at different concentrations) this assumption was pretty good. However, when the anions or cations (or both) are clearly different (such as type 2 and 3 systems) this assumption would lead to erroneous results. Furthermore, we assumed that the junction was sharp. The junction of course has some finite width set by the diffusion of the species.

For the analysis of the general situation let's picture that the entire device can be portioned into an infinite number of infinitesimal sections (elements) that vary in composition from that of pure phase α to that of phase β. Transporting charge across one such element involves of course all the species in the element and so in a particular element and a particular species i, $t_i/|z_i|$ moles of species i moves for every mole of charge moved through the element. Movement of species is associated with a change in electrochemical free energy change in the element. The net flux of electrochemical energy associated with movement of $t_i/|z_i|$ moles of species i is $(t_i/z_i)\, d\bar{\mu}_i$. Summing this over all the species in the element, we obtain the differential free energy for the element:

$$(16.8.21) \qquad d\bar{G} = \sum_i \frac{t_i}{z_i} d\bar{\mu}_i$$

Integrating this across the entire device from pure phase α to pure β phase,

$$(16.8.22) \qquad \int_\alpha^\beta d\bar{G} = \sum_i \int_\alpha^\beta \frac{t_i}{z_i} d\bar{\mu}_i$$

As we have just seen in the simple, type 1 system example because of null-current condition and therefore equilibrium in the system the overall electrochemical free energy change in the system is zero, and so the integral on the left-hand side is equal to zero, and so

$$(16.8.23) \qquad \sum_i \int_\alpha^\beta \frac{t_i}{z_i} d\bar{\mu}_i = 0$$

Next, we expand the electrochemical potential using the properties of electrochemical potential as we have done for the simple, type 1 system, and noting that in the systems of interest the standard electrochemical potentials for any species, μ_i^0 in both phases are the same. (An exception to this would be if one phase was an aqueous solution and another was an organic solvent, for example; here we are concerned only with aqueous solutions). We obtain

$$(16.8.24) \qquad \sum_i \int_\alpha^\beta \frac{t_i}{z_i} RT d\ln(a_i) + \left(\sum_i t_i\right) F \int_\alpha^\beta d\varphi = 0$$

Since the sum of all transference numbers at a particular location is unity,

$$(16.8.25) \qquad V_J = \varphi^\beta - \varphi^\alpha = -\frac{RT}{F} \sum_i \int_\alpha^\beta \frac{t_i}{z_i} d\ln(a_i)$$

This is the general expression of liquid junction potential. While this relation is quite useful, it is fairly difficult to evaluate since the concentration profile of ions across an arbitrary junction is not known. Therefore, either the entire situation including the hydrodynamics, (i.e. Navier-Stokes equations), advection-diffusion and current conservation equations must be solved to obtain the concentration (and so the potentials) or simplifying assumptions must be made about the junction and species concentration profiles to use (16.8.25). One example of such assumption is that (a) the solution is dilute enough that the activities of each species are equal to the concentration of each species; (b) the concentration profile of each species at the junction is a linear profile between the two phases. Using these assumptions we can evaluate the integral in (16.8.25) and obtain

$$(16.8.26) \qquad V_J = \frac{\sum_i \mu_i \left(c_i^\beta - c_i^\alpha \right)}{\sum_i z_i \mu_i \left(c_i^\beta - c_i^\alpha \right)} \frac{RT}{F} \ln \left(\frac{\sum_i z_i \mu_i c_i^\alpha}{\sum_i z_i \mu_i c_i^\beta} \right)$$

Where, as a clarification, here μ_i are the electrophoretic mobilities, and c_i are the species concentrations in the pure phase, with the superscript indicating the phase. (We have also assumed that the individual species electrophoretic mobilities are the same in all phases). This sometimes known as the Henderson equation (Henderson (1907)). This equation can be quite a useful first approximation for estimating the electric field in a diffusiophoresis situation.

Another general relation for diffusiophoresis can be obtained by carefully noting that we have a null-current condition. We have previously obtained a current conservation equation, (16.1.22), which we rewrite here for convenience:

$$(16.8.27) \qquad \mathbf{i} = \mathbf{u}F \sum_k z_k c_k - F \sum_k D_k z_k \nabla c_k - \frac{eF\nabla\varphi}{k_B T} \sum_k z_k^2 D_k c_k$$

We note as a reminder, $n_k = N_A c_k$ and $F = eN_A$. As we have previously assumed that there is no appreciable movement of the solvent in the system in the direction of the anticipated junction potential electric field, we set the first term of (16.8.27) to zero. Thus rearranging

$$(16.8.28) \qquad E = -\nabla\varphi = \frac{\sum_k D_k z_k \nabla c_k}{\frac{e}{k_B T} \sum_k z_k^2 D_k c_k}$$

To obtain useful information from (16.8.28) we need to obtain concentration profiles for all the ions. To do this in general, we would solve the ion conservation equation (here written in a form without reactions):

$$(16.8.29) \qquad \frac{\partial c_i}{\partial t} - \nabla \cdot \mathbf{i}_i = 0$$

where $\mathbf{i}_i$ is the current due an individual ion

$$(16.8.30) \qquad \mathbf{i} = \mathbf{u}Fz_i c_i - FD_i z_i \nabla c_i - \frac{eF\nabla\varphi}{k_B T} z_i^2 D_i c_i$$

together with Poisson's equation

$$(16.8.31) \qquad \varepsilon\varepsilon_0 \nabla \cdot E = F \sum_i z_i c_i$$

and the appropriate boundary concentrations of the ions as well as their initial distribution. In practice this is normally solved numerically.

As we mentioned before, diffusiophoresis is electrophoresis of particles (including ions) in the junction potential electric field. However, we have also mentioned before that electrophoresis and electroosmosis is the same phenomena where in electrophoresis the solid is free to move, and in electroosmosis the solid is fixed and the fluid moves. Thus just like there is diffusiophoresis, there is of course diffusioosmosis. Solvent with a concentration gradient of solute placed in a capillary or a porous media will setup a junction potential and this in turn will setup a flow of solvent. This flow of solvent is often termed diffusiosmosis but sometimes also termed capillary osmosis.

CHAPTER 17

Induced charge electrokinetic phenomena

Before jumping head on into polarization electrokinetic phenomena where we do not work with charges directly, but rather with dipoles, let's first discuss induced charge electrokinetic phenomena. In induced charge phenomena a particle or a surface becomes polarized but we treat the charge density that develops due to polarization similar as we have done in the free charge electrokinetic phenomena section. Thus this phenomena is a good bridge between free charge electrokinetics and polarization electrokinetics. In this section we will discuss two common types of induced charge electrokinetics: induced charge electroosmosis (ICEO) and induced charge electrophoresis.

The induced charge electrokinetic phenomena we are concerned about here arises when there is a (typically sharp) gradient in electrical conductivity inside our space of interest upon which an electric field is applied. This is typically due to an object with (typically much) higher conductivity placed in a lower conductivity fluid. This situation can be treated by either considering the charges that build up at the location of the high conductivity gradient as we do here, or by directly considering the force on the media due to this gradient using the electrical body force equation (Korteweg-Helmholtz equation (14.0.43), see Section 14) as we do in the AC electrothermal flow section below.

Free charge occurs at locations of gradients in conductivity due to Gauss's law. Imagine two slabs next to each other with different conductivities (see Figure 17.0.1). If an electric field is applied to them in a direction normal to their interface due to current continuity there is a jump in electric field and so a divergence of electric field and so due to Gauss's law charge build up at the interface.

In situations we are concerned here typically the low conductivity material is a liquid and the high conductivity is a metallic particle. Therefore, not only there is a jump in conductivity, there is also a jump in charge carrier type at the interface. In the liquid the dominant charge carriers are ions, and in the metal they are electrons. Thus for current to flow across the interface ions must "turn into" electrons or electrons must "turn into" ions at the interface. The only way this happens is if reduction-oxidation (red-ox) reactions, i.e. chemical reactions involving a transfer of an electron, occur at the interface. Sometimes these reactions are referred to as electrochemical reactions or electrode reactions (because they often occur at electrodes). However, for these reactions to occur, like with all reactions, a certain energy barrier has to be overcome. Actually there are two energy barriers (see Figure 17.0.2 for a typical reaction energy diagram). The small energy barrier is the total energy of the reaction - this is the thermodynamic energy change of the reaction and for red-ox reactions this is the electrochemical potential. Sometimes this energy barrier is negative, i.e. there is no barrier at all - the reaction is thermodynamically favorable. Then there is a large barrier - the activation energy

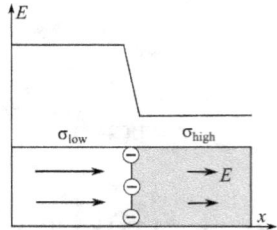

FIGURE 17.0.1. Sketch of charge build up at an interface between
two media having different electrical conductivities due to Gauss's
law and current continuity. When an electric field is applied normal
to the interface between the two blocks, due to current continuity
and Ohm's law the electric field in the low conductivity section is
high and that in the high conductivity section is low (as illustrated
by the graph in the sketch). At the interface there is a jump
in electric field (the width of the jump is greatly exaggerated for
clarity). Since now the electric field has a negative divergence
(arrows shrink in the direction of the arrows), the charge at the
interface is negative, by Gauss's law.

barrier. The reactants need to have this energy for the products to form, although
a good portion of that energy may be returned to the environment and so recycled
for the next reactant molecule. For electrochemical reaction overpotential scales
with the activation energy. Actually the overpotential is the potential difference
between the reaction's electrochemical (thermodynamic) potential and the potential
at which the reaction is appreciable. Note for the reaction to occur, the reactants
on average just need an amount of energy just a bit more than the total energy
of the reaction. This is because the energy among the reactant molecules is not
uniformly distributed - some molecules will have momentarily a very large amount
of energy and some much less than the total energy of the reaction. The ones that
have more energy than the activation energy will "jump" over the barrier, consume
the energy of the total energy of the reaction and return the rest of their energy
to the environment, so that some other reactant molecule can do the same. The
reaction will proceed, albeit very slowly. For the reaction to occur at an appreciable
rate, most of the reactant molecules should have an amount of energy about that
of the activation energy.

Thus for the reaction to occur at the interface and for current to flow there
must be a potential drop at the interface equal to the sum of the overpotential and
the electrochemical potential. At the very least, for some infinitesimal current to
flow, the potential drop must be equal to the electrochemical potential. Below this
threshold potential, no electrochemical reactions can occur at the interface, and so
no charge can be transferred across the interface, and so no current can flow through
the interface. Thus the interface reminds us of the current-voltage behavior of a
varistor.

In a realistic situations in which we are interested in there are two interfaces:
an interface into which the electric field enters the object, and an interface through
which the electric field exists the object. For example, let's consider a metal cylinder
in a globally uniform electric field. Let's assume the electric field is low enough such

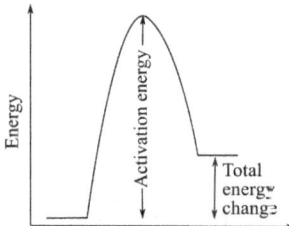

FIGURE 17.0.2. Typical energy diagram for a chemical reaction. The reaction will proceed if the reactants have an amount of energy equal to the total energy change of the reaction, but albeit the rate of such a reaction will be very low. For an appreciable reaction to occur the reactants must have an energy roughly equal to the activation energy.

that the electrochemical reactions cannot occur. Let's begin with the situation where the cylinder is in a liquid electrolyte and the electric field is initially zero. At time zero, we suddenly slam on our electric field from left to right as in Figure 17.0.3. At this time, for a brief time, the electric field is non-zero everywhere in our domain, including inside the metal cylinder (Figure 17.0.3a). Inside the metal cylinder the electric field forces the electrons to migrate to the left, making the left side of the cylinder have negative charge and the right side positive charge. Once the electrons reached their positions on the left side of the cylinder, they cannot be go any further and so the charge cannot go any further, as we assumed the electric field is low enough so that no electrochemical reactions occur on the surface of the cylinder. This charge on the surface of the cylinder as well as the external electric field drives ions of opposite charge (to that on the surface) towards the surface and so a double layer forms on the surface (Figure 17.0.3b). Once the double layer has formed, no current flows through the cylinder, and the electric field then goes around the cylinder rather than through the cylinder (Figure 17.0.3b). The tangential component of this electric field applies a force on the ions in the double layer, forcing them to move in the direction from the equator to the poles. Notice, that if the electric field was switched so it now points from right to left, the sign of the charge on the two sides of the particles will also flip, but the direction of the charge's motion, i.e. from the equator to the poles will remain the same. This charge of course drags liquid with it, and so there is a flow from the equator to the poles. This is induced charge electroosmotic flow around a metal cylinder.

17.1. Quantitative view of induced charge electroosmosis

So far we have discussed induced charge electroosmotic flow around an uncharged conducting cylinder from a qualitative point of view. Let's now discuss this from a quantitative point of view. As we noted in the quantitative picture, initially the cylinder behaves as a perfectly conducting cylinder, with electric field lines passing through the cylinder (Figure 17.0.3a). As the cylinder becomes polarized at later times, the cylinder behaves as if it expels all of the electric field lines, as current inside the cylinder is zero (Figure 17.0.3b). Thus, let's start with the problem of a uniform electric field applied perpendicularly to the axis of a perfectly

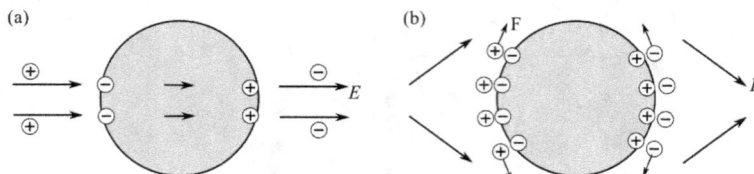

FIGURE 17.0.3. Charge induced around a metallic cylinder in aqueous electrolyte. (a) Just when the electric field is turned on the electric field is present everywhere including in the cylinder, causing electrons to migrate left, leaving the right side positive. In our situation, the electric field is low enough so that no electrochemical reactions occur at the surface of the cylinder. Thus the electrons stop there, and attract positive ions from the liquid. Thus a double layer forms on the surface of the cylinder. At this moment also, current no longer flows through the cylinder and electric field there goes to zero (b). The electric field now goes around the cylinder rather than through it. The tangential component of the electric field forces ions in the double layer to move in the direction from the equator to the poles, generating a flow from the equator to the poles.

conducting cylinder of infinite length as shown in Figure 17.0.3a. Since we are talking about a cylinder it is convenient to use polar coordinates. The field at infinity is unperturbed and so we write that

(17.1.1) $$\varphi = -E_0 r \cos \theta \text{ at } r \to \infty$$

Since the cylinder is a conductor, the surface of the conductor is an equipotential. For convenience, we set the potential at the surface to be zero

(17.1.2) $$\varphi = 0 \text{ at } r = a$$

where, as usual, a is the radius of the cylinder. Since there are no free charges in this situation, to obtain the potential and then electric field distribution we solve Laplace's equation in polar coordinates. Laplace's equation for these boundary conditions has a well known solution

(17.1.3) $$\varphi_0 (r, \theta) = -E_0 \left(1 - \frac{a^2}{r^2} \right) r \cos \theta$$

To find the electric field around the sphere, we recall that $\mathbf{E} = -\nabla \varphi$ and obtain

$$\mathbf{E} = E_0 \cos \theta \left(1 + \frac{a^2}{r^2} \right) \hat{\mathbf{r}} - E_0 \sin \theta \left(1 - \frac{a^2}{r^2} \right) \hat{\theta}$$

We plot this result in Figure 17.1.1a. At later times, the cylinder becomes polarized (and current inside the cylinder goes to zero) and the cylinder expels all of the electric field lines. Thus, the boundary condition at the cylinder surface becomes

(17.1.4) $$\mathbf{E} \cdot \hat{\mathbf{r}} = 0 \text{ at } r = a$$

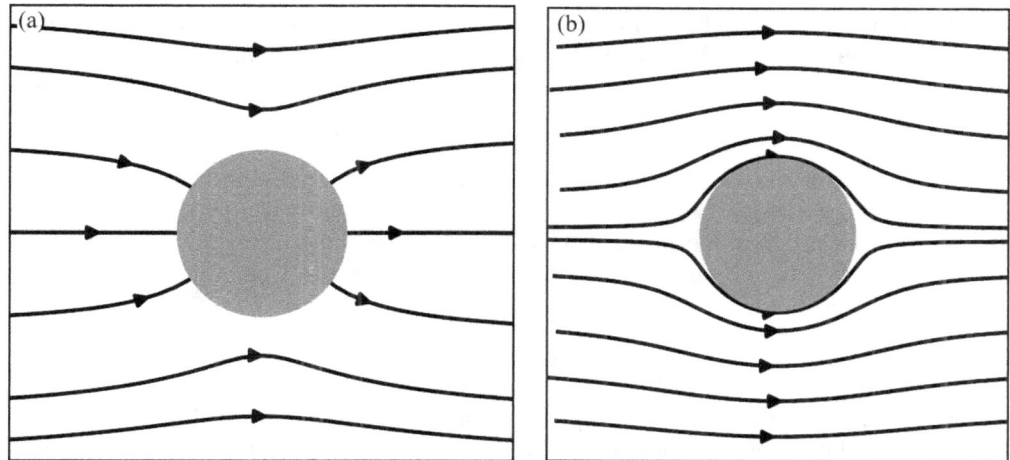

FIGURE 17.1.1. Electric field lines around an infinite ideally polarizable conducting cylinder. (a) Electric field at time $t = 0$ when a DC electric field has just been applied. The electric field penetrates into the particle and causes electrons to move the left side of the particle, making the left side negatively charged and leaving the right side positively charged. (b) Electric field lines at a long time (steady state). No current flows inside the cylinder and electric field inside the cylinder is zero.

i.e., the no-flux boundary condition. With this boundary condition, Laplace's equation also has a well known solution,

$$(17.1.5) \qquad \varphi_f(r, \theta) = -E_0 \left(1 + \frac{a^2}{r^2} \right) r \cos \theta$$

for which the electric field outside the sphere is

$$(17.1.6) \qquad \mathbf{E} = E_0 \cos \theta \left(1 - \frac{a^2}{r^2} \right) \hat{\mathbf{r}} - E_0 \sin \theta \left(1 + \frac{a^2}{r^2} \right) \hat{\theta}$$

We plot this result in Figure 17.1.1b.

To find the resulting zeta potential of the surface we subtract surface potential at the later times from the initial potential

$$(17.1.7) \qquad \zeta = \varphi_0|_{r=a} - \varphi_f|_{r=a}$$

You can think about this as a conservation of energy of the system: initially the system just had an electrostatic potential given by (17.1.3) while at the later times the surface has a potential that is the sum of the zeta potential and the electrostatic potential given by (17.1.5); but the energy at the surface must be conserved and so must be the potential. Thus

$$(17.1.8) \qquad \zeta = 2aE_0 \cos \theta$$

Knowing the zeta potential and the electric field from equation (17.1.6) allows us to calculate the fluid velocity. Assuming that our cylinder has thin double layers compared to the size of the cylinder ($\kappa a \gg 1$) we can use the Helmholtz-Smoluchowski result, (16.7.35), that we originally derived for spherical particles.

Recall that we derived the Helmholtz-Smoluchowski by assuming that the curvature of particle negligible and the surface of the particle was a plane. Thus, using the same assumption we would obtain the same result if the particle was not a sphere but a cylinder. Additionally, we derived the result for the motion of the particle. If we instead hold our solid plane fixed, and allow the fluid above it to flow, then we would obtain the same result, only with the sign inverted. We would then obtain the velocity directly at the surface (actually at the slip plane). This would make it appear that the fluid has a slip condition rather than a no-slip condition. Thus the slip velocity on the surface of a cylinder given the Helmholtz-Smoluchowski assumptions is

$$(17.1.9) \qquad\qquad u_s = -\frac{\varepsilon\varepsilon_0\zeta}{\mu}E_t$$

where E_t is the electric field component tangential to the surface. From (17.1.6) the tangential component of the electric field at the surface is just the theta component evaluated at $r = a$,

$$(17.1.10) \qquad\qquad E_t = -2E_0\sin\theta$$

Combining (17.1.9) with (17.1.8) and (17.1.9) we obtain that the slip velocity on the surface of the cylinder is

$$(17.1.11) \qquad\qquad u_s = \frac{4a\varepsilon\varepsilon_0 E_0^2\cos\theta\sin\theta}{\mu}\hat{\theta}$$

where, as a reminder, ε is the relative permittivity of the liquid. Combining this boundary condition at the surface of the cylinder with the boundary condition at infinity,

$$(17.1.12) \qquad\qquad \mathbf{u} = 0 \text{ at } \mathrm{r} \to \infty$$

we can utilize Taylor's approach (Taylor (1966)) to obtain the velocity field around the cylinder to be

$$(17.1.13) \qquad u = \frac{2a^2\left(a^2 - r^2\right)}{r^3}\frac{\varepsilon\varepsilon_0 E_0^2}{\mu}\cos 2\theta\hat{\mathbf{r}} + \frac{2a^4}{r^3}\frac{\varepsilon\varepsilon_0 E_0^2}{\mu}\sin 2\theta\hat{\theta}$$

We plot the resulting velocity streamlines in Figure 17.1.2. Notice that as expected, the flow is from the equator to the poles of the cylinder. Although we have considered the ICEO flow specifically for a circular geometry, the same analysis holds for similar shapes: the flow will be in the direction from the equator to the poles. If the shape is itself symmetric (including in its properties or if its properties are homogeneous) the flow will also be symmetric. If the shape is not symmetric, so will be the flow. One way to break this symmetry is to "add" charge to the shape, as we will discuss next.

17.2. ICEO around a charged conducting cylinder

Up till now we have assumed that our cylinder has zero native charge. This is a fairly unrealistic assumption. Surfaces with zero charge (especially at any pH) are very rare, and a pH at which a surface has zero charge is typically referred to as a point of zero charge. Hence at all other points the surface is charged! Thus we now consider a realistic case where initially the cylinder surface has a finite zeta potential ζ_0. Since Laplace's equation is linear, and we already arbitrarily set the potential at the surface of the cylinder to zero for the initial (early) time, we

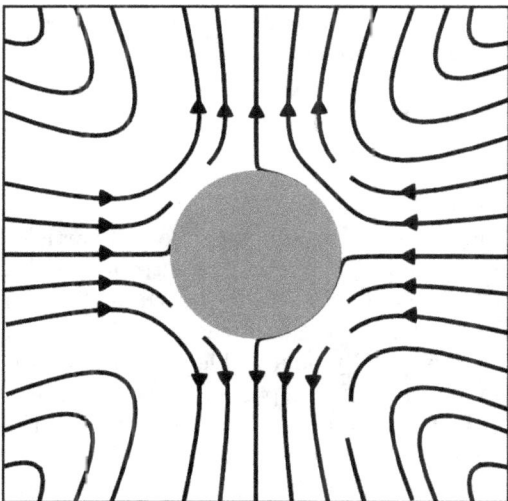

FIGURE 17.1.2. Steady state induced charge electroosmotic streamlines around an infinite ideally polarizable conducting cylinder for an electric field directed from left to right. Notice the quadrapolar nature of the flow. The electric field brings fluid from the equator, drags along the cylinder surface and ejects it at the poles.

can just as well set the potential to be ζ_0 and it will linearly add to the solution. Hence the total zeta potential will be the sum of zeta potential due to polarization, calculated in (17.1.7) and ζ_0

$$(17.2.1) \qquad\qquad \zeta = 2aE_0\cos\theta + \zeta_0$$

Again, combining (17.1.9) with (17.2.1) and (17.1.10) we obtain that the slip velocity on the surface of the cylinder is

$$(17.2.2) \qquad \mathbf{u}_{s.q} = \mathbf{u}_s - \frac{2\varepsilon\varepsilon_0\zeta_0\sin\theta}{\mu}\hat{\theta} = \frac{\varepsilon\varepsilon_0\sin\theta}{\mu}\left(4aE_0^2\cos\theta - 2\zeta_0\right)\hat{\theta}$$

where the subscript q indicates that the cylinder is charged. Now the surface velocity is asymmetrical. Since the Stokes simplification of the Navier-Stokes is also linear, the velocity field around the cylinder is a superposition of the velocity due to an uncharged polarized cylinder (17.1.13) and a non-polarizable cylinder with a constant zeta potential.

Let's now consider the electrophoresis of such a cylinder. Again, while this is not realistic, for simplicity we assume that the shape is 2D. Stone and Samuel has shown that for a 2D shape such as a cylinder the electrophoretic velocity is related to the surface velocity by

$$(17.2.3) \qquad\qquad U = -\frac{1}{2\pi}\int_0^{2\pi}\mathbf{u}_s\left(\theta\right)d\theta$$

If we insert our velocity from (17.1.13) into the integral in (17.2.3) then we see the component due to the polarization, i.e., (17.2.3) is always zero, while the component due to the charge is that as we found previously in the electrophoresis section, (16.7.35). Namely, the velocity is equal to

$$(17.2.4) \qquad\qquad U = \frac{\varepsilon\varepsilon_0\zeta}{\mu}E_0$$

Levich (1962) has obtained similar results for spherical particles. Ideally polarizable particles that have no charge will not undergo electrophoresis! Their velocity will be zero, in both AC or DC fields. This is due to the symmetry of the flow around them - the flow is symmetric and so it will keep the particle at rest. However, lucky for us, as we mentioned above, ideally polarizable particles with no net surface charge almost never exist in nature. Thus, particles will undergo electrophoresis, just due to their charge, which causes an asymmetry in the zeta potential distribution around the particle, and so an asymmetry in flow, and so a net flow in the reference frame of the particle (or net motion) in a particular direction.

17.3. ICEO time scales

So far, we have discussed what happens to the flow around our cylinder at long times when the cylinder has time to polarize and the double layer has time to form. But what happens before that? And how long is "a long time?" We will attempt to answer these questions in this section. We begin by finding the potential field around the particle, and to do so guess its form to be

$$(17.3.1) \qquad\qquad \varphi(r,\theta,t) = -E_0\left(1 + g(t)\frac{a^2}{r^2}\right)r\cos\theta$$

where $g(t)$ is an unknown function we need to obtain. We guess this form for the time dependence of the potential because it satisfies the solutions for the potential for both short times (by setting $g(t=0) = -1$) and for long times (by setting $g(t \to \infty) = 1$). This also gives us the boundary conditions for $g(t)$. Furthermore, we can obtain a relation for the total charge that has to accumulate between $t = 0$ and $t \to \infty$ from the value of the zeta potential. Recall from the electrophoresis section, equation (16.7.42), that for low surface potentials and thin double layers for a spherical particle

$$(17.3.2) \qquad\qquad \frac{Q_s}{4\pi a^2} = \varepsilon\varepsilon_0\kappa\zeta$$

Thus, we can infer that for a cylinder, the surface charge density, σ_s would also be

$$(17.3.3) \qquad\qquad \sigma_s = \varepsilon\varepsilon_0\kappa\zeta$$

The flux of the surface charge to the surface, (i.e. the current density) is given by the product of the conductivity of the liquid and the electric field normal to the surface.

$$(17.3.4) \qquad\qquad j = -\frac{d\sigma_s}{dt} = -\sigma\,E_n|_{r=a}$$

Here the negative sign in front of the time derivative of the surface charge is due to the fact that the charge traveling to the surface is opposite in sign to that attached to the surface. Here σ is the conductivity of the solution, not to be confused with the surface charge density, σ_s. We find the electric field by taking the negative gradient of the potential, (17.3.1)

$$(17.3.5) \qquad \mathbf{E} = E_0 \cos\theta \left(1 - g(t)\frac{a^2}{r^2}\right)\hat{\mathbf{r}} - E_0 \sin\theta \left(1 + g(t)\frac{a^2}{r^2}\right)\hat{\theta}$$

Evaluating the radial component of the electric field at $r = a$ in (17.3.5) we obtain

$$(17.3.6) \qquad -\frac{d\sigma_s}{dt} = -\sigma E_0 \cos\theta\,(1 - g(t))$$

We simplify this using (17.3.5)

$$(17.3.7) \qquad \frac{d\zeta}{dt} = \frac{\sigma E_0}{\varepsilon\varepsilon_0\kappa}\cos\theta\,(1 - g(t))$$

Furthermore, we can obtain another relation for the zeta potential from (17.1.7) and note that we originally set $\varphi_0|_{r=a} = 0$ for convenience. Using the new relation for the potential, (17.3.1)

$$(17.3.8) \qquad \zeta = E_0 \left(1 + g(t)\frac{a^2}{r^2}\right)r\cos\theta$$

Next, taking the derivative with respect to time,

$$(17.3.9) \qquad \frac{d\zeta}{dt} = \frac{dg}{dt}\frac{a^2}{r^2}E_0 r\cos\theta$$

Evaluating at $r = a$ and combining this with (17.3.7)

$$(17.3.10) \qquad \frac{dg}{dt} = \frac{\sigma}{\varepsilon\varepsilon_0\kappa a}(1 - g)$$

Integrating this ODE subject to the initial condition $g(t = 0) = -1$, we obtain

$$(17.3.11) \qquad g(t) = 1 - 2\exp(-t/\tau)$$

where

$$(17.3.12) \qquad \tau = \kappa a \frac{\varepsilon\varepsilon_0}{\sigma}$$

We have already seen this time scale before - the same scaling occurs in AC electroosmotic flow (see equation (16.6.10) and Table 4). This is not surprising at all - in both cases this time scale represents the time for a double layer to form. It is often termed "double layer relaxation" time scale. This time scale can also be rewritten in terms of the diffusion constant of the ions. Assuming that in our situation both the positive and negative ions have roughly the same mobilities, and that we have a 1:1 electrolyte, and employing the Einstein-Smoluchowski relation, we can obtain that the conductivity of the solution

$$(17.3.13) \qquad \sigma = \frac{2c_0 N_A e^2 D}{k_B T}$$

Then using the same assumptions and the definition of Debye length, we obtain that the time scale for the double layer to form

$$(17.3.14) \qquad \tau = \frac{\kappa^{-1}a}{D}$$

Interestingly this time scale is not the time scale for the ions to diffuse the length of the Debye length (i.e. the "length" of the double layer),

$$(17.3.15) \qquad \frac{\left(\kappa^{-1}\right)^2}{D}$$

or even the time scale for the ions to diffuse the radius of the cylinder,

$$(17.3.16) \qquad a^2/D$$

In fact, it is the geometric mean of the two! Geometric means are used to compare different items having widely different scales, and indeed the cylinder size is at a very different scale then Debye length, especially that we here have made a realistic assumption that the cylinder size is much larger than the Debye length. The geometric mean "normalizes" the scales being averaged so no scale dominates the weighting and a given percentage change in any of the items has the same effect on the geometric mean. For further discussion on double layer relaxation time scale, see MacDonald (1970). Having found $g(t)$ we can now substitute this into (17.3.8) and obtain the slip velocity at the cylinder surface as we have done before.

$$(17.3.17) \qquad \mathbf{u}_s = 2\frac{\varepsilon\varepsilon_0 E_0^2 a}{\mu}\sin 2\theta(1 - \exp\left(-t/\tau\right))^2\hat{\theta}$$

We see that the slip velocity increases in time as the double layer grows and saturates at long times to the value we have derived previously ($2\cos\theta\sin\theta = \sin 2\theta$). We can obtain the slip velocity for a charged cylinder in the same way and here we just prove the result:

$$(17.3.18) \qquad \mathbf{u}_{s,q} = \mathbf{u}_s - 2\frac{\varepsilon\varepsilon_0\zeta}{\mu}\sin\theta\left(1 - \exp\left(-t/\tau\right)\right)\hat{\theta}$$

Comparing the free charge (second) term to the polarization (first term) we see that the slip velocity due to free charge grows quicker with time. However, the polarization term typically dominates at longer times for higher fields as it depends on the electric field squared.

17.4. ICEO in an AC field

Having discussed the relevant time scale for ICEO, let's now move on to the most practical implementation of ICEO: ICEO in an AC field. As we touched on above, using DC fields limits us to low electric fields due to the fact that we would like to avoid electrochemical reactions. We would like to avoid electrochemical reactions not only for the reason that that situation is much easier to analyze, but mostly because electrochemical reactions either deposit (or remove) ions into (from) solution, often changing solution properties (including bio or reaction compatibility); or electrochemical reactions generate gas, which can introduce its own flows. Thus it is often advantageous to avoid electrochemical reactions. We can do this in part by applying an AC field at a sufficiently high frequency. At a sufficiently high frequency, if the kinetics of the electrochemical reaction are slow, the reaction will not occur to an appreciable extent. Even if the kinetics of the electrochemical reaction are fast, at the first half cycle of the sinusoid, for example, the reduction reaction occurs, but since the products do not have much time to diffuse, at the second half cycle oxidation reaction occurs and on the net it appears that no reactions took place.

To obtain the slip velocity for a cylinder in an AC field, we first assume that the field is of the form

(17.4.1)
$$\mathbf{E} = E_0 \exp\left(i\omega t\right) \hat{\mathbf{x}}$$

We then solve Laplace's equation, subject to the same boundary conditions as before, and find the potential distribution around the cylinder. As before, we then evaluate the potential at the surface of the cylinder and find that the zeta potential is

(17.4.2)
$$\zeta = 2E_0 a \cos\theta \, \mathrm{Re}\left(\frac{\exp\left(i\omega t\right)}{1 + i\omega\tau}\right)$$

We then use the same procedure as above to find the slip velocity

(17.4.3)
$$\mathbf{u}_s = 2\frac{\varepsilon\varepsilon_0 E_0^2 a}{\mu}\sin 2\theta\left(\mathrm{Re}\left(\frac{\exp\left(i\omega t\right)}{1 + i\omega\tau}\right)\right)^2 \hat{\theta}$$

We observe that the velocity oscillates with the frequency of the electric field. However, typically that frequency is too high to observe and to be of consequence, and so we look for a net component of the slip velocity. Thus, we time average the slip velocity over the cycle of the electric field,

(17.4.4)
$$\langle\mathbf{u}_s\rangle = \frac{\varepsilon\varepsilon_0 E_0^2 a}{\mu}\frac{1}{1 + \omega^2\tau^2}\sin\theta\hat{\theta}$$

We can see that for low frequencies, when $\omega \ll (1/\tau)$ the slip velocity is maximum, and the slip velocity begins to decay significantly when $\omega \geqslant (1/\tau)$. In the low frequency limit the double layer develops in phase with the applied electric field, while as the frequency increases the double layer does not have time to develop and so the net ICEO is not as efficient. Due to the assumptions made, this analysis is valid for $\omega \ll \sigma/\varepsilon\varepsilon_0$. For a more detailed and systematic derivation of ICEO phenomena see Squires and Bazant (2004, 2006).

17.5. ICEO and dielectric contamination

In the previous sections we have explored ICEO with absolutely clean metal surfaces. However, under real working conditions microfluidic devices often become contaminated either during manufacturing (e.g., adhesive leaks, surface blocking) or especially during use. For example, when using microfluidic devices with protein rich samples (blood, saliva, cell lysate), the proteins may coat the metal surfaces and form a dielectric layer. Sometimes it is also desirable to pre-coat microfluidic devices with dummy protein such as bovine serum albumin (surface blocking) so that proteins of interest do not attach to the walls detrimentally to the protein's function. Thus ICEO on metal surfaces with a dielectric layer is of practically importance.

To analyze ICEO with a dielectric layer, let's imagine that the metal layer is coated with a dielectric with a dielectric constant ε_d with a thickness λ_d above which sits the double layer with the Debye length $\kappa^{-1} = \lambda_D$ and whose permittivity is assumed to be that of the liquid (as we have assumed everywhere in this text) and equal to ε_w (see Figure 17.5.1). As before, for convenience we set the potential at the conductor's surface to be zero, while outside of the double layer to be φ_f. We approximate that the potential distribution in both the dielectric and inside

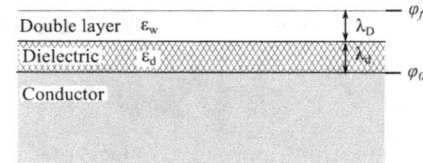

FIGURE 17.5.1. Schematic of our model for dielectric contamination of a conductive surface. We assume that the dielectric contamination is a uniform layer of thickness λ_d and relative permittivity ε_d.

the double layer is a linear distribution. Thus we can write that the potential drop across the dielectric and the double layer is

$$(17.5.1) \qquad E_d\lambda_d + E_D\lambda_D = \varphi_f$$

where E_d and E_D are the electric fields inside the dielectric and double layer respectively. Next, we can draw a Gaussian surface around the interface between the double layer and the dielectric layer. We expect that no free charge develops at this interface. Under this assumption, using Gauss's law, we can write

$$(17.5.2) \qquad \varepsilon_d E_d = \varepsilon_w E_D$$

Combining the above two equations, we can eliminate the unknown electric field in the dielectric,

$$(17.5.3) \qquad \left(1 + \frac{\varepsilon_w}{\varepsilon_d}\frac{\lambda_d}{\lambda_D}\right) E_D\lambda_D = \varphi_f$$

We can furthermore simplify this, by recalling that we assumed that $E_D\lambda_D = \zeta$ and using the relation for the potential outside the double layer that we have derived previously, (17.1.5), evaluated at $r = a$. We obtain

$$(17.5.4) \qquad \zeta = \frac{2aE_0\cos\theta}{\left(1 + \frac{\varepsilon_w}{\varepsilon_d}\frac{\lambda_d}{\lambda_D}\right)}$$

From this relation we can firstly observe that the dielectric layer reduces the overall zeta potential and thus reduces the slip velocity and flow around the cylinder, by the amount proportional to the denominator. Furthermore, considering that for most commonly encountered organic materials (plastics, adhesives) the dielectric constant is between 2 and 7, and the dielectric constant of water is around 80, for the effect of the dielectric coating on ICEO to be minimal, the thickness of the dielectric layer must be smaller than the Debye length. Since for typical solutions the Debye lengths are pretty small (see Table 4) this condition is even difficult to satisfy with thin films of bovine serum albumin which have a thickness of around 5 nm (Phan et al. (2015)). However, if we have the luxury to protect the metal surface with a coating of our choice, and we are interested in high ICEO, we should choose to coat it with a high-k dielectric, some of which have dielectric constants $> 10,000$. Thus, we see that it is quite easy to eliminate ICEO (and similarly AC EO) by coating the metal layer with a thick, low relative permittivity dielectric. Thus for ICEO (and AC EO) it is important to keep the metal surfaces clean.

TABLE 1. Steady state zeta potential and induced charge electroosmosis velocity profile for a cylinder and a sphere.

	Cylinder	Sphere
Steady state zeta potential	$2E_0 a \cos\theta$	$(3/2) E_0 a \cos\theta$
Radial flow velocity, u_r	$\frac{2a^2\left(a^2-r^2\right)}{r^3} \frac{\varepsilon\varepsilon_0 E_0^2}{\mu} \cos 2\theta$	$\frac{9a^3\left(a^2-r^2\right)}{16r^4} \frac{\varepsilon\varepsilon_0 E_0^2}{\mu} \left(1+3\cos 2\theta\right)$
Azimuthal flow velocity, u_θ	$\frac{2a^4}{r^3} \frac{\varepsilon\varepsilon_0 E_0^2}{\mu} \sin 2\theta$	$\frac{9a^5}{8r^4} \frac{\varepsilon\varepsilon_0 E_0^2}{\mu} \sin 2\theta$

17.6. ICEO in other geometries and uses of ICEO

In the previous sections for we have considered a cylindrical geometry, which can represent a metal or metal coated post in a microfluidic chip. Here we give results for a sphere, as derived by Gamayunov et al. (1986) and quantitatively describe several geometries of interest.

Let's begin with the spherical geometry. Results similar to which we derived here for a circular cylinder were derived by Gamayunov et al. for a sphere. We tabulate the steady-state zeta potential and the radial and azimuthal components of the fluid velocity around a fixed sphere in Table 1. As we have mentioned before, the net migration velocity of a purely uncharged polarizable sphere is zero; but that of a charged polarizable sphere similar to that of a charged cylinder we derived earlier. The time scale for the double layer to develop is the same for the cylinder and the sphere.

The next interesting geometry is a metal rectangle embedded in a channel floor (see Figure 17.6.1). This geometry can be easily created by depositing a metal rectangle onto the floor of the channel using standard lithographic techniques, printing, or even screen printing. Just like in the case of a sphere (Figure 17.0.3) when the electric field is first turned on (or switched direction for AC) the electric field lines penetrate into the metal rectangle (Figure 17.6.1a). Just like with the sphere, once the rectangle becomes polarized, no current flows through the rectangle and the field inside the rectangle becomes zero - the field lines become "expelled" from the rectangle (Figure 17.6.1b). The electric field acts on the two oppositely charged double layers formed on the opposite sides of the rectangle to create a flow consisting of two counter rotating vortices. This geometry can be used as a micromixer, especially for bringing reagents for surface reactions, such as those found in microarrays. An array of such rectangles is especially useful for mixing surface reactions (Figure 17.6.2a). This geometry is quite convenient because, unlike with AC EO counter rotating vortex flows on surfaces can be generated by applying a field with electrodes on the end of the channel, and so we do not need to route traces to a large number of electrodes inside the channel, as we do for AC EO.

We can create another mixer by implementing an array of metal or metal coated posts in a microchannel (Figure 17.6.2b). This geometry is somewhat harder to implement, and may interfere with other processes in the channel (e.g., it might undesirably filter particles) than the one in the previous paragraph. However, this geometry generates a large number of vortices in the bulk of the channel rather than at the surface. This may be used for mixing for bulk reactions. Such a mixer may be especially advantageous for situations where appropriate AC or DC field is already present in the microfluidic system and a simple bulk mixer is needed.

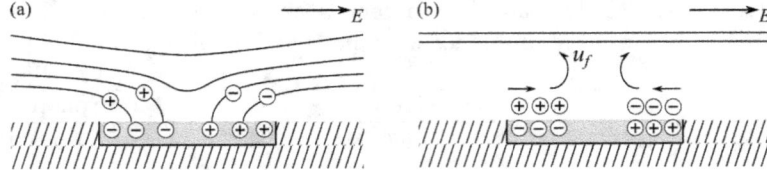

FIGURE 17.6.1. Electric field lines at early times (a) and at late times (steady state, b) for a metal rectangular strip embedded in an insulating wall, parallel to which an electric field is applied. When the electric field is first turned on (or switched direction for AC) the electric field lines penetrate into the metal rectangle. However, once the rectangle becomes polarized, no current flows through the rectangle and the field inside the rectangle becomes zero - the field lines become "expelled" from the rectangle.

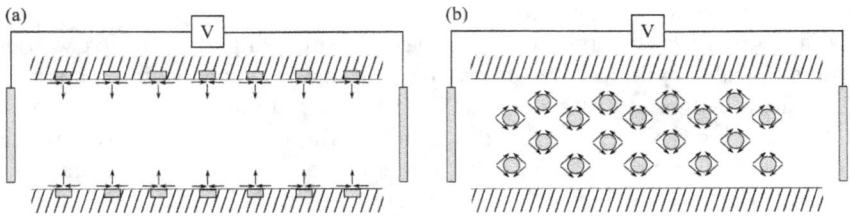

FIGURE 17.6.2. Induced charge electroosmotic flow surface mixers (a) and bulk mixers (b). The arrows represent the direction of the local flow velocity. These mixers can be driven at DC or AC frequency.

In addition to mixing, we can use ICEO to pump. We list a few pumping geometries in Figure 17.6.3. All of these pumps are unidirectional, i.e. their flow direction cannot be changed by altering the direction of the electric field, unlike that of DC (regular) electroosmosis. These pumps can be driven at both AC or DC. The first geometry, the least obtrusive one, consists of metal or more realistically metal coated asymmetrically positioned triangles on the channel wall (Figure 17.6.3a). The triangles are positioned such that the flow of one leading face is larger than that of the trailing face and so the net flow is in the direction of the flow of the leading face.

The next two pump configurations are more obtrusive and consists of an array of objects with broken fore-aft symmetry (Figure 17.6.3b) or left-right symmetries (Figure 17.6.3c) with respect to the direction of electric field. In the first case, this asymmetry generates a net flow parallel to the electric field, while in the second case this generates a net flow orthogonal to the electric field. While we drew triangles in Figure 17.6.3, any shape that has the appropriate symmetry broken (e.g., tear drop, air foil) will work, but of course with varying efficiencies.

The next set of pumps pump flow at channel junctions: L junction (Figure 17.6.3d), T junction (Figure 17.6.3e), and X (cross) junction (Figure 17.6.3f). These are especially useful if appropriate AC or DC field is already present in the microfluidic device and some pumping action or an extra pressure boost is necessary in the device. For the L and X junctions, simply placing a metal or metal coated post at

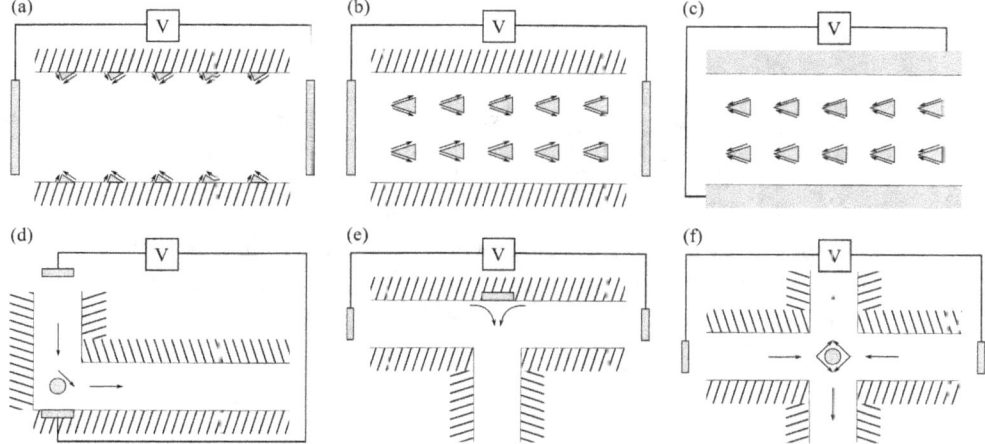

FIGURE 17.6.3. Induced charge electroosmotic pumps. The arrows represent the local flow velocity. All of these pumps are unidirectional, meaning the geometry, not the electric field direction controls the direction of the flow (unlike that for DC (regular) electroosmotic pumps). However, these pumps can be driven by both DC or AC field. The arrows represent the local flow velocity.

the center of the junction generates pumping action. Generation of a T junction pump is more involved, as it involves embedding a metal strip inside one of the channel walls.

17.7. Induced charge electrophoresis

As we have discussed above induced charge electrophoresis (ICEP) is not possible for objects both having left-right and fore-aft symmetries. This due to the quadrapolar nature of the ICEO flow around the particle. Spherical perfectly polarizable particles that do not have free charge do not undergo ICEP. However, if we break at least one of the symmetries, the particle will undergo ICEP. We have already discussed breaking such symmetry by adding free charge to a spherical particle. The other three ways are by coating part of the particle with a dielectric to alter zeta potential distribution on the particle surface; by creating a homogeneous particle, e.g., a Janus particle (Gangwal et al. (2008)); or by changing the particle geometry. See Squires and Bazant (2006) for a detailed discussion of these cases. Here we will briefly touch upon just the last case.

If the particle has a broken for-aft symmetry with respect to the electric field (Figure 17.7.1a) then there is a higher charge concentration at the sharp end compared to the dull end, and so effectively higher ICEO flow coming of the sharp end. Thus since more flow comes of the sharp end than the dull end, the particle is propelled in the direction pointing from the sharp end towards the dull end, in the axis of the electric field. If the particle has a broken top-bottom symmetry with respect to the electric field, (Figure 17.7.1b) the particle motion is even stranger. The corners that are polarized are themselves symmetric. However, the electric field has a larger component aligned with the surface on the dull end than on the sharp end. Thus the charge is driven more efficiently on the dull end than on the

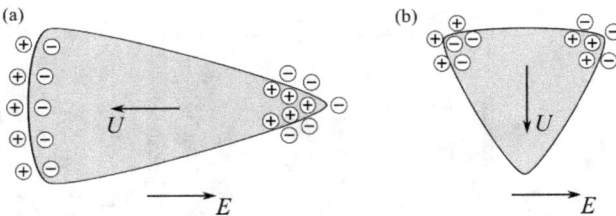

FIGURE 17.7.1. Schematic showing polarization and resulting particle velocities for particles undergoing induced charge electrophoresis. Notice that particle velocity depends on particle shape and its orientation to the electric field.

sharp end, and so there is more flow of the dull end than the sharp end. Thus, the particle is propelled in the direction pointing from the dull end to the sharp end, orthogonal to the direction of the electric field!

While it is somewhat laborious to calculate the ICEP velocities for particles of arbitrary shape, the ICEP velocity scales as

$$(17.7.1) \qquad\qquad U \sim \frac{\varepsilon\varepsilon_0 E_0^2 a}{\mu}$$

where a is the characteristic size of the particle. Comparing this with the scaling electrophoresis velocity we derived in the electrophoresis section,

$$(17.7.2) \qquad\qquad U \sim \frac{\varepsilon\varepsilon_0 \zeta E_0}{\mu} = \frac{\varepsilon\varepsilon_0 E_D \lambda_D E_0}{\mu}$$

we immediately see two distinctions. Firstly, the ICEP velocity scales as the square of the applied electric field, while electrophoresis velocity scales as the first power. Secondly the characteristic length scale for ICEP is the size of the particle, while that for electrophoresis is the size of the double layer (the Debye length). This allows ICEP to separate particles based on their size. Furthermore, ICEP can also be used to simply separate particles by their shape, something that is usually very difficult to do. This is particularly useful for sorting out defective particles, for example from the desired perfectly spherical particles. For details on calculating ICEP velocities of non-spherical particles see the work by Yariv (2005) and that by Squires and Bazant (2006).

ICEP can also cause rotation of polarizable particles, and align them with respect to the field. ICEO again acts on the particle, and rotates the particle. The rotation velocity scales as

$$(17.7.3) \qquad\qquad \Omega \sim \frac{\varepsilon\varepsilon_0 E_0^2}{\mu}$$

Interestingly, the rotation velocity is independent of the particle size (Yariv (2005)).

In dense suspensions of particles in which the particles are undergoing ICEP, the particles can hydrodynamically interact. For example, the quadrupolar flow around particles (for similar flow field, see Figure 17.1.2 and Figure 17.0.3) causes two symmetric particles to move towards each other in the direction along the field axis. However, the same flow causes two symmetric particles to move apart from each other in the direction normal to the field axis. Thus if left in such a field the suspension will turn into a spreading disk with the axis of the disk pointing in the direction of the electric field, while the radius of the disk pointing normal to the electric field.

In electric fields having a strong gradients polarizable particles show even more interesting behaviors. For example, an ideally polarizable sphere in a uniform gradient also does not move, as ICEP force on it exactly balances dielectrophoresis force! However, broken symmetries either in the field gradient or in the particle allow the particle to move despite the fact that ICEP force and dielectrophoresis force continue to oppose each other (but not equally) (Squires and Bazant (2006)).

CHAPTER 18

Polarization electrokinetic phenomena

18.1. AC Electrothermal flow

Let's begin our discussion with the motion of bulk fluid due to polarization effects. This fluid flow is termed electrothermal flow, because heating of the fluid causes local changes in permittivity and conductivity of the flow (as both are a function of temperature). As we have seen in Section 14, electrical fields coupled with gradients in permittivity produce electrical body forces in the media, and these forces produce fluid motion. Often heating and so temperature gradients are due to joule heating or dielectric heating from the electric field itself. From Section 14, the total electrical body force per unit volume in a fluid is

$$(18.1.1) \qquad \mathbf{f} = \rho\mathbf{E} - \frac{1}{2}|\mathbf{E}|^2\varepsilon_0\nabla\varepsilon + \frac{1}{2}\varepsilon_0\nabla\left(\rho_m\left(\frac{\partial\varepsilon}{\partial\rho_m}\right)|\mathbf{E}|^2\right)$$

where here we use ρ for electrical charge density, and ρ_m for the density of the fluid. In order to determine the magnitude and frequency dependence of fluid motion, we first need to re-cast equation (18.1.1) into a form in terms of gradients of permittivity and conductivity, combine this with appropriate electrical and thermal boundary conditions and geometry, and then couple it with a simplified version of Navier-Stokes momentum equation. We begin with expanding the equation using a perturbation approach. Since typically changes in permittivity and conductivity are small, we may write that the electric field

$$(18.1.2) \qquad \mathbf{E} = \mathbf{E}_0 + \mathbf{E}_1$$

where $\mathbf{E}_0$ is the applied field and $\mathbf{E}_1$ is the perturbation field, and where $|\mathbf{E}_1| \ll |\mathbf{E}_0|$. We now write Gauss's law

$$(18.1.3) \qquad \rho = \varepsilon_0\nabla\cdot(\varepsilon\mathbf{E}) = \varepsilon_0\left(\nabla\cdot(\varepsilon\mathbf{E}_0) + \nabla\cdot(\varepsilon\mathbf{E}_1)\right)$$

which we then expand to

$$(18.1.4) \qquad \rho = \varepsilon_0\left(\varepsilon\nabla\cdot\mathbf{E}_0 + \nabla\varepsilon\cdot\mathbf{E}_0 + \varepsilon\nabla\cdot\mathbf{E}_1 + \nabla\varepsilon\cdot\mathbf{E}_1\right)$$

We assume that the divergence of the applied field is zero ($\nabla\cdot\mathbf{E}_0 = 0$) and that any divergence of the field due to any charges is in the perturbation field; thus we eliminate the first term. We also eliminate the last term, by comparing it to the second term and noting that $|\mathbf{E}_1| \ll |\mathbf{E}_0|$. Thus

$$(18.1.5) \qquad \rho = \varepsilon_0\left(\nabla\varepsilon\cdot\mathbf{E}_0 + \varepsilon\nabla\cdot\mathbf{E}_1\right)$$

Next we substitute this expression for charge density into (18.1.1), while simultaneously dropping the compressibility term (third term in (18.1.1)), and so obtain

$$(18.1.6) \qquad \mathbf{f} = \varepsilon_0\left(\nabla\varepsilon\cdot\mathbf{E}_0 + \varepsilon\nabla\cdot\mathbf{E}_1\right)(\mathbf{E}_0 + \mathbf{E}_1) - \frac{1}{2}|\mathbf{E}_0 + \mathbf{E}_1|^2\varepsilon_0\nabla\varepsilon$$

Since $|\mathbf{E}_1| \ll |\mathbf{E}_0|$, we drop the terms multiplying $\mathbf{E}_1$ and obtain

$$(18.1.7) \qquad \mathbf{f} = \varepsilon_0 \left(\nabla \varepsilon \cdot \mathbf{E}_0 + \varepsilon \nabla \cdot \mathbf{E}_1 \right) \mathbf{E}_0 - \frac{1}{2} |\mathbf{E}_0|^2 \varepsilon_0 \nabla \varepsilon$$

Since $\mathbf{E}_0$ is the field we apply, it can be easily calculated for a particular geometry and boundary conditions. On the other hand, we need to find the perturbed field, or for this particular problem, just its divergence. To do this we employ the charge conservation equation: change in the amount of charge at a particular location with time is a result of charge either advected by the fluid flow, or transported in by the electric field (electrical conduction current). We can write this in equation form as

$$(18.1.8) \qquad \frac{\partial \rho}{\partial t} + \nabla \cdot (\rho \mathbf{u}) + \nabla \cdot (\sigma \mathbf{E}) = 0$$

The relative magnitude of the importance of the amount of charge advected by the flow to that transported by the electric field (conduction) is sometimes referred to as the electric Reynolds number

$$(18.1.9) \qquad Re_e = \frac{u_0}{l_0} \frac{\varepsilon \varepsilon_0}{\sigma}$$

Where u_0 and l_0 is the characteristic velocity scale and length scale. For typical microfluidic applications the electric Reynolds number is much less than unity, and so the electrical conduction is much more dominant than electrical convection. Thus, we simplify the charge conservation equation to

$$(18.1.10) \qquad \frac{\partial \rho}{\partial t} + \nabla \cdot (\sigma \mathbf{E}) = 0$$

Substituting in the perturbed electric field (18.1.2) (and keeping only the large terms, like in derivation of (18.1.5)) and our expression for change density (18.1.5), we obtain

$$(18.1.11) \qquad \varepsilon_0 \frac{\partial}{\partial t} \left(\nabla \varepsilon \cdot \mathbf{E}_0 + \varepsilon \nabla \cdot \mathbf{E}_1 \right) + \nabla \sigma \cdot \mathbf{E}_0 + \sigma \nabla \cdot \mathbf{E}_1 = 0$$

For convenience we assume that the electric field is a harmonic time varying electric field of a single frequency ω, i.e. $E \sim \exp(i\omega t)$, and note that

$$(18.1.12) \qquad \frac{\partial}{\partial t} \exp(i\omega t) = i\omega \exp(i\omega t)$$

and so obtain that

$$(18.1.13) \qquad \varepsilon_0 i\omega \left(\nabla \varepsilon \cdot \mathbf{E}_0 + \varepsilon \nabla \cdot \mathbf{E}_1 \right) + \nabla \sigma \cdot \mathbf{E}_0 + \sigma \nabla \cdot \mathbf{E}_1 = 0$$

which we then rearrange to find

$$(18.1.14) \qquad \nabla \cdot \mathbf{E}_1 = -\frac{(\nabla \sigma + \varepsilon_0 i\omega \nabla \varepsilon) \cdot \mathbf{E}_0}{\sigma + \varepsilon \varepsilon_0 i\omega}$$

Now we are ready to substitute this into our expression for the force (18.1.7); however, since our field is oscillating in time, our force will be as well. This is not useful, and so we first time average (18.1.7) with a time scale larger than the inverse of our frequency ω but smaller than the observation time and so obtain

$$(18.1.15) \qquad \langle \mathbf{f} \rangle = \frac{1}{2} \operatorname{Re} \left[\varepsilon_0 \left(\nabla \varepsilon \cdot \mathbf{E}_0 + \varepsilon \nabla \cdot \mathbf{E}_1 \right) \mathbf{E}_0^* - \frac{1}{2} |\mathbf{E}_0|^2 \varepsilon_0 \nabla \varepsilon \right]$$

where the asterisk stands for complex conjugate. Substituting in the expression for the divergence of the perturbation field, we obtain

$$(18.1.16) \qquad \langle \mathbf{f} \rangle = \frac{1}{2} \mathrm{Re} \left[\left(\frac{\varepsilon_0 \left(\sigma \nabla \varepsilon + \varepsilon \nabla \sigma \right) \cdot \mathbf{E}_0}{\sigma + \varepsilon \varepsilon_0 i \omega} \right) \mathbf{E}_0^* - \frac{1}{2} |\mathbf{E}_0|^2 \varepsilon_0 \nabla \varepsilon \right]$$

Now, all it is left is to relate the gradient of permittivity and conductivity to the gradient of temperature and we have a force relation that we can insert into a simplified Navier-Stokes equation; and together with a heat transfer equation we can obtain the fluid motion. We can relate the gradient of permittivity and conductivity to the gradient of temperature as

$$(18.1.17) \qquad \nabla \varepsilon = \frac{\partial \varepsilon}{\partial T} \nabla T$$

$$(18.1.18) \qquad \nabla \sigma = \frac{\partial \sigma}{\partial T} \nabla T$$

and so we can rewrite (18.1.16) as

$$(18.1.19) \qquad \langle \mathbf{f} \rangle = \frac{1}{2} \mathrm{Re} \left[\frac{\sigma \varepsilon \varepsilon_0 \left(\alpha - \beta \right)}{\sigma + i \omega \varepsilon \varepsilon_0} \left(\nabla T \cdot \mathbf{E}_0 \right) \mathbf{E}_0^* - \frac{1}{2} \varepsilon \varepsilon_0 \alpha |\mathbf{E}_0|^2 \nabla T \right]$$

$$(18.1.20) \qquad \alpha = \frac{1}{\varepsilon} \frac{\partial \varepsilon}{\partial T}$$

$$(18.1.21) \qquad \beta = \frac{1}{\sigma} \frac{\partial \sigma}{\partial T}$$

The first term in (18.1.19) arose from the force on the free charge (Coulomb force), $\rho \mathbf{E}$ term in (18.1.1), while the second term arose from the polarization force term $1/2 |\mathbf{E}|^2 \varepsilon_0 \nabla \varepsilon$ in (18.1.1). The first term depends on the dot product of the temperature gradient and the nominal electric field, and is in the direction of electric field. If electric field and temperature gradient are orthogonal, then this term is zero, and the force becomes independent of the frequency of the field. In practice one may achieve this when passing current through a solution in a capillary, with the solution transferring its heat to the capillary walls. In this situation an axially directed electric field sets up a radially directed temperature gradient. Note that for a long capillary filled with uniform conductivity buffer, far from its ends, there should be minimal axial temperature gradient. Thus there should be no Coulomb force in this situation. Now let's consider a tapering capillary. By conservation of current, current must be the same in all sections of the capillary. Since the resistance in the narrow part will be higher and so the temperature there should be higher than in the wide region. Another way to achieve a temperature gradient like this is to create a gradient in thermal insulation along the axis of the capillary (even in a uniform cross-section capillary), with one end more insulated then another. Thus, now that the temperature gradient and the electric field are aligned and we can pump fluid along such capillaries with this method! As we will see shortly this term is just as valid at zero frequency and so one can expect this flow with DC fields. Thus one should be quite careful about temperature gradients when performing electroosmotic flow, streaming potential, or related experiments (Sections 16.1- 16.3), lest you get a contribution from AC electroosmotic flow.

The second term (polarization term) on the other hand is in the direction of the temperature gradient. If the temperature gradient and the electric field are aligned, then we can rearrange (18.1.19) and rewrite it then in a more convenient form as

$$(18.1.22) \qquad \langle \mathbf{f} \rangle \sim \frac{1}{2}\varepsilon\varepsilon_0 |\mathbf{E}_0|^2 \nabla T \left[\frac{\alpha - \beta}{1 + (\omega/\omega_0)^2} - \frac{\alpha}{2} \right]$$

$$(18.1.23) \qquad \omega_0 = \sigma/\varepsilon\varepsilon_0$$

where ω_0 is the fluid's relaxation frequency ($1/\omega_0 = \tau_0$, the charge relaxation time). We plot the function in the square brackets as a function of the scaled frequency in Figure 18.1.1. The term in the square brackets is zero when

$$(18.1.24) \qquad \omega/\omega_0 = \sqrt{1 - 2\beta/\alpha}$$

We have this zero point because at this point two forces, the Coulomb force that arose from the first term (frequency dependent term, which came from $\rho\mathbf{E}$) balances the second term, the polarization force which came from $1/2|\mathbf{E}|^2\varepsilon_0\nabla\varepsilon$. As we move from the limit where one dominates to where the other dominates, the flow changes directions.

However, always operating near this frequency in order to avoid electrothermal flow is not practical and so we consider the two limits: where the scaled frequency is large and when it is small. When the scaled frequency is small, the term in the square brackets reduces to $\alpha/2 - \beta$. Here Coulomb force term dominates.

When the scaled frequency is large, the term in the square brackets reduces $-\alpha/2$. Here polarization force dominates. Since typically α and β have opposite signs, the limit of large frequency gives smaller electrothermal flow then the small frequency limit. In fact, we would expect α to be negative since when temperature increases, random molecular motion increases, and so it is harder to align molecules and so polarization of the medium is less and so the dielectric constant of the medium is less. On the other hand, typically we would expect β to be positive as with increasing temperature electrophoretic mobility typically increases, as well as acid, base, and salt dissociation, and so conductivity of a solution should typically increase with temperature. For example, for a typical aqueous solution $\alpha = -0.46\% \, \mathrm{K}^{-1}$ (Haynes (2017)). For most solvents, α is between $-0.24\% \, \mathrm{K}^{-1}$ (bromoform) to $-1.5\% \, \mathrm{K}^{-1}$ (hydrocyanic acid). For other notable solvents, α is $-0.62\% \, \mathrm{K}^{-1}$ (ethanol), $-0.61\% \, \mathrm{K}^{-1}$ (methanol), $-0.71\% \, \mathrm{K}^{-1}$ (2-propanol), and $-0.91\% \, \mathrm{K}^{-1}$ (1-octanol) (Haynes (2017)). β of course varies strongly with the makeup of the electrolyte, but, for example, for 0.01 M KCl, $\beta = +2.2\% \, \mathrm{K}^{-1}$ (Forsythe (1969)). This value of β is fairly representative of commonly encountered inorganic salts, with many ranging between 2 and 3% K^{-1} (Forsythe (1969)). However, for organic buffers, some buffers may exhibit much stronger conductivity dependence on temperature and thus higher β. For example, this would occur when the dissociation reaction of the buffer is strongly favorable at higher temperatures. One should seek such buffers if strong electrothermal flow is desired. We list some typical temperature coefficients in Table 1. Finally, for aqueous solution of KCl, using equation (18.1.24), we find that electrothermal flow is zero when ω/ω_0 is roughly 4. For typical solvents and electrolytes this value varies between 1 and 10. Furthermore, for aqueous solution of KCl electrothermal flow in the low scaled frequency limit is roughly 11 times higher velocity (as force and velocity are proportional) than in the high scaled frequency limit.

TABLE 1. Typical temperature coefficients for various liquids, including aqueous 0.01M aqueous solutions.

Solution	Temperature coefficient, (%/°C)
Ultrapure water*	4.55
Pure water**	2.3-7.4
Sugar syrup*	5.64
Acids**	1.0 - 1.6
0.01 M H_2SO_4	1.25
0.01 M HCl	1.62
Bases**	1.8 - 2.2
0.01 M KOH	1.94
Salts**	2.2 - 3.0
0.01 M KCl	2.21
0.01 M NaCl	2.38
0.01 M KNO_3	2.16
0.01 M $NaNO_3$	2.26

From *Ionode.com (2015), ** from Mettler-Toledo (2013), the rest from Forsythe (1969).

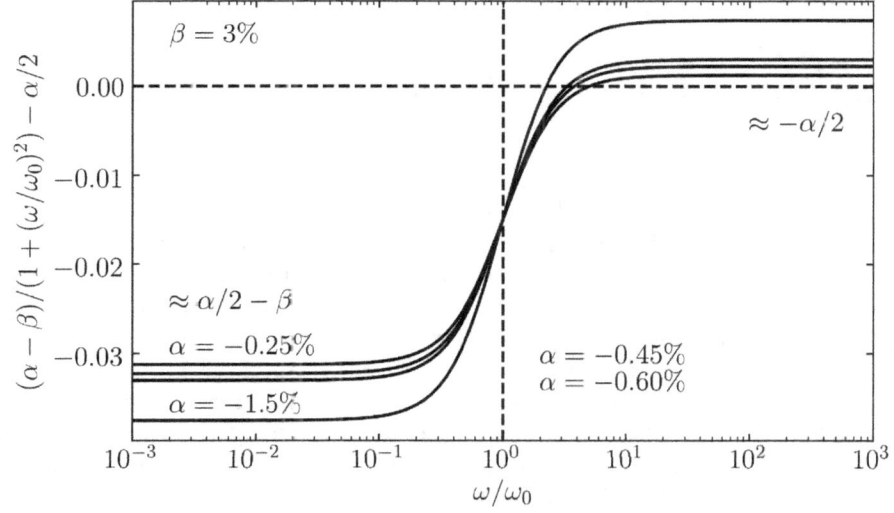

FIGURE 18.1.1. Electrothermal flow body force scaled by media permittivity, magnitude of electric field squared and temperature gradient as a function of the frequency scaled by the relaxation frequency of the solution. Electrothermal flow velocity is proportional to the electrothermal body force.

18.1.1. Electrothermal flow above planar parallel electrodes. In order to get a feel for electrothermal flows in a realistic device, let's examine the case of parallel planar electrodes that we have examined in AC electroosmotic flow (Section

16.6) and that we will examine shortly as the simplest case of DEP electrodes in Section 18.2.9. Consider the geometry of these DEP electrodes (Figure 18.2.6a and b). In this geometry we assume that the distance between the electrodes is much smaller than the width of the electrode (the thickness of the electrodes is negligible). This is quite realistic, as many such electrodes may be designed to be 100 μm in width, 5 μm apart, and only 100 nm thick. We apply an AC potential to the electrodes of the form

$$(18.1.25) \qquad \varphi = V_0 \exp{(i\omega t)}$$

To find the electric field in the domain above the electrodes, we solve Laplace's equation in polar coordinates, subject to the boundary condition that the above given potential is fixed at the electrodes while the potential tends to zero at infinity. This problem has a well-known solution (Zahn (1979)), where

$$(18.1.26) \qquad \mathbf{E}\,(r,\theta) = \frac{2}{\pi}\frac{V_0}{r}\hat{\theta}$$

The thermal power dissipation per unit volume is due to motion of ions in solution and for a sinusoidally oscillating electric field the time averaged power dissipation is

$$(18.1.27) \qquad \langle P \rangle = \frac{1}{2}\sigma\,(\mathbf{E}\cdot\mathbf{E}^*)$$

So for our electric field this is

$$(18.1.28) \qquad \langle P \rangle = \frac{2}{\pi^2}\frac{\sigma V_0^2}{r^2}$$

The heat transport equation in polar coordinates is

$$(18.1.29) \qquad \frac{k}{r}\frac{\partial}{\partial r}\left(r\frac{\partial T}{\partial r}\right) + \frac{k}{r^2}\frac{\partial^2 T}{\partial\theta^2} = -\frac{2}{\pi^2}\frac{\sigma V_0^2}{r^2}$$

where k is the thermal conductivity of the solution. Here we have already substituted in the power dissipation term from (18.1.28) and assumed that (a) the situation is quasistatic, i.e. temperature at a given point does not change on the time scale we are interested in, and (b) that the heat transfer due to advection is negligible. We have made these two simplifying assumptions to give an analytical, but a crude approximation for this problem. These assumptions may not be appropriate for many electrothermal problems. However, carrying on, we further assume that the electrodes serve as "cold plates" (i.e. their temperature never changes, no mater the heat flux into them) and are at ambient temperature. Furthermore, far away from the electrodes, the temperature is also ambient. With these boundary conditions, (18.1.29) has a solution of

$$(18.1.30) \qquad T = \sigma V_0^2 \left(\frac{\theta}{\pi k} - \frac{\theta^2}{\pi^2 k} \right)$$

And the temperature gradient is

$$(18.1.31) \qquad \nabla T = \frac{1}{\pi k}\frac{\sigma V_0^2}{r}\left(1 - \frac{2\theta}{\pi}\right)\hat{\theta}$$

Substituting this result into (18.1.22), we find that the force on the fluid is

$$(18.1.32) \qquad \langle \mathbf{f} \rangle = \frac{2}{\pi^3}\frac{\varepsilon\varepsilon_0\sigma V_0^4}{k r^3}\left(1 - \frac{2\theta}{\pi}\right)\left[\frac{\alpha - \beta}{1 + (\omega/\omega_0)^2} - \frac{\alpha}{2}\right]\hat{\theta}$$

To obtain the flow profile we would solve the Naiver-Stokes equation, which assuming the flow is low Reynolds number and steady, can be simplified to the Stokes equation

(18.1.33)
$$\mu\nabla^2\mathbf{u} - \nabla p + \mathbf{f} = 0$$

However, for our purposes here to get a very crude approximation, we assume that the pressure gradients are negligible and so obtain the scaling that

(18.1.34)
$$u \sim |\langle\mathbf{f}\rangle| l_0^2/\mu$$

Where l_0 is the characteristic dimension, e.g., the electrode width. While our result is fairly approximate, it does capture valuable scaling. For example, we see that the velocity is proportional the dielectric constant and conductivity of the solution; inversely proportional to its thermal conductivity and viscosity; and scales as the fourth power of the applied voltage. This fourth power scaling comes from product of the second power scaling for power dissipation and so production of the temperature gradient and second power scaling for the electric field dependence of the force ($|\mathbf{E}_0|^2$ in (18.1.22)). This fourth power scaling will not always be evident in electrothermal flows, as the temperature gradient will not always scale as the square of the voltage. An example of this is when the gradient is more strongly influenced (set up) by the boundary conditions then by the joule heating of the flow, as was analyzed here.

18.1.2. Scaling for AC heating of a solution.

Here we develop a scaling for the heating of a solution subject to an AC field by a pair of parallel plate electrodes (Figure 18.1.2) to understand the effect of frequency on electrical heating of a solution. We model the solution and the electrodes as a series of a parallel plate capacitor representing the double layer on the first electrode, a resistor, representing the solution, and another parallel plate capacitor, representing the double layer on the second electrode. For simplicity, we assume that the two electrodes and so their double layers are the same. The power dissipated in the entire system is given by the real part of the total power,

(18.1.35)
$$P_{dis} = \mathrm{Re}\,(VI^*) = \mathrm{Re}\left(|V|^2\big/Z^*\right)$$

We take the real part of the total power, as reactive components (ideal capacitors and inductors) dissipate zero power. The double layer capacitance can be approximated by (16.6.4)

(18.1.36)
$$C_{DL} = \frac{\varepsilon\varepsilon_0}{\kappa^{-1}} A$$

where A is the electrode area. The impedance of the double layer is then

(18.1.37)
$$Z_{DL} = \frac{1}{j\omega} \frac{\kappa^{-1}}{\varepsilon\varepsilon_0 A}$$

For the purposes of scaling we assume our solution consists of a monovalent symmetric electrolyte and so Debye length is given by (15.3.22) as

(18.1.38)
$$\kappa^{-1} = \sqrt{\frac{\varepsilon\varepsilon_0 RT}{2F^2 c}}$$

where c is the concentration of electrolyte. Thus, the double layer impedance becomes

$$(18.1.39) \qquad Z_{DL} = \frac{1}{A} \frac{1}{j\omega} \frac{1}{\varepsilon\varepsilon_0} \sqrt{\frac{\varepsilon\varepsilon_0 RT}{2F^2 c}}$$

The impedance of the solution scales as the inverse of its conductivity,

$$(18.1.40) \qquad Z_{sol} = R_{sol} = \frac{l}{A\sigma}$$

where l is the spacing between electrodes. From (18.2.86) for a monovalent symmetric electrolyte,

$$(18.1.41) \qquad \sigma = Fc\left(|\mu_+| + |\mu_-|\right)$$

Thus, the solution impedance scales as

$$(18.1.42) \qquad Z_{sol} = R_{sol} = \frac{l}{AFc\left(|\mu_+| + |\mu_-|\right)}$$

Combining the double layer impedances with the solution impedance, we obtain that the total impedance is

$$(18.1.43) \qquad Z = 2\frac{1}{A} \frac{1}{j\omega} \frac{1}{\varepsilon\varepsilon_0} \sqrt{\frac{\varepsilon\varepsilon_0 RT}{2F^2 c}} + \frac{l}{AFc\left(|\mu_+| + |\mu_-|\right)}$$

or

$$(18.1.44) \qquad Z = \frac{1}{AF}\left(\frac{1}{j\omega}\sqrt{\frac{2RT}{\varepsilon\varepsilon_0 c}} + \frac{l}{c\left(|\mu_+| + |\mu_-|\right)}\right)$$

The total power dissipation from (18.1.35) becomes

$$(18.1.45) \qquad P_{dis} = \mathrm{Re}\left(\frac{|V|^2 AFc}{\left(\frac{\sqrt{c}}{j\omega}\sqrt{\frac{2RT}{\varepsilon\varepsilon_0}} + \frac{l}{(|\mu_+|+|\mu_-|)}\right)}\right)$$

Simplifying this further,

$$(18.1.46) \qquad P_{dis} = \frac{|V|^2 AFc \frac{|\mu_+|+|\mu_-|}{l}}{\frac{c}{\omega^2}\frac{2RT}{\varepsilon\varepsilon_0}\left(\frac{|\mu_+|+|\mu_-|}{l}\right)^2 + 1}$$

We see that at high frequencies, where

$$(18.1.47) \qquad \omega \gg \sqrt{c\frac{2RT}{\varepsilon\varepsilon_0}\left(\frac{|\mu_+| + |\mu_-|}{l}\right)^2} = \omega_{ch}$$

(where we define ω_{ch} as the critical heating frequency) the dissipated power becomes independent of frequency and simplifies to

$$(18.1.48) \qquad P_{dis,\mathrm{high}\ \omega} = |V|^2 AFc\frac{(|\mu_+| + |\mu_-|)}{l} = |V|^2\frac{A}{l}\sigma$$

For low frequencies, where

$$(18.1.49) \qquad \omega \ll \sqrt{c\frac{2RT}{\varepsilon\varepsilon_0}\left(\frac{|\mu_+| + |\mu_-|}{l}\right)^2} = \omega_{ch}$$

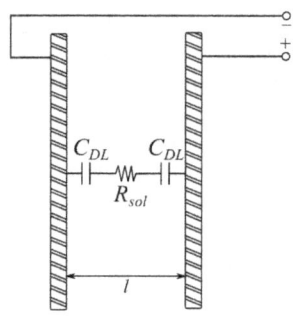

FIGURE 18.1.2. Schematic for analysis of heating of a solution between parallel plate electrodes. The electrodes have an area A and are spaced apart a distance l. The solution is treated as an equivalent electric circuit consisting of a capacitor, a resitor, and another capacitor. The capacitors represent the electrical double layers, while the resistor represents the ionic transport in the solution bulk.

dissipated power becomes independent of solution concentration and simplifies to

$$(18.1.50) \qquad P_{dis,\text{low } \omega} = \frac{|V|^2 A F \omega^2}{\frac{2RT}{\varepsilon\varepsilon_0} \frac{|\mu_+| + |\mu_-|}{l}}$$

For a solution with typical electrophoretic mobilities of $30 \times 10^{-9}\,\mathrm{m^2/(V \cdot s)}$, and for a device with a length scale of 250 µm, ω_{ch} is equal to 200 Hz, 640 Hz, 2 kHz, and 6.4 kHz for 0.1 mM, 1 mM, 10 mM, and 100 mM solutions respectively. For the same solution and a device with a length scale of 100 µm, ω_{ch} is equal to 500 Hz, 1.6 kHz, 5 kHz, and 16 kHz for 0.1 mM, 1 mM, 10 mM, and 100 mM solutions respectively. For the same solution and a device with a length scale of 25 µm, ω_{ch} is equal to 2 kHz, 5.40 kHz, 20 kHz, and 64 kHz for 0.1 mM, 1 mM, 10 mM, and 100 mM solutions respectively.

18.1.3. Other electrothermal and similar flows. Here we focused on a condition where the gradient in solution properties were generated by a temperature gradient produced via electrical Joule heating of the solution. However, we can produce temperature gradients in a solution not only via Joule heating but with many other means, including laser or light heating selective parts of the solution, heating solution boundaries, via exothermic or endothermic chemical reactions, and via microwave heating of the solution. All of these will generate electrothermal flow when a non-uniform electric field is applied to such a solution. Furthermore, we can produce gradients in conductivity and permittivity via non-thermal means. For example, we can generate conductivity gradient via a photodecomposition of a molecule that decomposes into ionic compounds, by selectively illuminating a solution with appropriate light. When a non-uniform electric field is applied to such a solution a flow akin to electrothermal flow will occur.

18.2. Dielectrophoresis

Dielectrophoresis is a term coined by Herbert Pohl (Pohl (1978)) to describe the force exerted by a non-uniform electric field on a small (infinitesimal) polarized but uncharged particle. Dielectrophoretic force is a ponderomotive force as it typically scales as the gradient of the norm of the electric field squared, as we have derived in Section 14.3. We here repeat this result for convenience

$$(18.2.1) \qquad \mathbf{F} = 2\pi\varepsilon_0\varepsilon_f a^3 \frac{\varepsilon_p - \varepsilon_f}{\varepsilon_p + 2\varepsilon_f} \nabla\left(\mathbf{E}^2\right) = 2\pi\varepsilon_0\varepsilon_f a^3 K\nabla\left(\mathbf{E}^2\right)$$

This force is typically weaker than the Coulombic force (force on charged particles), but it has a number of advantages. Namely, it is (a) insensitive to the field polarity, (b) can be used to separate particles based on their polarizability relative to the fluid, as a function of field frequency. Its main weakness is that it is proportional to the gradient of the norm of the electric field squared, which is very difficult to create a uniform field of, and therefore it is very difficult to create a uniform force field for separations. However, the insensitivity to field polarity allows separation devices to have AC fields and thus avoid dreaded electrode reactions (and the associated pH changes, gas evolution, and metal ion generation) that electrophoretic (i.e. Coulombic) separation devices have to endure and work around. The ability to separate particles based on their polarizability as a function of field frequency allows the separations to be fairly flexible and separate a wide range of particles. The steady state dielectrophoretic velocity which arises when dielectrophoretic force is completely opposed by the Stokes' drag ($\mathbf{F} = 6\pi\mu\mathbf{U}a$, equation (1.3.12)) is

$$(18.2.2) \qquad u_{DEP} = \frac{\varepsilon_0\varepsilon_f}{3\mu}a^2 K\nabla\left(\mathbf{E}^2\right)$$

This velocity, u_{DEP} scales as the square of the particle radius, proportional to the polarization coefficient K, proportional to the fluid relative permittivity ε_f, and inversely as the fluid viscosity, μ (as expected). The fact that velocity scales as the square of the particle radius sometimes makes it difficult to separate particles with widely differing sizes based on their polarization coefficient, even when the polarization coefficient is of the same order. However, this dependence makes it easy to separate particles by size. Nevertheless, separations based on polarization coefficient and the ability to tune these separations by altering the surrounding fluid properties and electric field is the greatest convenience of dielectrophoresis based separations.

In Section 14.3 we have already derived dielectrophoretic force for the most basic case: a lossless dielectric sphere in a lossless dielectric medium in a weakly spatially varying electric field. In the next sections we will dive deeper into dielectrophoresis and explore more general cases of more generally varying field, real particles (with conduction), and shelled particles (an example of which are biological cells). Here we will first consider dielectrophoresis of various particles (with increasing level of "realism" or complexity). Secondly, we will consider the effect of the media on dielectrophoresis. Thirdly we will investigate electrode and insulator geometries for producing electric fields useful for dielectrophoresis.

18.2.1. Lossless dielectric sphere and medium in axisymmetric DC field. To construct a solution for a lossless dielectric sphere in a lossless dielectric medium in an axisymmetric but non-uniform field, we being with the simplest non-uniform field there is, that of a point charge. Imagine a dielectric sphere positioned in the center of a coordinate system as shown in Figure 18.2.1. The sphere has a radius a and relative permittivity ε_p and is surrounded by a fluid with relative permittivity ε_f in a field of a point charge q located a distance ξ from the origin.

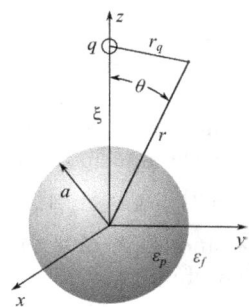

FIGURE 18.2.1. Dielectric sphere radius a and relative permittivity ε_p centered at the origin of the coordinate system surrounded by a fluid with permittivity ε_f in a field of a point charge q located a distance ξ from the origin.

The general solution for the axisymmetric electrostatic potential at an arbitrary point (distance r_q from charge q and distance r from the origin in Figure 14.3.1) is

$$(18.2.3) \qquad \varphi_{out} = \frac{q}{4\pi\varepsilon_0\varepsilon_f r_q} + \sum_{n=0}^{\infty} \frac{A_n P_n\left(\cos\theta\right)}{r^{n+1}}$$

$$(18.2.4) \qquad \varphi_{in} = \sum_{n=0}^{\infty} B_n r^n P_n\left(\cos\theta\right)$$

for the potential outside the sphere (subscript out, $r \geqslant a$) and inside the sphere (subscript in, $r < a$); here $P_n(x)$ are the Legendre polynomials (see (14.3.13) for the first few); A_n and B_n are coefficients to be found from the boundary conditions. The first term on the right-hand side of (18.2.3) is due to the imposed field due to the point charge and the second term is due to induced (linear) multipolar terms. To simplify (18.2.3), we expand the $1/r_q$ using Legendre polynomials as well:

$$(18.2.5) \qquad \frac{1}{r_q} = \frac{1}{\xi}\sum_{n=0}^{\infty}\left(\frac{r}{\xi}\right)^n P_n\left(\cos\theta\right) \text{ for } a \leqslant r \leqslant \xi$$

The boundary conditions for this problem are as in the previous problem, namely that at the surface of the particle, $r = a$, the electrostatic potential must be continuous, thus

$$(18.2.6) \qquad \varphi_{out}\left(r = a, \theta\right) = \varphi_{in}\left(r = a, \theta\right)$$

Secondly, the normal component of the displacement flux must be continuous across the interface, due to Gauss's law:

$$(18.2.7) \qquad -\varepsilon_0\varepsilon_p\left.\frac{\partial\varphi_{in}}{\partial r}\right|_{r=a,\theta} = -\varepsilon_0\varepsilon_f\left.\frac{\partial\varphi_{out}}{\partial r}\right|_{r=a,\theta}$$

Thus applying these boundary conditions to (18.2.3) and (18.2.4) and inserting (18.2.4) into (18.2.3), we obtain that

$$(18.2.8) \qquad A_n = \frac{-q}{4\pi\varepsilon_0\varepsilon_f}\frac{n\left(\varepsilon_p - \varepsilon_f\right)}{n\varepsilon_p + \left(n + 1\right)\varepsilon_f}\frac{a^{2n+1}}{\xi^{n+1}}$$

$$(18.2.9) \qquad B_n = \frac{q}{4\pi\varepsilon_0} \frac{2n+1}{n\varepsilon_p + (n+1)\,\varepsilon_f} \frac{1}{\xi^{n+1}}$$

Inserting (18.2.8) into (18.2.3) and comparing the induced (second right-hand) term to the expression for the general induced potential due to a multipole, (14.3.15) we see that the general expression for the multipolar moment for this situation is

$$(18.2.10) \qquad p^{(n)} = 4\pi\varepsilon_0\varepsilon_f A_n$$

To make our result more general and applicable to any axisymmetric field and not only that of a point particle, it is useful to rewrite the various moment expressions in terms of the electric field due to the point charge and its axial derivatives measured at the center of the dielectric sphere. Using Coulomb's law, we obtain that the field due to a point charge at the center of the dielectric sphere is

$$(18.2.11) \qquad \mathbf{E}\,(z=0) = -\left(\frac{q}{4\pi\varepsilon_0\varepsilon_f\xi^2}\right)\hat{\mathbf{z}}$$

and the partial derivatives with respect to z are

$$(18.2.12) \qquad \frac{\partial^n \mathbf{E}}{\partial z^n} = -\left(\frac{(n+1)!q}{4\pi\varepsilon_0\varepsilon_f\xi^{n+2}}\right)\hat{\mathbf{z}}$$

Thus, combining equations (18.2.8), (18.2.11), (18.2.12), and (18.2.10) we obtain that the linear multipolar moment in term of the electric field and its derivatives on the central axis is

$$(18.2.13) \qquad p^{(n)} = \frac{4\pi\varepsilon_0\varepsilon_f K^{(n)} a^{2n+1}}{(n-1)!} \frac{\partial^{n-1} E_z}{\partial z^{n-1}}$$

where

$$(18.2.14) \qquad K^{(n)} = \frac{\varepsilon_p - \varepsilon_f}{n\varepsilon_p + (n+1)\,\varepsilon_f}$$

for $n = 1, 2, 3...$ is the generalized Clausius-Mossotti function (polarization coefficient) for a dielectric sphere in a dielectric medium. Notice for $n = 1$ it reduces to the standard Clausius-Mossotti function of (14.3.26). Notice that the higher order moments are induced by the derivatives in the axial electric field. For example, the linear quadrupole is induced by the first derivative, the linear octupole induced by the second derivative, and so on. While (18.2.13) was derived for a field of a point charge, it is applicable for any axial axisymmetric electric field. This can be proven by invoking the superposition principle and constructing the desired axisymmetric electric fields from a distribution of point charges. Thus the equation for force as a function of dipole moment, (14.3.4), the force on a lossless dielectric sphere in a lossless dielectric medium in an axisymmetric but nonuniform field is

$$(18.2.15)$$

$$\mathbf{F} = \mathbf{p}\cdot\nabla\mathbf{E} = 4\pi\varepsilon_0\varepsilon_f\left(\frac{K^{(1)}a^3}{0!}E_z + \frac{K^{(2)}a^5}{1!}\frac{\partial^1 E_z}{\partial z^1} + \frac{K^{(3)}a^7}{2!}\frac{\partial^2 E_z}{\partial z^2}...\right)\hat{\mathbf{z}}\cdot\nabla\mathbf{E}$$

Notice, that as n increases, the polarization coefficient $K^{(n)}$ decreases. However, if we create an electric field with large higher derivatives (compared to lower derivatives), we can negate the decrease in $K^{(n)}$ as n increases, and create a force that varies very strongly with the particle radius, and thus achieve separation based on particle size.

We can also apply this equation to the two extreme cases: a perfectly conducting sphere in a dielectric medium and a dielectric sphere in a perfectly conducting

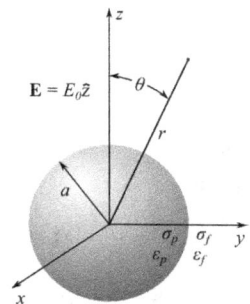

FIGURE 18.2.2. Real sphere radius a and relative permittivity ε_p, conductivity σ_p in a medium with relative permittivity ε_f and conductivity σ_f with an imposed uniform electric field, E_0 along the z-axis.

medium. The first can be a model for a metal sphere in water or dielectric oil, and the second can be a model for, for example, a silica or polystyrene sphere in mercury. Note that inside a perfect conductor the electric field is zero (it is a perfect conductor, it needs no finite field to drive charge); to have zero electric field, it needs to cancel the imposed field, and so have infinite polarization and so infinite permittivity. Of course we will not encounter perfect conductors, however a metal particle is such a better conductor then the surrounding water, it may be modeled as such. Hence, for a perfectly conducting sphere in a dielectric medium $\varepsilon_p/\varepsilon_f \to \infty$ and $K^{(n)} = 1/n$. For a dielectric sphere in a perfectly conducting medium, $\varepsilon_p/\varepsilon_f \to 0$ and $K^{(n)} = 1/(n+1)$.

18.2.2. Real dielectric particle and medium in uniform AC field. Having gotten a feel for forces on an ideal dielectric particle in a dielectric fluid, we now move on to real particles. Real particles posses both finite conduction and polarization phenomena, and the associated energy dissipation (loss) mechanisms. When such are present the induced dipole moment of a particle exhibits a time delay when an electric field is suddenly applied or a phase lag when the electric field is sinusoidal in time.

Consider a sphere shown in Figure 18.2.2 which has radius a and relative permittivity ε_p, conductivity σ_p and is immersed in a medium with relative permittivity ε_f and conductivity σ_f. The applied electric field is uniform with a magnitude of E_0 but unlike previously, the field is AC with a radian frequency ω. We can write the electric field thus as

(18.2.16) $$\mathbf{E}(t) = \mathrm{Re}\left[E_0 \exp\left(j\omega t\right)\right] \hat{\mathbf{z}}$$

The solution to the electric field is similar to that previously obtained,

(18.2.17) $$\varphi_{out}(r,\theta) = -E_0 r \cos\theta + \underline{A}\frac{\cos\theta}{r^2}$$

(18.2.18) $$\varphi_{in}(r,\theta) = -\underline{B}r\cos\theta$$

with the unknown coefficients to be obtained from the boundary conditions, $\underline{A}$ and $\underline{B}$ being complex. We will use the underline to designate complex variables. The boundary conditions for this problem are: firstly at the surface of the particle, $r = a$, the electrostatic potential must be continuous, thus

(18.2.19) $\varphi_{out}\left(r = a, \theta\right) = \varphi_{in}\left(r = a, \theta\right)$

and, secondly, there is current continuity at the boundary, including the possibility for accumulating free surface charge:

(18.2.20) $J_{r,f} - J_{r,p} + \dfrac{\partial \sigma_e}{\partial t} = 0$ at $r = a$

where J is the ohmic current, the subscript r designates the radial component, and the second subscript designates the material. Here σ_e is the electric charge density, not to be confused with the conductivity of one of the materials. The ohmic current is given by

(18.2.21)

$$J_{r,f} = \sigma_f \left.\dfrac{\partial \varphi_{out}}{\partial r}\right|_{r=a,\theta}$$

$$J_{r,p} = \sigma_p \left.\dfrac{\partial \varphi_{in}}{\partial r}\right|_{r=a,\theta}$$

while the electric free charge density, using Gauss's law, is given by

(18.2.22) $\sigma_e = \varepsilon_0 \left(\varepsilon_f \dfrac{\partial \varphi_{out}}{\partial r} - \varepsilon_p \dfrac{\partial \varphi_{in}}{\partial r}\right)$ at $r = a$

In this analysis, we will employ electric fields with exponential time dependence, $\exp\left(j\omega t\right)$ and thus assume that all variables have the same time dependence. Since the time derivative of $\exp\left(j\omega t\right)$ is $j\omega \exp\left(j\omega t\right)$ we will exploit the shortcut, substituting $j\omega$ for $\partial / \partial t$. Performing this and substituting in (18.2.22) and (18.2.21) into (18.2.20) we simplify (18.2.20) to

(18.2.23)

$$\sigma_f \left.\dfrac{\partial \varphi_{out}}{\partial r}\right|_{r=a,\theta} - \sigma_p \left.\dfrac{\partial \varphi_{in}}{\partial r}\right|_{r=a,\theta} + j\omega\varepsilon_0 \left(\varepsilon_f \left.\dfrac{\partial \varphi_{out}}{\partial r}\right|_{r=a,\theta} - \varepsilon_p \left.\dfrac{\partial \varphi_{in}}{\partial r}\right|_{r=a,\theta}\right) = 0$$

$$\left(\sigma_f + j\omega\varepsilon_0\varepsilon_f\right) \left.\dfrac{\partial \varphi_{out}}{\partial r}\right|_{r=a,\theta} - \left(\sigma_p + j\omega\varepsilon_0\varepsilon_p\right) \left.\dfrac{\partial \varphi_{in}}{\partial r}\right|_{r=a,\theta} = 0$$

and dividing the lower equation (18.2.23) by $j\omega$ we can define a complex dielectric permittivity as

(18.2.24) $\underline{\varepsilon} \equiv \varepsilon_0\varepsilon + \sigma/j\omega$

and write (18.2.23) as

(18.2.25) $\underline{\varepsilon}_f \left.\dfrac{\partial \varphi_{out}}{\partial r}\right|_{r=a,\theta} = \underline{\varepsilon}_p \left.\dfrac{\partial \varphi_{in}}{\partial r}\right|_{r=a,\theta}$

Compare this result with (14.3.22) for the analogous DC case. By inspection, the complex coefficients for the boundary conditions are

(18.2.26) $\underline{A} = \dfrac{\underline{\varepsilon}_p - \underline{\varepsilon}_f}{\underline{\varepsilon}_p + 2\underline{\varepsilon}_f} a^3 E_0$

(18.2.27) $\underline{B} = \dfrac{3\underline{\varepsilon}_f}{\underline{\varepsilon}_p + 2\underline{\varepsilon}_f} E_0$

Comparing the second term of (18.2.17), that is due to the induced dipole to the potential from a dipole (14.3.17), we again see that

$$\text{(18.2.28)} \qquad \mathbf{\underline{p}} = 4\pi\varepsilon_0\varepsilon_f\underline{K}a^3\mathbf{E_0}$$

where the dipole moment is complex, and so is the polarization coefficient (Clausius-Mossotti factor):

$$\text{(18.2.29)} \qquad \underline{K} = \frac{\underline{\varepsilon}_p - \underline{\varepsilon}_f}{\underline{\varepsilon}_p + 2\underline{\varepsilon}_f}$$

Since the polarization coefficient is complex, its magnitude and phase are a function of the frequency of the field. The phase angle here represents a lag between the applied electric field and the induced moment. This lag is due to the fact that it takes a finite time to build up surface charge, σ_e on the surface of the particle. However, note that the ε_f in (18.2.28) is not complex, as it comes from (14.3.17)! Furthermore, both the permittivities and conductivities can be functions of frequency, but not of time, at least on the time scale of the oscillation of the field.

Using the complex notation that we have just introduced, including the complex form of the polarization coefficient (Clausius-Mossotti factor), we can rewrite the force equation (18.2.1) as

$$\text{(18.2.30)} \qquad \mathbf{F} = 2\pi\varepsilon_0\varepsilon_f a^3\underline{K}\nabla\left(\mathbf{E}^2\right)$$

From this equation we see that the force oscillates on the same time scale as the electric field, namely with the time scale of $1/\omega$. Typically this time scale is very short, and practically we are not interested in changes to the force on this time scale. Hence, it is useful for us to take the time average of (18.2.30) over a time slightly longer than this time scale. Performing this averaging we obtain,

$$\text{(18.2.31)} \qquad \langle\mathbf{F}\rangle = 2\pi\varepsilon_0\varepsilon_f a^3\,\text{Re}\left[\underline{K}\right]\nabla\left(E_{RMS}^2\right)\frac{\nabla\left(\mathbf{E}^2\right)}{|\nabla\left(\mathbf{E}^2\right)|}$$

where we designate the time averaging with $\langle\rangle$. Notice that in obtaining this force equation we did not rely on the particular form of $\underline{K}$. Hence we can use this force equation with $\underline{K}$ from (18.2.29) as well as with other Clausius-Mossotti factors we obtain later.

18.2.3. Real dielectric particle and medium in uniform transient field.
Having analyzed the problem of a real particle in a real medium in an oscillating field, we turn our attention to a similar problem, but instead of an oscillating field, a field that is suddenly turned on. While such a field is rarely of practical importance we analyze this problem here to gain insights of what occurs to the particle as the field is changing on a time scale less than the time scale of a full cycle of an oscillation. Thus, let's again consider the particle in Figure 18.2.2, however now the particle is subjected to a field such that at $t = 0_+$ a uniform electric field is suddenly turned on:

$$\text{(18.2.32)} \qquad \mathbf{E}\left(t\right) = E_0u\left(t\right)\hat{\mathbf{z}}$$

where $u\left(t\right)$ is a step function. The solutions to the electroquasistatic potential both inside and outside the sphere are the same as before for this geometry, however now the coefficients responsible for the boundary conditions, A and B are functions of time:

(18.2.33)
$$\varphi_{out}(r,\theta) = -E_0 r \cos\theta + A(t) \frac{\cos\theta}{r^2}$$

(18.2.34)
$$\varphi_{in}(r,\theta) = -B(t) r \cos\theta$$

At the surface of the particle, $r = a$, the boundary conditions are, firstly, that the electrostatic potential must be continuous,

(18.2.35)
$$\varphi_{out}(r=a,\theta) = \varphi_{in}(r=a,\theta)$$

and, secondly, there is current continuity across the boundary, including the possibility for accumulating free surface charge:

(18.2.36)
$$\sigma_f \frac{\partial \varphi_{out}}{\partial r}\bigg|_{r=a,\theta} - \sigma_p \frac{\partial \varphi_{in}}{\partial r}\bigg|_{r=a,\theta} + \frac{\partial \sigma_e}{\partial t} = 0$$

where equations (18.2.20) and (18.2.21) have already been combined to give the above boundary condition. Inserting (18.2.33), (18.2.34) into the boundary condition equations and performing the appropriate differentiation, we obtain the following set of coupled ordinary differential equations:

(18.2.37)
$$a^{-3} A(t) + B(t) = E_0 u(t)$$

(18.2.38)
$$2a^{-3} A(t) \cos\theta - \frac{\sigma_p}{\sigma_f} B(t) \cos\theta + \frac{1}{\sigma_f} \frac{\partial \sigma_e}{\partial t} = -E_0 u(t) \cos\theta$$

(18.2.39)
$$2a^{-3} A(t) \cos\theta - \frac{\varepsilon_p}{\varepsilon_f} B(t) \cos\theta + \frac{\sigma_e}{\sigma_f} = -E_0 u(t) \cos\theta$$

As previously, we are interested in finding $A(t)$, since as before, $p = 4\pi\varepsilon_0\varepsilon_f A(t)$. Following Jones (1979) we find $A(t)$ and find that
(18.2.40)
$$p = 4\pi\varepsilon_0\varepsilon_f a^3 E_0 \left(\left(\frac{\sigma_p - \sigma_f}{\sigma_p + 2\sigma_f} \right) (1 - \exp[-t/\tau_{MW}]) + \left(\frac{\varepsilon_p - \varepsilon_f}{\varepsilon_p + 2\varepsilon_f} \right) \exp[-t/\tau_{MW}] \right)$$

where

(18.2.41)
$$\tau_{MW} = \varepsilon_0 \frac{\varepsilon_p + 2\varepsilon_f}{\sigma_p + 2\sigma_f}$$

is the Maxwell-Wagner time constant associated with the so called "relaxation time" or the time to accumulate free charge σ_e at the surface of the boundary between the sphere material and the fluid. Let's carefully observe the exponential terms in (18.2.40). We see that as the exponential term multiplying the permittivities term decays as time becomes large (relative to the time constant) the exponential term multiplying the conductivities goes to unity; for short times (time near zero relative to the time constant) the opposite happens: the permittivity term has a large effect on the dipole and the conductivity term has little effect. Thus at short times the particle and fluid combination behave as perfect dielectrics, while at large times both behave as perfect conductors. Thus we can write the approximation

(18.2.42)
$$p(t) = 4\pi\varepsilon_0\varepsilon_f a^3 E_0 \begin{cases} \left(\dfrac{\varepsilon_p - \varepsilon_f}{\varepsilon_p + 2\varepsilon_f} \right) & t \ll \tau_{MW} \\[3mm] \left(\dfrac{\sigma_p - \sigma_f}{\sigma_p + 2\sigma_f} \right) & t \gg \tau_{MW} \end{cases}$$

This agrees with our intuition. On a short time scale ($t \ll \tau_{MW}$) free charge has not yet accumulated and so the sphere appears as if it is an insulating dielectric sphere - as if conduction current does not exist (recall discussion of boundary conditions in the previous section). On long time scales ($t \gg \tau_{MW}$) surface charge had plenty of time to accumulate and now the dipole is governed by DC conduction. What is considered short time or long time is up to the ratio of the combination of permittivities of both the particle and the media to the conductivities of both the particle and the media. This makes sense as if the conductivity is small compared to permittivity, the time constant is large and almost for any reasonable time appears as the short time scale, and the situation appears as a dielectric particle in a dielectric media. Thus the Maxwell-Wagner time scale is the time scale where the particle behavior goes from that of a dielectric to that of a conductor.

Observing the equation (18.2.42) more closely we see that there could be a situation that, for example, the conductivity of the particle is larger than the fluid, while the permittivity of the particle is smaller than that of the fluid. Hence in this situation, as we cross the Maxwell-Wagner time, going from short times to long times, the dipole movement is first negative and then is positive. In other words, the particles change direction of how they move with respect to the field! Now imagine, we have several such particles, but all with slightly different Maxwell-Wagner time constants. Each will "decide" to change its direction relative to the field at a different time!

We can use this to build a continuous particle separator which we sketch in Figure 18.2.3 for situations where the Maxwell-Wagner time constant is large (preferably seconds). The particles (exaggerated size for clarity) are pumped at a steady flow rate through a channel with the structure shown in Figure 18.2.3, in which a DC electric field is applied. In the frame of reference of the particle, for fast enough flow velocity and high enough Maxwell-Wagner time constants it appears that when the particle enters part of the apparatus with electric field, it is as if the electric field is suddenly turned on for the particle. (Note, the time for the jump in electric field to make this jump appear as a step function just needs to be much smaller than the Maxwell Wagner time constants). For example, the particle properties are such that the particles to be separated first experience a force towards high gradient of electric field at short times and away from the high gradient regions at larger times. Since our set of particles to be separated has a different Maxwell-Wagner time constant, they will "decide" to move to the low gradient region at different times and so spend different amount of time moving towards that region. Hence, as shown in Figure 18.2.3, they will be separated into bands. For this separation technique to be practical the Maxwell-Wagner time constant has to be fairly large (order seconds preferably). This can be achieved when the fluid is fairly insulating, e.g., silicone transformer oil, $\sigma_f = 10^{-12} S/m$, $\varepsilon_f = 2.7$, and the particle is fairly insulating as well, e.g., polystyrene $\sigma_p = 10^{-12} S/m$, $\varepsilon_p = 2.6$, for which $\tau_{MW} = 24$ s.

Let's also consider the case where the charge relaxation time of the suspending medium and the particle are equal. Charge relaxation time is defined as

(18.2.43)
$$\tau_R = \varepsilon_0 \varepsilon / \sigma$$

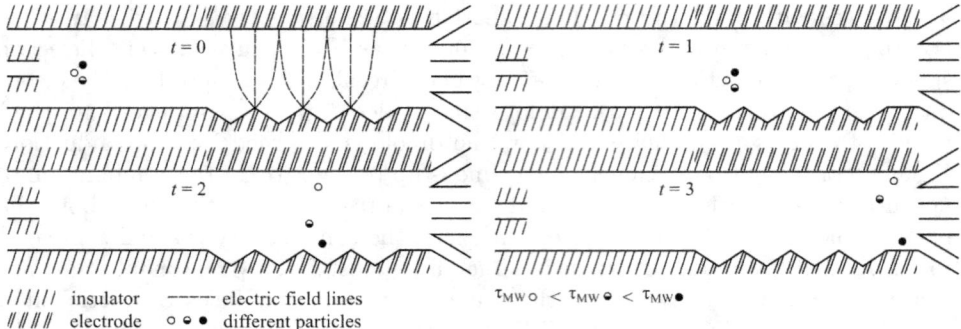

$\tau_{MW\circ} < \tau_{MW\ominus} < \tau_{MW\bullet}$

////// insulator   ----- electric field lines
//// electrode   ∘ ⊖ ● different particles

FIGURE 18.2.3. Sketch of a transient field Maxwell-Wagner time constant separator. The first panel shows the device at an initial time ($t = 0$) where the particles entered from the channel to the left and are traveling in the field free region. We only sketch representative electric field lines in the device in the first panel for clarity. Due to the shape of the electrodes, there is high electric field gradient at the bottom electrode, and low electric field gradient at the top electrode. Initially all the particles are attracted to the high field gradient region. As charge accumulates on their surface however, they begin to act as conductors and are now repelled more from the high field gradient region. As drawn, the accumulation of charge occurs fastest on the non-shaded particle, followed by the half shaded particle and lastly the fully shaded particle. Thus the non-shaded particle spends more time closer to the top wall and exists out the corresponding channel; this is followed by the half shaded particle (exits out of middle output channel) and then finally the fully shaded particle, which does not have much time to travel upward and so exists out of the lowest channel. While only 3 particles are drawn, this device can be operated in a fully continuous mode and of course with a larger set of particles.

and is analogous to the RC time constant for a parallel plate capacitor with a leaky (finite conductivity) dielectric insert. For this, case

$$(18.2.44) \qquad \varepsilon_p/\sigma_p = \varepsilon_f/\sigma_f$$

and the Maxwell-Wagner time constant becomes

$$(18.2.45) \qquad \tau_{MW} = \varepsilon_0 \frac{\varepsilon_p}{\sigma_p}$$

As the relaxation times are matched, it can be shown that no charge accumulates at the interface between the particle and the fluid.

18.2.4. Crossover frequency for a real particle and fluid. Before we move onto shelled particles, let's examine in detail what happens to a real dielectric particle as the frequency of the AC field is varied using the tools we obtained exploring particle response to a step change in electric field in the previous section. Firstly, let's expand and rearrange equation (18.2.29) for the polarization coefficient as

(18.2.46)
$$K(\omega) = \left(\frac{\sigma_p - \sigma_f}{\sigma_p - 2\sigma_f} \right) \left[\frac{1 + j\omega (\tau_0 - \tau_{MW}) + \omega^2 \tau_0 \tau_{MW}}{1 + \omega^2 \tau_{MW}^2} \right]$$

where

(18.2.47)
$$\tau_{MW} = \varepsilon_0 \frac{\varepsilon_p + 2\varepsilon_f}{\sigma_p + 2\sigma_f}$$

is the familiar Maxwell-Wagner time scale, and we define another time scale

(18.2.48)
$$\tau_0 = \varepsilon_0 \frac{\varepsilon_p - \varepsilon_f}{\sigma_p - \sigma_f}$$

Since we are concerned with the real portion of the force, we are interested in the real portion of the polarization coefficient

(18.2.49)
$$\text{Re} \left[K(\omega) \right] = \left(\frac{\sigma_p - \sigma_f}{\sigma_p - 2\sigma_f} \right) \left[\frac{1 - \omega^2 \tau_0 \tau_{MW}}{1 - \omega^2 \tau_{MW}^2} \right]$$

We are most interested in the frequency at which the real part of the polarization coefficient changes sign as this is this the point where the force changes direction from being towards high gradient of electric field region to being away from the high gradient region. This frequency is termed as the crossover frequency, ω_{xo}. Polarization coefficient is zero when the conductivity of the particle and the fluid are the same for any frequency. Polarization coefficient is also zero when

(18.2.50)
$$1 + \omega^2 \tau_0 \tau_{MW} = 0$$

or

(18.2.51)
$$\omega_{xo} = \sqrt{-1/\tau_0 \tau_{MW}}$$

(18.2.52)
$$\omega_{xo} = \frac{1}{\varepsilon_0} \sqrt{-\frac{\sigma_p + 2\sigma_f}{\varepsilon_p + 2\varepsilon_f} \left(\frac{\sigma_p - \sigma_f}{\varepsilon_p - \varepsilon_f} \right)}$$

Note that the time scale τ_0 can of course be negative, and in fact has to be for a crossover to occur. Thus, for the particle - fluid system to have a crossover, either $\varepsilon_p < \varepsilon_f$ or $\sigma_p < \sigma_f$ but not both simultaneously. Also notice that the conductivity term in parentheses in (18.2.46) controls the direction of the crossover. If that term is positive, then polarization coefficient crosses from being positive at low frequencies to being negative at high frequencies. If it is negative, then the polarization coefficient crosses from being negative at low frequencies to being positive at high frequencies.

Since the crossover behavior, i.e., the change of direction of particle travel as a function of frequency, is easy to observe, crossover frequency is easy to measure experimentally. Measuring crossover frequency while varying solution conductivity, permittivity, or both allows us to calculate both particle conductivity and permittivity. Since there are two unknowns (permittivity and conductivity) we need at least two different solution conditions to construct two versions of (18.2.52) to find these. It is preferable to use more conditions and use least squares fit and fit the data to (18.2.52) to find the unknown particle permittivity and conductivity.

18.2.5. Real dielectric shell in nearly uniform AC field. We now turn our attention to dielectrophoresis of real multilayer spherical particles. Such particles are very widely encountered - examples include biological cells (both animal and plant) as well as many coated particles. We will approach this problem in a similar manner as we have done in the previous sections. Consider a spherical shell, outer radius a_1 and permittivity and conductivity ε_{p1} and σ_{p1} respectively. Inside the shell is a spherical core with radius a_2 and permittivity and conductivity ε_{p2} and σ_{p2} respectively. Imposed on this particle is an external electric field $\mathbf{E}(t) = \text{Re}\,[E_0 \exp(j\omega t)]\,\hat{\mathbf{z}}$. Thus the situation is similar to that in Figure 18.2.2, with the exception that the particle is a core-in-shell particle. For this situation, the governing equation, Laplace's equation has the following solution for the three regions:

$$(18.2.53) \qquad \varphi_1 = \left(-E_0 r + \frac{A}{r^2}\right)\cos\theta, \; r > a_1$$

$$(18.2.54) \qquad \varphi_2 = \left(-\underline{B}r + \frac{C}{r^2}\right)\cos\theta, \; a_1 > r > a_2$$

$$(18.2.55) \qquad \varphi_3 = -\underline{D}r\cos\theta, \; r < a_2$$

where A, B, C, D are complex constants to be determined from the boundary conditions. We have two interfaces, and as before two boundary conditions at each:

$$(18.2.56) \qquad \varphi_1 = \varphi_2 \text{ at } r = a_1$$

$$(18.2.57) \qquad \varepsilon_f \frac{\partial\varphi_1}{\partial r} = \varepsilon_{p1}\frac{\partial\varphi_2}{\partial r} \text{ at } r = a_1$$

$$(18.2.58) \qquad \varphi_2 = \varphi_3 \text{ at } r = a_2$$

$$(18.2.59) \qquad \varepsilon_{p1}\frac{\partial\varphi_2}{\partial r} = \varepsilon_{p2}\frac{\partial\varphi_3}{\partial r} \text{ at } r = a_2$$

To determine the effective dipole moment we only need to know the field outside of the particle, i.e., that described by equation (18.2.53) as we compare the induced field of the sphere in our situation (second term on right in (18.2.53)) with that of an ideal dipole. Thus we need to only find the constant A. We find this as usual by inserting the solutions (18.2.53) through (18.2.59) into our boundary conditions and solving the resulting equations. We obtain

$$(18.2.60) \qquad \underline{A} = \underline{K}a_1^3 E_0$$

$$(18.2.61) \qquad \underline{B} = -\frac{3\underline{\varepsilon}_f\gamma^3}{\left(\underline{\varepsilon}_p + 2\underline{\varepsilon}_f\right)\left(\gamma^3 - \underline{K}\right)}E_0$$

$$(18.2.62) \qquad \underline{C} = -\frac{3\underline{\varepsilon}_f\underline{K}a^3}{\left(\underline{\varepsilon}_p + 2\underline{\varepsilon}_f\right)\left(\gamma^3 - \underline{K}\right)}E_0$$

$$(18.2.63) \qquad \underline{D} = -\frac{3\underline{\varepsilon}_f\left(1 - \underline{K}\right)\gamma^3}{\left(\underline{\varepsilon}_p + 2\underline{\varepsilon}_f\right)\left(\gamma^3 - \underline{K}\right)}E_0$$

where

$$(18.2.64) \qquad \gamma = a_1/a_2$$

$$(18.2.65) \qquad \underline{K} = \frac{\underline{\varepsilon}_p - \underline{\varepsilon}_f}{\underline{\varepsilon}_p + 2\underline{\varepsilon}_f}$$

$$(18.2.66) \qquad \underline{\varepsilon}_p = \underline{\varepsilon}_{p1} \left(\frac{\gamma^3 + 2\left(\frac{\underline{\varepsilon}_{p2} - \underline{\varepsilon}_{p1}}{\underline{\varepsilon}_{p2} + 2\underline{\varepsilon}_{p1}}\right)}{\gamma^3 - \left(\frac{\underline{\varepsilon}_{p2} - \underline{\varepsilon}_{p1}}{\underline{\varepsilon}_{p2} + 2\underline{\varepsilon}_{p1}}\right)} \right)$$

We have specifically written the solution in this form, where we have used an effective total particle complex permittivity (defined in (18.2.66)) so that our solution has the form of the solution for the simple (non-shelled) particle. Since we are able to do this, evidently, it appears that the electrostatic potential outside the sphere is identical to one for a solid particle with a complex permittivity $\underline{\varepsilon}_p$ given by (18.2.66). This implies that we can apply this analysis recursively, where now the core inside the particle has another core. In this case, now $\underline{\varepsilon}_{p2}$ is an effective total particle complex permittivity of that inner core-shell system, defined analogously to (18.2.66). We could of course continue this indefinitely, for as many core-shells as we need to have. Additionally, we could have obtained the same result for the effective total particle complex permittivity using the Maxwell-Garnett effective medium model.

As before, comparing the second term of (18.2.53), that is due to the induced dipole to the potential from a dipole (14.3.17), we again see that

$$(18.2.67) \qquad \mathbf{p} = 4\pi\varepsilon_0\varepsilon_f \underline{K} a_1^3 \mathbf{E_0}$$

where $\underline{K}$ is given by (18.2.65). Note that ε_f in (18.2.67) is not complex, as it comes from (14.3.17). Since now we have two interfaces, we can have two Maxwell-Wagner time scales for each charge build up at each interface. Unfortunately, simple explicit algebraic expressions for $\underline{K}(\omega)$ as well as expressions for the zeros of the real $\underline{K}(\omega)$, the crossover frequencies, become very unwieldy for shelled particles. Thus, it is convenient to find the crossover frequencies numerically, especially for particles with large number of layers. As before, we may use the measurement of crossover frequencies for different media conditions together with this model for particle dielectric behavior to measure the electrical properties of the particle layers without disturbing the particle!

18.2.6. Real dielectric particle with a thin shell in AC field. Let's also consider a special case of a particle with a single thin shell in an AC field. This is an important case for example for biological cells where the membrane thickness is ~ 10 nm while cell diameter is ~ 10 μm, as well as some industrial particles coated with a conductive layer to increase particle conductivity. There are two limits for the thin shelled particle: (1) series admittance limit, where a finite electrostatic potential is supported over the thin shell; and (2) the shunt admittance limit, where finite current flows around the layer tangential to the surface of the particle. The first limit is especially applicable to cells, many of which regularly maintain order 70 mV potential across their membranes. The second limit is typically of interest for particles with thin surface coating of conductive material.

Let's consider the series admittance limit first. Consider a shelled particle having $a_1 - a_2 = \delta$ where $\delta \ll a_1$. Thus, $a_1 \approx a$. Furthermore, the conductivity of the membrane layer, σ_{1p} is much less than the conductivity of the particle core σ_{2p}. This is quite a realistic situation for a cell, where the cytoplasm conductivity is much higher than that of a lipid membrane. For convenience, we define surface capacitance of the membrane $c_m \equiv \varepsilon_0 \varepsilon_{1p}/\delta$, which as units of F/m^2 and surface transconductance $g_m \equiv \sigma_{1p}/\delta$, which has units of S/m^2. As before, upon the particle we impose an electric field

(18.2.68)
$$\mathbf{E}(t) = \mathrm{Re}\left[E_0 \exp(j\omega t)\right] \hat{\mathbf{z}}$$

The solution to the electric field is similar to that previously obtained for a homogeneous particle,

(18.2.69)
$$\varphi_{out}(r,\theta) = -E_0 r \cos\theta + \underline{A}\frac{\cos\theta}{r^2},\ r > a$$

(18.2.70)
$$\varphi_{in}(r,\theta) = -\underline{B}r \cos\theta,\ r < a$$

However, the boundary conditions are slightly different. They consist of current continuity across the interface:

(18.2.71)
$$\underline{\varepsilon}_f \frac{\partial \varphi_{out}}{\partial r} = \underline{\varepsilon}_{p2} \frac{\partial \varphi_{in}}{\partial r}\ \text{at}\ r = a$$

and the finite potential drop across the interface due to the finite surface capacitance and transconductance

(18.2.72)
$$\varphi_{out} - \varphi_{in} = -\frac{j\omega\varepsilon_0\underline{\varepsilon}_f}{j\omega c_m + g_m} \frac{\partial \varphi_{out}}{\partial r}$$

Solving this set of equations we obtain

(18.2.73)
$$\underline{A} = \underline{K}a_1^3 E_0$$

(18.2.74)
$$\underline{K} = \frac{\underline{\varepsilon}_p - \underline{\varepsilon}_f}{\underline{\varepsilon}_p + 2\underline{\varepsilon}_f}$$

where

(18.2.75)
$$\underline{\varepsilon}_p = \frac{\underline{c}_m a \underline{\varepsilon}_{p2}}{\underline{c}_m a + \underline{\varepsilon}_{p2}}$$

and where we have defined a complex membrane capacitance in a manner analogous to the complex permittivity

(18.2.76)
$$\underline{c}_m = c_m + g_m/j\omega$$

This way, the particle complex permittivity, $\underline{\varepsilon}_p$ in (18.2.75) is identical to the admittance of two series connected circuit elements $\underline{\varepsilon}_{p2}$ and $\underline{c}_m a$ for a standard circuit modeling a thin membrane particle (Figure 18.2.4a). This circuit models the membrane as a parallel resistor and capacitor (together having admittance $\underline{c}_m a$) and models the core (e.g., cytoplasm) of the particle also as a resistor and capacitor (together having admittance $\underline{\varepsilon}_{p2}$). Unfortunately, once again, the crossover frequency relation is difficult to write explicitly and so the crossover frequency should be obtained numerically.

Next, let's consider the shunt admittance limit. Here the surface layer on a particle supports the flow of ohmic or displacement current tangent to the surface, and so some (or most) current incident normal to the particle will flow (be "shunted") through this conductive layer around the particle. An example of such particle might be hollow gold shells, or gold coated silica particles. Again, consider a shelled particle having $a_1 - a_2 = \delta$ where $\delta \ll a_1$. Thus, $a_1 \approx a$. As before, upon the particle we impose an electric field

(18.2.77)
$$\mathbf{E}(t) = \mathrm{Re}\left[E_0 \exp(j\omega t)\right] \hat{\mathbf{z}}$$

and obtain the solution to the electric field that is similar to that previously obtained,

(18.2.78)
$$\varphi_{out}(r,\theta) = -E_0 r \cos\theta + \underline{A}\frac{\cos\theta}{r^2}, \; r > a$$

(18.2.79)
$$\varphi_{in}(r,\theta) = -\underline{B}r\cos\theta, \; r < a$$

However, once again, the boundary conditions are slightly different. Firstly, now the membrane does not support any potential drop, and so the potential is continuous across the membrane:

(18.2.80)
$$\varphi_{out} = \varphi_{in} \text{ at } r = a$$

Secondly, there is conservation of charge, that is current in and out must balance charge accumulation at interfaces. Because now current has a strong tangential direction around the particle, the relation appears more complicated to account for flow of current in this direction:

(18.2.81)
$$j\omega\left(-\underline{\varepsilon}_f\frac{\partial\varphi_{out}}{\partial r} + \underline{\varepsilon}_{p2}\frac{\partial\varphi_{in}}{\partial r}\right)\hat{\mathbf{r}} + \frac{1}{a\sin\theta}\frac{\partial(\sin\theta\underline{K}_\theta)}{\partial\theta}\hat{\theta} + \frac{1}{a\sin\theta}\frac{\partial(\underline{K}_\phi)}{\partial\theta}\hat{\phi} = 0$$

where K_θ and K_ϕ are the zenithal and azimuthal components of the surface current density. While the relation (18.2.81) looks complicated, all it states is that the total current is split into three components: one that goes into and through the particle (first term), one that goes around the particle in the zenithal direction, and one that goes around the particle in the azimuthal direction (second and third terms); there is no charge accumulation. The total current is both an ohmic current and a displacement current - hence the relation is complex. We now assume that the surface current is proportional to the tangential electric field and so

(18.2.82)
$$\underline{K}_\theta = -j\omega\underline{\varepsilon}_\Sigma\frac{1}{r}\frac{\partial\varphi_{out}}{\partial\theta}\bigg|_{r=c}\hat{\theta}$$
$$\underline{K}_\phi = 0$$

where we define the complex surface permittivity as usual $\underline{\varepsilon}_\Sigma \equiv \varepsilon_0\varepsilon_\Sigma + \sigma_\Sigma$. Here ε_Σ (Farads) is the surface permittivity, which accounts for any electric field induced out-of-phase motion of the ionic charge layer (the Stern layer, see Section (15.3)). Here σ_Σ (Simens) is the surface conductivity of the membrane surrounding the particle as well as the double layer. We again solve for $\underline{A}$, compare the resulting potential to the potential of an ideal dipole, and then obtain the dipole moment. As before, the dipole moment is the same as in the previous problem, and the polarization coefficient is

(18.2.83)
$$\underline{K} = \frac{\underline{\varepsilon}_p - \underline{\varepsilon}_f}{\underline{\varepsilon}_p + 2\underline{\varepsilon}_f}$$

where now

(18.2.84)
$$\underline{\varepsilon}_p = \underline{\varepsilon}_{p2} + 2\underline{\varepsilon}_\Sigma/a$$

As in the previous problem, we can model the situation with an equivalent circuit of a capacitor in parallel with a resistor (together $\underline{\varepsilon}_p$, representing the core of the particle) in parallel with another parallel capacitor resistor pair (these together $2\underline{\varepsilon}_\Sigma/a$, representing the shunting layer), (Figure 15.3.16b). Thus, (18.2.84) gives the total admittance of this circuit. We can simplify (18.2.84) further, for the case of

(a) Series admittance model   (b) Shunt admittance model

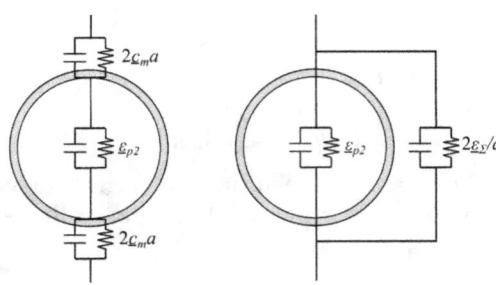

FIGURE 18.2.4. Models for real dielectric particles with a single thin shell: (a) series admittance model, where a finite electrostatic potential is supported over the thin shell and (b) shunt admittance limit, where finite current flows around the layer tangential to the surface of the particle.

a particle with a highly conductive coating ("membrane layer"), where $\varepsilon_{p2} \gg \varepsilon_\Sigma/a$. Then (18.2.84) simplifies to

$$(18.2.85) \qquad \underline{\varepsilon}_p = \varepsilon_0 \varepsilon_{p2} + \frac{\sigma_{p2} + 2\sigma_\Sigma/a}{j\omega}$$

18.2.7. Altering aqueous solution permittivity. In the previous sections we have seen how the solution and particle properties influence the polarization coefficient which in turn influences the direction of particle travel (up or down electric field gradient) at a given field frequency. In this section we will consider how we can control this. For a majority of the particles we are interested in separating, we cannot change their conductivity or permittivity without changing the particles themselves (although exceptions to this do exist). Largely then what is left is to alter solution conductivity and permittivity. Altering solution conductivity is fairly easy. Solution conductivity scales as the product of the concentration of ions in the solution and their electrophoretic mobility:

$$(18.2.86) \qquad \sigma = F \sum_i c_i \mu_i$$

where F is Faraday's constant, c_i is the concentration of ion i and μ_i is its mobility. Electrophoretic mobility has almost no concentration dependence in dilute solutions and has some concentration dependence for concentrated solutions. Typically solution conductivity at zero frequency is expressed by Debye-Huckel-Onsager relation

$$(18.2.87) \qquad \frac{\sigma}{c} = \Lambda^0 - \left(A + B\Lambda^0\right)\sqrt{c}$$

where Λ^0 is the limiting molar conductivity, i.e., conductivity of the solution per mole of solution extrapolated to infinite dilution; A and B are constants dependent on temperature. Λ^0 and A and B are tabulated and can be looked up in standard references such as CRC handbook of chemistry and physics. Temperature has a strong effect on conductivity of solutions and for many solutions can vary up to six-fold over a range from 0 to 100°C. However, much of this is due to the exponential decrease of viscosity of solutions with temperature. The product of viscosity and limiting molar conductivity varies within less than 30% for many solutions of strongly ionized electrolytes for a temperature range from 0 to 100°C and to a first approximation can be assumed to be constant. For weak acids and bases the dissociation constant usually also has a strong temperature dependence and so must be accounted for when calculating the net mobility of the species (dissociated and non-dissociated forms together) and so the conductivity of the solution. However, electrical conductivity of ionic solutions does have some frequency dependence (Debye-Falkenhagen effect) and some dependence on the applied electric field (Wien effect). Qualitatively, both of these effects are described by the presence of a counterion atmosphere around each ion (see Section (15)). As the ion travels there is an excess of counterions behind it and a deficit of counterions in front of it. Thus there is an asymmetry in the ionic atmosphere around the ion. This ionic atmosphere retards the ion's travel and so lowers its observed velocity and so its mobility, and so the overall observed conductivity of the solution. When the field is turned off, the atmosphere rearranges to become symmetric. The time for the atmosphere to rearrange is termed the relaxation time; it is roughly the time for an ion to move the length of the ionic atmosphere, i.e. roughly the Debye length. Debye and Falkenhagen (1928) has shown this to be

$$(18.2.88) \qquad \tau_{DF} = 2 \left(\frac{\mu_+}{z_+} + \frac{\mu_-}{z_-} \right)^{-1} \frac{1}{k_B T \kappa^2}$$

where μ is the electrophoretic mobility, z is valance, and κ^{-1} is the Debye length and the subscripts denote cation and anion. Another, simpler, expression for Debye-Falkenhagen time scale is

$$(18.2.89) \qquad \tau_{DF} = \frac{\varepsilon_0 \varepsilon}{\Lambda_\infty c} \approx \frac{\varepsilon_0 \varepsilon}{\sigma(0)}$$

where Λ_∞ is the equivalent conductance at infinite dilution,

$$(18.2.90) \qquad \Lambda_\infty \equiv N_A z (\mu_+ + \mu_-)$$

for a symmetric electrolyte. N_A is Avogadro's number and, $\sigma(0)$ is the DC conductivity of the solution. This time is order 10^{-6} s, 10^{-7} s, 10^{-8} s, and 10^{-9} s for 0.1 mM, 1 mM, 10 mM, and 100 mM 1:1 symmetric electrolyte respectively. When the electric field frequency is so high that the ion moves back and forth faster than this relaxation time, the asymmetry in the ionic atmosphere never develops. Therefore, the ion's motion is not retarded, its apparent mobility is higher, and the overall conductivity of the solution is higher. Conductivity as a function of frequency is given by

$$(18.2.91) \qquad \frac{\sigma(\omega) - \sigma(0)}{\varepsilon} = \frac{A \omega \tau_{DF}}{1 + (\omega \tau_{DF})^2}$$

$$(18.2.92) \qquad A = \frac{z^2 e^2}{6 \varepsilon_0 \varepsilon k_B T} \kappa \Lambda_\infty$$

Debye-Falkenhagen model is based on the Debye-Huckel model (see Section (15.3)) and so shares the same limitations, such as the assumption of a dilute solution. Furthermore, while Debye-Falkenhagen time scale overlaps with the time scales (frequencies) used in DEP, especially for low ionic strength solutions, for many ions the change in conductivity is small. For example, for 0.1 mM KCl the Debye-Falkenhagen frequency is around 300 kHz - well within typical frequencies used for

DEP, but the coefficient A/Λ_∞ in (18.2.92) is only about 0.4%! For 10 mM KCl solution, Debye-Falkenhagen frequency is about 30 MHz - well outside of typical DEP frequencies, but the coefficient A/Λ_∞ in (18.2.92) is now about 4%. Thus, for most electrolytes and for most DEP experiments, we can ignore this effect of additional conductivity. For further discussion on Debye-Falkenhagen effect, see Anderson (1994).

Similar to the Debye-Falkenhagen effect is the Wien effect. In the Wien effect, the electric field is so high that the ion travels so fast that the ionic atmosphere does not have time to form. To determine how high the field has to be we compare the time for the ion to travel one Debye length to the Debye-Falkenhagen time scale. The former time is

$$(18.2.93) \qquad\qquad t = \frac{1}{\kappa \, |\mu| \, E}$$

For univalent symmetric electrolytes, the Debye length scales as the inverse of the square root of the concentration, while the Debye-Falkenhagen time scale scales inversely with concentration. For example, for KCl, where the mobility of the potassium ion and chloride ion are about equal, and have a magnitude of about $80 \times 10^{-9} \, \mathrm{m^2/V \cdot s}$, for 0.1 mM solution the Debye length is about 30 nm and the Debye-Falkenhagen time is about 3 μs. Thus for this solution Wein effect should be felt above an electric field of 100 kV/m or about 1 V per 10 μm. This is a fairly reasonable field for DEP. On the other hand, for a 10 mM solution, the Debye length is about 3 nm and the Debye-Falkenhagen time is about 30 ns, and so the Wien effect should be felt above an electric field of 1 MV/m or about 10 V per 10 μm, again not unreasonable field for DEP. However, even at these high fields, the increase in conductivity is typically less than 1%, even for multivalent electrolytes. Wien effect becomes somewhat appreciable (1-10% change in conductivity) at 10 MV/m fields and larger, but this increase in conductivity does not increase much as the field increases further. For further discussion on the Wien effect see Eckstrom and Schmelzer (1939). In conclusion, we can easily manipulate solution conductivity by changing the ion concentration, and to a lesser effect choosing ions with different mobility. For an isothermal DEP separation, we should typically not worry about the effect of field strength or field frequency on conductivity.

Like solution conductivity, we may also change solution permittivity. In general, the permittivity of pure (deionized) water has no appreciable frequency dependence below 100 MHz. The temperature dependence of permittivity of pure water is given by Malmberg and Maryott (1956) to be

$$(18.2.94) \qquad \varepsilon = 87.740 - 0.40008T + 9.398 \times 10^{-4} T^2 - 1.410 \times 10^{-6} T^3$$

where T is temperature in degrees Celsius. The permittivity of a solution is given by

$$(18.2.95) \qquad\qquad \varepsilon = \varepsilon_s + \delta c$$

where ε_s is the solvent permittivity, δ is the permittivity increment, and c is the concentration of a solution. We tabulate some of the permittivity increments in Table 2. More permittivity increments and a discussion of effect on frequency on permittivity can be found in Hasted (1973). However, one should keep in mind that permittivity increments are not as widely tabulated as for example limiting molar conductivities or electrophoretic mobilities. Furthermore, notice that while we can easily change conductivity of an aqueous solution by a six orders of magnitude (roughly from 5×10^{-6} S/m to 20 S/m), we can barely change the permittivity of the solution by an order of magnitude. Further notice that many of the substances with a high dielectric increment are acids and so contribute to conductivity as well as permittivity of the solution. However, we should also notice that adding large, >100 mM, concentration of ions to tune solution conductivity also affects solution permittivity. Furthermore, many non-acids that have high dielectric increment (e.g., hexaglycine) are poorly soluble in water. All this makes it much more difficult to tune solution permittivity than it is to tune its conductivity. Lastly, remember that dissipated power scales as V^2/Z where V is the applied potential and Z is the impedance of the system (here our solution). Typically a solution between electrodes is modeled as a resistor and capacitor in parallel. The impedance for this circuit is given by

$$(18.2.96) \qquad |Z| = \frac{1}{\sqrt{\left(1/R\right)^2 + \left(\omega C\right)^2}}$$

Thus the higher the conductivity and/or permittivity of the solution, the less its overall impedance and the more power is dissipated in the solution. This heating either boils the solution - "catastrophic failure" or changes the solution conductivity and permittivity enough to have the field have a force on the solution (see Section 14 and Section 18.1). This causes additional flows, (electrothermal flows) which in turn typically interfere with separations.

18.2.8. DEP in deep electrode geometries. Having now thoroughly discussed the K (polarization coefficient) portion of equation (18.2.1), let's now discuss the $\nabla\left(\mathbf{E}^2\right)$ portion, which is generally set by the electrode geometry. We will start first with the two dimensional geometries and them move onto the more realistic planar geometries, which result in 3D fields. In these 2D geometries

$$(18.2.97) \qquad \nabla\left(\mathbf{E}^2\right) = \nabla\left(\sqrt{E_x^2 + E_y^2}^2\right) = \nabla\left(E_x^2 + E_y^2\right)$$

where

$$(18.2.98) \qquad \begin{aligned} E_x^2 &= \left(\frac{\partial V}{\partial x}\right)^2 \\ E_y^2 &= \left(\frac{\partial V}{\partial y}\right)^2 \end{aligned}$$

Thus

$$(18.2.99) \quad \nabla\left(\mathbf{E}^2\right) = 2\left(\frac{\partial V}{\partial y}\frac{\partial^2 V}{\partial x \partial y} + \frac{\partial V}{\partial x}\frac{\partial^2 V}{\partial x^2}\right)\hat{\mathbf{i}} + 2\left(\frac{\partial V}{\partial y}\frac{\partial^2 V}{\partial y^2} + \frac{\partial V}{\partial x}\frac{\partial^2 V}{\partial x \partial y}\right)\hat{\mathbf{j}}$$

To obtain the gradient of the norm of the electric field squared (which we will refer to as the normalized force) we calculate the potential in the 2D geometry of interest using the Laplace equation subject to the appropriate potential and insulation boundary conditions and then apply equation (18.2.99).

In Figure 18.2.5 we plot the normalized force (proportional to the positive dielectrophoretic force) in simple 2D geometries with the arrow size proportional to the magnitude of the force. This is proportional to a force experienced by a particle undergoing positive dielectrophoresis (pDEP). While we plot a limited number of geometries, we can extend learning from these to many other realistic 2D geometries.

TABLE 2. Dielectric increments of ions and polar molecules in water at 25°C.

Substance	$\delta \left[M^{-1} \right]$	Substance	$\delta \left[M^{-1} \right]$
H^+	-17	Glycine	23
Li^+	-11	Glycine dipeptide	71
Na^+	-8	Glycine tripeptide	113
K^+	-8	Glycine tetrapeptide	159
Rb^+	-7	Glycine pentapeptide	215
Mg^{2+}	-24	Glycine hexapeptide	234
Ba^{2+}	-22	Glycine heptapeptide	290
La^{3+}	-35	Lysylglutamic acid	345
F^-	-5	d-Arginine	62
Cl^-	-3	l-Aspartic acid	28
I^-	-7	d-Glutamic acid	26
OH^-	-13	l-Asparagine	28
SO_4^{2-}	-7	l-Glutamine	21
Acetonitrile	-1.74	dl-Proline	21
Ribose	-2.72	α-Alanine	23
Galactose	-3.28	β-Alanine	35
Sucrose*	-8.2	dl- α-Valine	25
Glucose	-4.27	l- α-Leucine	25
Tetrahydrofuran	-4.85	Taurine	41
Pyrazine	-6.4	Creatine	32
Phenol	-6.6	γ-Aminobutyric acid	51
Ethanol**	-4.6	γ-Aminovaleric acid	55
Methanol**	-2.1	δ-Aminovaleric acid	63
Glycerol*	-1.9	ε-Aminocaproic acid	78
Benzoic acid	-67	Urea	3.4

Data from: Hasted (1973). * from: Arnold et al. (1993); ** from: Akerlof (1932). For salts, dielectric increments add: e.g., $\delta_{NaCl} = -11$.

In all of these geometries the electrodes are located on the top and bottom walls and the left and right walls are electrically insulated. All of the geometries are scaled such that their horizontal span is equal to unity.

The simplest geometry we consider is a linear wedge (Figure 18.2.5a). In this geometry the electric field increases from left to right as the wedge tapers, but the rate at which the electric field increases is constant. Thus there is a linearly increasing pDEP force from left to right. The geometry in Figure 18.2.5 b is another tapered wedge, with again the electric field linearly increasing from left to right and a pDEP force from left to right. This geometry has horizontal symmetry and so of course does the resulting force field. Thus it is unnecessary to calculate the force field below the horizontal line of symmetry - we could have just calculated the force in the upper part (i.e. Figure 18.2.5a) and mirrored it.

In Figure 18.2.5c we have a converging-diverging geometry formed by a part of a cylinder and a plane. Positive DEP force points toward the converging region where both the norm of the electric field and its gradient is highest. Right at the center of the converging region the force points towards the cylindrical surface. In

Figure 18.2.5d we have a mirrored geometry of Figure 18.2.5c over the plane surface. As we expect, the force is also mirrored. Geometries in Figure 18.2.5c and d also have vertical symmetry. Thus the resulting force field is also vertically symmetric, and so it is enough to calculate this force field in only one of the symmetric parts

In Figure 18.2.5e we have another geometry formed from two cylindrical surfaces. In contrast to Figure 18.2.5d where the region of interest was in the convex part of the geometry, in Figure 18.2.5e the region of interest is in the concave part of the geometry. The electrode walls diverge from the center and converge towards the ends. Thus the pDEP force also diverges from the center and converges in the converging sections. At the line of vertical symmetry the force is directed towards the center. Generally, at such lines of symmetry, the force is directed away from a concave region and towards the convex region.

In Figure 18.2.5f, g, and h we plot the pDEP force in three wedges: a hyperbolic wedge, a parabolic wedge, and an exponential wedge. In the hyperbolic wedge the force is fairly non-uniform, with the force near the hyperbolic surface going from high to low to high as we go from left to right, but the force at the plane surface monotonically increasing. In the parabolic wedge the force is more uniform, but again generally non-monotonic. It goes from low at the left edge, to high in the center to low again towards the right edge. In the exponential wedge the force monotonically increases from left to right. Using the symmetry rules outlined in the previous paragraphs these wedges can be built up into more complex geometries.

18.2.9. DEP in common, planar electrode geometries. Having discussed the simpler two dimensional geometries, let's turn to the much more commonly used planar geometries that typically result in three dimensional DEP force fields. We illustrate some of the common such geometries in Figure 18.2.6. The reason why these geometries are so popular is that they are fairly easy to fabricate especially in an academic setting. To fabricate these you just need to put down a metal layer and lithographically pattern it. Their drawback however is that the dielectrophoretic force field produced decays rapidly from the electrode surface and for some geometries is highly non-uniform near the electrode surface.

Let's begin with the simplest of such planar electrode geometries: parallel electrode strips, Figure 18.2.6a. These electrodes can be assumed to be infinitely long (compared to the gap between them and to the height above them of interest) and so the problem becomes two dimensional. The other electrode geometries to first order can be thought of long electrode strips but just "bunched up" and with some nonuniformities in the force field. For this two dimensional problem we follow the analysis of Morgan et al. (2001). For this two dimensional problem, Laplace's equation has an analytical solution for the potential

$$(18.2.100) \qquad \varphi(x,y) = \sum_{n=0}^{\infty} A_n \cos(k_n x) \exp[-k_n y]$$

$$(18.2.101) \qquad k_n = \frac{(2n+1)\pi}{2d}$$

$$(18.2.102) \qquad d = (d_1 + d_2)/2$$

$$(18.2.103) \qquad A_n = \frac{16V_0 d}{\pi^2 d_2 (2n+1)^2} \cos\left(\frac{(2n+1)\pi d_1}{4d}\right)$$

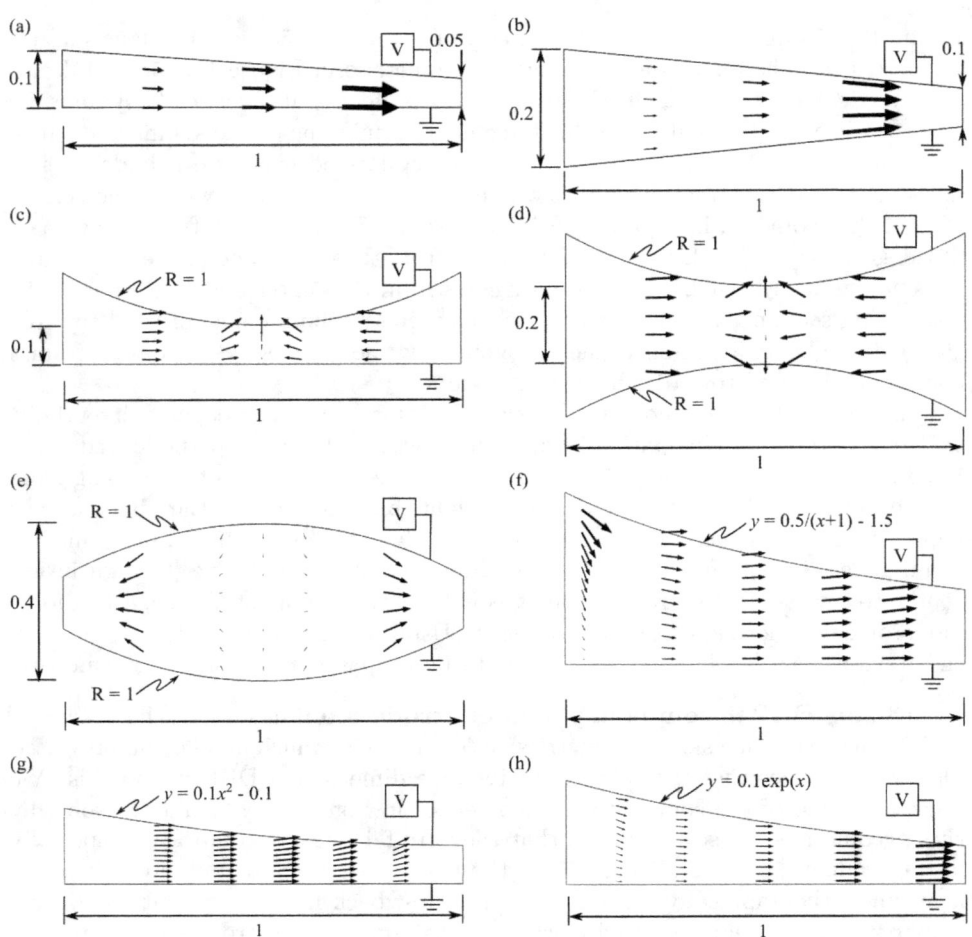

FIGURE 18.2.5. Dielectrophoretic force field in simple 2D geometries with the arrow size is proportional to the force. The arrows point in the direction of positive dielectrophoresis force.

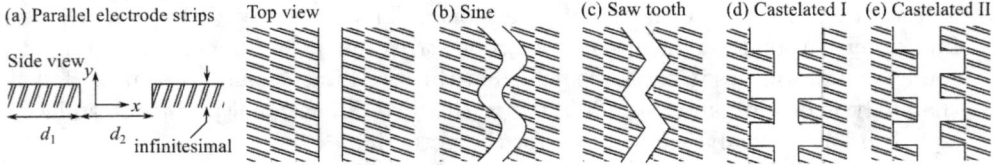

FIGURE 18.2.6. Schematic of common planar electrode geometries. (a) Side view: the side view of common electrode geometries. The height of planar electrodes is typically order 100 nm, when the electrode spacing is typically order 10 μm; thus the height of such electrodes can be assumed to be infinitesimal. Top view: top view of a parallel electrode strip configuration.(b)-(e) Configurations with increasing number of points of non-uniform electric field per area.

where V_0 is the potential difference between the electrodes, d_1 is the electrode width and d_2 is the electrode gap. Note that the height of the electrode is assumed to be infinitesimal compared to the gap width or electrode width. For the case where the electrode width and gap are the same, $d_1 = d_2 = d$

$$(18.2.104) \qquad A_n = \frac{16 V_0}{\pi^2 (2n+1)^2} \cos\left(\frac{\pi}{4}(2n+1)\right)$$

The solution for $\nabla\left(\mathbf{E}^2\right)$ is also an infinite series, but if we desire a solution for $y > d$ and so can assume d is small, we can approximate $\nabla\left(\mathbf{E}^2\right)$ with the first term in the series:

$$(18.2.105) \qquad \nabla\left(\mathbf{E}^2\right) = -\frac{64}{\pi}\frac{V_0^2}{d_2^2 d}\cos^2\left(\frac{\pi d_1}{4d}\right)\exp\left[-\pi y/d\right]\hat{\mathbf{y}}$$

which simplifies to

$$(18.2.106) \qquad \nabla\left(\mathbf{E}^2\right) = \frac{32}{\pi}\frac{V_0^2}{d^3}\exp\left[-\pi y/d\right]\hat{\mathbf{y}}$$

for when the gap width and electrode width are the same: $d_1 = d_2 = d$. Solution in (18.2.106) is especially interesting as it shows clear scaling relationship between the force (scaled by the prefactor of $\nabla\left(\mathbf{E}^2\right)$ in equation (18.2.1); we will call this force here "scaled force") and gap as a function of the distance from the electrode surface and applied potential. Just from observation of (18.2.106), we can see several interesting things. Firstly, as we would expect, the scaled force scales as the square of the applied potential. Secondly, the scaled force for $y > d$ is approximately unidirectional, in the direction away from the planar electrodes; the force orthogonal to this direction is negligible. This is great as it simplifies predicting the motion of particles in such a force field, however, unfortunately, this force decays exponentially with distance from the electrode surface. For ease of understanding the relationship between scaled force, electrode gap, and distance from the electrode, we plot the salient part of (18.2.106) in Figure 18.2.7. We can clearly see that to achieve high forces at a considerable distance from the electrode it is best to choose a larger electrode spacing. Furthermore, notice that the scaled force exponentially decays with distance away from the electrode, for $y > d$. This exponential decay of force on the particle is indeed the main drawback of planar electrode geometries. Particles several gap spacing away from the electrodes experience orders of magnitude less force than those a few gap spacings closer. Thus particles further away from the electrodes may effectively never experience any effect of DEP separation, while particles very near the electrodes may experience a force enough to damage (e.g. shear) the particle!

To analyze the case where $y < d$ we resort to numerical means. For simplicity, we consider the case where the gap width and electrode width are the same($d_1 = d_2 = d$) and numerically evaluate (18.2.100) calculating the scaled force via the definition

$$(18.2.107) \quad \nabla\left(\mathbf{E}^2\right) = 2\left(\frac{\partial\varphi}{\partial x}\frac{\partial^2\varphi}{\partial x^2} + \frac{\partial\varphi}{\partial y}\frac{\partial^2\varphi}{\partial x\partial y}\right)\hat{\mathbf{x}} + 2\left(\frac{\partial\varphi}{\partial x}\frac{\partial^2\varphi}{\partial y\partial x} + \frac{\partial\varphi}{\partial y}\frac{\partial^2\varphi}{\partial y^2}\right)\hat{\mathbf{y}}$$

We plot the scaled force, scaled by the square of the applied potential in Figure 18.2.8. We see that near the electrode surface the force is highly non-uniform. The force vectors curve into the center of the gaps between the electrodes. Zooming

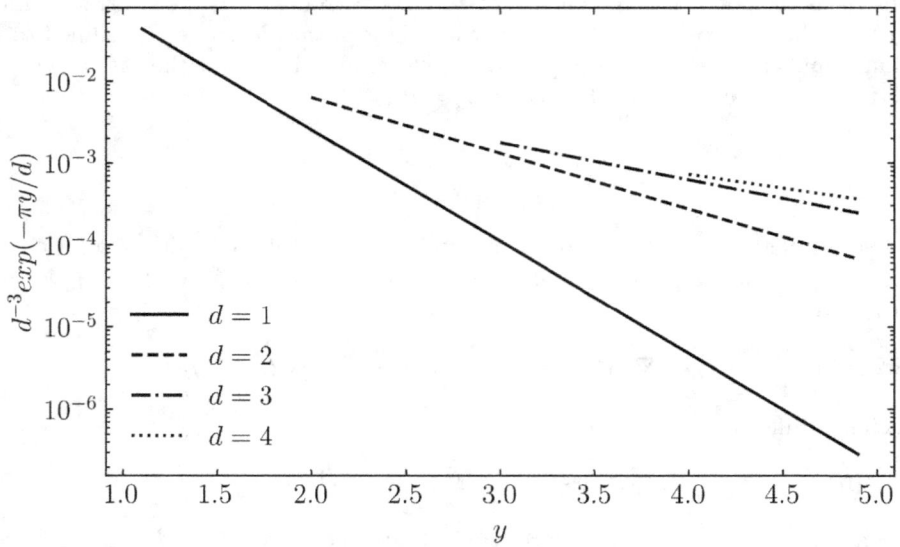

FIGURE 18.2.7. Relationship between scaled force as a function of gap width d and distance from the electrode surface y.

into a single gap, Figure 18.2.9, we observe that the force is the highest near the electrode edge and it pushes particles experiencing positive dielectrophoresis into the center of the gap. Particles that are on top of the electrodes experiencing positive dielectrophoresis (and for which Stokes' drag law applies) are first slowly swept towards the edge of the gap then are accelerated abruptly as they pass the edge and then slow down as they reach the center of the gap. Thus particle behavior near the gap is fairly complicated, and this is another drawback of planar electrode geometries for DEP. Lastly, to further illustrate together the fact that force is highest at the electrode edge and the fact that the force decays exponentially as the distance from the electrode surface increases, we plot the log of the magnitude of this scaled force in Figure 18.2.10.

We will not discuss in great detail the other planar electrode geometries. To the first approximation the force field produced by these other geometries is the same as that for the parallel strips configuration with the condition that the parallel strip is snaked around the curve of the geometries. Of course the corners and especially the sharp points in the geometry generate high non-uniformity, but these can be treated as second order effects. One important outcome of this snaking is the ability of for example for geometries in Figure 18.2.6d and e to surround the particles from three edges in the valley of the castellation and thus corral (push) the particles undergoing positive dielectrophoresis to the center of this valley. Particles undergoing negative dielectrophoresis at the same time would be pushed away from the edges of the electrodes towards the main electrode material, and thus the negative DEP particles would gather in the centers of the electrode square teeth. The effect of sharp corners in geometries in Figure 18.2.6b-e is that the sharp corners tend to pull particles towards those sharp corner regions.

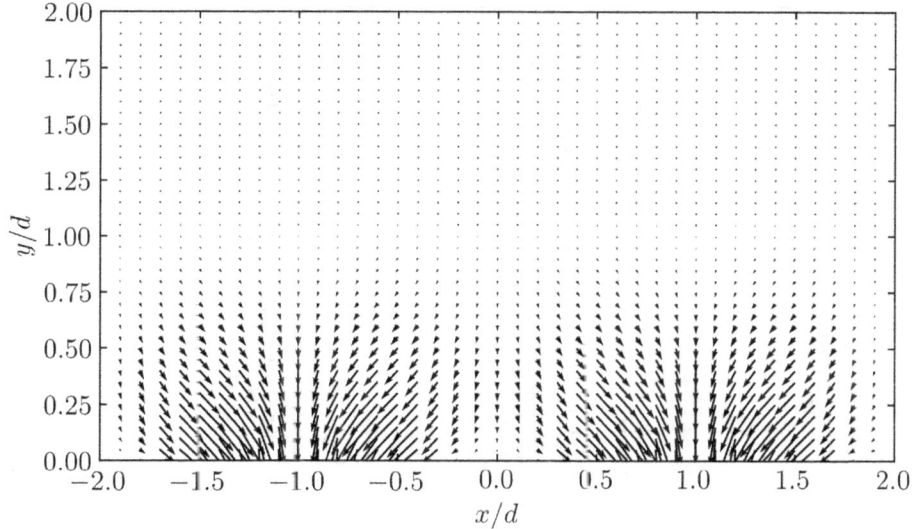

FIGURE 18.2.8. Vector plot of $\nabla\left(\mathbf{E}^2\right)/V_o^2$ for d equal to unity. The electrodes span from -2.5 to -1.5; -0.5 to 0.5; and 1.5 to 2.5. Thus there are two gaps: one from -1.5 to -0.5 and one from 0.5 to 1.5. Near the surface of the planar electrode the positive dielectrophoretic force is directed into the center of the gap and away from the center of the electrode. Far away from the electrode the positive dielectrophoretic force is nearly vertical and directed downward towards the electrodes. The magnitude of this force exponentially decays as the distance from the electrode floor.

18.2.10. Alternative field shaping: Insulator-based dielectrophoresis. In addition to shaping the electric field by clever placement of electrodes, we can also shape the electric field by clever placement of insulating material. The main advantage of this approach is that it is often much simpler to shape and manipulate insulating material than it is to carefully place electrodes and wire them to the outside of the chip. Thus this approach is especially useful in designing low cost dielectrophoresis based devices. A typical insulator dielectrophoresis device consists of an array of posts; at the end of such an array electrodes are placed and constrictions between the posts shape the electric field. A numerical simulation of a DEP force in a typical constriction is shown in Figure 18.2.11. This constriction represents a unit cell of a larger structure of posts, with electrodes at two distant ends supplying the electric field. The simplest incarnation of this simply a tube packed with chromatography beads (e.g. silica) with electrodes applying electric field at both ends. Constrictions between the beads force the electric field lines to go through the narrow constrictions which have high electric field gradient and thus high local force, which in turn collects smaller particles experiencing positive dielectrophoresis in this area. Thus, a simple application of an appropriate frequency electric field to a porous filter can help the filter trap particles much smaller than the filter pore size. Insulator-based dielectrophoresis can be applied to enhance

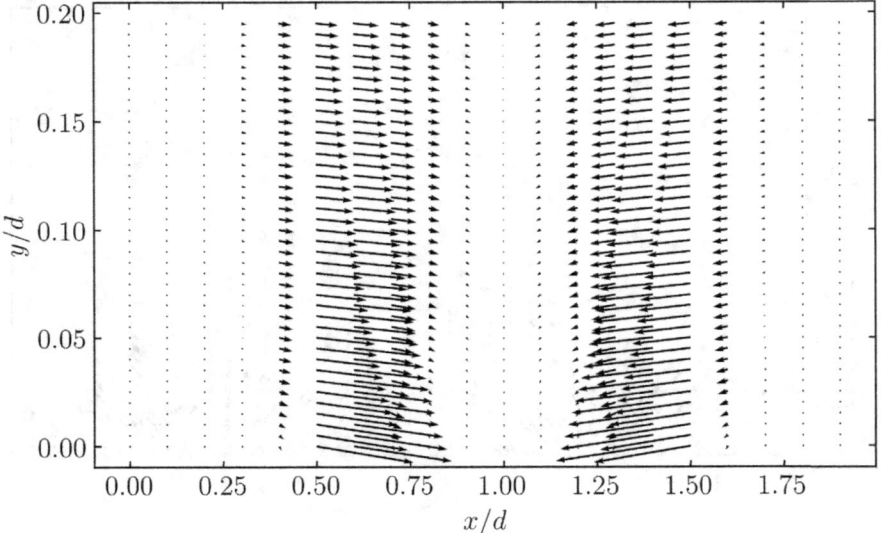

FIGURE 18.2.9. Zoomed in vector plot of $\nabla\left(\mathbf{E}^2\right)/V_0^2$ for d equal to unity. The electrodes span from -0.5 to 0.5 and 1.5 to 2.5. Thus we have zoomed into a single gap from 0.5 to 1.5. Positive dielectrophoretic force is towards the center of the gap. This force is the highest near the electrode edge and it pushes particles into the center of the gap. Particles that are on top of the electrodes experiencing positive dielectrophoresis (and for which Stokes' drag law applies) are first slowly swept towards the edge of the gap then are accelerated abruptly as they pass the edge and then slow down as they reach the center of the gap.

filtration and potentially enhance chromatography of particles. One also needs to watch for the unwanted effects of insulator-based dielectrophoresis in electrochromatography or any other places where an electric field is applied through porous media.

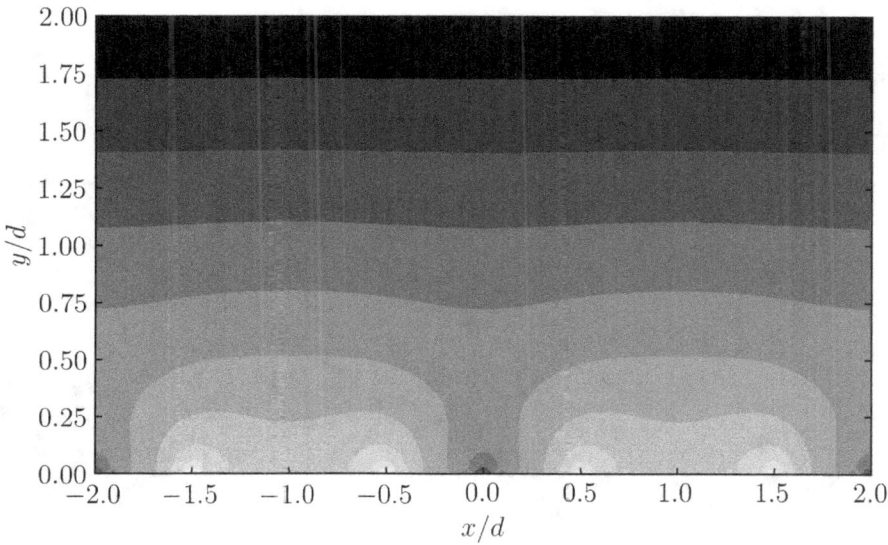

FIGURE 18.2.10. Contour plot of the natural logarithm of the magnitude of $\nabla \left(\mathbf{E}^2 \right)/V_0^2$ for d equal to unity. The lighter areas indicate greater force magnitude. The electrodes span from -2.5 to -1.5; -0.5 to 0.5; and 1.5 to 2.5. Thus there are two gaps: one from -1.5 to -0.5 and one from 0.5 to 1.5. The areas of greatest force are the edges of the electrodes.

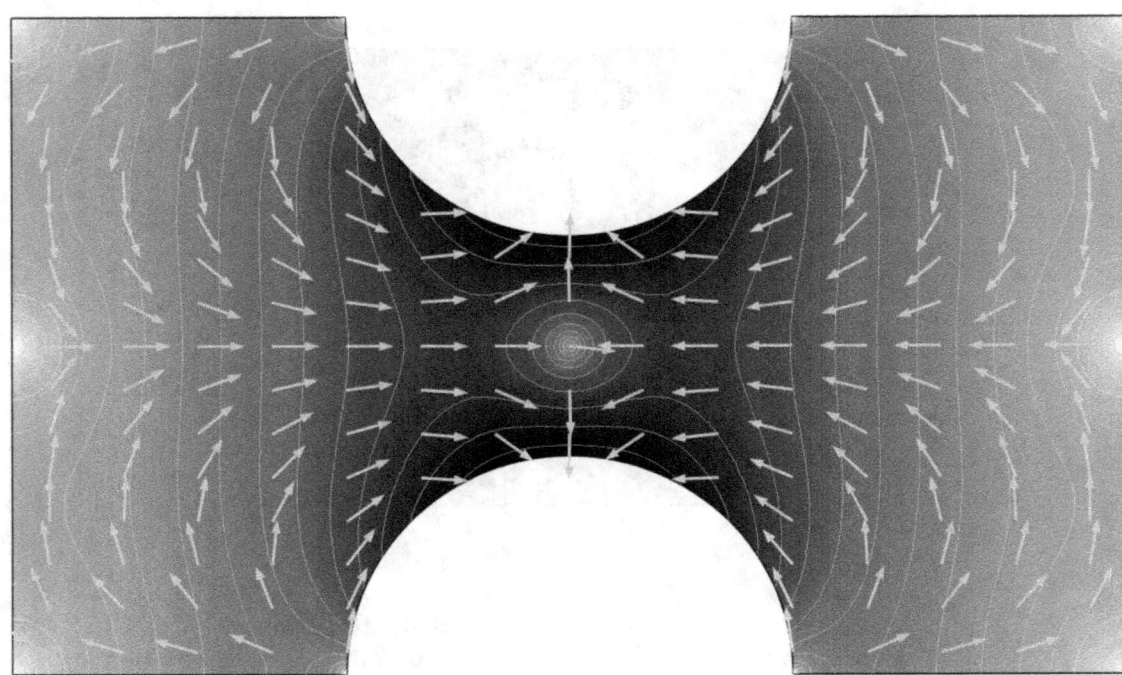

FIGURE 18.2.11. Dielectrophoretic force field in a constriction. The electrodes are located at the left and right most boundaries of the domain. The other walls of the domain are set to be perfectly insulating. The gap is equal to the radius of the semi-circles forming the constriction. The arrows show the direction of the dielectric force field (positive dielectrophoresis), while the contour shading show the magnitude of the force (logarithmic scale). There is roughly three orders of magnitude difference between the force in the lowest contours (near the electrodes) vs. the highest force contours (on the surface of the bumps in the constriction). The (positive dielectrophoresis) force is directed into the gap and onto the surface of the circular bump. This can be used to enhance capture of particles by affinity probes (e.g., antibodies) attached to the surface of pillars in a pillar array, as particles (e.g., protein of interest) would be pushed towards the antibodies. Furthermore, dielectrophoretic force would work to exclude particles undergoing negative dielectrophoresis from the gap. This can be used to enhance filtering capabilities of a post array, allowing the array to exclude much smaller particles than would be excluded mechanically. Such filter is also less prone to clogging then conventional mechanical filters.

CHAPTER 19

Particle rotation

Having explored various particle translation mechanisms, here we turn to particle rotation mechanisms. We will analyze two particle rotation mechanisms: the Born-Lertes rotation (Born (1920), Lertes (1921)) and the Quincke rotation (Quincke (1896)) In Born-Lertes rotation, particles rotate in a rotating electric field, and the rotation is not synchronous – the angular velocity of the particle is different from that of the field. In Quince rotation, particles rotate in a static electric field, given a "kick" to rotate in one or the other direction, when the electric field is greater than some threshold field. In both effects, the rotational axis is perpendicular to the imposed electric field.

Let's begin with the Born-Lertes effect. Consider a homogeneous particle in a four electrode setup in Figure 19.0.1. This setup is filled with a fluid with permittivity ε_f, conductivity σ_f, and viscosity μ. The particle has a permittivity ε_p and conductivity σ_p. The setup consists of two pairs of electrode plates, with each pair connected to a separate channel of a power supply. One channel of this power supply is 90° phase shifted with respect to the other channel, but both channels are at the same frequency, ω (radians/s). This configuration produces a rotating circularly polarized electric field in the center between the electrodes, rotating with

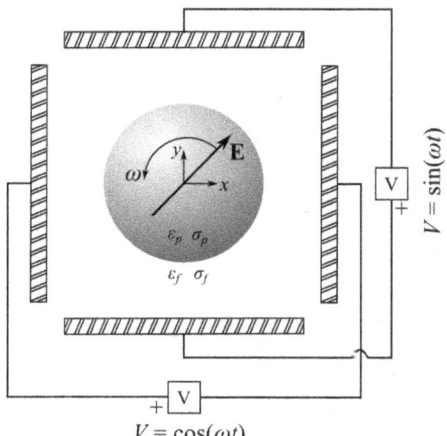

FIGURE 19.0.1. Schematic of a four electrode setup with two independent, 90° phase offset power supply channels that produces a counter clockwise rotating electric field with rotation speed ω. To rotate a particle via the Born-Lertes effect, we place a particle into the center of this electrode structure.

the frequency ω,

(19.0.1) $$\mathbf{E}(t) = E_0 \left(\hat{\mathbf{x}} \cos \omega t + \hat{\mathbf{y}} \sin \omega t \right)$$

or in complex notation

(19.0.2) $$\mathbf{E}(t) = E_0 \left(\hat{\mathbf{x}} - j\hat{\mathbf{y}} \right) \exp \left(j\omega t \right)$$

Substituting this into (18.2.28) we find that the effective dipole moment induced in the stationary particle induced by the rotating field is

(19.0.3) $$\underline{\mathbf{p}} = 4\pi\varepsilon_0\varepsilon_f \underline{K} a^3 E_0 \left(\hat{\mathbf{x}} - j\hat{\mathbf{y}} \right) \exp \left(j\omega t \right)$$

From (14.3.7), the torque on a particle is

(19.0.4) $$\mathbf{T} = \mathbf{p} \times \mathbf{E}$$

Substituting in (19.0.3) and (19.0.2),

(19.0.5) $$\mathbf{T} = \mathrm{Re}\left[4\pi\varepsilon_0\varepsilon_f \underline{K} a^3 E_0 \left(\hat{\mathbf{x}} - j\hat{\mathbf{y}} \right) \exp \left(j\omega t \right) \right] \times \mathrm{Re}\left[E_0 \left(\hat{\mathbf{x}} - j\hat{\mathbf{y}} \right) \exp \left(j\omega t \right) \right]$$

Taking the time average over a time slightly longer than $1/\omega$,

(19.0.6) $$\langle \mathbf{T} \rangle = \frac{\mathrm{Re}\left[\mathbf{p} \times \mathbf{E} \right]}{2\exp\left(j\omega t \right)} = 4\pi\varepsilon_0\varepsilon_f a^3 E_0 \left| \underline{K} \left(\hat{\mathbf{x}} - j\hat{\mathbf{y}} \right) \right| E_0 \sin \alpha \hat{\mathbf{z}}$$

(19.0.7) $$\alpha = -\sin^{-1}\left(\frac{\mathrm{Im}\left[\underline{K} \right]}{\left| \underline{K} \right|} \right)$$

where α is the lag angle between the rotating electric field and the effective dipole moment. Combining (19.0.6) and (19.0.7), and simplifying

(19.0.8) $$\langle \mathbf{T} \rangle = -4\pi\varepsilon_0\varepsilon_f a^3 E_0^2 \, \mathrm{Im}\left[\underline{K} \right] \hat{\mathbf{z}}$$

This is the time averaged torque on a spherical particle. Compare this to the relation for the dielectrophoretic force on a particle, (18.2.31), which we rewrite here for convenience as,

(19.0.9) $$\langle \mathbf{F} \rangle = 2\pi\varepsilon_0\varepsilon_f a^3 \, \mathrm{Re}\left[\underline{K} \right] \nabla \left(E_{RMS}^2 \right) \frac{\nabla \left(\mathbf{E}^2 \right)}{\left| \nabla \left(\mathbf{E}^2 \right) \right|}$$

We see that while the force on the particle scales with the real part of the Clausius-Mossotti factor, the torque on a particle scales with the imaginary part of the Clausius-Mossotti factor. Furthermore, while the force scales as the gradient of the square of the root mean square of the electric field (and thus a non-uniform field is necessary for a non-zero force), the torque scales as the square of the field magnitude. This implies that a particle can experience a torque in a completely uniform electric field!

We obtained the Clausius-Mossotti factor for a particle and fluid with finite permittivity and conductivity in Section 18.2.4, equation (18.2.46). Substituting in this relation into (19.0.9), we obtain that the torque on such particle is

(19.0.10) $$\langle \mathbf{T} \rangle = -4\pi a^3 E_0^2 \varepsilon_0 \varepsilon_f \omega \frac{\left(\tau_0 - \tau_{MW} \right)}{\left(1 + \omega^2 \tau_{MW}^2 \right)} \frac{\left(\sigma_p - \sigma_f \right)}{\left(\sigma_p - 2\sigma_f \right)} \hat{\mathbf{z}}$$

Instead of the time scale τ_0, we can similarly define more intuitive charge relaxation time scales for each of the individual phases,

$$
(19.0.11) \qquad \begin{aligned} \tau_f &= \frac{\varepsilon_0 \varepsilon_f}{\sigma_f} \\ \tau_p &= \frac{\varepsilon_0 \varepsilon_p}{\sigma_p} \end{aligned}
$$

Using this notation, the torque on the particle is

$$
(19.0.12) \qquad \langle \mathbf{T} \rangle = \frac{-6\pi a^3 E_0^2 \varepsilon_0 \varepsilon_f \omega \tau_{MW} \left(1 - \tau_f/\tau_p\right)}{\left(1 + 2\varepsilon_f/\varepsilon_p\right)\left(1 + \sigma_f/2\sigma_p\right)\left(1 + \omega^2 \tau_{MW}^2\right)} \hat{\mathbf{z}}
$$

In this form we see that the sign of the torque depends on the ratio of the charge relaxation time scales of the individual phases. If $\tau_p < \tau_f$ then the particle rotates with the electric field, and if $\tau_p > \tau_f$ then the particle rotates in the direction opposite of the field rotation. We can understand this by considering the distribution of free charge on the surface of the particle (Figure 19.0.2). The effective polarization moment vector and the associated free charge always rotate synchronously with the electric field. When $\tau_p < \tau_f$ then the sign of the induced charge on the particle is opposite to that of the charge on the electrodes, resulting in electrodes pulling the particle around (Figure 19.0.2a). When $\tau_p > \tau_f$ then the sign of the induced charge on the particle is the same to that of the charge on the electrodes, resulting in electrodes pushing the particle around (Figure 19.0.2b). The lag of charge on the particle relative to that on the electrodes is given by (19.0.7).

Let's now generalize this situation to one where the particle already is rotating with an angular speed Ω and is subjected to a rotating electric field described above. In the frame of reference of the particle, the effective electrical frequency is shifted from ω to $\omega - \Omega$, so we rewrite our torque expression as

$$
(19.0.13) \qquad \langle \mathbf{T} \rangle = \frac{-6\pi a^3 E_0^2 \varepsilon_0 \varepsilon_f \left(\omega - \Omega\right) \tau_{MW} \left(1 - \tau_f/\tau_p\right)}{\left(1 + 2\varepsilon_f/\varepsilon_p\right)\left(1 + \sigma_f/2\sigma_p\right)\left(1 + \left(\omega - \Omega\right)^2 \tau_{MW}^2\right)} \hat{\mathbf{z}}
$$

We assume that the Reynolds number of this motion is small, so we are in the creeping flow limit. Thus we assume that the inertia of the particle is negligible and so the torque on the particle is balanced by Stokes' drag torque (analogous to Stokes' drag force),

$$
(19.0.14) \qquad T_\mu = 8\pi \mu a^3 \Omega
$$

Combining (19.0.13) and (19.0.14), we obtain
(19.0.15)

$$
\Omega \tau_{MW} \left(\frac{-8\mu \left(1 + 2\varepsilon_f/\varepsilon_p\right)\left(1 + \sigma_f/2\sigma_p\right)}{\tau_{MW} 6 E_0^2 \varepsilon_0 \varepsilon_f} \right) \frac{1}{\left(1 - \tau_f/\tau_p\right)} = \left(\frac{\left(\omega - \Omega\right)\tau_{MW}}{1 + \left(\omega - \Omega\right)^2 \tau_{MW}^2} \right)
$$

Notice that the steady state rotation speed is independent of the particle size and only depends on the electrical properties of the particle and the fluid, as well as the electric field strength and fluid viscosity. While we can obtain an exact analytical solution to this equation, the expression is quite cumbersome. Hence it is more instructive to consider the overall solution space graphically (Figure 19.0.3). We plot the left-hand side and the right-hand side of (19.0.15) separately and look for the intersections of the two curves to obtain the solution. We see when $\tau_p < \tau_f$ then equation (19.0.15) will have a single solution for Ω, and that $\Omega > 0$. On the

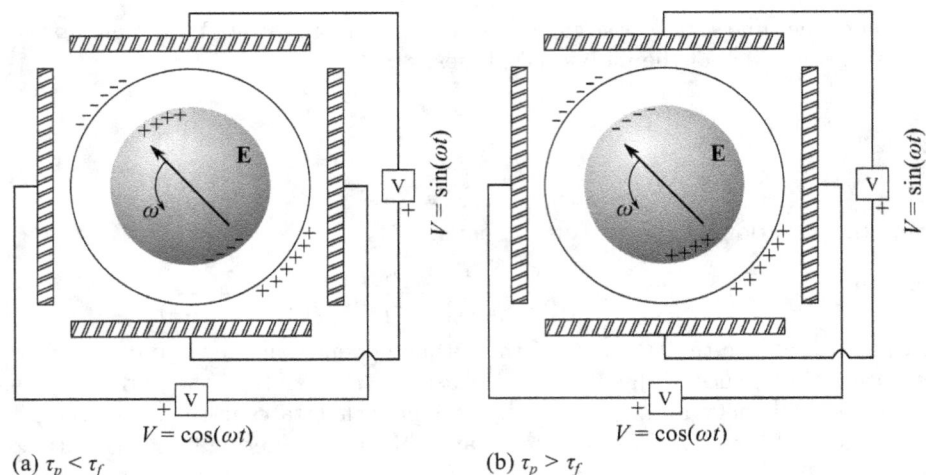

(a) $\tau_p < \tau_f$ (b) $\tau_p > \tau_f$

FIGURE 19.0.2. Schematic of a sphere in a rotating electric field, showing effective charges on the electrodes (large circle) and the induced charges on the sphere. (a) When the relaxation time for the particle is smaller than that for the surrounding fluid, $\tau_p < \tau_f$, the sign of the induced charge on the particle is opposite to that on the electrodes, resulting in a torque that turns the particle in the same direction as the direction of the rotating field. (b) When the relaxation time for the particle is larger than that for the surrounding fluid, $\tau_p > \tau_f$, the sign of the induced charge on the particle is the same as that on the electrodes, resulting in a torque that turns the particle in the direction opposite of the direction of the rotating field.

other hand, when $\tau_p > \tau_f$, then there will be one, two (very rare), or three solutions for Ω. One of these solutions will be always negative. For the three solution case, only two of the solutions are stable, as the solution is stable when

$$(19.0.16) \qquad \left(\left. \frac{\partial T_\mu}{\partial \Omega} \right|_{\Omega_{eq}} \right) \left(\left. \frac{\partial \langle T \rangle}{\partial \Omega} \right|_{\Omega_{eq}} \right) < 0$$

Thus, for a reproducible rotation of a particle, it is advisable to operate in the $\tau_p < \tau_f$ regime, or if we must operate in the $\tau_p > \tau_f$, we should pick a low enough electric field strength or high enough viscosity where only a single solution is possible.

Having obtained the torque and rotational velocity for a particle in a rotating electric field, now let's consider a particle in a stationary electric field – Quincke rotation. Let's imagine that we took a particle, and gave it a kick, so it started to rotate with an angular velocity $\delta\Omega$. If this particle was in a static, DC electric field, (e.g., an electric field between two electrode plates), then the torque expression (19.0.13), would just be

$$(19.0.17) \qquad \langle \mathbf{T} \rangle = \frac{-6\pi a^3 E_0^2 \varepsilon_0 \varepsilon_f \left(\delta\Omega \right) \tau_{MW} \left(1 - \tau_f / \tau_p \right)}{\left(1 + 2\varepsilon_f / \varepsilon_p \right) \left(1 + \sigma_f / 2\sigma_p \right) \left(1 + \left(\delta\Omega \right)^2 \tau_{MW}^2 \right)} \hat{\mathbf{z}}$$

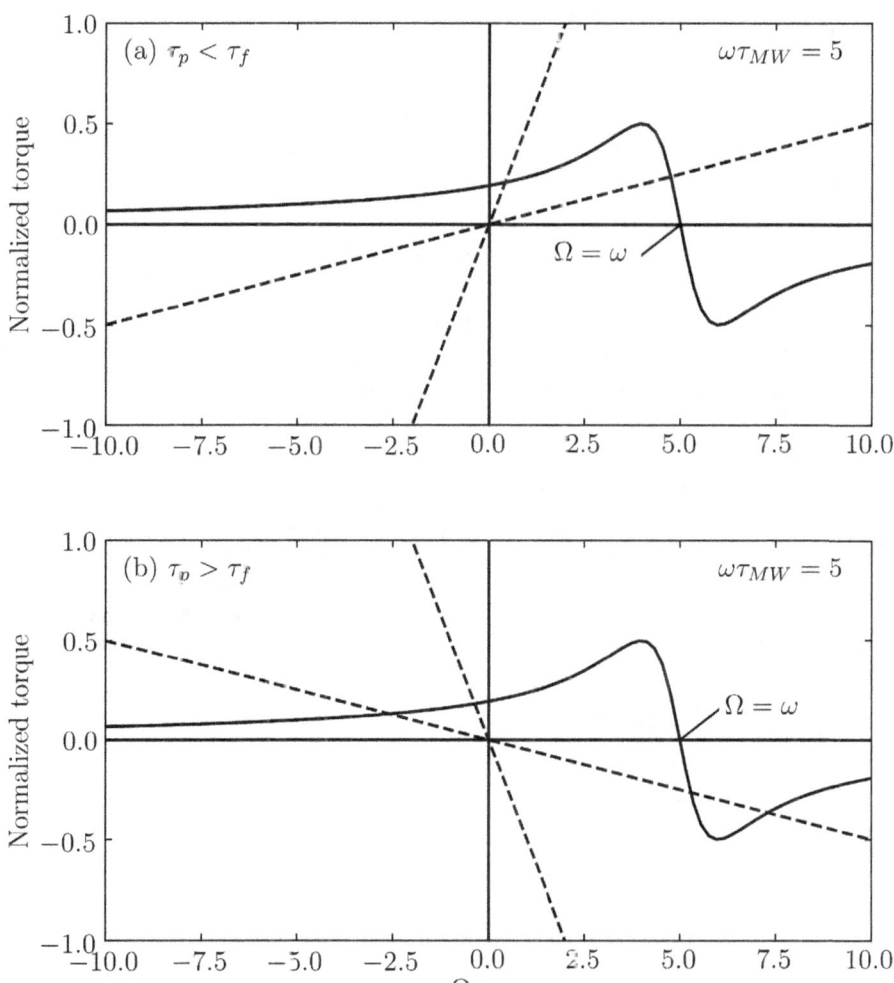

FIGURE 19.0.3. Plot of left-hand side (dashed) and right-hand side (solid) of (19.0.15), for when the factor multiplying $\Omega\tau_{MW}$ in left-hand side is positive, (a), $\tau_p < \tau_f$; and when this factor is negative, (b), $\tau_p > \tau_f$. We plot the left-hand side for a high value and low value of this factor.

where we have set Ω to $\delta\Omega$ and ω to zero, as the field is DC. Thus, due to our kick, in the frame of reference of the particle, the particle is experiencing a rotating electric field! Thus, the particle is experiencing an electrical torque. But the particle will also be experiencing a Stokes' drag torque. Hence the rotational version of Newton's second law, conservation of angular momentum, would be

$$(19.0.18) \qquad I\frac{d\left(\delta\Omega\right)}{dt} = \langle \mathbf{T} \rangle \left(\delta\Omega\right) - T_\mu \left(\delta\Omega\right)$$

Where $I = 8\pi\rho_p a^5/15$ is the moment of inertia of the spherical particle. Substituting (19.0.17) and (19.0.14) into (19.0.18) we get an ODE that is difficult to integrate. However, we can linearize the ODE if we assume that $\delta\Omega\tau_{MW} \ll 1$. Since our kick $\delta\Omega$ is supposed to be small, this is quite reasonable. We thus obtain that,

$$(19.0.19) \qquad I\frac{d\left(\delta\Omega\right)}{dt} = \left(\frac{-6\pi a^3 E_0^2 \varepsilon_0 \varepsilon_f \tau_{MW}\left(1 - \tau_f/\tau_p\right)}{\left(1 + 2\varepsilon_f/\varepsilon_p\right)\left(1 + \sigma_f/2\sigma_p\right)} - 8\pi\mu a^3\right)\delta\Omega$$

This ODE integrates to an exponential function, where the angular speed grows or decays exponentially with time. We see that the angular speed will grow with time if the term in the parenthesis on the right-hand side is positive, which is only possible when $\tau_p > \tau_f$. Furthermore, the crossover from decay in angular speed, to growth of angular speed occurs when the term in the parenthesis is zero, i.e., when the electric field is above a critical threshold,

$$(19.0.20) \qquad E_{0,crit} = \left(1 + \frac{\sigma_p}{2\sigma_f}\right)\sqrt{\frac{8\mu\sigma_f\sigma_p}{3\varepsilon_0\varepsilon_f\left(\tau_p - \tau_f\right)}}$$

This agrees with experimental observations of Quincke rotation, where particles only rotate if $\tau_p > \tau_f$ and when the electric field is above a critical value.

CHAPTER 20

Electrocapillary phenomena

In this section we will discuss phenomena that combines the effects of electrokinetics and capillarity. Our major focus will be on the phenomena in "digital microfluidics" (DMF) (sometimes known as electrowetting-on-dielectric (EWOD)), specifically the movement and splitting of droplets. To do this we will first examine fundamental electrocapillary experiments: electrowetting of a sessile drop, Quincke's bubble experiment, and Pellat's parallel plate experiment. We will then derive scaling rules for designing DMF devices and the key figures of merit for these devices.

20.1. Simple experiments in electrocapillary phenomena

To begin our study of electrocapillary phenomena we consider several classical electrocapillary experiments: electrowetting of a sessile drop demonstrating change in apparent contact angle upon application of an electric field, Quincke's bubble experiment demonstrating electrocapillary pressure and Pellat's demonstration of capillary rise between parallel plate electrodes.

20.1.1. Electrowetting of a sessile drop. We begin our study of electrocapillary phenomena by studying a simple experiment. Consider a drop sitting on top of an electrode (Figure 20.1.1), which we will call the working electrode. For practical purposes the electrode may be coated with a thin layer of a hydrophobic dielectric (e.g. PTFE film). Another, thin wire electrode (which we will call the counter electrode) is inserted into the droplet body as shown in Figure 20.1.1. When we apply an electric field between these two electrodes, we observe that the drop flattens - the contact angle of the drop decreases. To understand what is happening, let's consider the simplest case of a liquid drop directly on the electrode surface. Assume that we have chosen a liquid (e.g. silicon oil) such that does not undergo electrode reactions at the voltages we are exploring. Instead, when the electric field is applied the working electrode charges and counterions to that charge travel from the bulk of the liquid to shield that charge. Hence our familiar double layer builds up at the liquid-electrode interface (here driven by the electric field, in addition to that induced by the immobile charged groups on the solid). From Gibbs' interfacial thermodynamics, the accumulation of charges leads to a reduction in the interfacial tension (here the liquid-solid interfacial tension). This reduction of interfacial tension, γ_{sl}, is proportional to the amount of charge at the interface, and the change in potential of that charge.

$$(20.1.1) \qquad d\gamma_{sl}^{eff} = -\rho_e dV$$

Here ρ_e is the surface charge density (electrical charge per unit area) and V is voltage. (Consider that by changing the potential of that charge, you are pumping

461

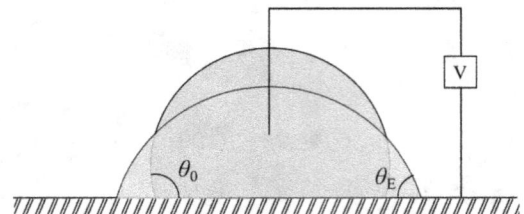

FIGURE 20.1.1. Schematic of a sessile drop on top of a flat electrode ("working electrode"), with a "counter electrode" inserted into the drop such that the liquid is always in contact with the electrodes. Initially (background) the contact angle is high. When we apply a field across the drop, the drop flattens (foreground), and the contact angle decreases.

energy into (or out of) that charge – you are pumping energy into (our out of) the interface. This in turn changes the electrostatic forces between the solid and the liquid, i.e. γ_{sl}).

To obtain an estimate for the charge in the induced double layer we will use the simplest model for the double layer – the parallel plate capacitor (otherwise known as the Helmholtz model, see Section 15.3). Hence we will assume that the ions are located a fixed distance d_H from the surface of the electrode and so the double layer has a capacitance (per area) given by

$$(20.1.2) \qquad\qquad C_H = \frac{\varepsilon_l \varepsilon_0}{d_H}$$

where ε_l is just the bulk liquid permittivity. Since this is just a parallel plate capacitor, the charge per area on it is then,

$$(20.1.3) \qquad\qquad \rho_e = C_H V$$

Substituting this into (20.1.1) and integrating we obtain,

$$(20.1.4) \qquad\qquad \gamma_{sl}^{eff}(V) = \gamma_{sl} - \int_{V_{pzc}}^{V} C_H V \, dV$$

where γ_{sl} is the solid-liquid interfacial tension when there is no charge at the interface – at the point of zero charge, really γ_{sl}^{pzc}. In practice this is taken as just the solid-liquid interfacial tension when there is no electric field is applied between the electrodes. We know that that unless the pH of the liquid is adjusted perfectly such that the interface is at point of zero charge, the interface will normally not be at the point of zero charge and the surface will have some finite zeta potential, and so there will be a difference between γ_{sl} and γ_{sl}^{pzc}. However, we assume that this difference is not significant. Furthermore, V is the voltage on the Helmholtz parallel plate capacitor. However, for practical purposes we equate this voltage with the voltage drop between the working and the counter electrodes, and in doing so we are assuming that there is negligible Ohmic loss in the conducting fluid. This assumption is not a bad one when the double layer is fully charged and no ion current flows through the bulk. For AC voltage, this is true when the AC frequency is much smaller than the inverse of the relaxation time of the double layer (see Section

17.3). Performing the integration in (20.1.4) we obtain,

$$(20.1.5) \qquad \gamma_{sl}^{eff}(V) = \gamma_{sl} - \frac{C_H}{2}(V - V_{pzc})^2$$

The Young's equation for when there is no electric field applied between the electrodes is

$$(20.1.6) \qquad \gamma_{sg} - \gamma_{sl} = \gamma_{lg} \cos \theta_0$$

where θ_0 is the equilibrium contact angle when no field is applied. When an electric field is applied, we assume that the solid-gas and liquid-gas surface tensions do not change, and only the solid-liquid surface tension does, hence the Young's equation becomes

$$(20.1.7) \qquad \gamma_{sg} - \gamma_{sl}^{eff}(V) = \gamma_{lg} \cos \theta_E$$

where θ_E is the contact angle when the electric field is applied. Subtracting (20.1.6) from (20.1.7) and combining the result with (20.1.5), we obtain

$$(20.1.8) \qquad \cos \theta_E = \cos \theta_0 + \frac{C_H}{2\gamma}(V - V_{pzc})^2$$

where γ is the liquid-gas surface tension. Relation (20.1.8) is known as the Lippmann-Young law and is named after Gabriel Lippmann who studied this around 1875 (Lippmann (1875)). This relation has been experimentally tested, and the square dependence on voltage (up to a saturation voltage) verified. C_H is typically not known and is taken as a fitting parameter.

In practice, it is more common to place a dielectric later (or several) between the liquid and the working electrode. This forms a chain of capacitors in series and so the equivalent capacitance (per area) of this chain should be used in the Lippmann-Young law (20.1.8). For capacitors in series,

$$(20.1.9) \qquad \frac{1}{C_{eq}} = \frac{1}{C_H} + \frac{1}{C_{D1}} + \frac{1}{C_{D2}}$$

where the subscripts $D1$ and $D2$ refer to the first and second dielectric layers. For most systems the double layer capacitance is significantly larger than the capacitance of the dielectrics and so it can be neglected from the equivalent capacitance,

$$(20.1.10) \qquad \frac{1}{C_{eq}} \approx \frac{1}{C_{D1}} + \frac{1}{C_{D2}}$$

We can see that two series parallel plate capacitors, can be replaced by a single effective parallel plate capacitor with

$$(20.1.11) \qquad (d/\varepsilon)_{eff} = d_1/\varepsilon_1 + d_2/\varepsilon_2$$

Furthermore, the native zeta potential of surfaces is typically less than 100 mV, and so the voltage to achieve point of zero charge is also of this order. Meanwhile the voltages typically used in electrowetting are typically tens to hundreds of volts. Hence V_{pzc} can be neglected in (20.1.8). Taking this into account, we can rewrite Lippmann-Young relation as

$$(20.1.12) \qquad \cos \theta_E = \cos \theta_0 + \frac{C_{eq}}{2\gamma}V^2$$

The second term on the right-hand side is dimensionless and is sometimes referred to as the electrowetting number.

20.1.2. Quincke's bubble experiment. Having examined the change in contact angle in a sessile drop, let's examine another classical electrocapillary experiment - Quincke's bubble experiment (Figure 20.1.2). In our discussion of this experiment, we follow the analysis of Jones et al. (2003). For this experiment we place two parallel plate electrodes separated by a small gap d into a bath of liquid. In the top electrode we drill a small hole and connect it to a manometer and to a system that can push another fluid into the gap. We use this system to push a bubble of another fluid (e.g., air) into the gap as shown in Figure 20.1.2. We then shut of the valve to the fill fluid, and open the valve to a manometer to measure the pressure in this fluid. We assume that pressure head due to gravity is small, as the gap size is small. We measure this pressure in this fluid as a function of the voltage applied to the electrodes. Depending on the permittivity of this fluid and the surrounding liquid, the pressure either increases or decreases depending on the difference in permittivity of the liquid and the fluid. We can explain this based on change in contact angle at the interface between the fluid and the liquid, but in this section we will take another approach. If we look closely at the setup of this experiment, this situation reminds us of the capacitor experiments we considered in Section 14. In fact, we will study this situation using the Maxwell stress tensor approach. Recall from Section 14 that the electrical body force is given by

$$(20.1.13) \qquad \mathbf{f} = \rho_f \mathbf{E} - \frac{\varepsilon_0 E^2}{2} \nabla \varepsilon + \nabla \left(\frac{\varepsilon_0 E^2}{2} \rho \frac{\partial \varepsilon}{\partial \rho} \right)$$

and the associated Maxwell stress tensor is

$$(20.1.14) \qquad T_{mn} = \varepsilon \varepsilon_0 E_m E_n - \frac{\delta_{mn} \varepsilon \varepsilon_0}{2} E_k E_k + \frac{\delta_{mn} \varepsilon_0}{2} \rho \frac{\partial \varepsilon}{\partial \rho}$$

Recall that the first term in (20.1.13) is due to free charge, the second term due to a gradient in permittivity, and the third term due to the compressibility of the media. In (20.1.14) the third term is from the force due to compressibility of the fluid, whereas the first and second terms are from the "material inhomogeneity force." In this case we will ignore the compressibility of the fluid inside the bubble. If this fluid is a liquid, this is quite reasonable (here we will still refer to this as a "bubble" even though a droplet may be more appropriate). This is less reasonable assumption for a gas bubble, but if the pressures developed are not too high and $\partial \varepsilon / \partial \rho$ is also not too high, then this can be a reasonable assumption. Furthermore, we note that we don't expect any free charge in the bulk of the fluid, but anticipate free charge build up at the interfaces (and for AC voltages, at least at low frequencies when the double layers have time to relax and build up charge at the interfaces). Thus, in our case the Maxwell stress tensor, without electrostriction simplifies to

$$(20.1.15) \qquad T_{mn} \approx \varepsilon \varepsilon_0 E_m E_n - \delta_{mn} \frac{\varepsilon \varepsilon_0}{2} E_k E_k$$

To determine the net force on an object enclosed in a control volume, we take the surface integral around the surface enclosed by that volume,

$$(20.1.16) \qquad F = \oint_{\Sigma} T_{mn} \mathbf{n}_n dA$$

where Σ is the enclosing surface, and $\mathbf{n}_n$ is the unit normal on the nth face of Σ. The convenience of this approach is that we can choose any convenient control volume, as we only need to evaluate the electric field on its surface. Therefore we choose a control volume such that the weird fringing electric field at the fluid-liquid

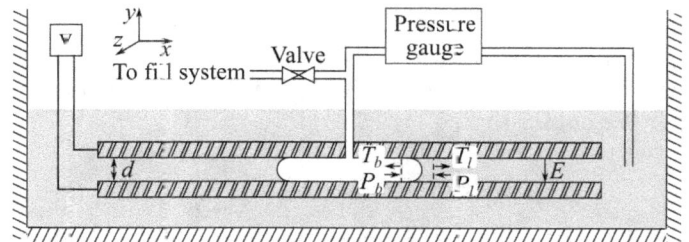

FIGURE 20.1.2. Schematic of Quincke's bubble experiment. We immerse two parallel plate electrodes into a bath of liquid. The top electrode has a small hole drilled into it allowing us to pump in another fluid into the between-electrode-space as needed, creating a bubble or a drop. This fill system is also connected to a pressure gauge allowing us to monitor the pressure in the bubble or the drop. This allows us to monitor the pressure inside the bubble as electric field is applied between the electrodes. To help us calculate the forces on the liquids due to the electric field via the Maxwell stress approach, we draw a control volume (dashed lines) that encompasses both the bubble fluid and the bath liquid. For detailed, zoomed in view of the control volume, see Figure 20.1.3.

interface is inside the control volume and we don't have to worry about knowing it (Figure 20.1.2). Furthermore, for the control volume we chose, we don't need to worry about knowing the charge build up at the fluid-liquid interface. Lastly, we don't even need to worry about what happens at the apparent three-phase contact line – as this too is inside the control volume. The beauty of the Maxwell stress tensor approach is you shove everything you don't know inside the control volume.

For our particular situation, the electric field is only in one dimension, the y direction (at the walls of the control volume), as we place the walls of the control volume far enough away from the interface such that the fringe fields are negligible. Taking this into account, we can simplify the Maxwell stress tensor (20.1.15) further, to

$$(20.1.17) \qquad T_{mn} = -\delta_{mn} \frac{\varepsilon \varepsilon_0}{2} E_k E_k$$

Thus, we see that the electrical (Maxwell) force on the control volume is normal to the vertical faces of the control volume. Consider the control volume in Figure 20.1.2. P_b and P_l are the hydrostatic pressures inside the bubble and the liquid respectively. T_b and T_l are the Maxwell surface stresses acting on the vertical faces of the control volume from the bubble side and the liquid side respectively. Combining the pressures and stresses on the control volume and assuming the control volume does not accelerate we obtain,

$$(20.1.18) \qquad T_b - P_b = T_l - P_l$$

With the manometer we are measuring the differential pressure between the bubble and the liquid, $P_b - P_l = \Delta P_E$, which comparing with (20.1.18) we see is

$$(20.1.19) \qquad \Delta P_E = T_b - T_l$$

To evaluate each of the stresses we use (20.1.17). To obtain the electric field in each region we made sure to draw the walls of the control volume far enough from the interface so the field is uniform in the region, hence we can estimate the field simply as V/d. We obtain that

$$(20.1.20) \qquad \Delta P_E = -\frac{1}{2}\left(\varepsilon_b - \varepsilon_l\right)\varepsilon_0\left(\frac{V}{d}\right)^2$$

This gives us an equation to predict the pressure changes in the bubble as a function of voltage. Notice again the quadratic dependence on the voltage as in the previous section.

Above we obtained a pressure relation for the simplest Quincke's bubble apparatus, that with bare electrodes, but a more practical apparatus would have the electrodes be coated with the dielectric. This does not change the analysis much, but does change the electric field and the forces. We show a schematic of this more realistic Quincke's bubble apparatus in Figure 20.1.3a, including the new control volume in a more zoomed in view onto the fluid-liquid interface. In Figure 20.1.3b, we show the equivalent circuit model, which allows us to calculate voltages at different interfaces and therefore obtain the relevant electric fields. In this apparatus the fluid has both a finite conductivity σ_f and finite permittivity ε_f. The dielectric has a thickness d_d, which is much smaller than the electrode gap thickness d. The capacitance (per unit area) of the dielectric is

$$(20.1.21) \qquad C_d = \varepsilon_d \varepsilon_0 / d_d$$

The capacitance per unit area of the liquid phase is

$$(20.1.22) \qquad C_l = \varepsilon_l \varepsilon_0 / d$$

The capacitance per unit area of the fluid phase is

$$(20.1.23) \qquad C_f = \varepsilon_f \varepsilon_0 / d$$

And the conductance (inverse of resistance) per unit area of the fluid phase is

$$(20.1.24) \qquad G_f = \sigma_w / d$$

Notice that we have neglected the double layer capacitance for both the liquid and the fluid. As we have mentioned in the section above, the double layer capacitance is typically much larger than the capacitance of the dielectric and because of the capacitor in series arrangement, the double layer capacitance is then negligible. We apply an AC voltage in the form of

$$(20.1.25) \qquad V\left(t\right) = \mathrm{Re}\left[\sqrt{2}V\exp\left(j\omega t\right)\right]$$

where V is the root mean square (RMS) magnitude of the applied voltage. Working through the circuit in Figure 20.1.3b we can find the voltages and then assuming that the electric field is uniform, the electric fields. We find that the electric field in the dielectric bordering the liquid phase is

$$(20.1.26) \qquad E_{dl} = \frac{C_l}{2C_l + C_d}\frac{V}{d_d}$$

The electric field in the liquid phase (at the wall of the control volume),

$$(20.1.27) \qquad E_l = \frac{C_d}{2C_l + C_d}\frac{V}{d}$$

The electric field in the dielectric bordering the fluid phase is,

$$(20.1.28) \qquad E_{df} = \text{Re}\left[\frac{j\omega C_f + G_f}{j\omega\left(2C_f + C_d\right) + 2G_f}\frac{V}{d_d}\right]$$

And the electric field in the fluid phase is (at the wall of the control volume),

$$(20.1.29) \qquad E_f = \text{Re}\left[\frac{j\omega C_d}{j\omega\left(2C_f + C_d\right) + 2G_f}\frac{V}{d}\right]$$

Now we calculate the pressure on the control volume as before using (20.1.17)

$$\langle\Delta P\rangle = -\frac{1}{2}\varepsilon_0\left(\varepsilon_f E_f^2 - \varepsilon_l E_l^2\right)$$

Where the angular brackets indicate a time average over a time longer than the period of oscillation of the electric field. Similarly, we can find electrical force (per unit length into the page) on the control volume as before, since the force per unit length is this pressure multiplied by the height of the control volume,

$$(20.1.30) \qquad \langle F_x\rangle = -\frac{1}{2}\varepsilon_0\left(\varepsilon_f E_f^2 - \varepsilon_l E_l^2\right)d$$

By observation of (20.1.29), we see that there is a critical frequency

$$(20.1.31) \qquad \omega_c = \frac{2G_f}{2C_f + C_d}$$

above which versus below which the behavior of the electric field in the fluid and therefore the pressure generated by electric fields is different. For example, for $\omega \ll \omega_c$ the fluid behaves like a perfectly conducting medium and the electric field inside the fluid phase is zero. The voltage drop is concentrated in the liquid phase. When on the other hand, $\omega \gg \omega_c$ the fluid behaves like a dielectric and whole circuit behaves like a capacitive voltage divider. Thus, in these two limits, the pressure is

$$(20.1.32) \qquad \langle\Delta P\rangle = -\frac{\varepsilon_d\varepsilon_0}{4d_d d}V^2 \text{ when } \omega \ll \omega_c$$

$$(20.1.33) \qquad \langle\Delta P\rangle = -\frac{\left(\varepsilon_f - \varepsilon_l\right)\varepsilon_0}{2d^2}V^2 \text{ when } \omega \gg \omega_c$$

Notice that in the low frequency regime, the properties of the dielectric dominate, while in the high frequency regime, the effect of the dielectric disappears and the relation for pressure collapses to that for the case where dielectric is not present (compare (20.1.33) with (20.1.20)). The low frequency limit is sometimes referred to as the electrowetting on dielectric (EWOD) limit, while the high frequency limit is sometimes referred to as the liquid dielectrophoresis limit. As we will discuss this below, it is often advantageous to operate digital microfluidics system in the low frequency limit as fluid motion becomes independent of fluid conductivity and permittivity. This is particularly advantageous as it allows us to move and mix droplets with different conductivity with the same electrical parameters. As an example, for an example fluid of phosphate buffered saline, with a conductivity of 2 S/m, relative permittivity of 80 and an example device with a dielectric having d_d/ε_d of 100 nm and electrode gap of 20 µm, we find that the critical frequency is 1 GHz. For an example solution of DI water (equilibrated with atmosphere, without special handling) with a conductivity of 10^{-4} S/m, the critical frequency goes down to 63 kHz.

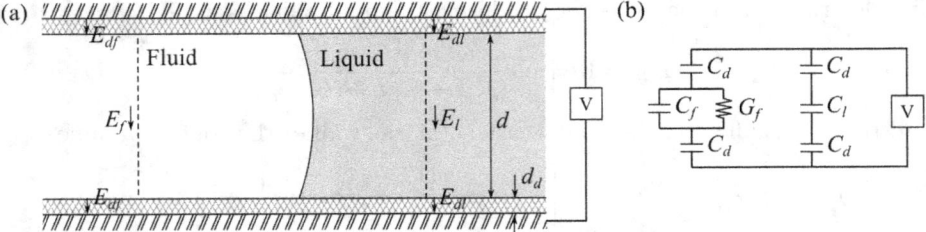

FIGURE 20.1.3. (a) Zoom in onto the interface between the introduced fluid and the bath liquid in the Quincke's bubble experiment, including the details of the control volume for obtaining the electrical forces on the two fluids. (b) An equivalent circuit for the control volume. The left side of the circuit represents the introduced fluid (subscript f), while the right side of the circuit represents the bath liquid (subscript l); subscript d represents the double layer formed at the liquid (fluid) solid interfaces.

20.1.3. Pellat's parallel plate experiment. Having explored Quincke's bubble experiment, we now consider a very similar experiment – Pellat's parallel plate experiment. In our discussion of this experiment, we follow the analysis of Jones et al. (2004). In this experiment, we turn the Quincke's apparatus on the side so that the gravity vector points parallel to the plates (and forget about the hole for introducing another fluid). In this apparatus, we observe the height of rise of the fluid as a function of the voltage applied to the plate. We show a schematic of this apparatus, including the control volume around the fluid air interface in Figure 20.1.4a; in Figure 20.1.4b we show the associated circuit diagram. Notice that the circuit is the same as for the Pellat's experiment, with the exception that the names of the different phases has been changed. With the new names, the electrical fields are:

$$(20.1.34) \qquad E_{da} = \frac{C_a}{2C_a + C_d} \frac{V}{d_d}$$

$$(20.1.35) \qquad E_a = \frac{C_d}{2C_a + C_d} \frac{V}{d}$$

$$(20.1.36) \qquad E_{df} = \mathrm{Re}\left[\frac{j\omega C_f + G_f}{j\omega\left(2C_f + C_d\right) + 2G_f} \frac{V}{d_d}\right]$$

$$(20.1.37) \qquad E_f = \mathrm{Re}\left[\frac{j\omega C_d}{j\omega\left(2C_f + C_d\right) + 2G_f} \frac{V}{d}\right]$$

Where the subscripts d imply the dielectric, and the second subscript a and f imply neighboring air or fluid. The critical frequency is again,

$$(20.1.38) \qquad \omega_c = \frac{2G_f}{2C_f + C_d}$$

The electrical forces on the walls of the control volume are the same as in Quincke's apparatus. However, now we have additional forces due to gravity. Thus, the force equation becomes

$$(20.1.39) \qquad \langle F_y \rangle = \frac{1}{2}\varepsilon_0 \left(\varepsilon_f E_f^2 - \varepsilon_a E_a^2 \right) wd - \rho_f hgwd$$

where h is the height of the control volume from the level of fluid in the bath. Here we have assumed that the density of air is negligible compared to the density of the fluid. w is the length of the plates into the board. When the fluid reaches its equilibrium height for a given voltage the net force $\langle F_y \rangle$ is zero. Thus, solving for the height of the interface we find,

$$(20.1.40) \qquad h = \frac{1}{2}\frac{\varepsilon_0}{\rho_f g} \left(\varepsilon_f E_f^2 - \varepsilon_a E_a^2 \right)$$

Once again, we can take a look at the low and high frequency limits. We find that the height-of-rise h in these limits is

$$(20.1.41) \qquad h = \frac{\varepsilon_d \varepsilon_0}{4\rho_f g d_d d} V^2 \text{ when } \omega \ll \omega_c$$

$$(20.1.42) \qquad h = \frac{(\varepsilon_f - \varepsilon_a)\varepsilon_0}{2\rho_f g d^2} V^2 \text{ when } \omega \gg \omega_c$$

We can approach this experiment also from the point of view of changing contact angle, rather than looking at it from the point of view of Maxwell stress. The capillary rise height for liquid between two plates is

$$(20.1.43) \qquad h = \frac{2\gamma \cos\theta}{d\rho_f g}$$

Substituting in Young-Lippmann relation (20.1.12) and noting that C_{eq} in this case is equal to C_d and noting that we are using RMS voltage defined as per (20.1.25) we find that

$$(20.1.44) \qquad h = \frac{2\gamma \cos\theta_0}{d\rho_f g} + \frac{\varepsilon_d \varepsilon_0}{4 d_d d\rho_f g} V^2$$

We see that the additional height-of-rise due to the electric field (the second term in (20.1.44)) predicted by using Young-Lippmann relation is exactly the same as that in (20.1.41) predicted using the more general Maxwell stress approach. It shouldn't be surprising that the two results agree – both were derived from using a conservation of energy approach. Hence the two approaches are equivalent. In this case, the Maxwell stress approach proved to be more insightful as it predicts a frequency dependence for the height-of-rise and predicts that there are two regimes.

20.2. Scaling of digital microfluidic devices

In this section we will explore the design space for digital microfluidic (DMF) devices, sometimes referred to as electrowetting-on-dielectric (EWD or EWOD) microfluidics. These devices consist of an array of electrodes that actuate individual droplets to move, merge, and split these. With these (and a few additional auxiliary) operations this microfluidic scheme allows the individual droplets to behave as individual enclosed chemical reactors. This in turn allows the microfluidic system to perform a large number of individual chemical reactions in parallel, as well as perform complex serial reaction operations. The mechanism of actuation of these droplets are based on the forces we have considered in the previous section – Maxwell stress at the interface between the droplet and the surrounding fluid (commonly,

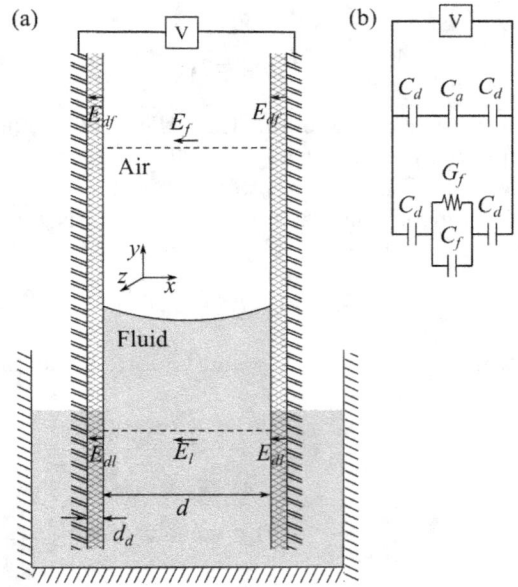

FIGURE 20.1.4. (a) Schematic of Pellat's apparatus with the gap between the plates exaggerated for clarity. For this experiment, we immerse parallel plate electrodes coated with a dielectric into a dielectric bath and wait for the liquid to rise to an equilibrium height. We then apply an electric field between the plates and observe the bath fluid rise further as an electric field is applied. To calculate the forces on the fluids we employ the control volume shown. (b) An equivalent circuit for situation inside the control volume in Pellat's apparatus, with the top part of the circuit representing air (subscript a) and the bottom part representing the bath fluid (subscript f); subscript d represents the double layer formed at the fluid-solid interfaces.

air, but sometimes silicon oil). As we have seen in the previous section, this is largely equivalent to attributing the actuation to the modulation of the interfacial tension between the liquid of the drop and the solid comprising he dielectric above the electrode. In this section, we will consider the droplet actuation from the later point of view, and in our approach follow the analysis of Song et al. (2009). While this point of view may not always yield as much insight as the Maxwell stress tensor approach, it is the approach most commonly used in the DMF literature and so we will use it also, as to help the reader interface with this literature. Here we will obtain scaling arguments and rules of thumb helpful for design of DMF devices.

In a DMF device we apply an electric field on an electrode (coated with a dielectric layer) beneath one part of the droplet. This creates an imbalance in the liquid-dielectric surface tension, causing the droplet to move. (Equivalently, there is an imbalance in the Maxwell stress between one part of the droplet vs. another and this causes the droplet to move.) In this section we will consider a DMF device where the droplets are sandwiched between an array of electrodes and a ground electrode, the so called "covered configuration", as this is the most common configuration (Figure 20.2.1). (The other two less common configurations are the "catena configuration" where the ground is provided by thin wires stretched above the electrode array, and the "open configuration" which does not have a ground plane, and just consists of an array of working electrodes on top of which the droplets move). In the closed configuration, we place droplets between two parallel plates coated with a dielectric. This dielectric typically consists of a separate dielectric layer and a separate low contact angle hysteresis (low roll off angle) surface, that is actually in contact with the drop. This low hysteresis surface is often a hydrophobic surface, especially when aqueous drops are used. The space around the drops is typically filled with air or (less commonly) silicon oil. The electrode array beneath this combined dielectric layer is connected to AC or DC power supply through a switching array, such that each electrode can be actuated individually. The droplet begins its journey positioned on the electrode, such that the droplet size and the electrode size roughly match (Figure 20.2.1). In typical DMF devices the Bond number is small and so gravitational forces have negligible effect on both the droplet interface or in the motion of the droplet. Hence the shape of a droplet at rest is approximately that of a sphere cut by two planes. To move a droplet onto a neighboring electrode, we apply a voltage (that exceeds a threshold voltage) to neighboring electrode. This changes (lowers) the contact angle of the droplet on this side, creating a force imbalance on the droplet and accelerating the droplet towards this neighboring electrode.

20.2.1. Scaling for droplet motion. To determine droplet velocity, let's first consider the scaling of forces on the droplet, starting with the electrical actuation force. The force per unit length of the triple line, on the apparent triple line on the part of the drop above the actuated electrode scales as

$$(20.2.1) \qquad f_R \sim \gamma_{sg} - \gamma_{lg} \cos \theta_0 - \gamma_{sl} \left(V \right)$$

whereas the force per unit length on the other side of the droplet, above the non-actuated electrode is

$$(20.2.2) \qquad f_L \sim \gamma_{sl} \left(0 \right) + \gamma_{lg} \cos \theta_0 - \gamma_{sg}$$

The net driving force per unit length then scales as the sum of these two forces,

$$(20.2.3) \qquad f_m \sim f_A + f_B \sim \gamma_{sl} \left(0 \right) - \gamma_{sl} \left(V \right)$$

Note that the amount of the droplet over the actuated area changes as the droplet moves on the neighboring electrode. Integrating of the driving force f_m over the contact line of a droplet on an electrode with a characteristic length L gives us an estimate of the net actuation force on the droplet,

$$(20.2.4) \qquad F_m \sim f_m L \sin \phi$$

Note that the angle ϕ changes as the droplet traverses and so does the net force on the droplet. In addition, we have to account for hysteresis – the difference between the receding and advancing contact angle, α – as the front of the droplet is advancing and the rear of the droplet is receding. Thus, at the rear of the drop we will see a contact angle of $\theta \left(0 \right) - \alpha$ and at the front of the drop, $\theta \left(V \right) + \alpha$. Typically α is between 1.5-2° for water droplet in silicon oil on SiOC surface, and 7-9° for water droplet in air on PTFE surface (Berthier (2012)). Note that in this analysis we treat the contact angle hysteresis as a simple processes represented by a single number, and this is a simplifying assumption. Furthermore, we don't distinguish

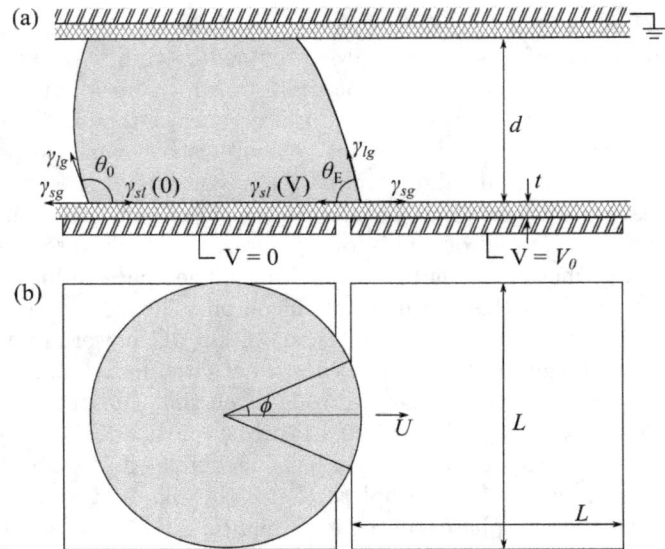

FIGURE 20.2.1. Schematic of a traditional, "covered configuration" of a digital microfluidics device. In this configuration a drop of liquid is sandwiched between a ground electrode (coated with a dielectric) and an array of working electrodes (again coated with a dielectric). Typically, in a resting configuration, all working electrodes are grounded. When we want to move a droplet to a neighboring electrode we apply a voltage to that electrode (DC or AC). This decreases the contact angle on that side of the droplet, effectively making the surface above that electrode appear more wetting, thus enticing the droplet to move over this electrode. (a) side view; (b) top view of this configuration.

between the dynamic and static contact angle, again as a simplification. Also note that in our Maxwell stress analysis in the Quincke apparatus, we neglected the effect of hysteresis altogether. Neglecting hysteresis has the effect that we cannot predict the experimentally observed threshold voltage – a voltage above which the drop just begins to move, when it just begins to overcome the differences in the advancing and receding contact angle. Thus taking hysteresis into account the force per unit length scales as

$$(20.2.5) \qquad f_m \sim \gamma_{lg} \cos\left(\theta\left(V\right) + \alpha\right) - \gamma_{lg} \cos\left(\theta\left(0\right) - \alpha\right)$$

Note that if the neighboring electrode voltage is zero, the droplet will not move there and so there is no hysteresis – hence in that case α is also zero, so then the force collapses to zero, as expected. We can simplify the above equation by using the Young-Lippmann equation, and obtain

$$(20.2.6) \qquad f_m \sim \cos\alpha \frac{C_{eq}}{2} V^2 - \gamma \sin\alpha \left(\sin\theta\left(V\right) + \sin\theta\left(0\right)\right)$$

and using (20.2.4) we obtain

(20.2.7) $F_m \sim \left(\cos \alpha \dfrac{C_{eq}}{2} V^2 - \gamma \sin \alpha \left(\sin \theta \left(V \right) + \sin \theta \left(0 \right) \right) \right) L \sin \phi$

where $C_{eq} = \varepsilon \varepsilon_0 / t$ for a single layer dielectric; see (20.1.11) for a multilayer dielectric. This equation gives us insight into both the effect of contact angle hysteresis and surface tension. Notice the role of contact angle hysteresis – it robs us of the actuation force for a given voltage. Thus for a low voltage DMF device we want to choose surfaces with as low hysteresis as possible. Furthermore, we want to ensure that all of our droplets have roughly the same hysteresis, or we will need to compensate for this by adjusting the actuation voltage for different droplets. This is less of an issue for dilute solutions but can be an issue for concentrated solutions, or solutions heavily laden with particles and/or beads. Furthermore, notice the role of surface tension – it again robs us of the actuation force. Hence, often surfactant is added to the droplets to decrease the surface tension. However, we have to be careful with adding surfactant and make sure that the surfactant does not interfere with the chemical processes we are using DMF for or change the contact angle of the dielectric surface.

Having obtained a scaling for the driving force on the droplet, let's now turn to the drag forces the droplet experiences. The droplet experiences two types of drag forces: drag from the surrounding fluid, and drag from the plates. We can obtain a very rough scaling for the drag from the surrounding medium for a droplet moving at relatively low velocities (such that its Reynolds number is low), from the Stokes' drag on a sphere,

(20.2.8) $F_{DS} \sim 3 \pi \mu_f L U$

Where μ_f is the viscosity of the surrounding fluid, U is the characteristic velocity of the drop and L is the characteristic length scale of the electrode and so of the same order as the diameter of the drop. Note that this is very approximate, as this relation is meant to describe a drag on a rigid sphere far away from walls, and our drop is neither rigid, spherical, and is in fact in contact with the walls. To obtain a scaling for the drag force from the plates, we can imagine that the drop experiences Poiseuille flow type drag, and write the drag in the similar form as,

(20.2.9) $F_{DP} \sim 2 C_v \dfrac{\mu_d U L^2}{d}$

where μ_d is the viscosity of the drop, d is the spacing (gap) between the plates, and C_v is an empirical constant and ranges between 2 and 15 and depends on d/L. When the velocity in the drop profile is roughly parabolic, i.e., the maximum velocity occurs in the center of the drop, C_v can be taken as 6. These two drag forces balance the actuation force,

(20.2.10) $F_m = F_{DS} + F_{DP}$

Substituting the individual relations into (20.2.10) and solving for the droplet velocity we obtain

(20.2.11) $U \sim \dfrac{\left(\cos \alpha \frac{C_{eq}}{2} V^2 - \gamma \sin \alpha \left(\sin \theta \left(V \right) + \sin \theta \left(0 \right) \right) \right) \sin \phi}{12 \mu_f \frac{d}{L} + 2 C_v \mu_d \frac{L}{d}}$

Notice that the velocity scales linearly with the square of the voltage. We also see that the denominator of the velocity scales either as d/L or as L/d depending if the surrounding fluid viscosity is dominant or if the droplet viscosity is dominant. For an aqueous droplet surrounded by air, since the viscosity of water is about 500 times that of air, the denominator scales as L/d,

$$(20.2.12) \qquad U_{w-in-air} \sim \frac{\left(\cos\alpha \frac{C_{eq}}{2} V^2 - \gamma\sin\alpha\left(\sin\theta\left(V\right) + \sin\theta\left(0\right)\right)\right)\sin\phi}{2C_v\mu_d L/d}$$

We see that for such a system we want to minimize L/d. However, practically, we cannot make L/d even close unity, as then the droplet will not be in contact with both the working electrode and the ground electrode.

Observing (20.2.11), we see that there is a voltage at which the droplet velocity, as predicted by this equation is zero. This is the minimum voltage to get the droplet to move against contact angle hysteresis, and is termed the threshold voltage. Setting U in (20.2.11) to zero, we find,

$$(20.2.13) \qquad V_T = \sqrt{\frac{2\gamma}{C_{eq}}\tan\alpha\left(\sin\theta\left(V\right) + \sin\theta\left(0\right)\right)}$$

For convenience we rewrite (20.2.11) in terms of this threshold voltage,

$$(20.2.14) \qquad U \sim \frac{C_{eq}\sin\phi\cos\alpha\left(V^2 - V_T^2\right)}{2\left(12\mu_f d/L + 2C_v\mu_d L/d\right)}$$

where the hysteresis terms have been folded into V_T.

20.2.2. Scaling for droplet splitting. Next important operation in DMF after droplet movement is droplet splitting. To split a droplet we actuate the two electrodes neighboring the electrode currently housing the droplet (Figure 20.2.2). This lowers the contact angles θ_{b2} on both sides of the droplet, and so pulls the droplet in both directions onto the neighboring electrodes. This in turn reduces the radius of curvature r_2, which then creates a meniscus over the center electrode, as there has to be conservation of liquid. The neck generated (with radius R_1) grows and narrows, and droplet splits in two (Figure 20.2.2). Examining this process from a static point of view, the static balance of capillary forces on the droplet (from geometry in Figure 20.2.2) is such that the droplet splits when

$$(20.2.15) \qquad 1/R_1 = 1/R_2 - (\cos\theta_{b2} - \cos\theta_{b1})/d$$

By applying voltage to the neighboring electrodes we control the contact angles θ_{b1} and θ_{b2}. Thus substituting the Young-Lippmann equation accounting for hysteresis into (20.2.15)

$$(20.2.16) \qquad 1/R_1 = 1/R_2 - C_{eq}\frac{\cos\alpha}{2d\gamma}\left(V^2 - V_T^2\right)$$

To obtain the splitting voltage we need to evaluate characteristic scales for R_1 and R_2. From geometry in Figure 20.2.2, we see R_1 scales as L and R_2 scales as $L/2$. Thus, the splitting voltage scales as

$$(20.2.17) \qquad \left(V^2 - V_T^2\right) \sim \frac{d}{L}\frac{2\gamma}{C_{eq}\cos\alpha}$$

Droplet splitting operations typically require a larger voltage than droplet movement operations. Analogous to droplet splitting is the droplet dispensing operation

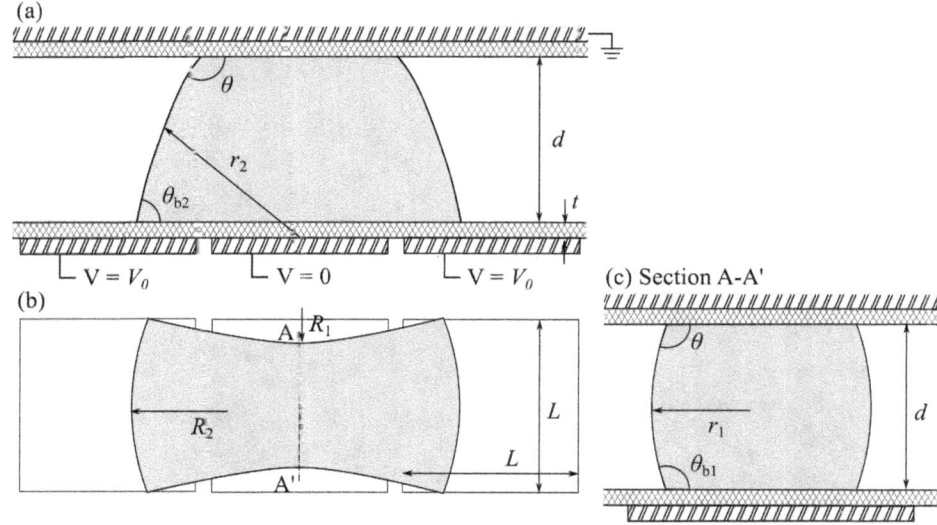

FIGURE 20.2.2. Schematic of droplet splitting in a traditional, "covered configuration" digital microfluidic device. (a) side view; (b) top view, (c) cross-section view through section A-A'. Once again, in a resting configuration, all working electrodes are grounded. To split a droplet, we apply a voltage to the neighboring electrodes, which makes the surface above these electrodes appear more wetting. The droplet is thus stretched into a shape in (b) (shape, and sharpness greatly exaggerated for clarity). The droplet necks above the grounded, central electrode and as the neck narrows, the droplet splits into two.

- splitting of a small droplet from a much larger droplet which resides on a large, "reservoir" electrode. Due to the size of the reservoir electrode, droplet dispensing typically requires lower voltage than droplet splitting.

Appendix

CHAPTER 21

Appendix A: Review of tensors

In this text we use tensors and results from tensor calculus to study hydrodynamic and electric fields. In this appendix we will briefly review properties of tensors, and refer the reader to the text by Aris (2012) for a more detailed discussion. Tensors are geometric objects that describe linear relations. We commonly encounter three types of tensors: zeroth rank (or order) tensor - the familiar scalar; the first rank tensor - the familiar vector; and the second rank tensor, which is often represented by the familiar 2D matrix. Even higher rank tensors of course exist (represented by higher dimension matrices) but we will not discuss them in detail here.

To get an intuitive understanding of tensors let's begin with tensors we are most familiar with: scalars, and vectors. Notice that a scalar has no direction. A vector has a single direction. So a second rank tensor has two directions. In this text we often use second rank tensors to describe stress on an element of a fluid. Stress inherently has two directions: the direction of the face it is applied to and the direction in which it is applied (Figure 21.0.1). For example, σ_{21} is the stress acting on the face that is oriented with its normal in direction x_2 and with the stress acting in direction x_1.

A second rank tensor is said to be symmetric if for all of its components $A_{ij} = A_{ji}$. A second rank tensor is said to be antisymmetric if for all of its components $A_{ij} = -A_{ji}$. Notice that by this definition for antisymmetric tensors, the components that must have the same value i.e., A_{11}, A_{22}, A_{33} must be zero. In other words, the diagonal components of an antisymmetric tensor are zero!

With second rank tensors we can continue to use the Einstein summation notation. That is if an index appears twice in a term, we imply a summation with

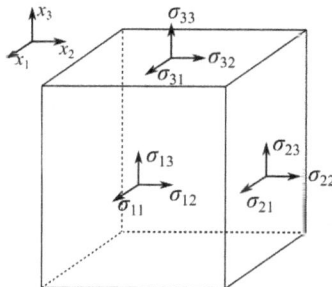

FIGURE 21.0.1. Sketch of a Cauchy stress tensor, a second order tensor. The first index in the stress indicates the direction of the normal of the face, and the second index indicates direction of the stress on that face.

respect to that index of the dimensions of the space. We apply this rule always when an index is repeated in a term. For example, for 3D space,

$$(21.0.1) \qquad A_{ii} = A_{11} + A_{22} + A_{33}$$

$$(21.0.2) \qquad u_i T_{ji} = u_1 T_{j1} + u_2 T_{j2} + u_3 T_{j3}$$

The transpose of a second rank tensor $\mathbf{A}$ is defined as $\mathbf{A}^T$ such that

$$(21.0.3) \qquad A_{ij} = A_{ji}$$

We can see from this definition that the transpose of a symmetric second rank tensor is itself.

The trace of a second rank tensor $\mathbf{A}$ is defined as a scalar A_{ii} or

$$(21.0.4) \qquad \text{trace } \mathbf{A} = A_{ii} = A_{11} + A_{22} + A_{33}$$

Addition of tensors is term by term,

$$(21.0.5) \qquad \mathbf{A} + \mathbf{B} = \mathbf{C} \Rightarrow A_{ij} + B_{ij} = C_{ij}$$

and so the tensors must be both the same rank and also the same size.

We can define quite a number of different products of tensors. The first class of products and the simplest product is the element wise product, otherwise known as Hadamard (or Schur) product

$$(21.0.6) \qquad \mathbf{A} \circ \mathbf{B} = \mathbf{C} \Rightarrow A_{ij} B_{ij} = C_{ij}$$

For this product the tensors must be both the same rank and also the same size. The result is a tensor of the same rank and size. Hadamard product is commutative, associative, and distributive over addition,

$$(21.0.7) \qquad \mathbf{A} \circ \mathbf{B} = \mathbf{B} \circ \mathbf{A}$$

$$(21.0.8) \qquad \mathbf{A} \circ (\mathbf{B} \circ \mathbf{C}) = (\mathbf{A} \circ \mathbf{B}) \circ \mathbf{C}$$

$$(21.0.9) \qquad \mathbf{A} \circ (\mathbf{B} + \mathbf{C}) = \mathbf{A} \circ \mathbf{B} + \mathbf{A} \circ \mathbf{C}$$

The second class of products is the direct product. For a direct product of two tensors we multiply the components of the two tensors together, pair by pair. The result is a tensor whose rank is the sum of the ranks of the tensors being multiplied. However, this product should not be confused with element wise multiplication. When A and B are scalars we get the familiar scalar product

$$(21.0.10) \qquad AB = C$$

The direct product of two zeroth rank tensor is a zeroth rank tensor. When A is a scalar and $\mathbf{B}$ is a second order tensor,

$$(21.0.11) \qquad A\mathbf{B} = \mathbf{C} \Rightarrow AB_{ij} = C_{ij}$$

and so a product of a zeroth rank tensor and a second order tensor is a second order tensor. When $\mathbf{A}$ is a vector and $\mathbf{B}$ is a vector,

$$(21.0.12) \qquad \mathbf{AB} = \mathbf{C} \Rightarrow A_i B_j = C_{ij}$$

The direct product of two first rank tensors is a second rank tensor. This product is sometimes referred to as a dyadic product, and is symbolized as

$$(21.0.13) \qquad \mathbf{A} \otimes \mathbf{B} = \mathbf{C} \Rightarrow A_i B_j = C_{ij}$$

When $\mathbf{A}$ is a vector and $\mathbf{B}$ is a second rank tensor,

$$(21.0.14) \qquad \mathbf{AB} = \mathbf{C} \Rightarrow A_i B_{jk} = C_{ijk}$$

and so a product of a first rank tensor and a second rank tensor is a third rank tensor. Note that direct product is in general not commutative. Of course when taking direct product of two tensors we must use different indices, so that not to confuse multiplication with summation over repeated index. Thus, $\mathbf{AB} = A_i B_j$ and not $A_i B_i$.

The third class of products is the inner (dot) product. We perform the inner product between two tensors by first forming the direct product between these two tensors, and then setting the two nearest indicies (with one index coming from each tensor) equal to one another. This results in the repeated index, and we then perform summation according to the summation convention. For example, for first rank tensors,

$$(21.0.15) \qquad \mathbf{A} \cdot \mathbf{B} = C$$

we first write $A_i B_j$, then set $i = j$, and so get $A_i B_i = A_1 B_1 + A_2 B_2 + A_3 B_3$. The rank of tensors resulting from taking an inner product is equal to the sum of the ranks of the tensors that entered into the product minus 2. The summing over the two indices lowers the rank by two. Hence the rank of a tensor participating in an inner product must be greater than zero (it must have an index that it contributes to sum over). When $\mathbf{A}$ is a vector and $\mathbf{B}$ is a second rank tensor the inner product is

$$(21.0.16) \qquad \mathbf{A} \cdot \mathbf{B} = \mathbf{C} \Rightarrow A_i B_{jk} \delta_{ij} = A_i B_{ik} = C_k$$

where δ_{ij} is the Kronecker delta

$$(21.0.17) \qquad \delta_{ij} = \left\{ \begin{array}{l} 0 \text{ if } i \neq j \\ 1 \text{ if } i = j \end{array} \right.$$

Multiplication by the Kronecker delta with appropriate indices performs the operation of setting those indices to be the same. The inner product of a first and second rank tensors is a first rank tensor. If we swap the order (commute) $\mathbf{A}$ and $\mathbf{B}$

$$(21.0.18) \qquad \mathbf{B} \cdot \mathbf{A} = \mathbf{C}_1 \Rightarrow B_{jk} A_i \delta_{ki} = B_{jk} A_k = C_j$$

we obtain a different result. In general the inner product of two tensors is not commutative! The only exception to this is the inner (dot) product of two vectors is commutative. This is because each vector has only one index. When $\mathbf{A}$ and $\mathbf{B}$ are both second rank tensors the inner product is

$$(21.0.19) \qquad \mathbf{A} \cdot \mathbf{B} = \mathbf{C} \Rightarrow A_{ij} B_{kl} \delta_{jk} = A_{ij} B_{jl} = C_{il}$$

The inner product of a two second rank tensors is also a second rank tensor.

So far we have considered a single inner product - where we set a single index from each tensor to be equal. We can of course define double (and even higher multiple) inner products. For example, we perform a double inner product, we take the result of the single inner product, set the nearest pair of indices (the inner indices) from the two tensors equal, and perform the summation according to the summation notation. Because of the need for at least two indices from each tensor, only second or higher rank tensors can participate in double inner products. When **A** and **B** are both second rank tensors the double inner product is

$$(21.0.20) \qquad \mathbf{A} : \mathbf{B} = C \Rightarrow A_{ij}B_{kl}\delta_{jk}\delta_{il} = A_{ij}B_{jl}\delta_{il} = A_{ij}B_{ji} = C$$

where we have again used a Kronecker delta to first set the inner most indices equal, and then a second Kronecker delta to set the second inner most indices to be equal. The double inner product of two second rank tensors is a zeroth order tensor.

There are primarily two ways to represent second (and higher) rank tensors. The first way is by using the familiar matrix type array. For a second rank tensor, this is just a 2D matrix,

$$(21.0.21) \qquad \mathbf{A} = \begin{bmatrix} A_{11} & A_{12} & A_{13} \\ A_{21} & A_{22} & A_{23} \\ A_{31} & A_{32} & A_{33} \end{bmatrix}$$

We can represent higher order tensors with higher order matrices. The second, less popular, way to represent tensors is via basis dyads. The so called bases dyads are just direct products of two basis vectors (for a second rank tensor). For example, we can write the above second rank tensor as

$$(21.0.22) \quad \mathbf{A} = A_{11}\hat{\mathbf{x}}_1\hat{\mathbf{x}}_1 + A_{12}\hat{\mathbf{x}}_1\hat{\mathbf{x}}_2 + A_{13}\hat{\mathbf{x}}_1\hat{\mathbf{x}}_3$$
$$+ A_{21}\hat{\mathbf{x}}_2\hat{\mathbf{x}}_1 + A_{22}\hat{\mathbf{x}}_2\hat{\mathbf{x}}_2 + A_{23}\hat{\mathbf{x}}_2\hat{\mathbf{x}}_3$$
$$+ A_{31}\hat{\mathbf{x}}_3\hat{\mathbf{x}}_1 + A_{32}\hat{\mathbf{x}}_3\hat{\mathbf{x}}_2 + A_{33}\hat{\mathbf{x}}_3\hat{\mathbf{x}}_3$$

This convention is an attempt to maintain semblance to the conventional vector notation, where a vector **A** is written as

$$(21.0.23) \qquad \mathbf{A} = A_1\hat{\mathbf{x}}_1 + A_2\hat{\mathbf{x}}_2 + A_3\hat{\mathbf{x}}_3$$

We can represent higher order tensors with dyadic notation as well. For example, for a third order tensor the basis dyad would be a product of three basis vectors. The dyadic notation can be found in older literature, while it seems that the preference today is for the matrix notation. The matrix notation is potentially more popular because it resembles how tensors are represented in the common data structures in common modern computer languages.

Two common tensor operations that we will use are the extraction of a vector component of a tensor and taking a gradient of a vector field. As an example, let's take the stress tensor (Figure 21.0.1). The components of this stress tensor, σ_{ij} represent a stress on a surface that is oriented perpendicular to direction i applied in direction j. Let's say we want to extract the stress vector **S** on a surface oriented in a particular direction. To do this, we simplify take the inner product of the stress tensor with a unit vector pointed in that direction,

$$(21.0.24) \qquad \mathbf{S} = \mathbf{n} \cdot \sigma \Rightarrow n_i\sigma_{ij}$$

To take the gradient of the vector field, we take the direct product of the gradient operator with the vector,

$$(21.0.25) \qquad \nabla\mathbf{u} = \frac{\partial u_j}{\partial x_i}$$

Just like we can have scalar fields and vector fields, we can have fields of any rank tensor. A field just means that the quantity varies in general depending on position in space (and potentially in time). As an example a stress tensor in a fluid depends on the location of the fluid,

$$(21.0.26) \qquad \sigma = \sigma\left(x_1, x_2, x_3\right)$$

Bibliography

Abbott, J. R., N. Tetlow, A. L. Graham, S. A. Altobelli, E. Fukushima, L. A. Mondy, and T. S. Stephens (1991). Experimental observations of particle migration in concentrated suspensions: Couette flow. *Journal of rheology 35*(5), 773–795.

Abramowitz, M. and I. A. Stegun (1964). *Handbook of mathematical functions: with formulas, graphs, and mathematical tables*, Volume 55. Courier Corporation.

Accudynetest (2017). Critical surface tension and contact angle with water for various polymers.

Adam, N. K. (1941). *The Physics and Chemistry of Surfaces*. Oxford University Press.

Ajdari, A., N. Bontoux, and H. A. Stone (2006). Hydrodynamic dispersion in shallow microchannels: the effect of cross-sectional shape. *Analytical Chemistry 78*(2), 387–392.

Ajmani, R. S., E. J. Metter, R. Jaykumar, D. K. Ingram, E. L. Spangler, O. O. Abugo, and J. M. Rifkind (2000). Hemodynamic changes during aging associated with cerebral blood flow and impaired cognitive function. *Neurobiology of aging 21*(2), 257–269.

Akerlof, G. (1932). Dielectric constants of some organic solvent-water mixtures at various temperatures. *Journal of the American Chemical Society 54*(11), 4125–4139.

Al-Housseiny, T. T. and H. A. Stone (2013). Controlling viscous fingering in tapered Hele-Shaw cells. *Physics of Fluids 25*(9), 092102.

AlMomani, T., H. S. Udaykumar, J. S. Marshall, and K. B. Chandran (2008). Micro-scale dynamic simulation of erythrocyte-platelet interaction in blood flow. *Annals of biomedical engineering 36*(6), 905–920.

Altobelli, S. A., R. C. Givler, and E. Fukushima (1991). Velocity and concentration measurements of suspensions by nuclear magnetic resonance imaging. *Journal of Rheology 35*(5), 721–734.

Anderson, J. E. (1994). The Debye-Falkenhagen effect: experimental fact or friction? *Journal of Non-Crystalline Solids 172*, 1190–1194.

Angeli, P. and A. Gavriilidis (2008). Hydrodynamics of Taylor flow in small channels: a review. *Proceedings of the Institution of Mechanical Engineers, Part C: Journal of Mechanical Engineering Science 222*(5), 737–751.

Anna, S. L. and H. C. Mayer (2006). Microscale tipstreaming in a microfluidic flow focusing device. *Physics of Fluids 18*(12), 121512.

Aris, R. (1956). On the dispersion of a solute in a fluid flowing through a tube. *Proceedings of the Royal Society of London A: mathematical, physical and engineering sciences 235*(1200), 67–77.

Aris, R. (1959). On the dispersion of a solute by diffusion, convection and exchange between phases. *Proceedings of the Royal Society of London A: Mathematical, Physical and Engineering Sciences 252*(1271), 538–550.

Aris, R. (2012). *Vectors, tensors and the basic equations of fluid mechanics*. Courier Corporation.

Arkles, B. (2011). Hydrophobicity, hydrophilicity, and silane surface modification.

Arnold, W. M., A. G. Gessner, and U. Zimmermann (1993). Dielectric measurements on electro-manipulation media. *Biochimica et Biophysica Acta (BBA)-General Subjects 1157*(1), 32–44.

Arp, P. A. and S. G. Mason (1977). The kinetics of flowing dispersions: VIII doublets of rigid spheres (theoretical). *Journal of colloid and interface science 61*(1), 21–43.

Asmolov, E. S. (1999). The inertial lift on a spherical particle in a plane Poiseuille flow at large channel Reynolds number. *Journal of Fluid Mechanics 381*, 63–87.

Assael, M. J., A. E. Kalyva, K. D. Antoniadis, R. Michael Banish, I. Egry, J. Wu, E. Kaschnitz, and W. A. Wakeham (2010). Reference data for the density and viscosity of liquid copper and liquid tin. *Journal of Physical and Chemical Reference Data 39*(3), 033105.

Aussillous, P. and D. Quere (2000). Quick deposition of a fluid on the wall of a tube. *Physics of Fluids 12*(10), 2367–2371.

Baret, J. C. (2012). Surfactants in droplet-based microfluidics. *Lab on a Chip 12*(3), 422–433.

Bell, J. M. and F. K. Cameron (1906). The flow of liquids through capillary spaces. *The Journal of Physical Chemistry 10*(8), 658–674.

Bensimon, D., L. P. Kadanoff, S. Liang, B. I. Shraiman, and C. Tang (1986). Viscous flows in two dimensions. *Reviews of Modern Physics 58*(4).

Berger, S. A., L. Talbot, and L. S. Yao (1983). Flow in curved pipes. *Annual review of fluid mechanics 15*(1), 461–512.

Bergstrom, L. (1997). Hamaker constants of inorganic materials. *Advances in colloid and interface science 70*, 125–169.

Berker, R. (1963). Integration des equations du mouvement d'un fluide visqueux incompressible. *Handbuch der physik 2*, 1–384.

Berthier, J. (2012). *Micro-drops and digital microfluidics*. William Andrew.

Bhaskar, K. R., P. Garik, B. S. Turner, J. D. Bradley, R. Bansil, H. E. Stanley, and J. T. LaMont (1992). Viscous fingering of HCl through gastric mucin. *Nature 360*(6403), 458–461.

Bico, J. and D. Quere (2002). Rise of liquids and bubbles in angular capillary tubes. *Journal of colloid and Interface Science 247*(1), 162–166.

Bland, D. R. (1962). *Solutions of Laplace's equation*, Volume 14. Springer.

Blokhuis, E. M. and J. Kuipers (2006). Thermodynamic expressions for the Tolman length. *The journal of chemical physics 124*(7), 074701.

Bockris, J. O., A. K. N. Reddy, and G.-A. M. (2000). *Modern Electrochemistry 2A: Fundamentals of Electrodics*. Kluwer Academic, New York.

Bormashenko, E. (2008). Why does the Cassie-baxter equation apply? *Colloids and Surfaces A: Physicochemical and Engineering Aspects 324*(1), 47–50.

Born, M. (1920). Uber die beweglichkeit der elektrolytischen ionen. *Zeitschrift fur Physik 1*(3), 221–249.

Bosanquet, C. (1923). Lv. on the flow of liquids into capillary tubes. *The London, Edinburgh, and Dublin Philosophical Magazine and Journal of Science 45*(267), 525–531.

Boucher, E. and M. Evans (1975). Pendent drop profiles and related capillary phenomena. *Proc. R. Soc. Lond. A 346*(1646), 349–374.

Bretherton, F. P. (1961). The motion of long bubbles in tubes. *Journal of Fluid Mechanics 10*(2), 166–188.

Bretherton, F. P. (1962). The motion of rigid particles in a shear flow at low reynolds number. *Journal of Fluid Mechanics 14*(2), 284–304.

Brown, A. B. D., C. G. Smith, and A. R. Rennie (2000). Pumping of water with ac electric fields applied to asymmetric pairs of microelectrodes. *Physical review E 63*(1), 016305.

Cahill, B. P., L. J. Heyderman, J. Gobrecht, and A. Stemmer (2004). Electroosmotic streaming on application of traveling-wave electric fields. *Physical review E 70*(3), 036305.

Caupin, F., M. W. Cole S. Balibar, and J. Treiner (2008). Absolute limit for the capillary rise of a fluid. *EPL (Europhysics Letters) 82*(5), 56004.

Chan, P. H. and L. G. Leal (1979). The motion of a deformable drop in a second-order fluid. *Journal of Fluid Mechanics 92*(1), 131–170.

Chang, C. and R. L. Powell (1994). Effect of particle size distributions on the rheology of concentrated bimodal suspensions. *Journal of rheology 38*(1), 85–98.

Chapman, D. L. (1913). Li. a contribution to the theory of electrocapillarity. *The London, Edinburgh, and Dublin philosophical magazine and journal of science 25*(148), 475–481.

Chen, J. D. (1986). Measuring the film thickness surrounding a bubble inside a capillary. *Journal of Colloid and Interface Science 109*(2), 341–349.

Chen, J. D. (1988). Experiments on a spreading drop and its contact angle on a solid. *Journal of colloid and interface science 122*(1), 60–72.

Chen, J. M., P. C. Huang, and M. G. Lin (2008). Analysis and experiment of capillary valves for microfluidics on a rotating disk. *Microfluidics and Nanofluidics 4*(5), 427–437.

Chester, W., D. R. Breach, and I. Proudman (1969). On the flow past a sphere at low Reynolds number. *Journal of Fluid Mechanics 37*(4), 751–760.

Cidraprecisionservices (2017). Cyclic olefin copolymer (coc) and cyclic olefin polymer (cop) material properties.

Cini, R., G. Loglio, and A. Ficalbi (1972). Temperature dependence of the surface tension of water by the equilibrium ring method. *Journal of Colloid and Interface Science 41*(2), 287–297.

Ciric, M. (2009). Notes on constant mean curvature surfaces and their graphical presentation. *Filomat 23*(2), 97–107.

Clark, W. C., J. M. Haynes, and G. Mason (1968). Liquid bridges between a sphere and a plane. *Chemical Engineering Science 23*(7), 810–812.

Clift, R., J. R. Grace, and M. E. Weber (1979). *Bubbles, Drops, and Particles, Academic Press*. Functional histology: New York.

Concus, P. and R. Finn (1969). On the behavior of a capillary surface in a wedge. *Proceedings of the National Academy of Sciences of the United States of America 63*(2).

Coogan, M. D., M. Z. Brettler, C. A. Newsom, and P. Perkins (2010). *The New Oxford Annotated Bible: New Revised Standard Version: with the Apocrypha: an Ecumenical Study Bible.* Oxford University Press, USA.

Cunningham, A. B., J. E. Lennox, and R. J. Ross (Eds.) (2011). *Biofilms: The Hypertextbook.* hypertextbookshop.com.

Daniels, V. G., P. R. Wheater, and H. Burkitt (1979). *Functional histology: A text and colour atlas.* Edinburgh: Churchill Livingstone.

Davis, A. M. J. and M. E. O'Neill (1977). Separation in a slow linear shear flow past a cylinder and a plane. *Journal of Fluid Mechanics 81*(3), 551–564.

Davis, A. M. J., M. E. O'Neill, J. M. Dorrepaal, and K. B. Ranger (1976). Separation from the surface of two equal spheres in Stokes flow. *Journal of Fluid Mechanics 77*(4), 625–644.

Davis, J. A. (2008). *Microfluidic separation of blood components through deterministic lateral displacement.* Princeton University.

de Gennes, P. G., F. Brochard-Wyart, and D. Quere (2004). *Capillarity and Wetting Phenomena: Drops.* Bubbles, Pearls, Waves.

Dean, W. R. (1927). Note on the motion of fluid in a curved pipe. *The London, Edinburgh, and Dublin Philosophical Magazine and Journal of Science 4*, 20.

Debye, P. and H. Falkenhagen (1928). Dispersion of the conductivity and dielectric constants of strong electrolytes. *Physik. Z. 29*(121), 401.

Demond, A. H. and A. S. Lindner (1993). Estimation of interfacial tension between organic liquids and water. *Environmental science and technology 27*(12), 2318–2331.

Derby, B. (2010). Inkjet printing of functional and structural materials: fluid property requirements, feature stability, and resolution. *Annual Review of Materials Research 40*, 395–414.

Deryaguin, B. V., N. V. Churaev, and V. M. Muller (1987). *Surface forces.* Plenum Pub Corp.

Dong, M. and I. Chatzis (1995). The imbibition and flow of a wetting liquid along the corners of a square capillary tube. *Journal of colloid and interface science 172*(2), 278–288.

Doshi, M. R., P. M. Daiya, and W. N. Gill (1978). Three dimensional laminar dispersion in open and closed rectangular conduits. *Chemical Engineering Science 33*(7), 795–804.

Doyeux, V., T. Podgorski, S. Peponas, M. Ismail, and G. Coupier (2011). Spheres in the vicinity of a bifurcation: elucidating the Zweifach-fung effect. *Journal of Fluid Mechanics 674*, 359–388.

Drake-Holland, A. J. and M. I. Noble (2009). The important new drug target in cardiovascular medicine-the vascular glycocalyx. *Cardiovascular & Haematological Disorders-Drug Targets (Formerly Current Drug Targets-Cardiovascular and Hematological Disorders) 9*(2), 118–123.

Drazer, G., J. Koplik, B. Khusid, and A. Acrivos (2002). Deterministic and stochastic behaviour of non-Brownian spheres in sheared suspensions. *Journal of Fluid Mechanics 460*, 307–335.

Dussan, E. B. (1985). On the ability of drops or bubbles to stick to non-horizontal surfaces of solids part 2. Small drops or bubbles having contact angles of arbitrary size. *J. Fluid Mech 151*(1).

Eckstrom, H. C. and C. Schmelzer (1939). The Wien Effect: Deviations of Electrolytic Solutions from Ohm's Law under High Field Strengths. *Chemical Reviews 24*(3), 367–414.

Edvinsson, R. K. and S. Irandoust (1996). Finite element analysis of taylor flow. *AIChE Journal 42*(7), 1815–1823.

Einstein, A. (1905). Uber die von der molekularkinetischen theorie der warme geforderte bewegung von in ruhenden flussigkeiten suspendierten teilchen. *Annalen der Physik 322.*

Einstein, A. (1906). Eine neue Bestimmung der Molek?ldimensionen. *Annalen der Physik 19.*

Elliott, J. A. W. (2001). On the complete Kelvin equation. *Chemical Engineering Education 35*(4), 274–279.

Ethington, E. F. (1990). *Interfacial contact angle measurements of water, mercury, and 20 organic liquids on quartz, calcite, biotite, and Ca-montmorillonite substrates.* U. S.

Extrand, C. W. and Y. Kumagai (1997). An experimental study of contact angle hysteresis. *Journal of Colloid and interface Science 191*(2), 378–383.

Fahraeus, R. (1928). Die stromungsverhaltnisse und die verteilung der blutzellen im gefasystem. *Klin Wochenschr 7,* 100–106.

Fahraeus, R. and T. Lindqvist (1931). The viscosity of the blood in narrow capillary tubes. *American Journal of Physiology-Legacy Content 96*(3), 562–568.

Fairbrother, F. and A. E. Stubbs (1935). 119. studies in electro-endosmosis. part vi. the "bubble-tube" method of measurement. *Journal of the Chemical Society (Resumed),* 527–529.

Ferrini, F., D. Ercolani, D. Cindio, N. B., N. L., L., and S. Ranaudo (1979). Shear viscosity of settling suspensions. *Rheologica Acta 18*(2), 289–296.

Formaggia, L., A. Quarteroni, and A. Veneziani (Eds.) (2010). *Cardiovascular Mathematics: Modeling and simulation of the circulatory system.* Springer Science & Business Media.

Forsythe, W. E. (1969). *Smithsonian physical tables* Smithsonian Institution.

Fredenslund, A., R. L. Jones, and J. M. Prausnitz (1975). Group contribution estimation of activity coefficients in nonideal liquid mixtures. *AIChE Journal 21*(6), 1086–1099.

Fries, N. and M. Dreyer (2008). An analytic solution of capillary rise restrained by gravity. *Journal of colloid and interface science 320*(1), 259–263.

Friz, G. (1965). On the dynamic contact angle in the case of complete wetting. *Z. Angew. Phys 19*(4).

Furmidge, C. G. L. (1962). Studies at phase interfaces i. The sliding of liquid drops on solid surfaces and a theory for spray retention. *Journal of colloid science 17*(4), 309–324.

Gadala-Maria, F. and A. Acrivos (1980). Shear-induced structure in a concentrated suspension of solid spheres. *Journal of Rheology 24*(4), 799–814.

Gamayunov, N. I., V. A. Murtsovkin, and A. S. Dukhin (1986). Pair interaction of particles in electric field 1. Features of hydrodynamic interaction of polarized particles. *Colloid J. USSR (Engl. Transl.) ;(United States) 48*(2).

Gangwal, S., O. J. Cayre, M. Z. Bazant, and O. D. Velev (2008). Induced-charge electrophoresis of metallodielectric particles. *Physical review letters 100*(5), 058302.

Garstecki, P., H. A. Stone, and G. M. Whitesides (2005). Mechanism for flow-rate controlled breakup in confined geometries: a route to monodisperse emulsions. *Physical Review Letters 94*(16), 164501.

Geary, A., H. V. Lowry, and H. A. Hayden (1955). *Advanced Mathematics for Technical Students*. Longmans, Green and Company.

Giavedoni, M. D. and F. A. Saita (1997). The axisymmetric and plane cases of a gas phase steadily displacing a newtonian liquid: A simultaneous solution of the governing equations. *Physics of Fluids 9*(8), 2420–2428.

Giavedoni, M. D. and F. A. Saita (1999). The rear meniscus of a long bubble steadily displacing a Newtonian liquid in a capillary tube. *Physics of Fluids 11*(4), 786–794.

Gieras, J. F. (2013). Electrical ignition of fuel-air mixture in aircraft fuel tanks. *Przeglad Elektrotechniczny (Electrical Review) 89*(7), 17–24.

Gill, W. N., U. Guceri, and R. J. Nunge (1969). Laminar dispersion in diverging channels and concentric annuli. *Office of Saline Water, Research and Development Report No 443*.

Gorce, J. B., I. J. Hewitt, and D. Vella (2016). Capillary imbibition into converging tubes: Beating washburn's law and the optimal imbibition of liquids. *Langmuir 32*(6), 1560–1567.

Gouy, M. (1910). Sur la constitution de la charge électrique à la surface d'un électrolyte. *J. Phys. Theor. Appl. 9*(1), 457–468.

Grant, R. P. and S. Middleman (1966). Newtonian jet stability. *AIChE Journal 12*(4), 669–678.

Green, N. G., A. Ramos, A. González, H. Morgan, and A. Castellanos (2000). Fluid flow induced by nonuniform ac electric fields in electrolytes on microelectrodes. i. experimental measurements. *Physical review E 61*(4), 4011.

Gu, Y. and D. Li (2000). The zeta-potential of glass surface in contact with aqueous solutions. *Journal of Colloid and Interface Science 226*(2), 328–339.

Hadamard, J. S. (1911a). Motion of liquid drops (viscous). *Comp. Rend. Acad. Sci. Paris 154*, 1735–1755.

Hadamard, J. S. (1911b). Motion of liquid drops (viscous). *Comp. Rend. Acad. Sci. Paris 154*, 1735–1755.

Hampton, R. E., A. A. Mammoli, A. L. Graham, N. Tetlow, and S. A. Altobelli (1997). Migration of particles undergoing pressure-driven flow in a circular conduit. *Journal of Rheology 41*(3), 621–640.

Han, M., C. Kim, M. Kim, and S. Lee (1999). Particle migration in tube flow of suspensions. *Journal of rheology 43*(5), 1157–1174.

Happel, J. and H. Brenner (1983). *Low Reynolds Number Hydrodynamics: With Special Applications to Particulate Media, Kluwer*. Boston.

Harkins, W. D. and F. E. Brown (1919). The determination of surface tension (free surface energy), and the weight of falling drops: The surface tension of water and benzene by the capillary height method. *Journal of the American Chemical Society 41*(4), 499–524.

Harned, H. S. and R. L. Nuttall (1947). The diffusion coefficient of potassium chloride in dilute aqueous solution. *Journal of the American Chemical Society 69*(4), 736–740.

Harrison, D. A., D. Yan, and S. Blairs (1977). The surface tension of liquid copper. *The Journal of Chemical Thermodynamics 9*(12), 1111–1119.

Hasimoto, H. (1996). Lorentz's theorem on the Stokes equation. *Journal of engineering mathematics*, 215–224.

Hasted, J. B. (1973). *Aqueous dielectrics*. Chapman and Hall.

Haynes, W. M. (Ed.) (2014). *CRC handbook of chemistry and physics*. CRC press.

Haynes, W. M. (Ed.) (2017). *CRC handbook of chemistry and physics*. CRC press.

Henderson, P. (1907). *Zur thermodynamik der flussigkeitsketten*. Druck von WF Kaestner.

Higuera, F. J., A. Medina, and A. Linan (2008). Capillary rise of a liquid between two vertical plates making a small angle. *Physics of fluids 20*(10), 102102.

Hill, R. and G. Power (1956). Extremum principles for slow viscous flow and the approximate calculation of drag. *Q. J. Mech. Appl 9*, 313–319.

Hogg, A. J. (1994). The inertial migration of non-neutrally buoyant spherical particles in two-dimensional shear flows. *Journal of Fluid Mechanics 272*, 285–318.

Huckel, E. (1924). Die kataphorese der kugel. *Phys. Z 25*, 204–210.

Huh, D., H. Fujioka, Y. C. Tung, N. Futai, R. Paine, J. B. Grotberg, and S. Takayama (2007). Acoustically detectable cellular-level lung injury induced by fluid mechanical stresses in microfluidic airway systems. *Proceedings of the National Academy of Sciences 104*(48), 18886–18891.

Hund, S. J., M. V. Kameneva, and J. F. Antaki (2017). A quasi-mechanistic mathematical representation for blood viscosity. *Fluids 2*(1).

Husny, J. and J. J. Cooper-White (2006). The effect of elasticity on drop creation in T-shaped microchannels. *Journal of non-newtonian fluid mechanics 137*(1), 121–136.

Inglis, D. W., J. A. Davis, R. H. Austin, and J. C. Sturm (2006). Critical particle size for fractionation by deterministic lateral displacement. *Lab on a Chip 6*(5), 655–658.

Ionode.com (2015). Conductivity theory.

Irandoust, S. and B. Andersson (1989). Liquid film in taylor flow through a capillary. *Industrial and engineering chemistry research 28*(11), 1684–1688.

Jaros, M., V. Hruska, M. Stedry, I. Zuskova, and B. Gas (2004). Eigenmobilities in background electrolytes for capillary zone electrophoresis: IV computer program PeakMaster. *Electrophoresis 25*(18-19), 3080–3085.

Jeffrey, R. C. and J. R. A. Pearson (1965). Particle motion in laminar vertical tube flow. *Journal of Fluid Mechanics 22*(4), 721–735.

Jones, T. B. (1979). Dielectrophoretic force calculation. *Journal of Electrostatics 6*, 69–82.

Jones, T. B., J. D. Fowler, Y. S. Chang, and C. J. Kim (2003). Frequency-based relationship of electrowetting and dielectrophoretic liquid microactuation. *Langmuir 19*(18), 7646–7651.

Jones, T. B., K. L. Wang, and D. J. Yao (2004). Frequency-dependent electromechanics of aqueous liquids: electrowetting and dielectrophoresis. *Langmuir 20*(7), 2813–2818.

Joos, P., V. Remoortere, P., and M. Bracke (1990). The kinetics of wetting in a capillary. *Journal of colloid and interface science 136*(1), 189–197.

Joswiak, M. N., N. Duff, M. F. Doherty, and B. Peters (2013). Size-dependent surface free energy and Tolman-corrected droplet nucleation of TIP4p/2005 water. *The journal of physical chemistry letters 4*(24), 4267–4272.

Jurin, J. (1719). Ii. an account of some experiments shown before the royal society; with an enquiry into the cause of the ascent and suspension of water in capillary tubes. *Philosophical Transactions 30*(355), 739–747.

Kalinin, Y. V., V. Berejnov, and R. E. Thorne (2009). Contact line pinning by microfabricated patterns: effects of microscale topography. *Langmuir: the ACS journal of surfaces and colloids 25*(9).

Kalliadasis, S. and H. C. Chang (1994). Apparent dynamic contact angle of an advancing gas-liquid meniscus. *Physics of Fluids 6*(1), 12–23.

Kolb, W. B. and R. L. Cerro (1991). Coating the inside of a capillary of square cross section. *Chemical Engineering Science 46*(9), 2181–2195.

Kolb, W. B. and R. L. Cerro (1993a). Film flow in the space between a circular bubble and a square tube. *Journal of colloid and interface science 159*(2), 302–311.

Kolb, W. B. and R. L. Cerro (1993b). The motion of long bubbles in tubes of square cross section. *Physics of Fluids A: Fluid Dynamics 5*(7), 1549–1557.

Kreutzer, M. T. (2003). *Hydrodynamics of Taylor flow in capillaries and monolith reactors.* Ph. D. thesis, TU Delft.

Kreutzer, M. T., F. Kapteijn, J. A. Moulijn, and J. J. Heiszwolf (2005). Multiphase monolith reactors: chemical reaction engineering of segmented flow in microchannels. *Chemical Engineering Science 60*(22), 5895–5916.

Kreutzer, M. T., F. Kapteijn, J. A. Moulijn, C. R. Kleijn, and J. J. Heiszwolf (2005). Inertial and interfacial effects on pressure drop of Taylor flow in capillaries. *AIChE Journal 51*(9), 2428–2440.

Krieger, I. M. (1972). Rheology of monodisperse latices. *Advances in Colloid and Interface science 3*(2), 111–136.

Krieger, I. M. and T. J. Dougherty (1959). A mechanism for non-newtonian flow in suspensions of rigid spheres. *Transactions of the Society of Rheology 3*(1), 137–152.

Krishnan, G. P., S. Beimfohr, and D. T. Leighton (1996). Shear-induced radial segregation in bidisperse suspensions. *Journal of Fluid Mechanics 321*, 371–393.

Kroesser, F. W. and S. Middleman (1969). Viscoelastic jet stability. *AIChE Journal 15*(3), 383–386.

Kumar, A. and M. D. Graham (2012). Margination and segregation in confined flows of blood and other multicomponent suspensions. *Soft Matter 8*(41), 10536–10548.

Laborie, S., C. Cabassud, L. Durand-Bourlier, and J. M. Laine (1999). Characterisation of gas-liquid two-phase flow inside capillaries. *Chemical Engineering Science 54*(23), 5723–5735.

Landau, L. and B. Lebich (1942). Dragging of a liquid by a moving plate. *Acta Physiochemica U.R. S. S.*, 42–54.

Langbein, D. (2002). *Capillary Surfaces.* Springer Berlin Heidelberg.

Le Grand, N., A. Daerr, and L. Limat (2005). Shape and motion of drops sliding down an inclined plane. *Journal of Fluid Mechanics 541*, 293–315.

Leal, L. G. (2007). *Advanced transport phenomena: fluid mechanics and convective transport processes.* Cambridge University Press.

Leighton, D. and A. Acrivos (1987). The shear-induced migration of particles in concentrated suspensions. *Journal of Fluid Mechanics 181*, 415–439.

Leighton, D. and I. Rampall (1993). Measurement of the shear-induced microstructure of concentrated suspensions of non-colloidal spheres. *Particulate Two-Phase*

Flow 6.

Lertes, P. (1921). Untersuchungen uber rotationen von dielektrischen flussigkeiten im elektrostatischen drehfeld. *Zeitschrift fur Physik 4*(3), 315–336.

Levich, V. G. (1962). *Physicochemical hydrodynamics.* Prentice hall.

Levine, S., J. R. Marriott, G. Neale, and N. Epstein (1975). Theory of electrokinetic flow in fine cylindrical capillaries at high zeta-potentials. *Journal of Colloid and Interface Science 52*(1), 136–149.

Lippmann, G. (1875). *Relations entre les phenomenes electriques et capillaires (Doctoral dissertation).* Gauthier-Villars.

Liu, H., C. O. Vandu, and R. Krishna (2005). Hydrodynamics of taylor flow in vertical capillaries: flow regimes, bubble rise velocity, liquid slug length, and pressure drop. *Industrial and engineering chemistry research 44*(14), 4884–4897

Lorentz, H. A. (1907). Ein allgemeiner staz, die bewegung einer reibenden flssigkeit betreffend. *Nebst Einigen Anwendungen Desselben. Abh. Theor. Phys. Leipzig 1,* 23–42.

Lucas, R. (1918). Rate of capillary ascension of liquids. *Kolloid Z 23*(15), 15–22.

MacDonald, J. R. (1970). Double layer capacitance and relaxation in electrolytes and solids. *Transactions of the Faraday Society 66*, 943–958.

Malmberg, C. G. and A. A. Maryott (1956). Dielectric constant of water from 00 to 1000 c. *J. Res. Nat. Bur. Stand 56*, 369131–8.

Marchessault, R. N. and S. G. Mason (1960). Flow of entrapped bubbles through a capillary. *Industrial and Engineering Chemistry 52*(1), 79–84.

Marmur, A. (1988). Penetration of a small drop into a capillary. *Journal of Colloid and Interface Science 122*(1), 209–219.

Martel, J. M. and M. Toner (2013). Particle focusing in curved microfluidic channels. *Scientific Reports 3*(3340.).

McGeary, R. K. (1961). Mechanical packing of spherical particles. *Journal of the American ceramic Society 44*(10), 513–522.

McGrath, J., M. Jimenez, and H. Bridle (2014). Deterministic lateral displacement for particle separation: a review. *Lab on a Chip 14*(21), 4139–4158.

McHale, G., N. J. Shirtcliffe, and M. I. Newton (2004). Contact-angle hysteresis on super-hydrophobic surfaces. *Langmuir 20*(23), 10146–10149.

McLaren, C., B. Houwen, J. Koepke, R. Rowan, P. McKay, B. Ortner, and M. Bishop (1993). Analysis of red blood cell volume distributions using the icsh reference method. detection of sequential changes in distributions determined by hydrodynamic focusing. *Clinical and Laboratory Haematology 15*(3), 173–184.

Meissner, H. P. and A. S. Michaels (1949). Surface tensions of pure liquids and liquid mixtures. *Industrial and Engineering Chemistry 41*(12), 2782–2787.

Mettler-Toledo (2013). A guide to conductivity measurement: Theory and practice of conductivity applications.

Middleman, S. (1995). *Modeling axisymmetric flows: dynamics of films, jets, and drops.* Academic Press.

Moffatt, H. K. (1964). Viscous and resistive eddies near a sharp corner. *Journal of Fluid Mechanics 18*(1), 1–18.

Morgan, H., A. G. Izquierdo, D. J. Bakewell, N. G. Green, and R. A. (2001). The dielectrophoretic and travelling wave forces for interdigitated electrode arrays: analytical solution using fourier series. *Phys. D: Appl. Phys 34.*

Mortimer, R. G. (2000). *Physical chemistry.* Academic Press.

Mwasame, P. M., N. J. Wagner, and A. N. Beris (2016). Modeling the viscosity of polydisperse suspensions: improvements in prediction of limiting behavior. *Physics of Fluids 28*(6), 061701.

Nepomnyashchy, A. A., M. G. Velarde, and P. Colinet (2001). *Interfacial phenomena and convection*. CRC Press.

Olivares, W., T. L. Croxton, and D. A. McQuarrie (1980). Electrokinetic flow in a narrow cylindrical capillary. *The Journal of Physical Chemistry 84*(8), 867–869.

Orr, F. M., L. E. Scriven, and A. P. Rivas (1975). Pendular rings between solids: meniscus properties and capillary force. *Journal of Fluid Mechanics 67*(4), 723–742.

Ouali, F. F., G. McHale, H. Javed, C. Trabi, N. J. Shirtcliffe, and M. I. Newton (2013). Wetting considerations in capillary rise and imbibition in closed square tubes and open rectangular cross-section channels. *Microfluidics and nanofluidics 15*(3), 309–326.

Park, C. W. and G. M. Homsy (1984). Two-phase displacement in Hele Shaw cells: theory. *Journal of Fluid Mechanics 139*, 291–308.

Paulus, J. (1975). Platelet size in man. *Blood 46*.

Phan, H. T., S. Bartelt-Hunt, K. B. Rodenhausen, M. Schubert, and J. C. Bartz (2015). Investigation of bovine serum albumin (BSA) attachment onto self-assembled monolayers (SAMs) using combinatorial quartz crystal microbalance with dissipation (QCM-D) and spectroscopic ellipsometry (SE). *PloS one 10*(10).

Phillips, R. J., R. C. Armstrong, R. A. Brown, A. L. Graham, and J. R. Abbott (1992). A constitutive equation for concentrated suspensions that accounts for shear-induced particle migration. *Physics of Fluids A: Fluid Dynamics 4*(1), 30–40.

Pohl, H. A. (1978). *Dielectrophoresis the behavior of neutral matter in nonuniform electric fields*. Cambridge University Press. Cambridge.

Ponomarenko, A., D. Quere, and C. Clanet (2011). A universal law for capillary rise in corners. *Journal of Fluid Mechanics 666*, 146–154.

Pries, A. R., D. Neuhaus, and P. Gaehtgens (1992). Blood viscosity in tube flow: dependence on diameter and hematocrit. *American Journal of Physiology-Heart and Circulatory Physiology 263*(6).

Pries, A. R. and T. W. Secomb (2005). Microvascular blood viscosity in vivo and the endothelial surface layer. *American Journal of Physiology-Heart and Circulatory Physiology 289*(6).

Pries, A. R., T. W. Secomb, T. Gessner, M. B. Sperandio, J. F. Gross, and P. Gaehtgens (1994). Resistance to blood flow in microvessels in vivo. *Circulation research 75*(5), 904–915.

Probstein, R. F. (2005). *Physicochemical hydrodynamics: an introduction*. John Wiley and Sons.

Probstein, R. F., M. Z. Sengun, and T. C. Tseng (1994). Bimodal model of concentrated suspension viscosity for distributed particle sizes. *Journal of rheology 38*(4), 811–829.

Proudman, I. and J. R. A. Pearson (1957). Expansions at small Reynolds numbers for the flow past a sphere and a circular cylinder. *Journal of Fluid Mechanics 2*(3), 237–262.

Quere, D. (1997). Inertial capillarity. *EPL (Europhysics Letters) 39*(5).

Quincke, G. (1896). Electrische untersuchungen: XIV uber Rotationen im konstanten electrischen Felde. *Ann. Phys. Chem. 59*, 417–486.

Ramos, A., A. Gonzalez, A. Castellanos, N. G. Green, and H. Morgan (2003). Pumping of liquids with ac voltages applied to asymmetric pairs of microelectrodes. *Physical review E 67*(5), 056302.

Ramos, A., H. Morgan, N. G. Green, A. Gonzalez, and A. Castellanos (2005). Pumping of liquids with traveling-wave electroosmosis. *Journal of Applied Physics 97*(8), 084906.

Rand, R. and A. Burton (1964). Mechanical properties of the red cell membrane I. *Membrane stiffness and intracellular pressure. Biophys J 4*, 115–135.

Ransohoff, T. C. and C. J. Radke (1988). Laminar flow of a wetting liquid along the corners of a predominantly gas-occupied noncircular pore. *Journal of colloid and interface science 121*(2), 392–401.

Rayleigh, L. (1892). Xvi on the instability of a cylinder of viscous liquid under capillary force. *The London, Edinburgh, and Dublin Philosophical Magazine and Journal of Science 34*(207), 145–154.

Reynolds, O. (1886). On the theory of lubrication and its application to mr. beauchamp tower's experiments, including an experimental determination of the viscosity of olive oil. *Proceedings of the Royal Society of London*, 191–203.

Reyssat, M., L. Courbin, E. Reyssat, and H. A. Stone (1922). Imbibition in geometries with axial variations. *Journal of Fluid Mechanics 615*(264), 335–344.

Rice, C. L. and R. Whitehead (1965). Electrokinetic flow in a narrow cylindrical capillary. *The Journal of Physical Chemistry 69*(11), 4017–4024.

Rideal, E. K. (1922). Cviii. on the flow of liquids under capillary pressure. *The London, Edinburgh, and Dublin Philosophical Magazine and Journal of Science 44*(264), 1152–1159.

Rosen, M. J. and J. T. Kunjappu (2012). *Surfactants and interfacial phenomena.* John Wiley and Sons.

Rybczynski, W. (1911). On the translatory motion of a fluid sphere in a viscous medium. *Bull. Acad. Sci., Cracow, Series A 40.*

Schmid-Schonbein, G. W., K. L. Sung, H. Tozeren, R. Skalak, and S. Chien (1981). Passive mechanical properties of human leukocytes. *Biophysical Journal 36*(1), 243–256.

Schonberg, J. A. and E. J. Hinch (1989). Inertial migration of a sphere in Poiseuille flow. *Journal of Fluid Mechanics 203*, 517–524.

Segre, G. and A. Silberberg (1962). Behaviour of macroscopic rigid spheres in Poiseuille flow Part 2. *Experimental results and interpretation. Journal of fluid mechanics 14*(1), 136–157.

Shauly, A., A. Averbakh, A. Nir, and R. Semiat (1997). Slow viscous flows of highly concentrated suspensions. part II: particle migration, velocity and concentration profiles in rectangular ducts. *International Journal of Multiphase Flow 23*(4), 613–629.

Shauly, A., A. Wachs, and A. Nir (1998). Shear-induced particle migration in a polydisperse concentrated suspension. *Journal of Rheology 42*(6), 1329–1348.

Shauly, A., A. Wachs, and A. Nir (2000). Shear-induced particle resuspension in settling polydisperse concentrated suspension. *International journal of multiphase flow 26*(1), 1–15.

Shen, C. and J. M. Floryan (1985). Low Reynolds number flow over cavities. *The Physics of fluids 28*(11), 3191–3202.

Shevtsova, V. M., H. C. Kuhlmann, and H. J. Rath (1996). *Thermocapillary convection in liquid bridges with a deformed free surface.* Materials and fluids under low gravity.

Shi, X., M. P. Brenner, and S. R. Nagel (1994). A cascade of structure in a drop falling from a faucet. *Science 265*(5169), 219–222.

Smoluchowski, M. (1906). Zur kinetischen theorie der brownschen molekularbewegung und der suspensionen. *Annalen der Physik 326.*

Song, J., R. Evans, Y. Y. Lin, B. N. Hsu, and R. B. Fair (2009). A scaling model for electrowetting-on-dielectric microfluidic actuators. *Microfluidics and Nanofluidics 7*(1), 75–89.

Sousa, P. C., F. T. Pinho, M. A. Alves, and M. S. Oliveira (2016). A review of hemorheology: measuring techniques and recent advances. *Korea-Australia Rheology Journal 28*(1), 1–22.

Sprunt, E. S., T. B. Mercer, and N. F. Djabbarah (1994). Streaming potential from multiphase flow. *Geophysics 59*(5), 707–711.

Squires, T. M. and M. Z. Bazant (2004). Induced-charge electro-osmosis. *Journal of Fluid Mechanics 509*, 217–252.

Squires, T. M. and M. Z. Bazant (2006). Breaking symmetries in induced-charge electro-osmosis and electrophoresis. *Journal of Fluid Mechanics 560*, 65–101.

Starov, V. M., M. G. Velarde, and C. J. Radke (2007). *Wetting and spreading dynamics.* CRC Press.

Stellwagen, N. C., C. Gelfi, and P. G. Righetti (1997). The free solution mobility of DNA. *Biopolymers 42*(6), 687–703.

Stratton, J. A. (1941). *Electromagnetic Theory.* McGraw-Hill.

Sutera, S. P., V. Seshadri, P. A. Croce, and R. M. Hochmuth (1970). Capillary blood flow: II deformable model cells in tube flow. *Microvascular research 2*(4), 420–433.

Sutherland, W. (1905). Lxxv. a dynamical theory of diffusion for non-electrolytes and the molecular mass of albumin. *The London, Edinburgh, and Dublin Philosophical Magazine and Journal of Science 9*(54), 781–785.

Tandon, V., S. K. Bhagavatula, W. C. Nelson, and B. J. Kirby (2008). Zeta potential and electroosmotic mobility in microfluidic devices fabricated from hydrophobic polymers: 1 the origins of charge. *Electrophoresis 29*(5), 1092–1101.

Taneda, S. (1979). Visualization of separating Stokes flows. *Journal of the Physical Society of Japan 46*(6).

Tanner, L. H. (1979). The spreading of silicone oil drops on horizontal surfaces. *Journal of Physics D: Applied Physics 12*(9).

Tarleton, S. and R. Wakeman (2005). *Solid/liquid separation: principles of industrial filtration.* Elsevier.

Tate, T. (1864). Xxx. on the magnitude of a drop of liquid formed under different circumstances. *The London, Edinburgh, and Dublin Philosophical Magazine and Journal of Science 27*(181), 176–180.

Taylor, B. (1712). Concerning the Ascent of Water between two Glas Planes. *Phil. Trans. Roy. Soc. London 27*(538.).

Taylor, G. (1953). Dispersion of soluble matter in solvent flowing slowly through a tube. *In Proceedings of the Royal Society of London A: Mathematical, Physical* and Engineering Sciences (Vol. 219, No 219(1137), 186–203.

Taylor, G. (1960). Deposition of viscous fluid in a plane surface. *J. Fluid Mech 9*, 218–224.

Taylor, G. (1966). Studies in electrohydrodynamics. I. The circulation produced in a drop by electrical field. *In Proceedings of the Royal Society of London A: Mathematical, Physical and Engineering Sciences 291*(1425), 159–166.

Taylor, G. I. (1961). Deposition of a viscous fluid on the wall of a tube. *Journal of Fluid Mechanics 10*(2), 161–165.

Thio, T. H. G., S. Soroori, F. Ibrahim, W. Al-Faqheri, N. Soin, L. Kulinsky, and M. Madou (2013). Theoretical development and critical analysis of burst frequency equations for passive valves on centrifugal microfluidic platforms. *Medical and biological engineering and computing 51*(5), 525–535.

Tien, C. (2012). *Principles of filtration*. Elsevier.

Tien, C. and B. V. Ramarao (2011). *Granular filtration of aerosols and hydrosols*. Elsevier.

Touchard, G. (2001). Flow electrification of liquids. *Journal of Electrostatics 51*, 440–447.

Tripathi, S., Y. B. V. Kumar, A. Prabhakar, S. S. Joshi, and A. Agrawal (2015). Passive blood plasma separation at the microscale: a review of design principles and microdevices. *Journal of Micromechanics and Microengineering 25*(8), 083001.

Troian, S. M., E. Herbolzheimer, S. A. Safran, and J. F. Joanny (1989). Fingering instabilities of driven spreading films. *EPL (Europhysics Letters) 10*(1).

TSI301.com (2017). Comparison of tsi 301 and wd-40.

Tsori, Y. (2006). Discontinuous liquid rise in capillaries with varying cross-sections. *Langmuir 22*(21), 8860–8863.

Tsori, Y. (2007). Discontinuous meniscus location in tapered capillaries driven by pressure difference and dielectrophoretic forces. *Langmuir 23*(15), 8028–8034.

Turgeon, M. L. (2005). *Clinical hematology: theory and procedures*. Lippincott Williams & Wilkins.

Tyn, M. T. and T. W. Gusek (1990). Prediction of diffusion coefficients of proteins *Biotechnology and bioengineering 35*(4), 327–338.

Umbanhowar, P. B., V. Prasad, and D. A. Weitz (2000). Monodisperse emulsion generation via drop break off in a coflowing stream. *Langmuir 16*(2), 347–351.

van den Berg, B. M., M. Nieuwdorp, E. S. Stroes, and H. Vink (2006). Glycocalyx and endothelial (dys) function: from mice to men. *Pharmacological Reports 58*, 75.

Venerus, D. C. and D. N. Simavilla (2015). Tears of wine: new insights on an old phenomenon. *Scientific reports 5*.

Vladisavljevic, G. T. and R. A. Williams (2005). Recent developments in manufacturing emulsions and particulate products using membranes. *Advances in colloid and interface science 113*(1), 1–20.

Vogel, J., M. Sperandic, A. R. Pries, O. Linderkamp, P. Gaehtgens, and W. Kuschinsky (2000). Influence of the endothelial glycocalyx on cerebral blood flow in mice. *Journal of Cerebral Blood Flow & Metabolism 20*(11), 1571–1578.

Vulto, P., S. Podszun, P. Meyer, C. Hermann, A. Manz, and G. A. Urban (2011). Phaseguides: a paradigm shift in microfluidic priming and emptying. *Lab on a Chip 11*(9), 1596–1602.

Walther, F., P. Davydovskaya, S. Zürcher, M. Kaiser, H. Herberg, A. M. Gigler, and R. W. Stark (2007). Stability of the hydrophilic behavior of oxygen plasma activated SU-8. *Journal of Micromechanics and Microengineering 17*(3).

Walther, F., T. Drobek, A. M. Gigler, M. Hennemeyer, M. Kaiser, H. Herberg, T. Shimitsu, G. E. Morfill, and R. W. Stark (2010). Surface hydrophilization of su-8 by plasma and wet chemical processes. *Surface and Interface Analysis 42*(12-13), 1735–1744.

Washburn, E. W. (1921). The dynamics of capillary flow. *Physical review 17*(3).

White, F. M. (1991). *Viscous fluid flow.* MacGraw, New York.

Whitehead, A. N. (1889). Second approximations to viscous fluid motion. *QJ Math 23*, 143–152.

Woolf, L. A. (1960). Tracer diffusion of hydrogen ion in aqueous alkali chloride solutions at 25ř. *The Journal of Physical Chemistry 64*(4), 481–484.

Yamada, M., M. Nakashima, and M. Seki (2004). Pinched flow fractionation: continuous size separation of particles utilizing a laminar flow profile in a pinched microchannel. *Analytical chemistry 76*(18), 5465–5471.

Yang, X. M., Z. W. Zhong, E. M. Diallo, Z. H. Wang, and W. S. Yue (2014). Silicon wafer wettability and aging behaviors: Impact on gold thin-film morphology. *Materials Science in Semiconductor Processing 26*, 25–32.

Yao, S. and J. G. Santiago (2003). Porous glass electroosmotic pumps: theory. *Journal of Colloid and Interface Science 268*(1), 133–142.

Yariv, E. (2005). Induced-charge electrophoresis of nonspherical particles. *Physics of Fluids 17*(5), 051702.

Young, N. O., J. S. Goldstein, and M. J. Block (1959). The motion of bubbles in a vertical temperature gradient. *Journal of Fluid Mechanics 6*(3), 350–356.

Zahn, M. (1979). *Electromagnetic Field Theory: A Problem Solving Approach.* Wiley.

Zeleny, E. (2014). The unduloid.

Zhang, K. and A. Acrivos (1994). Viscous resuspension in fully developed laminar pipe flows. *International journal of multiphase flow 20*(3), 579–591.

Zimmermann, M. H. and J. A. Milburn (2012). *Transport in plants I: Phloem transport*, Volume 1. Springer Science & Business Media.

Zipursky, A., E. Bow, R. S. Seshadri, and E. J. Brown (1976). Leukocyte density and volume in normal subjects and in patients with acute lymphoblastic leukemia. *Blood 48*(3), 361–371.

Index